Nucleic Acid–Metal Ion Interactions

RSC Biomolecular Sciences

Editorial Board:

Professor Stephen Neidle (Chairman), *The School of Pharmacy, University of London, UK*
Dr Simon F Campbell CBE, FRS
Dr Marius Clore, *National Institutes of Health, USA*
Professor David M J Lilley FRS, *University of Dundee, UK*

This Series is devoted to coverage of the interface between the chemical and biological sciences, especially structural biology, chemical biology, bio- and chemo-informatics, drug discovery and development, chemical enzymology and biophysical chemistry. Ideal as reference and state-of-the-art guides at the graduate and post-graduate level.

Titles in the Series:

Biophysical and Structural Aspects of Bioenergetics
Edited by Mårten Wikström, *University of Helsinki, Finland*
Computational and Structural Approaches to Drug Discovery: Ligand–Protein Interactions
Edited by Robert M Stroud and Janet Finer-Moore, *University of California in San Francisco, San Francisco, CA, USA*
Exploiting Chemical Diversity for Drug Discovery
Edited by Paul A. Bartlett, *Department of Chemistry, University of California, Berkeley, USA* and Michael Entzeroth, *S*Bio Pte Ltd, Singapore*
Metabolomics, Metabonomics and Metabolite Profiling
Edited by William J. Griffiths, *University of London, The School of Pharmacy, University of London, London, UK*
Nucleic Acid–Metal Ion Interactions
Edited by Nicholas V. Hud, *School of Chemistry and Biochemistry, Georgia Institute of Technology, Atlanta, GA, USA*
Protein–Carbohydrate Interactions in Infectious Disease
Edited by Carole A. Bewley, *National Institutes of Health, Bethesda, Maryland, USA*
Protein Folding, Misfolding and Aggregation: Classical Themes and Novel Approaches
Edited by Victor Muñoz, *Department of Chemistry and Biochemistry, University of Maryland, MD, USA*
Protein-Nucleic Acid Interactions: Structural Biology
Edited by Phoebe A. Rice, *Department of Biochemistry & Molecular Biology, The University of Chicago, Chicago IL, USA* and Carl C. Correll, *Dept of Biochemistry and Molecular Biology, Rosalind Franklin University, North Chicago, IL, USA*
Quadruplex Nucleic Acids
Edited by Stephen Neidle, *The School of Pharmacy, University of London, London, UK* and Shankar Balasubramanian, *Department of Chemistry, University of Cambridge, Cambridge, UK*
Ribozymes and RNA Catalysis
Edited by David MJ Lilley FRS, *University of Dundee, Dundee, UK* and Fritz Eckstein, *Max-Planck-Institut for Experimental Medicine, Goettingen, Germany*
Sequence-specific DNA Binding Agents
Edited by Michael Waring, *Department of Pharmacology, University of Cambridge, Cambridge, UK*
Structural Biology of Membrane Proteins
Edited by Reinhard Grisshammer and Susan K. Buchanan, *Laboratory of Molecular Biology, National Institutes of Health, Bethesda, Maryland, USA*
Structure-based Drug Discovery: An Overview
Edited by Roderick E. Hubbard, *University of York, UK and Vernalis (R&D) Ltd, Cambridge, UK*
Therapeutic Oligonucleotides
Edited by Jens Kurreck, *Institute for Chemistry and Biochemistry, Free University Berlin, Berlin, Germany*

Visit our website at www.rsc.org/biomolecularsciences

For further information please contact:
Sales and Customer Care, Royal Society of Chemistry, Thomas Graham House, Science Park, Milton Road, Cambridge, CB4 0WF, UK
Telephone: +44 (0)1223 432360, Fax: +44 (0)1223 426017, Email: sales@rsc.org

Nucleic Acid–Metal Ion Interactions

Edited by

Nicholas V. Hud
School of Chemistry and Biochemistry, Georgia Institute of Technology, Atlanta, GA, USA

RSCPublishing

ISBN: 978-0-85404-195-4

A catalogue record for this book is available from the British Library

© Royal Society of Chemistry 2009

All rights reserved

Apart from fair dealing for the purposes of research for non-commercial purposes or for private study, criticism or review, as permitted under the Copyright, Designs and Patents Act 1988 and the Copyright and Related Rights Regulations 2003, this publication may not be reproduced, stored or transmitted, in any form or by any means, without the prior permission in writing of The Royal Society of Chemistry or the copyright owner, or in the case of reproduction in accordance with the terms of licences issued by the Copyright Licensing Agency in the UK, or in accordance with the terms of the licences issued by the appropriate Reproduction Rights Organization outside the UK. Enquiries concerning reproduction outside the terms stated here should be sent to The Royal Society of Chemistry at the address printed on this page.

Published by The Royal Society of Chemistry,
Thomas Graham House, Science Park, Milton Road,
Cambridge CB4 0WF, UK

Registered Charity Number 207890

For further information see our web site at www.rsc.org

Preface

Natural biochemical processes are continuously being discovered that involve RNA. Some of these processes, such as RNA interference, are now being exploited for biotechnology and medicinal applications. DNA has also proven in recent years to be more than a passive storehouse of information. For example, non-B-form DNA structures formed by G-rich DNA have been shown to participate in the regulation of gene expression, a discovery that presents new possibilities for drug targets in the genome. The current quest to understand how nucleic acids function at the most fundamental levels requires that we have a detailed understanding of nucleic acid–metal ion interactions. Due to the polyanionic nature of nucleic acids, DNA and RNA molecules are always associated with cations in living cells and these cations are essential for maintaining nucleic acid structure and function. The nature of these interactions varies greatly, from monovalent alkali metal ions that are primarily delocalized in a diffuse cloud around duplex DNA and RNA, to transition metals that are directly coordinated to the nucleotide bases. During the past decade, many important insights regarding nucleic acid–metal ion interactions have resulted from the application of spectroscopic techniques in the solution state and the determination of X-ray crystal structures. The present volume has been compiled to provide readers with an overview of these biophysical investigations. Our goal in writing each chapter has been two-fold. We have sought to produce a book that could serve as a reference source for researchers in the field, and also a text with sufficient background to serve as an initial starting point for students and researchers interested in entering the field or simply learning about this exciting topic.

As Editor, I am indebted to my co-authors for their generous contributions to this book. We have worked together to make the chapters as complementary as possible and to minimize overlap.

Contents

Chapter 1 Complexes of Nucleic Acids with Group I and II Cations
*Chiaolong Hsiao, Emmanuel Tannenbaum,
Halena VanDeusen, Eli Hershkovitz, Ginger Perng,
Allen R. Tannenbaum and Loren Dean Williams*

1.1	Introduction		1
	1.1.1	Modern Treasure Troves of Structural Information: Large RNAs	2
1.2	Nucleic Acid Folding		2
	1.2.1	Cations	2
	1.2.2	The RNA Folding Hierarchy	3
	1.2.3	Alternative RNA Folding Hierarchies	4
1.3	Coordination Chemistry		4
	1.3.1	Group I	4
	1.3.2	Group II	7
1.4	Experimental Methods for Determination of Cation Positions in X-ray Structures		24
	1.4.1	Group I	24
	1.4.2	Group II	25
1.5	The Two Binding-mode Formalism		25
	1.5.1	Thermodynamic/Kinetic Definition	25
	1.5.2	Structural Definition	26
	1.5.3	Computational Definition	26
	1.5.4	Breaking the Two Binding-mode Formalism	27
1.6	Reaction Coordinates for RNA Folding		27
	1.6.1	The Utility of 3D Databases for Determining Mechanism	27

RSC Biomolecular Sciences
Nucleic Acid–Metal Ion Interactions
Edited by Nicholas V. Hud
© Royal Society of Chemistry 2009
Published by the Royal Society of Chemistry, www.rsc.org

		1.6.2	Mg^{2+}–RNA Complexes Report on Folding Intermediates	28
	Appendix			28
	Acknowledgements			32
	References			32

Chapter 2 Coordinative Bond Formation Between Metal Ions and Nucleic Acid Bases
Bernhard Lippert

	2.1	Introduction		39
		2.1.1	Nucleic Acids and Metal Ions	39
		2.1.2	Types of Interactions	40
		2.1.3	Brief History	41
		2.1.4	Scope	42
	2.2	Nucleobases as Ligands		42
		2.2.1	Location of Donor Sites in DNA and RNAs	42
		2.2.2	Acid–Base Equilibria of Nucleobases	43
		2.2.3	'Conventional' Metal Binding Sites in Neutral Nucleobases	44
		2.2.4	Exocyclic Amino Groups: Not Used Unless Deprotonated	45
		2.2.5	Metal Coordination to Rare Nucleobase Tautomers	46
		2.2.6	Deprotonated Nucleobases	48
		2.2.7	Organometallic Nucleobase Complexes	50
		2.2.8	Chelates, Macrochelates and 'Indirect' Chelates	52
		2.2.9	Coordination Behavior of Di- and Multinuclear Metal Complexes	52
		2.2.10	Protonated Purine Bases as Ligands	54
		2.2.11	Other Modes	54
		2.2.12	Metal Migration	54
		2.2.13	Metals as Chameleons in Nucleic Acid Coordination	54
		2.2.14	Complex Stabilities	55
	2.3	Consequences of Metal Coordination		57
		2.3.1	Cross-link Formation in dsDNA	57
		2.3.2	Multistranded Nucleic Acids	59
		2.3.3	Subtle Effects of Metal Coordination	62
		2.3.4	Irreversible Alterations	66
	2.4	Summary and Outlook		68
	References			69

Chapter 3 Sequence-specific DNA–Metal Ion Interactions
Nicholas V. Hud and Aaron E. Engelhart

3.1	Introduction	75
3.2	A Brief Overview of Cation Binding Modes to DNA	77
3.3	Defining A-tracts, G-tracts and Generic DNA	79
3.4	Divalent Cation Localization	79
	3.4.1 Divalent Cations and the Major Groove	79
	3.4.2 Divalent Cations and the Minor Groove	86
3.5	Monovalent Cation Localization	92
	3.5.1 Monovalent Cations and the Major Groove	92
	3.5.2 Monovalent Cations and the Minor Groove	96
3.6	Testing Cation Size Selection of the A-tract Minor Groove Using a Large Monovalent Cation	103
3.7	Additional Experimental Evidence of Sequence-specific Binding of Cations to DNA	104
	3.7.1 Capillary Electrophoresis	104
	3.7.2 Hydration Measurements	104
3.8	Sequence-specific Cation Localization and DNA Fine Structure	106
3.9	Investigations of the Role of Cations in DNA Fine Structure Using Non-natural Bases	110
3.10	Conclusions and Perspectives	112
Acknowledgements		112
References		112

Chapter 4 Metal Ion Interactions with G-Quadruplex Structures
Aaron E. Engelhart, Janez Plavec, Özgül Persil and Nicholas V. Hud

4.1	Introduction	118
4.2	A Brief Overview of G-quadruplex Structures and Folding Topologies	121
4.3	X-ray Crystallographic Studies of Cation Coordination by G-quadruplexes	123
	4.3.1 X-ray Studies of Monovalent Cation Coordination	123
	4.3.2 X-ray Studies of Divalent Cation Coordination	128
4.4	NMR Studies of Cation Coordination by G-quadruplexes	130
	4.4.1 Solution-state NMR Studies of Monovalent Cation Coordination	130

| | | 4.4.2 Solid-state NMR Studies of Cation Coordination | 133 |

    4.5 Direct Observation of Movement and Exchange of Cations Within G-quadruplexes   134
    4.6 G-quadruplex Cation-dependent Stability   139
    4.7 Cations and G-quadruplex Polymorphism   141
    4.8 Non-biological Applications of G-quartet–Cation Interactions   145
    4.9 Conclusion   147
    Acknowledgements   147
    References   147

Chapter 5 Characterization of Nucleic Acid–Metal Ion Binding by Spectroscopic Techniques
Victoria J. DeRose

    5.1 Introduction   154
    5.2 NMR Spectroscopy   159
        5.2.1 (Indirect) Detection of Metal Interactions Using NMR Spectroscopy   159
        5.2.2 Spectroscopy of NMR-active Cation Probes   160
    5.3 EPR Spectroscopy   163
        5.3.1 EPR Spectroscopy of Oligonucleotide-bound Paramagnetic Ions   164
        5.3.2 Mn^{2+} and VO^{2+} Environments Probed by ENDOR and ESEEM Spectroscopies   164
    5.4 X-ray Absorption Spectroscopy and Small-angle X-ray Scattering   166
        5.4.1 X-ray Absorption Spectroscopy (XAS)   168
        5.4.2 Small-angle X-ray Scattering (SAXS) and ASAX   169
    5.5 Optical Methods: Lanthanide Luminescence   170
    5.6 Vibrational Spectroscopy: IR, Raman and Resonance Raman   173
    5.7 Conclusion   174
    Acknowledgements   175
    References   175

Chapter 6 Metal Ions and the Thermodynamics of RNA Folding
David P. Giedroc and Nicholas E. Grossoehme

    6.1 Introduction   180
    6.2 Physical Properties of RNA and Metal Ions Relevant to RNA Folding   181

		6.2.1	Implications of RNA as a Folded Polyanion in Ion-RNA Interactions	181

		6.2.1	Implications of RNA as a Folded Polyanion in Ion-RNA Interactions	181
		6.2.2	Physical Properties of Ions: Monovalent *versus* Di(multi)valent Ions	182
		6.2.3	RNA Folding is Generally Hierarchical	185
		6.2.4	Intrinsic Competition of Monovalent and Divalent Ions for 'Sites' on RNA	186
	6.3	Mg^{2+} Ions, Hydrodynamic Collapse and Dynamical Restriction		187
		6.3.1	Mg^{2+} Drives Hydrodynamic Collapse of Complex RNAs	188
		6.3.2	Dynamical Restriction of RNA Conformational Fluctuations by Mg^{2+} Ions	190
	6.4	Metal Ion Binding Models		192
		6.4.1	Operational Classes of Ions that Associate with RNA	192
		6.4.2	General Considerations and Complications	194
		6.4.3	Direct Binding Methods	197
		6.4.4	The Hill Model	200
		6.4.5	Diffuse Ion Association Models	204
		6.4.6	Linkage of Protonation to Mg^{2+} Binding and RNA Folding	205
	6.5	How Mg^{2+} Stabilizes RNA from Single-molecule Unfolding Studies Using Mechanical Force		207
	6.6	Conclusion and Perspective		212
	Acknowledgements			213
	References			213

Chapter 7   Metal Ions and RNA Folding Kinetics
Somdeb Mitra and Michael Brenowitz

	7.1	Introduction		221
		7.1.1	A Brief Overview	221
		7.1.2	RNA Folding is Hierarchical	222
		7.1.3	Why are Counterions Necessary for RNA Folding?	223
	7.2	Cation-induced RNA Folding Kinetics		226
		7.2.1	Effects of Ionic Strength on Net Energetic Stability of Initial, Intermediate and Final States	226
		7.2.2	Position and Plasticity of the Transition State	235
		7.2.3	Effects on the Initial Conformational Ensemble and Pathway Partitioning	246

7.3	Conclusion	253
	Acknowledgement	253
	References	254

Chapter 8   Metal Ions in RNA Catalysis
John K. Frederiksen, Robert Fong and Joseph A. Piccirilli

8.1	Introduction		260
8.2	Metal Ions and Their Ability to Facilitate RNA Catalysis in Biological Systems		262
8.3	Mechanistic Roles for Metal Ions in Ribozyme Reactions		266
	8.3.1	Lewis Acid Stabilization of the Leaving Group	267
	8.3.2	Nucleophile Activation	267
	8.3.3	Coordination of Non-bridging Phosphoryl Oxygens	268
	8.3.4	Induced Intramolecularity	269
	8.3.5	Indirect Catalytic Contributions	269
8.4	Experimental Techniques for Detecting Catalytic Metal Ion–Ligand Interactions in Ribozymes		270
	8.4.1	X-ray Crystallography	270
	8.4.2	Spectroscopic Methods – Electron Paramagnetic Resonance (EPR) and Nuclear Magnetic Resonance (NMR)	271
	8.4.3	Computational and Database Methods	272
	8.4.4	Nucleotide Analogue Interference Mapping (NAIM)	273
	8.4.5	Lanthanide Ion-induced Cleavage and Luminescence	274
	8.4.6	Heavy Atom Isotope Effects	275
	8.4.7	Metal Ion Rescue Experiments	276
	8.4.8	Metal Rescue in Practice: The *Tetrahymena* Group I Ribozyme Active Site	277
8.5	The Large Ribozymes: A 'Two-Metal Ion' Mechanistic Paradigm		283
	8.5.1	RNase P	284
	8.5.2	Group II Intron	288
	8.5.3	Spliceosome	293
8.6	The Small Ribozymes: Endonucleolytic Cleavage and General Acid–Base Catalysis		295
8.7	Conclusions		298
	References		299

Chapter 9 Polyanion Models of Nucleic Acid–Metal Ion Interactions
J. Michael Schurr

9.1	Introduction	307
	9.1.1 Background and Objectives	307
	9.1.2 Polyanions and Their Compensating Charges	309
	9.1.3 The Ion Atmosphere	309
	9.1.4 Equilibrium Polyion Properties and Phenomena	310
9.2	Statistical Mechanics, Configuration Integral and Potential of Mean Force	313
9.3	Ion Atmosphere Theory	315
	9.3.1 The Standard Primitive Model for a Polyanion and Its Small Ion Atmosphere	315
	9.3.2 The Non-linear Poisson–Boltzmann Equation	317
	9.3.3 Solutions of the NLPB Equation	318
9.4	Connection of NLPB Theory to Thermodynamic Properties	319
	9.4.1 Electrostatic Free Energy	319
	9.4.2 Polyanion Activity Coefficient and Polyanion–Salt Preferential Interaction Coefficient	320
	9.4.3 Osmotic Pressure and Osmotic Coefficient	323
9.5	Polyion Equilibria and Repulsive Interactions	323
	9.5.1 Effect of Salt Concentration on the Association of a Polycation with a Polyanion	323
	9.5.2 Effect of Salt Concentration on the Equilibrium Constant for a Conformational Transition	325
	9.5.3 Temperature Dependence of a Conformational Transition	326
	9.5.4 Proton Binding to an Acidic Ligand Adjacent to a Polyion	327
	9.5.5 Binding a Ligand to Different Conformations of the Same Polyion	328
	9.5.6 Absolute Equilibrium Constants	329
	9.5.7 Repulsive Forces Between Rod-like Polyanions	332
9.6	Counterion Condensation Theory	333
	9.6.1 Relation to NLPB Theory	333
	9.6.2 Thermodynamic and Other Properties	335
	9.6.3 Electrostatic Contribution to the Bending Rigidity and Persistence Length	336
	9.6.4 Effect of Oligocations to Bend DNA	336

	9.7	Attractions Between Like-charged Filaments and Condensed Phase Formation	337
		9.7.1 Simulations and Analyses	337
		9.7.2 The Tightly Bound Ion (TBI) Model	340
		9.7.3 Effects of Helical Symmetry, DNA Shape and Fixed-counterion Models	342
	References		344

Chapter 10 Metal Ion–Nucleic Acid Interactions in Disease and Medicine
Ana M. Pizarro and Peter J. Sadler

10.1	Introduction		350
10.2	Mutagenesis and Carcinogenesis		351
	10.2.1	Chromium	351
	10.2.2	Arsenic	352
	10.2.3	Other Metals	353
10.3	Antiviral Drugs		353
10.4	Metal-containing Anticancer Agents		358
	10.4.1	Platinum Drugs	358
	10.4.2	Titanium Agents	379
	10.4.3	Ruthenium Drugs	381
	10.4.4	Rhodium Complexes	389
	10.4.5	Gold Complexes	393
	10.4.6	Iron Helicates	395
10.5	Conclusions and Perspectives		396
References			398

Subject Index   417

CHAPTER 1
Complexes of Nucleic Acids with Group I and II Cations

CHIAOLONG HSIAO,[a] EMMANUEL TANNENBAUM,[b] HALENA VanDEUSEN,[a] ELI HERSHKOVITZ,[c,d] GINGER PERNG,[c] ALLEN R. TANNENBAUM[c,d] AND LOREN DEAN WILLIAMS[a]

[a] School of Chemistry and Biochemistry, Georgia Institute of Technology, Atlanta, GA 30332-0400, USA; [b] School of Biology, Georgia Institute of Technology, Atlanta, GA, 30332, USA; [c] School of Electrical and Computer Engineering, Georgia Institute of Technology, Atlanta, GA 30332–0250, USA; [d] School of Biomedical Engineering, Georgia Institute of Technology, Atlanta, GA 30332–0250, USA

1.1 Introduction

Recent structures of large RNAs, such as the P4–P6 domain of the *Tetrahymena* ribozyme,[1–5] and larger RNAs such as rRNAs,[6–14] combined with a general increase over time in the sophistication of diffraction experiments, show cations in diverse and sometimes unexpected environments. The interactions of nucleic acids with cations follow basic principles of coordination chemistry. The effects of cations on RNA stability and conformation demonstrate the endurance of these relatively simple principles.

One focus of this chapter is the coordination of Na^+, K^+, Ca^{2+} and Mg^{2+} by phosphates and nucleic acids. We describe coordination chemistry, electrostatic forces/energetics, conformational effects and ion-selective binding. We explain the origins of the specific requirement for Mg^{2+} in RNA folding and the tight coupling between Mg^{2+} binding and RNA conformation. We deconstruct the two binding-mode formalism. We describe crystallographic methods for determining cation positions. We propose a model of RNA folding that is consistent with Mg^{2+} coordination properties of RNA. Previous reviews are available on roles of metals in biology,[15–17] in polyelectrolyte theory[18–22] (see Chapter 9), in DNA structure[22–28] (see Chapter 3), in RNA folding[29–32] (see Chapters 6 and 7) and in RNA catalysis[33–36] (see Chapter 8).

1.1.1 Modern Treasure Troves of Structural Information: Large RNAs

Very large RNA assemblies are now available at high resolution. The largest and most accurate structures are used here in conjunction with smaller structures, down to the level of mononucleotides, to illustrate patterns of interaction of nucleic acids with cations. 23S-rRNA^{HM} refers to the 23S rRNA from the archaeon *H. marismortui*[7,37] (2.4 Å resolution, PDB entry 1JJ2), a halophile from the Dead Sea. 23S-rRNA^{TT} refers to the 23S rRNA from the eubacterium *T. thermophilus*[13] (2.8 Å resolution, PDB entry 2J01), isolated from a thermal vent. The fractional sequence identity of the 23S rRNAs from HM and TT is around 60%. RNA^{P4-P6} refers the 160 nucleotide domain of the self-splicing *Tetrahymena thermophila* intron (2.3 Å resolution, PDB entry 1HR2, this ΔC209 mutant[5] gives the best available resolution).

1.2 Nucleic Acid Folding

1.2.1 Cations

During protein folding, water molecules in contact with hydrophobic surfaces are released to bulk solvent. During nucleic acid folding, cations are sequestered from bulk solvent and held in close proximity to the polymer. Protein side-chains are multifarious, with a variety of shapes and chemical properties. The nucleic acid backbone is intricate, with many accessible rotameric states,[38] and carries charge.

Functional nucleic acids generally fold into compact and stable states of given conformation.[39,40] DNA can form quadruplexes,[41–46] triplexes,[47–49] i-motifs,[50,51] *etc*. Structured RNAs range in size from aptamers and tRNAs to ribosomes. However some functional nucleic acids, such as riboswitches, are conformationally polymorphic.[52,53] For our purposes, folded nucleic acid structures fall into three general classes: (i) *helical structures* such as A-form, B-form and triplexes, (ii) *quasi-globular structures* such as tRNA, with base–base tertiary interactions but no buried phosphates,[54,55] and quadruplexes,[41–46] and (iii) *true globular structures* such as the *Tetrahymena* ribozyme[40] and its P4–P6 domain,[1–5] with base–base tertiary interactions plus buried OP atoms (OP indicates a non-bridging phosphate oxygen). True globular structures have distinct 'insides' and 'outsides'. Folding of helices, quasi-globular structures and true globular structures increases proximities of phosphate groups and the electrostatic repulsion among them. Therefore, folding is intrinsically linked to association with cations. Phosphate–phosphate repulsion must be offset by attraction between phosphates and cations. Cations most strongly associate with regions of DNA and RNA in which phosphate groups assume greatest 'density'.

As will become clear in the following sections, Mg^{2+} stabilizes distinctive conformational and energetic states of nucleic acids. Mg^{2+} shares a special geometric and electrostatic complementarity with phosphate, with a specific coordination and thermodynamic fingerprint. These states are simply not accessible in the absence of Mg^{2+} (or Mn^{2+}), even when other cations are at high concentration. The thermodynamic and conformational consequences of first-shell OP interactions with Mg^{2+} are different to those for neutral ligands or for other cations, with lesser charge or greater size.

1.2.2 The RNA Folding Hierarchy

RNA folding is hierarchical.[56,57] Folding progresses through a series of intermediates that are commonly characterized by extents and types of base–base hydrogen bonding and stacking interactions. The unfolded state, the random coil, is a conformationally polymorphic and fluctuating ensemble with few local or long-range base–base interactions. Early intermediates contain double-stranded stems and hairpin loops, interspersed by single-stranded regions. These stems and loops are known collectively as secondary structural units. Late intermediates and the final folded state are stabilized by base–base tertiary interactions, between residues that are remote in the secondary structure. To a first approximation, secondary structure can be conceptually and experimentally separated from tertiary structure. Secondary structure forms before tertiary structure and is favorable in a broad range of ionic conditions. Tertiary structure is favored by divalent cations.[58,59] Although compact structures with base–base tertiary interactions can be achieved at very high concentrations of other cations, for true globular structures, the fully folded state is absolutely dependent on Mg^{2+}. It can be useful to make a distinction between the *tertiary structure of an RNA*, which is a description of short- and-long range base–base interactions, and *a folded RNA*, which is a description of three-dimensional positions of all atoms, including of course the phosphate groups.

A hierarchical model that focuses exclusively on base–base interactions is a useful but somewhat limited approximation. In true globular structures, ground states are stabilized by specifically associated Mg^{2+} ions, each with up to four first-shell OP ligands. The importance of $Mg^{2+}(OP)_3$ and $Mg^{2+}(OP)_4$ coordination complexes is discussed in later sections. Multidentate interactions of OP atoms with Mg^{2+} are generally local along the RNA backbone. A small and important subset of OP ligands of a common Mg^{2+} are remote in the secondary and primary structures, thus forming 'electrostatic tertiary interactions' (Mg^{2+} mediated linkages between remote OP groups). Extensive base–base tertiary interactions during folding do not necessary imply the formation of electrostatic tertiary interactions. At least some globular RNAs fold into compact (but non-native) structures, with extensive base–base tertiary interactions – in the absence of Mg^{2+}.[60] We do not know at present if the converse is true (*i.e.* are electrostatic tertiary interactions fully dependent on correct base–base tertiary interactions?). At any rate, to understand and describe fully the structure of a globular RNA, one can extend a conventional tertiary description of base–base interactions to include electrostatic tertiary interactions.

1.2.3 Alternative RNA Folding Hierarchies

To illuminate the underlying dependence of folding on cations, one can re-state the hierarchy of RNA folding using 'phosphate density'. In early folding steps, a subset of phosphate–phosphate distances decreases from $>7\,\text{Å}$ (P to P) in random coil to around 5.8–6.2 Å (in A-form helical regions and loops). In subsequent steps, a subset of P to P distances decreases further, to 5.0–4.6 Å. Associated with this group of short P to P distances are tightly packed anionic OP atoms, which are in van der Waals contact with each other ($d_{OP-OP} = 2.8$–$3.2\,\text{Å}$). This tight packing of anionic oxygen atoms is dependent on multidentate chelation of Mg^{2+} by OP atoms. Neither monovalent cations nor polyamines can substitute for Mg^{2+} in stabilizing structures with such short OP–OP contacts. During folding, some phosphates and associated Mg^{2+} ions become buried in the globular interior.

1.3 Coordination Chemistry

The binding of ligands to Group I and II cations is dictated by the chemical properties of the cations and of the ligands and, to a significant extent, by interactions between ligands.[15] Chelators, with covalently linked ligands, create cavities for ions and bind with greater affinity and selectivity than monomeric ligands. The length of the chelator linker is a critical component of stability. As the linker length increases, the entropic cost of assembling the ligands for joint coordination increases. Hud and Polak previously noted the chelation properties of DNA, calling it an ionophore.[27] Here we illustrate how the phosphate groups of nucleic acids commonly act as chelators of cations.

1.3.1 Group I

Group I cations prefer hard neutral ligands or one singly charged ligand plus additional neutral ligands. In their associations with nucleic acids, Group I cations are most commonly associated with non-anionic oxygens (*i.e.* oxygen atoms other than OP) as inner shell ligands.[29,61]

The monovalent cations [sodium (Na^+), potassium (K^+), rubidium (Rb^+), caesium (Cs^+), thallium (Tl^+) and ammonium (NH_4^+), excluding lithium (Li^+)] are characterized by relatively large ionic radii, low charge density and modest enthalpies of hydration (Table 1.1). The coordination chemistry of Li^+, with its

Table 1.1 Physical properties of cations.

Property	Li^+	Na^+	K^+	Rb^+	Cs^+	Tl^+	NH_4^+	Mg^{2+}	Ca^{2+}	Mn^{2+}
Ionic radius, Å[a]	0.60	0.95	1.33	1.48	1.69	1.49	–[b]	0.65	0.99	0.80
$\Delta H_{hydration}$/kcal mol^{-1}[c]	–127	–99	–77	–72	–66	–78	–78	–458	–358	–[e]
AOCN[d]	5.3	6.7	9.0	9.8	10.4	8.3	–[e]	5.98	7.3	5.98

[a] From Brown.[65]
[b] The radius of a non-spherical species such as NH_4^+ is not well defined. Rashin and Honig[160] estimated from geometric considerations that the effective radius of NH_4^+ is very nearly the same as the radius of K^+.
[c] From Rashin and Honig.[160]
[d] Average observed coordination numbers, from Brown.[65]
[e] Not reported.

small atomic radius and high charge density, is distinct from that of other Group I metals. Tl^+ and NH_4^+ are listed here along with the Group I metals because they are well-developed K^+ substitutes with useful spectroscopic and crystallographic signals. NH_4^+ positions are indicated by NOEs in solution.[42,62,63] Tl^+ positions are indicated in solution by NMR and in crystals by a distinctive X-ray scattering signal (anomalous scattering). Tl^+, K^+ and NH_4^+ have similar ionic radii and enthalpies of hydration (Table 1.1).

These monovalent ions, except Li^+, display irregular and variable coordination geometry.[64,65] The variability in coordination geometry is associated with non-covalency of interaction, weak ligand–ligand interactions and loose ligand–ligand packing. For a given monovalent ion, the number of first-shell ligands can vary from four to over 10. These properties are quantitated by 'average observed coordination numbers' (AOCN, Table 1.1) over a large number of structures within the Cambridge Structural Database as reported by Brown.[65]

1.3.1.1 Na^+

The ideal Na^+ to oxygen distance is 2.4 Å (Figure 1.1a). The distance between first-shell ligands of Na^+ is variable, depending on coordination number and coordination geometry. An octahedral arrangement of first shell oxygen ligands is loosely packed. The O to O distance is 3.4 Å, which is significantly greater than twice the van der Waals radius of oxygen (oxygen radius = 1.4 Å). Therefore, inner shell ligands of Na^+ are not crowded and the geometry of the Na^+ inner sphere is not determined by ligand–ligand interactions. An Na^+ ion with ideal octahedral geometry in association with the O6 position of a guanine of DNA with five water molecule inner-shell ligands[66] is shown in Figure 1.1a. As can be seen, the Na^+ to O distances average around 2.4 Å, whereas the distance between *cis* oxygen atoms (adjacent oxygen ligands) averages around 3.4 Å. DNA–cation interactions are discussed in Chapter 3.

1.3.1.2 K^+

The ideal K^+ to oxygen distance is around 2.7 Å. For an octahedral arrangement of first-shell oxygen ligands, the average O to O distance is over 4.0 Å.

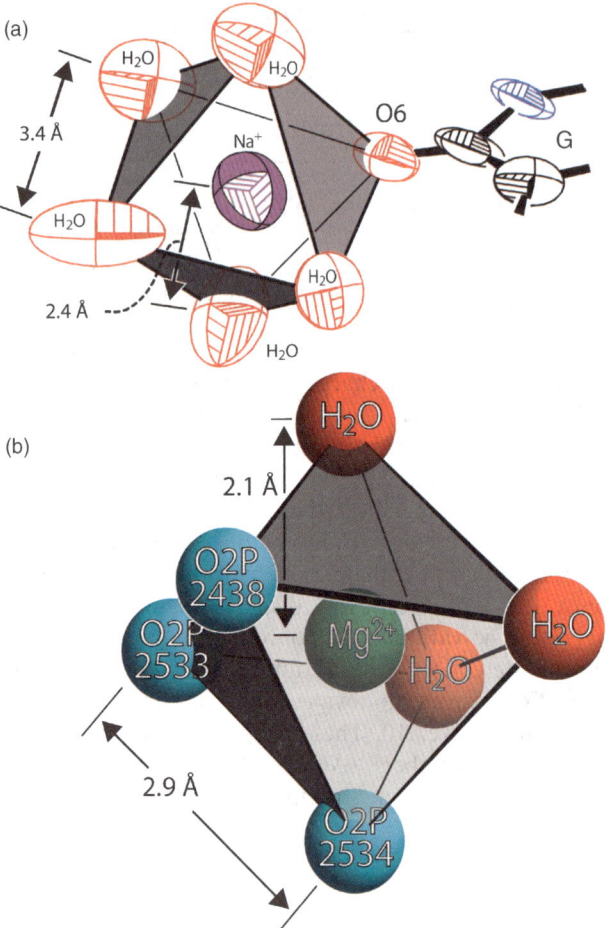

Figure 1.1 (a) An Na$^+$ ion bound to the O6 of a guanine of DNA (ion 487 from PDB entry 2DYW). The Na$^+$ ion (purple ellipsoid) is octahedral, with bonds to the floor of the major groove and five water molecules (hydrogens not shown). All first-shell oxygen ligands are represented by red ellipsoids. Na$^+$–oxygen bonds are around 2.4 Å in length. The first shell ligands are not in van der Waals contact with each other. (b) A trichelated Mg^{2+} ion (ion 8001 from 23S-rRNA^{HM}). This Mg^{2+} ion (green sphere) is octahedral, with three bonds to phosphate groups of RNA (cyan) and three bonds to water oxygens (red). Mg^{2+}–oxygen bonds are around 2.1 Å in length. Mg^{2+} coordination imposes oxygen–oxygen distances of around 2.9 Å in the first shell. For clarity, the radii of the spheres are reduced from the van der Waals radii of the atoms and have no physical significance.

Hence the inner-shell ligands of K$^+$ are of less defined geometry than those of Na$^+$. Specific K$^+$ binders, that exclude Na$^+$, are generally composed of stacked, planar arrangements of keto oxygens.[41,67,68] In these K$^+$-selective structures, the positions of the keto oxygens (O^{keto}) are fixed such that the enthalpy of dehydration is compensated by O^{keto}–K^+ interactions but not by O^{keto}–Na^+ interactions, which are too long.

1.3.2 Group II

1.3.2.1 Ca^{2+}

Ca^{2+} often shows irregular coordination geometry and coordination numbers greater than six. The ionic radius of Ca^{2+} is $ca.$ 0.99 Å. Compared with Mg^{2+}, Ca^{2+} is large, its charge density is low, its inner-shell ligands are loosely packed and the magnitude of is hydration enthalpy is small (Table 1.1). However, like Mg^{2+}, Ca^{2+} prefers a mix of anionic and neutral ligands.

1.3.2.2 Mg^{2+}

Mg^{2+}, from life's beginning, has been closely associated with some of the central players in biological systems – phosphates and phosphate esters.[69] Mg^{2+} shares a special geometric, electrostatic and thermodynamic relationship with phosphates and phosphate esters. In comparison with group I ions, Ca^{2+} or polyamines, Mg^{2+} has a greater affinity for OP atoms and binds to OP with well-defined geometry. Unlike other cations, Mg^{2+} brings OP atoms into direct contact with each other.

The ionic radius of Mg^{2+} is small ($ca.$ 0.65 Å), the charge density is high, the six ligands of an octahedral inner first shell are tightly packed and the magnitude of the hydration enthalpy is large (Table 1.1). The heat of hydration of Mg^{2+} is much greater than for other biological cations (Table 1.1). Mg^{2+} prefers two to four oxyanions (along with a complement of water molecules) over uncharged oxygens and nitrogens as inner-shell ligands.

The first coordination sphere of Mg^{2+} assumes octahedral geometry,[64,65,70,71] as shown in Figure 1.1b. The AOCN of Mg^{2+} is 5.98.[65] Because Mg^{2+} is small and highly charged, ligand–ligand crowding is one of the hallmarks of Mg^{2+} complexes, leading to highly restrained ligand–Mg^{2+}–ligand geometry and strong ligand–ligand repulsive forces. Although probably not germane to nucleic acid structure, four-coordinate Mg^{2+} is observed at high temperature in the gas phase.[72]

1.3.2.2.1 Hexaaqua Mg^{2+} Complexes. In hexaaqua complexes [$Mg^{2+}(H_2O)_6$ or Mg^{2+}_{aq}], oxygen–Mg^{2+} distances are 2.07 Å. The cis O···Mg···O angle is 90° and the cis O···O distance is 2.93 Å.[73–75] The $trans$ O···Mg···O angle is 180°. Adjacent oxygen atoms in Mg^{2+}_{aq} are in van der Waals contact. First-shell water molecules are strictly oriented such that their dipole moments are directed in towards the metal, with Mg^{2+}···O–H angles of 120–128°. This orientation prevents hydrogen bonding between water molecules in the Mg^{2+} first coordination shell.

1.3.2.2.2 ADP–Mg^{2+} Complexes. The nature of complexes formed by Mg^{2+} with ADP and ATP are highly predictive of Mg^{2+} interactions with other multidentate OP ligands such as RNA. Some of the ADP–Mg^{2+} and ATP–Mg^{2+} complexes identified in the Protein Data Bank are shown in Figure 1.2 and here we discuss the implications of these and related structures provided by X-ray crystallography. One can immediately appreciate that ADP and ATP are mono- and multidentate chelators of Mg^{2+}, contributing OP ligands to the inner coordination sphere. In this section, we focus on structural aspects of Mg^{2+} complexes with nucleotides. Thermodynamic aspects are discussed in a subsequent section.

Mg^{2+} interacts with non-bridging OP atoms, but not with bridging oxygens, base or sugar atoms of ADP/ATP. The OP atoms of ADP bind to Mg^{2+} by either monodentate interactions (40% of structures surveyed; Figure 1.2a and b, Table 1.2 and Table 1.3) or by bidentate chelation (60%, Figure 1.2c–e). Monodentate interactions occur exclusively by O$^\beta$P whereas bidentate chelation involves O$^\alpha$P and O$^\beta$P.

Chelation ring size is an important factor in modulating stability (see Section 1.3). The bidentate chelation complexes of Mg^{2+} with ADP/ATP are composed of six-membered rings consisting of atoms Mg^{2+}–O$^\alpha$P–P–O–P–O$^\beta$P–Mg^{2+} or (Mg^{2+}–O$^\beta$P–P–O–P–O$^\gamma$P–Mg^{2+}). Bidentate chelation by two OP atoms bound to a common phosphorus atom would require a chelation ring size of four and is not observed in ADP/ATP complexes. When Mg^{2+} is monochelated by ADP, at least one protein ligand is also found in the Mg^{2+} first shell. The protein ligand is invariably oxygen, but may be charged or neutral. Similarly, most bidentate ADP–Mg^{2+} complexes contain protein first-shell ligands. OP ligands are adjacent to each other within the octahedron of first-shell ligands of Mg^{2+}. All bidentate Mg^{2+}–ADP complexes assume *cis* (OP–Mg^{2+}–OP angle 90°), *cis–cis* or *cis–cis–cis* orientation of the ligands surrounding the Mg^{2+} depending on the number of first shell protein ligands. *Trans* bidentate Mg^{2+}–ADP complexes (OP–Mg^{2+}–OP angle 180°) are not observed and appear to be stereochemically prohibited.

1.3.2.2.3 ATP–Mg^{2+} Complexes. Mg^{2+} binds to ATP by bidentate (66%, Figure 1.2f–i) or tridentate (33%, Figure 1.2j) interactions. Monodentate ATP–Mg^{2+} complexes are not observed. The bidentate complexes are all *cis*, whereas the tridentate complexes are all *cis–cis*. The tridentate complexes, by definition, are bicyclic, with two six-membered rings. Each ring consists of six atoms (Mg^{2+}–O$^\alpha$P–P–O–P–O$^\beta$P–Mg^{2+}) or (Mg^{2+}–O$^\beta$P–P–O–P–O$^\gamma$P–Mg^{2+}). Bidentate complexes most commonly involve O$^\beta$P and O$^\gamma$P (88%). One O$^\alpha$P–O$^\beta$P bidentate complex is observed.

An O$^\alpha$P–Mg^{2+}–O$^\gamma$P bidentate complex would form an eight-membered ring. The importance of ring size in chelation complexes is underscored by the absence of O$^\alpha$P–Mg^{2+}–O$^\gamma$P bidentate complexes in the Protein Data Bank. The observed bidentate complexes contain 0–3 protein ligands (Table 1.2), which are all oxygens, with no obvious preference in the number or charge of protein ligands. First-shell protein ligands are not observed in tridentate ATP complexes.

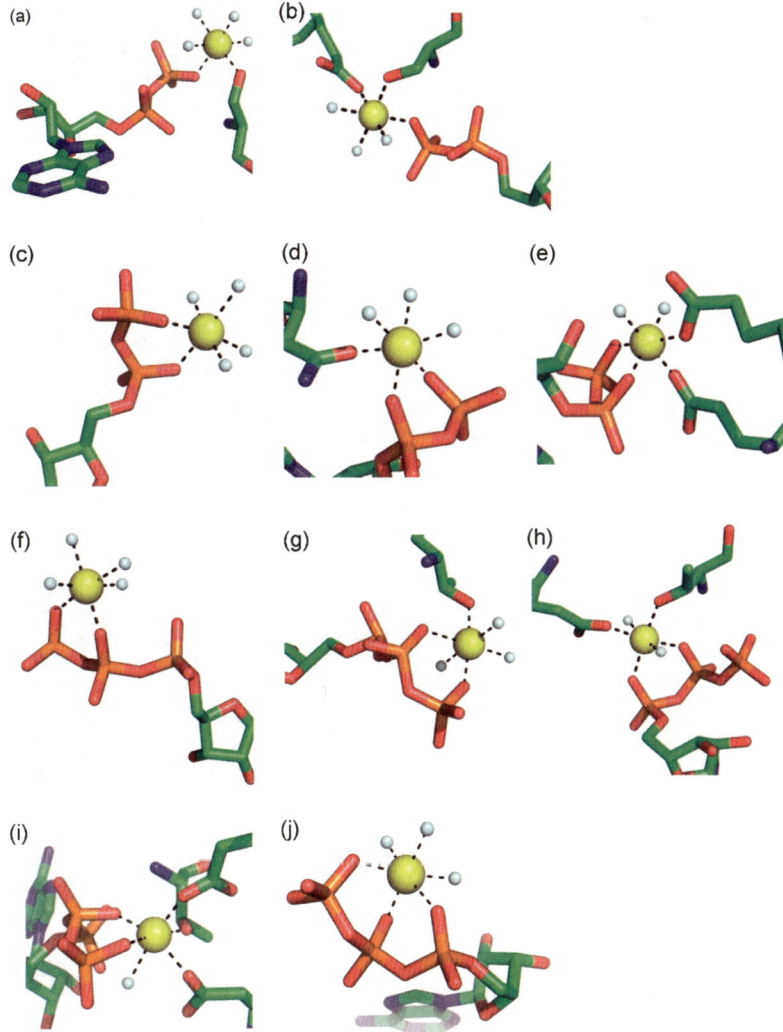

Figure 1.2 ADP and ATP as Mg^{2+} chelators. Mg^{2+} ions are represented by yellow spheres and water molecules by small white spheres. Nucleotides and protein ligands are represented by sticks. (a) Monodentate chelation by ADP ($O^\beta P$) with one protein ligand. (b) Monodentate chelation by ADP with two protein ligands. (c) Bidentate chelation by ADP ($O^\beta P$ and $O^\alpha P$) with no protein ligands. The OP ligands are *cis*. (d) Bidentate chelation by ADP with one protein ligand. (e) Bidentate chelation by ADP with two protein ligands. (f) Bidentate chelation by ATP ($O^\beta P$ and $O^\gamma P$) with no protein ligands. (g) Bidentate chelation by ATP plus one protein ligand ($O^\beta P$ and $O^\gamma P$). (h) Bidentate chelation by ATP plus three protein ligands ($O^\alpha P$ and $O^\beta P$). (i) Bidentate chelation by ATP ($O^\alpha P$ and $O^\beta P$) with two protein ligands. (j) Tridentate chelation by ATP. The OP ligands are *cis–cis*. All structures shown were extracted from the Protein Data Bank.

Table 1.2 Mg^{2+} interactions with ADP and ATP.[a]

PDB entry	No. of first-shell OP atoms	Total first-shell ligands[b]	Protein ligands	Orientation	Resolution/Å	Corresponding author
ADP						
1G6H	1	2	Ser	*cis*	1.6	Hunt
1L4Y	1	2	Ser	*cis*	2.0	Ji
1P5Z	1	3	Ser, Glu	*cis–cis*	1.6	Lavie
1SVL	1	2	Ser	*cis*	1.95	Chen
1BYQ	2	3	Asn	*cis–cis*	1.5	Pavletich
1KJQ	2	4	Glu, Glu	*cis–cis–cis*	1.05	Holden
1OHA	2	2		*cis*	1.9	Rubio
1PHP	2	3	Asp	*cis–cis*	1.65	Watson
1RL9	2	2		*cis*	1.45	Chapman
1Z2N	2	4	Asp, Asp	*cis–cis–cis*	1.2	Hurley
ATP						
1G5T	2	4	Glu, Thr	*cis–cis–trans*	1.8	Bauer
1A82	2	5	Glu, Asp, Thr		1.8	Schneider
1D4X	2	2		*cis*	1.75	Almo
1F2U	2	4	Ser, Gln	*cis–cis–trans*	1.6	Craig
1KAX	2	2		*cis*	1.7	McKay
1SVM	2	3	Ser	*cis–trans*	1.94	Chen
2IXE	2	3	Ser	*cis–trans*	2.0	Gaudet
2IYW	2	3	Ser	*cis–trans*	1.85	Bartunk
1J09	3	3		*cis–cis*	1.8	Yokoyama
1OBD	3	3[c]		*cis–cis*	1.4	Wilson
2A84	3	3		*cis–cis*	1.55	Eisenberg
2NT8	3	3[c]		*cis–cis*	1.68	Rayment

[a]Structures were obtained from the PDB. Structures were rejected if the Mg^{2+} ion is coordinated by anything other than the nucleotide, water or protein. OP–Mg^{2+} bond lengths greater than 2.25 Å or other bad geometries were excluded. Ten Mg^{2+}–ADP and 12 Mg^{2+}–ATP structures were obtained.
[b]The total number of first-shell ligands other than water molecules.
[c]These Mg^{2+} ions are four-coordinate, presumably because two water molecules were omitted during refinement.

Table 1.3 Mg^{2+} chelation by nucleotides[a] and by RNA.[b]

	No. of first-shell OP ligands			
	0 (hexaaqua)	1 (monodentate)	2 (bidentate)	3 (tridentate)
ADP	NA	4	6	0
ATP	NA	0	8	4
RNA	36	40	31	10

[a]Nucleotides with OP atoms in the Mg^{2+} first shell. Protein ligands are omitted.
[b]In 23S-rRNA^{HM}.

1.3.2.2.4 RNA–Mg²: OP Preferred. With RNA, Mg^{2+} associates preferentially with OP atoms over base and ribose atoms, just as it prefers OP atoms of ATP/ADP (above). Interactions with uncharged functional groups of sugars and bases of RNA are infrequent (Table 1.4).

The arrangement of three OP atoms and three water oxygens around one particular Mg^{2+} in a large globular RNA is shown in Figure 1.1b. The 23S-rRNA^{HM} is a tridentate chelator of this Mg^{2+} ion, contributing three OP ligands to the inner coordination shell.

Large rRNAs such as 23S-rRNA^{HM} are associated with many Mg^{2+} ions, which can be visualized by X-ray diffraction. Of the 117 Mg^{2+} atoms associated in some fashion with 23S-rRNA^{HM}, 98 are 'bonded' to the 23S rRNA, ($Mg^{2+}\cdots$RNA distance <2.6 Å), with first-shell RNA ligands. The most common Mg^{2+} ligands are water oxygens (407 ligands). Most RNA ligands are OP atoms (129 first-shell OP ligands). Mg^{2+} does not exhibit a preference for O1P (63 ligands) over O2P (66 ligands). RNA bases are infrequent ligands, with 12 O6 atoms, 10 N7 atoms, five O4 atoms and three O2 atoms within Mg^{2+} first shells. As expected from DNA structures,[61] RNA amino groups, (N6, N2 and N4) do not enter the first coordination shell of Mg^{2+}. The 5S rRNA associated with 23S-rRNA^{HM} interacts with one Mg^{2+}. There, two OP atoms chelate an Mg^{2+}.

1.3.2.2.5 Chelation Ring Size. In contrast to preferential bi- and tridentate chelation of Mg^{2+} by ADP and ATP, RNA generally prefers monodentate chelation of Mg^{2+} (Table 1.3). The primary reason for this preference is that RNA forms 10-membered chelation cycles with Mg^{2+} (Figures 1.3 and 1.4), which are less favorable than the six-membered chelation cycles of ADP/ATP (Figure 1.2). A second reason is that the relative positions of OP atoms in common RNA secondary structural elements such as A-form helices and tetraloops are not favorable for multidentate chelation of Mg^{2+}. In such canonical RNA conformations, the OP atoms are too far apart and are not optimally oriented for multidentate Mg^{2+} chelation. In regions of irregular RNA conformation, OP can assume the correct geometric disposition for Mg^{2+} chelation. For example, the RNA conformation in the $Mg^{2+}(OP)_2$ complex in Figure 1.3 deviates profoundly from that in A-form helical RNA.

The most frequent Mg^{2+} chelation motif in RNA is when OP atoms of adjacent residues chelate a common Mg^{2+} ($\Delta = 1$, Figure 1.4). Larger ring sizes are less frequent. $\Delta = 1$ is observed at higher frequency than in other

Table 1.4 Paleo-magnesium ions[a] in 23S-rRNA^{HM} and 23S-rRNA^{TT}.

Mg^{2+} ID[b]	Ligand type (OP, B, S, P)[c]	Mg^{2+} ID No.[d]	rRNA residue numbers[e]
Dual Mg^{2+} bicycles			
D1	(OP)$_3$	8001 (122)	2483 (2448), 2533 (2498), 2534 (2499)
D1'	(OP)$_2$B	8002 (123)	2483 (2448), 2534 (2499), 627^{Base} (570^{Base})
D2	(OP)$_3$	8003 (66)	876 (783), 877 (784), 2624 (2589)
D2'	(OP)$_2$	8013 (70)	877 (784), 2623 (2588)
D3	(OP)$_3$	8016 (86)	1504 (1395), 1678 (1603), 1679 (1604)
D3'	(OP)$_2$	8029 (85)	1503 (1394), 1679 (1604)
D4	(OP)$_3$	8005 (124)	1836 (1780), 1838 (1782), 1839 (1783)
D4'	(OP)$_3$	8007 (126)	832 (740), 1839 (1783), 1840 (1784)
Lone tridenate Mg^{2+}			
4	(OP)$_3$	8026 (130)	2608 (2573), 2609 (2574), 2610 (2575)
5	(OP)$_3$	8008 (127[f])	919 (826), 2464 (2427), 2465 (2428), na (2429)[f]
6	(OP)$_3$B	8033 (133)	1747 (1669), 1748 (1670), 2585 (2550), 1749^{Base} (1671^{Base})
9	(OP)$_3$	8006 (125)	821 (730), 822 (731), 854 (761)
10	(OP)$_3$	8081 (143)	1420 (1314), 1421 (1315), 1438 (1332)
Ancillary Mg^{2+} ions[g]			
a1	(OP)$_2$	8009 (272)	2611 (2576), 2612 (2577)
a2	(OP)$_2$	8023 (129)	2617 (2582), 2618 (2583)
a3	(OP)$_2$BP	8067 (422)	1845 (1789), 1846 (1790), 1884^{Base} (1828^{Base})
a4	(OP)$_2$	8015 (100)	844 (751), 1689 (1614)
a5	(OP)$_2$	8077 (141)	880 (787), 883 (790)

[a] Paleo-magnesium ions identified by coordination and are conserved in position and coordinating ligands of 23S-rRNA^{HM} and 23S-rRNA^{TT}.
[b] See Figure 1.7 for the locations of these ions in the 2D structure and Figure 1.8 for their locations in the 3D structure.
[c] OP indicates non-bridging phosphate; B indicates base atom; S indicates sugar (ribose) atom; P indicates protein atom.
[d] These are the numbers in the 1JJ2 coordinate file (23S-rRNA^{HM}). The numbers in parentheses refer to the numbers in the 2J01 coordinate file (23S-rRNA^{TT}).
[e] These are the rRNA residue numbers of the Mg^{2+} first shell in 23S-rRNA^{HM}. The numbers in parentheses refer to 23S-rRNA^{TT}, which follows the *E. coli* numbering scheme.
[f] This Mg^{2+} does not have conserved coordination in 23S-rRNA^{HM} [(OP)$_3$]. and 23S-rRNA^{TT} [(OP)$_4$].
[g] These ions are found in close association with paleo-Mg^{2+} ions and are conserved in the 23S-rRNA^{HM} and 23S-rRNA^{TT} structures.

bidentate complexes (Figure 1.4) because it offers (i) the shortest achievable linker and thus the lowest entropy penalty upon binding, (ii) charged oxygens (OP atoms) as first-shell Mg^{2+} ligands and (iii) favorable RNA bond rotamers. This 10-membered ring system (Mg^{2+}–OP–P–O5'–C5'–C4'–C3'–O3'–P–OP–Mg^{2+}), which requires *cis*-oriented OP atoms around the Mg^{2+}, is the elemental unit of Mg^{2+} chelation by RNA. Tri- and tetradentate Mg^{2+} complexes nearly always contain at least one of these 10-membered ring systems (Figures 1.5 and 1.6).

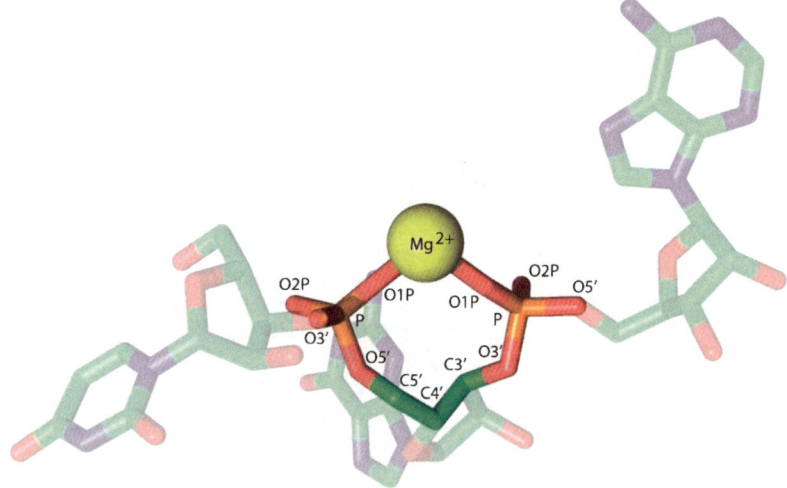

Figure 1.3 RNA as Mg^{2+} chelator. Shown here is a representative bidentate $Mg^{2+}(OP)_2$ complex. For clarity, the base atoms along with ribose atoms C1′, C2′ O4′ and O2′ are lightened. This bidentate complex forms a 10-membered chelation cycle ($\Delta = 1$). It can be seen that this trinucleotide is in a non-A-helical conformational state. This is Mg^{2+} 8009 from 23S-rRNA^{HM}, along with residues U(2610), G(2611) and A(2612).

1.3.2.2.6 Accuracy of 3D Structures. The Mg^{2+} positions in 23S-rRNA^{HM} are, as determined by various geometric criteria, highly credible, with a few exceptions. The Mg^{2+} to oxygen distance is an excellent metric in that the predicted and observed (in 23S-rRNA^{HM}) frequencies reach distinct maxima at 2.1 Å and fall to nearly zero by 2.6 Å. As indicated by relative OP positions, essentially all relevant Mg^{2+} ions were added correctly to the model. Other ribosome structures and smaller globular RNAs follow 23S-rRNA^{HM} in the general patterns of Mg^{2+} interaction [for example, compare Figure 1.5 (23S-rRNA^{HM}) and Figure 1.6 (RNA^{P4-P6} and 23S-rRNA^{TT})]. The same level of confidence does not apply to the monovalent cations of 23S-rRNA^{HM}, which in many cases display unorthodox coordination geometry such as fewer than three first-shell ligands and are judged to be less credible.

1.3.2.2.7 Why Mg^{2+}? The predominant mode of interaction of Mg^{2+} with RNAs large and small is in the form of Mg^{2+}_{aq} and $Mg^{2+}(OP)_1$ complexes (the five water molecules required to complete the hexacoordinate coordination sphere of $Mg^{2+}(OP)_1$ are not specified for the sake of brevity). Geometric considerations suggest that these types of Mg^{2+} ions can be substituted by other cations, such as Na^+, polyamines and cationic side-chains of proteins without substantial alteration of RNA conformation.

Specific requirements for Mg^{2+} in RNA folding derive from Mg^{2+} stabilization of distinctive conformational states of RNA. Mg^{2+}-specific states are characterized by bi-, tri- and tetradentate Mg^{2+} complexes with OP atoms

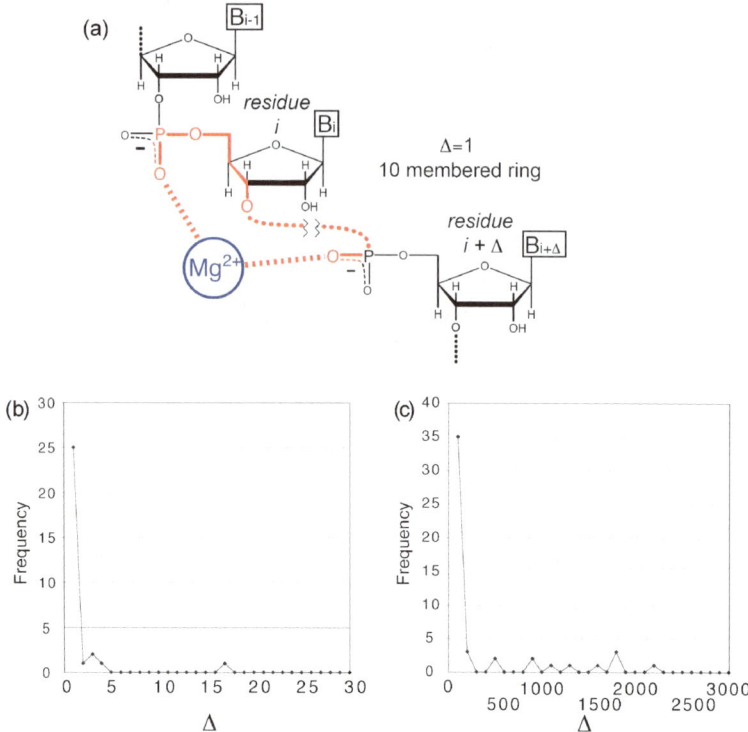

Figure 1.4 Bidentate Mg^{2+} binding is predominantly local. OP ligands are clustered along the RNA chain. (a) Definition of Δ, which is the distance in residues between two OP ligands that bind to a common Mg^{2+} ion. $\Delta = 1$ indicates binding by OPs of adjacent residues, giving a 10-membered chelation ring. (b) The observed frequency of occurrence of Δ values in 23S-rRNA^{HM}, expanded to show the distribution at small Δ. $\Delta = 1$ is preferred over all other Δ. (c) The observed frequency of occurrence of Δ values. Small Δ are preferred over large Δ.

[in $Mg^{2+}(OP)_2$, $Mg^{2+}(OP)_3$, $Mg^{2+}(OP)_4$ complexes]. A representative Mg^{2+}(OP)$_2$ complex in 23S-rRNA^{HM} is shown in Figure 1.3. The OP atoms in these multidentate complexes reach the global minimum in OP···OP distances in RNA. These OP atoms are in closer proximity and are more tightly restrained in position than in any other environment, such as when associated with larger ions such as K^+ or Na^+ or polyamines or cationic side-chains of proteins or when not directly associated with ions. Therefore, RNA conformation in the vicinity of these Mg^{2+} ions is dependent on and is specific for Mg^{2+}. Such tightly packed OP atoms, in association with Mg^{2+} ions, are found in all globular RNAs, including the P4–P6 domain of the tetrahymena ribozyme[1-5] and ribosomes.[6-14] Only Li^+ or Mn^{2+} can rival Mg^{2+} in driving the close packing of OP atoms. However, the first-shell ligands of Li^{2+} tend to assume tetrahedral rather than octahedral geometry.

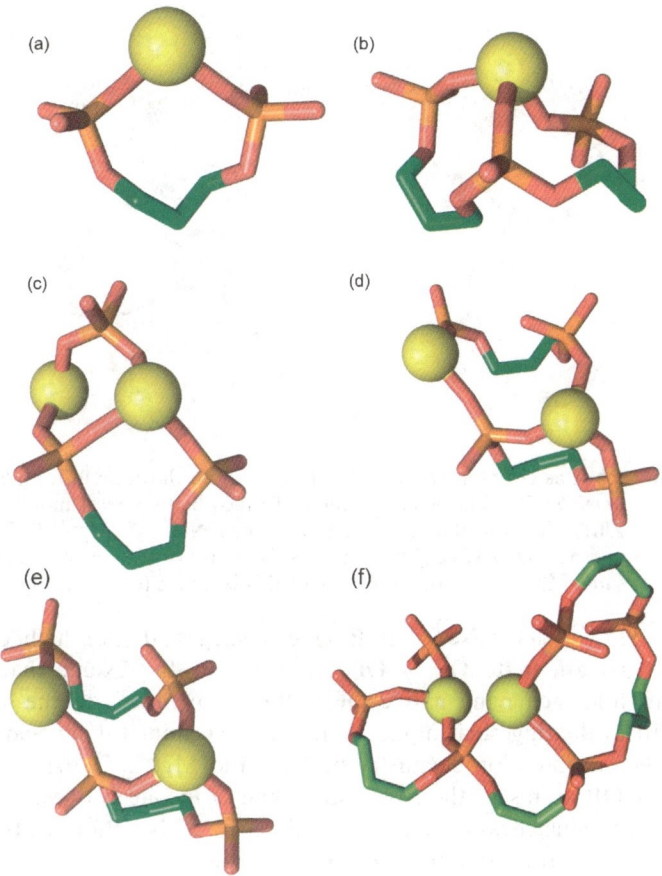

Figure 1.5 RNA as Mg^{2+} chelator. Shown here are some Mg^{2+} chelation complexes in 23S-rRNA^{HM}. The base atoms and ribose atoms C1', C2' O4' and O2' are omitted for clarity. (a) Bidentate chelation by neighboring OP atoms. This common motif is a 10-membered-ring with a *cis* orientation of OP ligands. This is a $\Delta = 1$ complex, indicating that the OP atoms are contributed by reside $i + 1$ and residue $i + 1$ (Mg^{2+} 8009, same as Figure 1.3). (b) Tridentate chelation by three OP atoms from neighboring residues ($\Delta_1 = 1$, $\Delta_2 = 1$, Mg^{2+} 8026). (c) Dual Mg^{2+} bicycle center D1. This is a bicyclic complex with an Mg^{2+}–OP–P–OP–Mg^{2+} bridge. One Mg^{2+} is involved in tridentate chelation by neighboring and a remote OP atoms (Mg^{2+} 8001, $\Delta_1 = 1$, $\Delta_2 = 50$). The tridentate complex is coupled to a bidentate complex (Mg 8002, $\Delta = 51$). (d) Dual Mg^{2+} bicycle center D3. This is a bicyclic complex with a Mg^{2+}–OP–P–OP–Mg^{2+} bridge. One Mg^{2+} is involved in tridentate chelation by two neighboring and one remote OP atoms (Mg^{2+} 8016, $\Delta_1 = 1$, $\Delta_2 = 175$). The tridentate complex is coupled to a bidentate complex (Mg^{2+} 8029, $\Delta = 176$). (e) Dual Mg^{2+} bicycle center D2. This is a bicyclic complex with an Mg^{2+}–OP–P–OP–Mg^{2+} bridge. One Mg^{2+} is involved in tridentate chelation by two neighboring and one remote OP atoms (Mg^{2+} 8003, $\Delta_1 = 1$, $\Delta_2 = 1747$). This complex is coupled to a bidentate complex (Mg^{2+} 8013, $\Delta = 1746$). (f) Dual Mg^{2+} tricycle center D4. This is a tricyclic complex with an Mg^{2+}–OP–P–OP–Mg^{2+} bridge. One Mg^{2+} is involved in tridentate chelation by two neighboring and one proximal OP atoms (Mg^{2+} 8005, $\Delta_1 = 1$, $\Delta_2 = 2$). This complex is coupled to another tridentate chelation complex, by two neighboring and one remote OP atoms (Mg^{2+} 8007, $\Delta_1 = 1$, $\Delta_2 = 1007$).

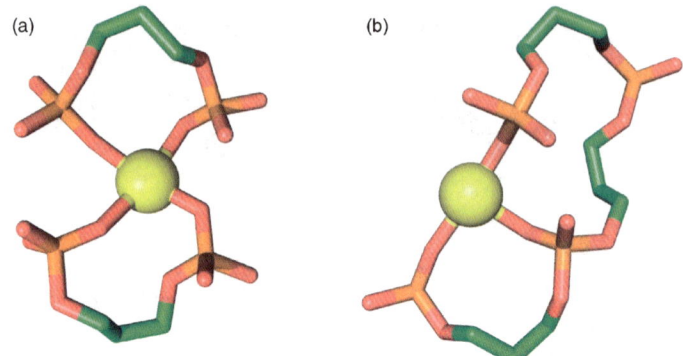

Figure 1.6 RNA as chelator, continued. (a) Tetradetate chelation in the *Thermus thermophilus* 70s ribosome. A complex formed by two $\Delta = 1$ motifs (Mg^{2+} 245, 2J01). (b) Tridentate Mg^{2+} chelation in RNA^{P4-P6} (Mg^{2+} 3). Neighboring and next-nearest neighboring phosphate groups ($\Delta_1 = 1$, $\Delta_2 = 2$) bind to the same Mg^{2+}. Note the similarity of this complex to that in Figure 1.5f.

1.3.2.2.8 Mg^{2+} Versus Na^+. It has been suggested that high concentrations of Na^+ attenuate OP···OP repulsion to the extent that globular RNAs can fold, achieving native OP···OP proximities, in the absence of Mg^{2+}.[60] In evaluating such models, one must account for unyielding differences in the coordination chemistry of Na^+ and Mg^{2+}. The close proximity of adjacent OP atoms in the Mg^{2+} first shell is inconceivable in Na^+ complexes at any concentration and the preference of Na^+ for neutral ligands would drive this cation to alternate sites.

It seems that globular RNAs can collapse in the presence of high [Na^+] to states stabilized by native-like base–base tertiary interactions, but lacking tightly packed OP atoms. One might expect base–base and electrostatic interactions to be somewhat independent of each other because they do not necessarily link the same secondary elements and do not, on a local level, necessarily act in concert. Where the phosphates come closest together the corresponding bases are remote from each other.

1.3.2.2.9 Why Chelation? When Chelation? Multidentate Mg^{2+} binding is commonly under control of local RNA conformation or, conversely, RNA conformation is coupled with multidentate Mg^{2+} binding. The relationship between local conformation and Mg^{2+} binding is important especially when OP atoms are contributed by adjacent residues ($\Delta = 1$, Figures 1.3 and 1.4). Distinctive conformational states are associated with such bidentate complexes. The 10-membered rings of RNA that chelate Mg^{2+} are not, with few exceptions, A-form helix or tetraloops. *Trans* bidentate complexes are favored electrostatically (by attenuated OP···OP repulsion) in comparison with *cis* bidentate complexes, but are disfavored by the entropic cost associated with the necessarily increased linker size. The minimum observed linker for a *trans* $Mg^{2+}(OP)_2$ complex in 23S-rRNA$^{HM/TT}$ is four residues.

What stabilizes cis Mg^{2+} $(OP)_2$ complexes in which OP atoms are forced into close proximity? First, $OP\cdots Mg^{2+}$ attraction offsets the $OP\cdots OP$ repulsion. Second, repulsion is attenuated by charge transfer from OP to Mg^{2+}. Third, the RNA backbone allows the formation of 10-membered chelation rings in the absence of rotameric restraints, i.e. in the absence of unfavorable bond rotations. The implications of multidentate Mg^{2+} coordination transcend thermodynamics. Close $OP\cdots OP$ proximities present kinetic barriers to changes in coordination state.[76] Chapters 6 and 7 contain detailed discussions of the effects of Mg^{2+} on the kinetics of RNA folding.

Many multidentate $Mg^{2+}(OP)_n$ complexes are found in 23S-rRNA^{HM}, as described by Klein et al.[11] As these authors noted, the most frequent chelation motif is $Mg^{2+}(OP)_2$, where phosphate groups from neighboring residues chelate a common Mg^{2+} (Figures 1.3–1.6). Twenty-five of these 10-membered bidentate chelation cycles are observed in 23S-rRNA^{HM}. The OP ligands within the 10-membered cycles can be either O1P or O2P atoms and are invariably in the cis orientation around the Mg^{2+}. For these 10-membered cycles, by definition $\Delta = 1$, where Δ is the distance, in number of residues, between the two OP groups (Figure 1.4). The tridentate complex shown in Figure 1.1b yields two different Δ values; 1 and 95. Nine of 10 $Mg^{2+}(OP)_3$ centers in 23S-rRNA^{HM} contain at least one $\Delta = 1$ complex. The $Mg^{2+}(OP)_3$ center in Figure 1.5b is composed of two $\Delta = 2$ complexes.

The chelating ring size patterns observed in 23S-rRNA^{HM} are general features of large RNAs, in which $Mg^{2+}(OP)_2$ $\Delta = 1$ complexes are observed alone and in combination with other chelation rings. A double $\Delta = 1$ complex, with tetradentate chelation, is observed linking the 16S RNA with the mRNA in the intact ribosome of Thermus thermophilus (PDB entry 2J01, Figure 1.6a). A homolog of the $Mg^{2+}(OP)_3$ $\Delta_1 = 1$, $\Delta_2 = 2$ complex that forms part of the D4 center of 23S-rRNA^{HM} (Figure 1.5f) is found in RNA^{P4-P6} (Figure 1.6b).

1.3.2.2.10 Bicycles, Tricycles. Tridentate Mg^{2+}–RNA complexes, although less frequent than monodentate and bidentate complexes, are especially important in structure, stability and function. Ten $\Delta = 1$ cycles are fused with secondary cycles to form bicyclic $Mg^{2+}(OP)_3$ structures in 23S-rRNA^{HM}. The bicycles fall into two classes; those composed of RNA and a single Mg^{2+} ion (a $\Delta = 1$, $\Delta = 1$ example is shown in Figure 1.5b) and those containing Mg^{2+}–O1P–P–O2P–Mg^{2+} linkages (Figure 1.5c–f). Four Mg^{2+}–O1P–P–O2P–Mg^{2+} linked bicycles are observed in 23S-rRNA^{HM}. In these dual Mg^{2+} bicycles, called here D1, D2, D3 and D4, both the O1P and O2P of a single phosphate group are first-shell Mg^{2+} ligands. In each of these centers, a tridentate $Mg^{2+}(OP)_3$ complex is paired with and mutually stabilizes a bidentate $Mg^{2+}(OP)_2$ complex. The exception is D4, in which two $Mg^{2+}(OP)_3$ complexes form a tricyclic structure. A dual Mg^{2+} bicycle (or tricycle) is essentially a single extended structural unit, of high rigidity and with stability greater than the sum of the parts. The high measure of similarity within subsets of these dual Mg^{2+} bicycles [compare Figure 1.5d (D3) with 1.5e (D2)] suggests that general rules for RNA conformation and Mg^{2+} interactions are discernible from them.

1.3.2.2.11 Paleo-magnesium Ions. Here we introduce the concept of the 'paleo-Mg^{2+} ion'. Paleo-Mg^{2+} ions play key roles in RNA folding, stability and function and are conserved in ribosomes over vast evolutionary timescales. They were identified initially by their RNA coordination and then validated by comparison between different rRNAs. Broadly, a paleo-Mg^{2+} ion is in either an $Mg^{2+}(OP)_3$ center, with at least three RNA first-shell OP ligands, or is a component of a dual Mg^{2+} bicycle or tricycle. Five ancillary $Mg^{2+}(OP)_2$ ions are located in close association with paleo-Mg^{2+} ions in 23S-rRNA$^{HM/TT}$. Figure 1.5b–f are examples of paleo-Mg^{2+} ions from 23S-rRNA^{HM}. The requirement for three OP first-shell ligands over base and sugar ligands is founded on the lower frequency and smaller contribution to stability of base/sugar ligands compared with OP ligands.

In sum, 13 paleo-Mg^{2+} ions associate with 23S-rRNA^{HM} (Figures 1.7 and 1.8, Table 1.4). Their importance in RNA folding, stability and evolution is underscored by the following:

(i) *Conservation in 3D.* The paleo-Mg^{2+} ions are highly conserved in position and mode of interaction between the 23S rRNAs of HM (archea) and TT (bacteria). There is essentially a 1:1 mapping of paleo-Mg^{2+}

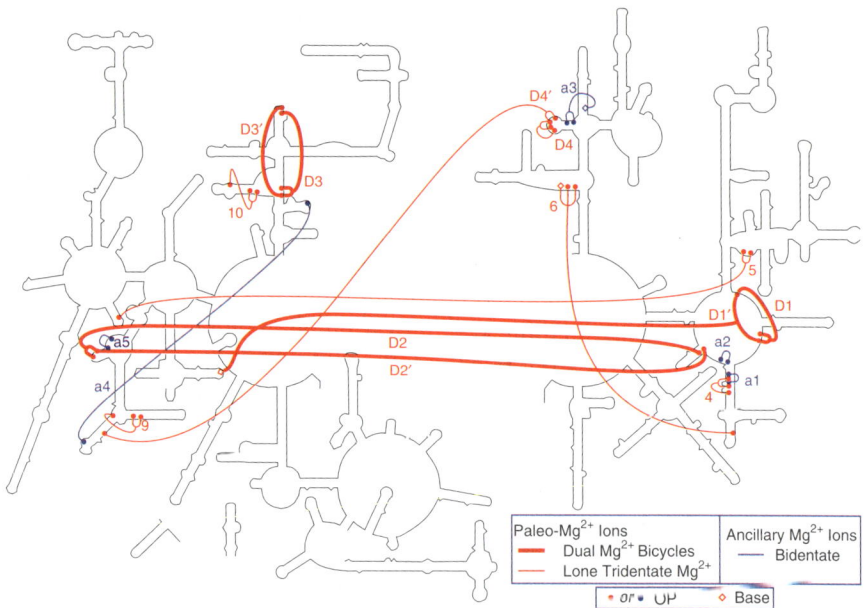

Figure 1.7 The secondary structure of 23S-rRNA^{HM}. The locations of paleo-Mg^{2+} ions are depicted in red. The thick red lines indicate the dual Mg^{2+} bicycles (D1, D2, D3 and D4). The thin red lines indicate the isolated $Mg^{2+}(OP)_3$ complexes. The ancillary $Mg^{2+}(OP)_2$ ions, which cluster on the 2D map with paleo-Mg^{2+} ions, are shown in blue. The circles indicate the OP atoms that contact Mg^{2+} ions. Diamonds indicate base atoms that contact Mg^{2+} ions.

Complexes of Nucleic Acids with Group I and II Cations

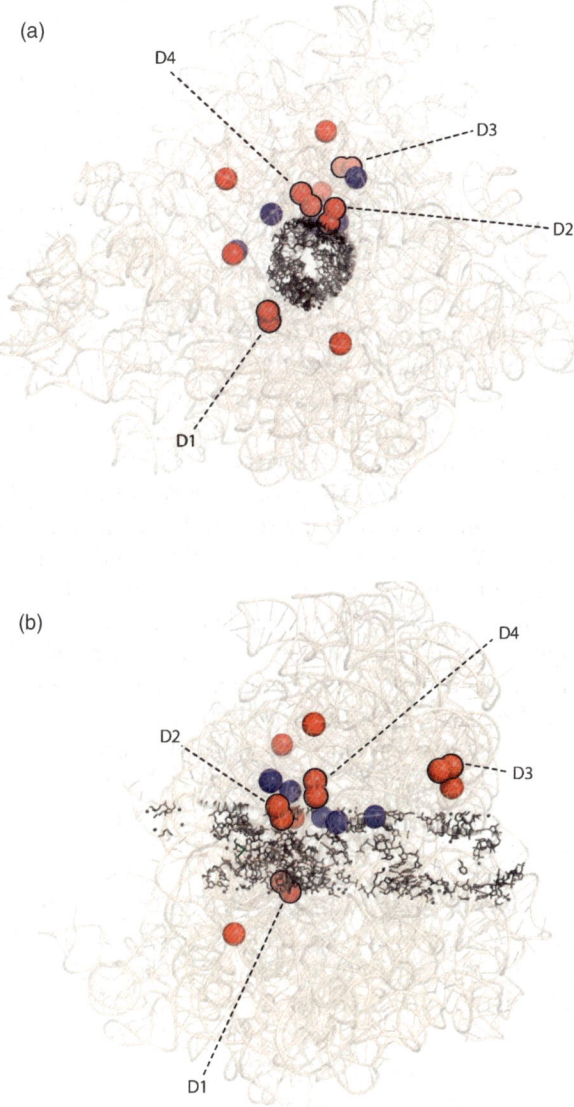

Figure 1.8 (a) View into the peptidyl transfer center of 23S-rRNA[HM]. The 13 paleo-Mg^{2+} ions are shown as red spheres. The Mg^{2+} ions of the four dual Mg^{2+} bicycles/tricycles (D1, D2, D3 and D4) are indicated. Ancillary Mg^{2+} ions are shown in blue. The RNA atoms lining the peptide exit tunnel are accented in black. (b) This view, looking across the peptide tunnel, is rotated by 90° relative to the top panel. The radii of the Mg^{2+} ions are increased over their normal ionic size for clarity and have no physical significance. The proteins and 5S-rRNA are omitted for clarity.

ions, of the surrounding RNA conformation and Mg^{2+} coordination geometry between the two 3D structures, which are separated by billions of years of evolution.

(ii) *Effect on 3D structure.* It is apparent that $Mg^{2+}(OP)_3$ complexes alone and especially dual Mg^{2+} bicycles and tricycles form unique structural entities with rigidity, stability and forced dispositions of functional groups that cannot be approximated by RNA alone or in conjunction with other ions.

(iii) *Conservation in 2D.* Paleo-Mg^{2+} ions generally associate with the most conserved 2D elements in rRNA and link these conserved 2D elements by electrostatic tertiary interactions. These elements are remote in secondary structure. The relationship of paleo-Mg^{2+} ions to secondary structural elements of 23S-rRNA^{HM} is shown in Figure 1.7. The secondary elements linked by paleo-Mg^{2+} ions are conserved between bacteria and archea (see the next item), in the proposed secondary structures of eukaryotic[77] and some mitochondrial rRNAs,[78] and in a proposed minimal 23S-rRNA.[79]

(iv) *Conservation of sequence.* The base sequence of the RNA surrounding paleo-Mg^{2+} ions is highly conserved between the 23S rRNAs of HM and TT. Where the sequences do differ, only purine to purine or pyrimidine to pyrimidine substitutions are observed.

(v) *Role in function.* The locations of paleo-Mg^{2+} ions appear to lend critical support to function (Figure 1.8). Ten paleo-Mg^{2+} ions form a loose ring around the peptidyl transfer center. Three paleo-Mg^{2+} ions are located by the exit of the peptide tunnel. Dual Mg^{2+} bicycles and tricycles D1, D2 and D4 flank the peptidyl transfer center, whereas D3 is located by the exit of the peptide tunnel.

(vi) *Absence of protein ligands.* Paleo-Mg^{2+} ions are not coordinated by protein ligands (Table 1.4), suggesting ancestry prior to development of the ribosomal proteins.

(vii) *Linkage with RNA conformation.* Paleo-Mg^{2+} ions, by nature of their 10-membered chelation cycles, impose constraints on RNA conformation and topology. The relationship between RNA conformation and Mg^{2+} chelation is discussed below.

1.3.2.2.12 Mg^{2+} Avoids RNA Motifs. Chelation of Mg^{2+} is coupled to RNA conformation. Mg^{2+} ions select against canonical conformations such as A-form helices and tetraloops. As noted by Moore[80] and others,[81] folded RNA is largely composed of a relatively small number of motifs such as A-helices,[82] tetraloops,[83–86] E-loop motifs,[87–91] and kink-turns.[37,92,93] RNA motifs are essentially equivalent to RNA secondary structural units, that form early in RNA folding processes. We have used multi-resolution data-mining approaches to extend the definition of RNA motifs, to allow for deletions, insertions, strand clips and topology switches.[94] Formation of RNA motifs (secondary structure) is not Mg^{2+} dependent.

One can observe that Mg^{2+} inner-shell complexes are apparently incompatible with RNA tetraloops. In Figure 1.9, first-shell OP interactions with Mg^{2+}

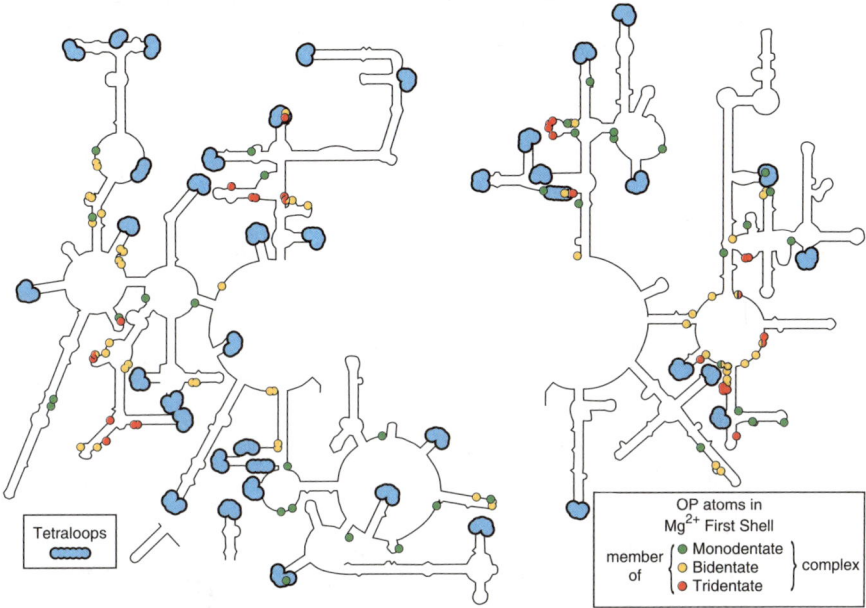

Figure 1.9 Multidentate Mg^{2+} binding selects against tetraloops. The secondary structure and Mg^{2+} contacts of the 23S rRNA from HM are shown. Tetraloop positions, as determined by Hsiao et al.,[94] are indicated in blue. Mg^{2+}–OP interactions are indicated by circles. OP atoms that interact with Mg^{2+} via monodentate interactions are green circles. OP atoms that interact with Mg^{2+} via bidentate interactions are yellow circles. OP atoms that interact with Mg^{2+} via tridentate interactions are red circles.

are mapped on to the 23S-rRNA^{HM} secondary structure, as are the locations of tetraloops.[94] The sites of first-shell Mg^{2+} coordination do not in general correspond to the locations of the tetraloops.

This preference is supported by statistical results. Automated methods[94–96] allow one to count conformational states and determine their populations (frequencies), locations and sequences. One can seek correlations between frequency of occurrence of conformational states and locations of Mg^{2+} ions. 23S-rRNA^{HM} can be partitioned into conformational states and grouped by frequency. For example, various conformational states can be differentiated by their torsion fingerprints. States with many occupants are motifs. States with few occupants represent non-motif RNA (conformational-deviants). To help determine if Mg^{2+} interacts preferentially with motifs or with non-motif RNA requires a comparison with a random binding model (see Appendix) that assumes no preference of Mg^{2+} binding to any residue or conformation over another. We have used this method to calculate populations and then compared the calculated with the observed populations. The results, illustrated in Figure 1.10 and tabulated in Table A1 in the Appendix, indicate that Mg^{2+} binds preferentially to non-motif RNA. The observed extent of binding to non-motif RNA exceeds that predicted by the random model. The observed extent

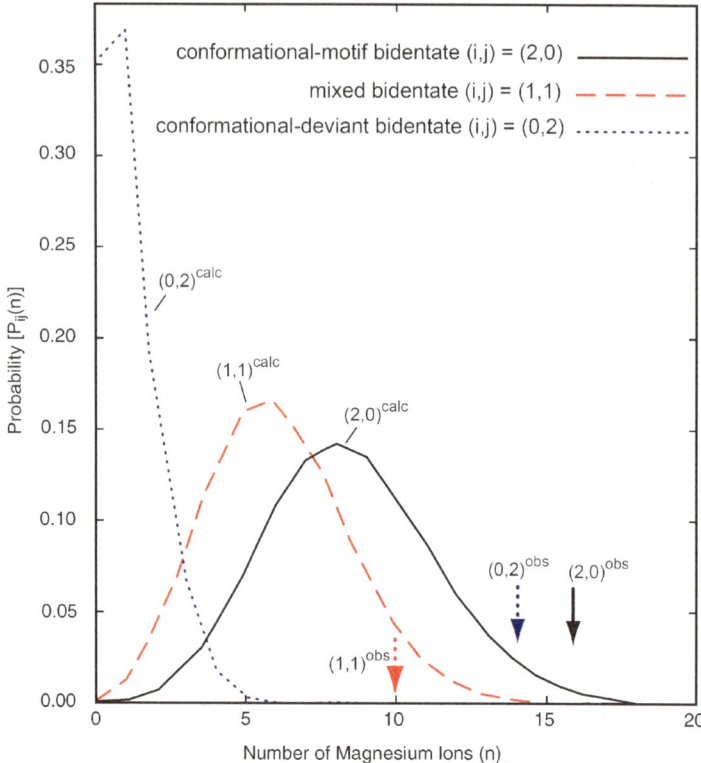

Figure 1.10 Mg^{2+} ions that form multiple bonds to RNA prefer conformational-deviants. The observed populations of various binding states are indicated by vertical arrows. The predicted propensity envelopes, obtained from a random binding model that assumes no preference of Mg^{2+} ion binding for canonical *versus* deviant conformations are shown as solid, dashed and dotted lines. $p_{ij}(n)$ is the probability of n Mg^{2+} ions occupying state (i,j). Predicted envelopes are shown for Mg^{2+} ions in state (2,0) with two bonds to canonical RNA and none to a conformational-deviant, state (1,1) with one bond to canonical and one bond to non-canonical RNA and (0,2) with two bonds to conformational-deviants and none to canonical RNA. The observed populations of states (2,0) and (1,1) are within or very close to the predicted binding envelopes. The observed population in the (0,2) state is far greater than predicted by the random binding envelope.

of binding to motif RNA, such as A-form helices and tetraloops, falls below that predicted by a random binding model.

1.3.2.2.13 RNA Associated with Mg^{2+} is Dispersed in Conformational Space.
The diversity of RNA conformation in the vicinity of $Mg^{2+}(OP)_2$, is illustrated in Figure 1.11. The conformation of RNA acting as bidentate Mg^{2+} ligands is different from the conformation of other RNA, including RNA in

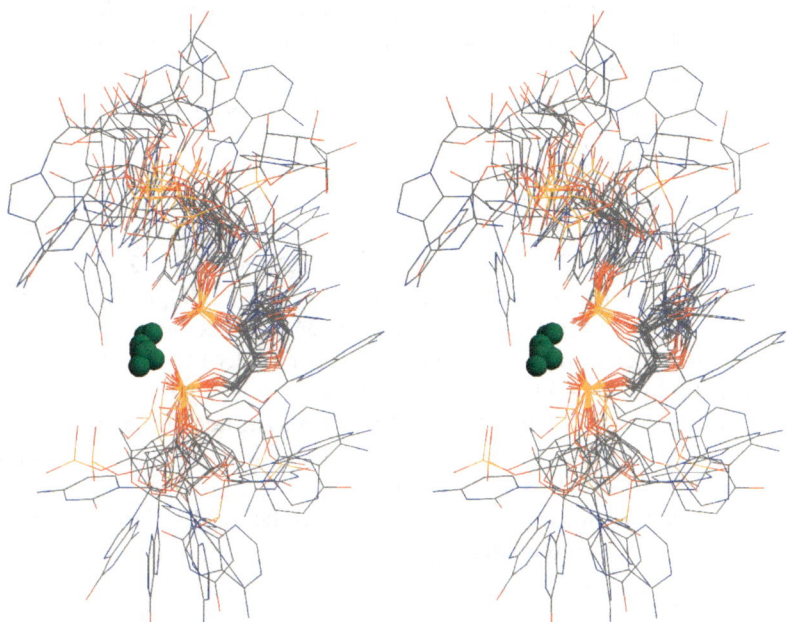

Figure 1.11 Mg^{2+} ions that form multiple bonds to RNA prefer conformational deviants. Fragments of RNA that bind directly to Mg^{2+} were superimposed. Each of these RNA fragments bind to Mg^{2+} using the same two atoms (O1 of residue i and OP2 of residue $i+1$). The backbone atoms of the two residues linking to the Mg^{2+} ions were used for the superimposition. Mg^{2+} ions are represented by green spheres. Other Mg^{2+} ligands (water molecules) are omitted for clarity. From 23S-rRNA[HM].

the vicinities of other bidentate Mg^{2+} ions. This result is consistent with the observation of Klein *et al*. that Mg^{2+} ions tend to bind to "highly idiosyncratic binding sites in 23S RNA that are unlike any previously reported".[11] However, there are reports that in some cases Mg^{2+} ions bind to specific motifs such as non-Watson–Crick base pairs in the E-loop of the 5S rRNA.[87]

1.3.2.2.14 Mg^{2+}–OP Energetics. Cations interact with nucleic acids by both electrostatic and non-electrostatic interactions. Stability is influenced by solvent screening and other ions and by entropic effects.[26,97–100] Entropic gain by water release may, in some cases, be a dominant factor in binding, especially for Mg^{2+} (see below). High-level calculations underscore the importance of non-electrostatic contributions, such as polarization and charge transfer. For example, calculations by Rulisek and Šponer suggest that inner-shell binding of Mg^{2+} to the N7 position of a guanine significantly reduces the energy of the outer-shell binding of the same Mg^{2+} to a phosphate group, due to polarization and charge transfer from Mg^{2+} to guanine.[98] NLPB, molecular dynamics simulation and other computational approaches generally assume that electrostatic interactions are the only significant contribution to energy of binding. Some level of error should be expected from these approximations.

The thermodynamic fingerprint for OP interactions with Mg^{2+} is anomalous in comparison with OP interactions with other divalent cations. For example, the standard enthalpy of interaction of ATP in aqueous media with Mg^{2+} is small and positive.[101] Other divalent cations (Ca^{2+}, Mn^{2+}, Zn^{2+}, *etc.*) give negative enthalpies for this reaction. Thus, in water enthalpic factors are less favorable for forming ATP complexes with Mg^{2+} than with other divalent cations. This effect is related to the large heat of hydration of Mg^{2+}. The negative enthalpy of ATP–Mg^{2+} interaction is offset by a larger negative entropy of interaction of Mg^{2+} with H_2O. The magnitude of the entropic effect is much greater for Mg^{2+} than for other divalent cations. Thus, the association of ATP with Mg^{2+} is driven by entropy – arising from the release of water molecules. This thermodynamic paradigm is specific to Mg^{2+} and is applicable to aqueous Mg^{2+}-OP interactions in general.

1.4 Experimental Methods for Determination of Cation Positions in X-ray Structures

The solvent/ion environment in the vicinity of nucleic acids is difficult to fit unambiguously to X-ray diffraction data. Solvent positions and species identified by X-ray diffraction should be considered to be approximations.[102] Monovalent cations in particular present non-trivial analytical challenges. Sodium ions (Na^+), potassium ions (K^+), rubidium ions (Rb^+), caesium ions (Cs^+), ammonium ions and water molecules, and even polyamines and divalent cations, compete for overlapping sites adjacent on DNA and RNA.[24,103,104] A variety of species coexist and co-localize, giving partial and mixed occupancies. In addition, Na^+ and NH_4^+ scatter X-rays with nearly the same power as water or partially occupied K^+. Therefore, it is not generally correct to exclude K^+ in favor of other species based on thermal factors. One can interchangeably fit H_2O, NH_4^+, Na^+, K^+, *etc.*, to many solvent peaks simply by adjusting or fitting occupancy levels.

1.4.1 Group I

Tl^+ was initially investigated as a K^+ mimic in biological systems by R.J.P. Williams *et al.*[15,105] and others.[106–111] Tl^+ can effectively replace K^+ in diol dehydratase, pyruvate kinase, some phosphatases and other enzymatic systems. In X-ray diffraction experiments, the anomalous signal of Tl^+ renders it a beacon that circumvents the necessity for interpretation of subtle differences in coordination geometry and scattering power of K^+ *versus* Na^+ *versus* H_2O. More recently, Tl^+ has been used as a K^+ substitute in the catalytic mechanisms of sodium–potassium pumps,[112] fructose-1-6-bisphosphatase[113] and pyruvate kinase.[114] Tl^+ has been shown to stabilize guanine quadruplexes in a manner analogous to K^+ and ammonium.[45,115,116] We have used Tl^+ as a probe for K^+ in association with B-DNA[104,117] and in DNA–drug complexes.[118] Doudna and co-workers used Tl^+ as a probe for K^+ in the tetrahymena ribozyme P4–P6 domain.[1] Draper and co-workers used Tl^+ as a K^+ probe in the structure of a fragment of the 23S rRNA.[119] Correll *et al.* used Tl^+ as a K^+ probe in the structure of the sarcin/ricin loop of the 23S rRNA.[86] Tl^+ was used by Caspar and co-workers to determine counterion positions adjacent to insulin[120,121] and by Gill and Eisenberg to determine the positions of ammonium ions in the binding pocket of glutamine synthetase.[122] Rb^+ has also been used, with some success, as a K^+ substitute in DNA structures[123,124] and in ribosomal structures.[11]

1.4.2 Group II

In contrast to monovalent cations, Mg^{2+} ions can often be identified by coordination geometry. As noted above, Mg^{2+} is surrounded by an octahedron of first-shell ligands, generally oxygen atoms, with ligand to Mg^{2+} distances of 2.1 Å and ligand to ligand distances of 2.9 Å (see Figure 1.1b). No other species found in proximity to DNA/RNA has this geometric fingerprint. Mg^{2+} ions that appear to have coordination numbers of less than six are disordered and/or partially occupied. Mg^{2+} can generally be substituted by manganese (Mn^{2+}),[70,125–128] which gives a useful anomalous signal.[2,129] Cobalt hexamine is a useful NMR probe for fully hydrated magnesium.[130–135]

1.5 The Two Binding-mode Formalism

Cation associations with nucleic acids are commonly ascribed to one of two modes, site-bound or non-specific.[18,19,29,35,58,97,119,136–142] This formalism is, at its core, a thermodynamic construct. Yet one constantly strives to interpret and extend this model in terms of three-dimensional structures and theoretical models. In fact, the definitions of the site-bound and non-specific modes are variable, depending on the application and context. It is important to understand that that the various definitions are not necessarily self-consistent; an ion can be site bound by one definition and diffusely bound by another. It seems clear that further progress will be aided by concise definitions of terms. Here we list what we believe are the consensus definitions of site-bound and non-specific ions in various contexts.

1.5.1 Thermodynamic/Kinetic Definition

Non-specific cations are found in diffuse 'ion clouds' and 'ion-atmospheres'. Ions on the outer reaches of an atmosphere interact weakly with the nucleic acid, whereas the inner cations interact strongly. Their integrated contribution to stability is large. These ions retain mobility and are exchangeable for a broad array of alternative ion types. Alternatively, site-bound cations localize in confined volumes with highly favorable enthalpies but at the cost of an entropic penalty (note that the contributions from water release can switch the signs of both effects). Site-bound ions show significantly reduced mobility compared with bulk aqueous ions and cannot be replaced by alternative ionic species.

1.5.2 Structural Definition

Non-specific cations retain water molecules as inner-sphere ligands. These cations are not closely associated with specific nucleic acid functional groups. Non-specific cations are not dependent on specific geometric dispositions of DNA/RNA functional groups. Site-binding is associated with inner-sphere coordination and especially with multidentate interactions (chelation). Site bound ions have lost one or more first-shell water molecules. Multidentate complexes, particularly with Mg^{2+}, require specific geometric dispositions of the RNA/DNA functional groups and can impose special conformational restraints on RNA/DNA. Multidentate Mg^{2+} interactions can stabilize conformational states that are otherwise unfavorable. *Cis*-bidentate Mg^{2+} binding brings OP atoms into van der Waals contact. Draper has discussed an intermediate binding mode called 'water-positioned' cations.[32]

1.5.3 Computational Definition

The non-specific and site-bound categories have specific definitions and practical consequences in NLPB theory,[20,22] in which electrostatic energies are computed from detailed atomic structures. Chapter 9 discusses computational approaches to understanding long-range electrostatic forces. Chapter 6 discusses the application of NLPB to Mg^{2+} and RNA folding. Here it is sufficient to know that NLPB is a continuum solvent model for treating electrostatic interactions. The electrostatic free energy is computed for empty sites plus the free cations and again for filled sites. Non-specific ions are treated non-atomistically, as a continuum that is distinct from the RNA/DNA, with intrinsically incorporated entropy. Specifically bound ions are generally treated as discrete entities. The entropy of site binding is not incorporated automatically. A specific binding site is distinguished from a region of diffuse binding by a small confinement volume and a large occupation probability per unit volume. A specific binding site is restricted to two occupational states, namely occupied or empty, with no partial occupancy permitted. A bound cation is regarded as fixed and part of the fixed charge array of the polyion, rather than part of the diffuse ion atmosphere.

1.5.4 Breaking the Two Binding-mode Formalism

The two binding-mode formalism may provide reasonably satisfactory descriptions of the interactions of Mg^{2+} with nucleic acids. However, for group I cations we believe that this formalism is not generally useful and is probably misleading. For group I cations the distinction between site- and non-specific binding is commonly blurred. At one limit, monovalent cations fully conform to the conventional site-bound description in association with G-quadruplexes,[41–46] with Draper's 58 nucleotide rRNA fragment[119] and with the AA platform of the RNA tetraloop receptor.[1] At the other limit, a subset of fully hydrated group I cations is held in loose association with double helical RNA and DNA.

However, group I ions commonly exhibit intermediate behavior, with a mix of site-bound and non-specific characteristics. Ions appear site bound in that they form inner-shell complexes, with up to four nucleic acid ligands.[66,103,104,123,124] Some of these ions appear non-specific in that they retain significant mobility, switching readily between coordination environments, with residence times from 10 ns to 100 µs,[143] and are exchangeable between a broad array of alternative ion types, including polyamines.[61,103,104,123,144]

DNA ligands of group I ions are uncharged base oxygens and nitrogens, which fall in overlapping arrays along the floors of helical grooves. Therefore, the activation energies and the changes in free energies for the transitions from one site to the next are nominal. Group I cations can transit along the major groove, from say N7/O6 to N7/O6 of adjacent guanines, or along the minor groove B-form A-tract without much of an activation barrier or change in free energy (discussed in Chapter 3). The free energies required to alter coordination geometry and coordination number are small in magnitude. Nucleic acids offer competing isoenergetic binding sites with varying numbers and geometries of first-shell ligands. In sum, differences in coordination chemistry suggest a continuum between site-bound and diffuse modes for Na^+ and K^+ and a clear demarcation for Mg^{2+}.

1.6 Reaction Coordinates for RNA Folding

1.6.1 The Utility of 3D Databases for Determining Mechanism

Crystal structures, when averaged, can provide excellent predictions of solution behavior. It has been observed that relative populations over a large number of crystal structures reflect populations and relative energies in solution.[145,146] Structural databases allow determination of averages and deviations of bond and hydrogen bond lengths, bond angles and dihedrals.[147,148] Structural databases also allow the determination of coordination sphere geometry,[73–75] reaction coordinates and transition pathways.[76,149–153]

Burgi and Dunitz used data-mining of crystal structures to determine reaction coordinates for simple organic reactions.[149,150,154] Similarly, reaction coordinates for conformational transition reaction coordinates[155] and along folding reaction coordinates[152] have been determined for biological polymers.

1.6.2 Mg^{2+}–RNA Complexes Report on Folding Intermediates

As noted in previous sections, in ground-state crystal structures of large RNAs OP atoms of Mg^{2+}(OP)$_2$ complexes tend to be from neighboring residues (*i.e.* $\Delta = 1$ is most probable). Further, RNA in Mg^{2+}(OP)$_2$ complexes is conformationally polymorphic [*i.e.* Mg^{2+}(OP)$_2$ complexes are conformational deviants].

Does one expect to capture such complexes within a large folded RNA? Yes, *if* Mg^{2+}(OP)$_2$ complexes form preferentially with single-stranded regions of RNA folding intermediates. Adjacent residues along a *single-stranded (i.e. flexible) RNA chain* achieve close proximity with greatest probability,[156] and are conformationally most polymorphic. Adjacent residues along a *double-stranded (i.e. relatively rigid) RNA chain* achieve close proximity with lower probability and are conformationally homogeneous.

Hence, the combined data suggest that much of Mg^{2+} binding to folding intermediates (i) is local, (ii) occurs in flexible, single-stranded regions, (iii) dampens flexibility and decreases the available number of conformational states and (iv) is fast on the time-scale of large-amplitude RNA conformational change. However, a subset of Mg^{2+} ions stitch together RNA elements that are remote in primary sequence (*i.e.* are distant along the backbone). During RNA folding, such electrostatic tertiary interactions may form after many base–base tertiary interactions. A summary of this RNA folding model is shown in Figure 1.12. Support for this general mechanism is provided by results of Woodson and co-workers,[157] who concluded that Mg^{2+} ions dampen the dynamics of RNA folding intermediates. Chapter 6 discusses restriction of RNA conformational fluctuations by Mg^{2+} ions. The mechanism here is consistent with preferential binding of Mg^{2+} to ssRNA over dsRNA as observed experimentally in solution,[158] even though that preference is counter to the predictions of polyelectrolyte theory.[18,159]

Appendix

Statistical Analysis of Mg^{2+}–RNA Interactions: Conformational Motifs *Versus* Conformational Deviants

An Mg^{2+} with *i* bonds to conformational motif(s) and *j* bonds to conformational deviant(s) [*i.e.* Mg^{2+}(RNA)$i+j$] is said to be in state (*i,j*). With a geometric algorithm we determined n_{ij}^{obs}, the number of observed Mg^{2+} ions in 23S-rRNA^{HM} in each state (*i, j*). The results, summarized in Table A1, indicate, for example, that 25 Mg^{2+} ions are observed in state (0,0) [*i.e.* Mg^{2+}(RNA)$_{n=0}$]. These Mg^{2+} ions are fully hydrated and do not form bonds to RNA in either canonical or non-canonical conformations. Thirty-three Mg^{2+} ions form a single bond to RNA [Mg^{2+}(RNA)$_{n=0}$]. Twenty of those Mg^{2+} ions bond to RNA in

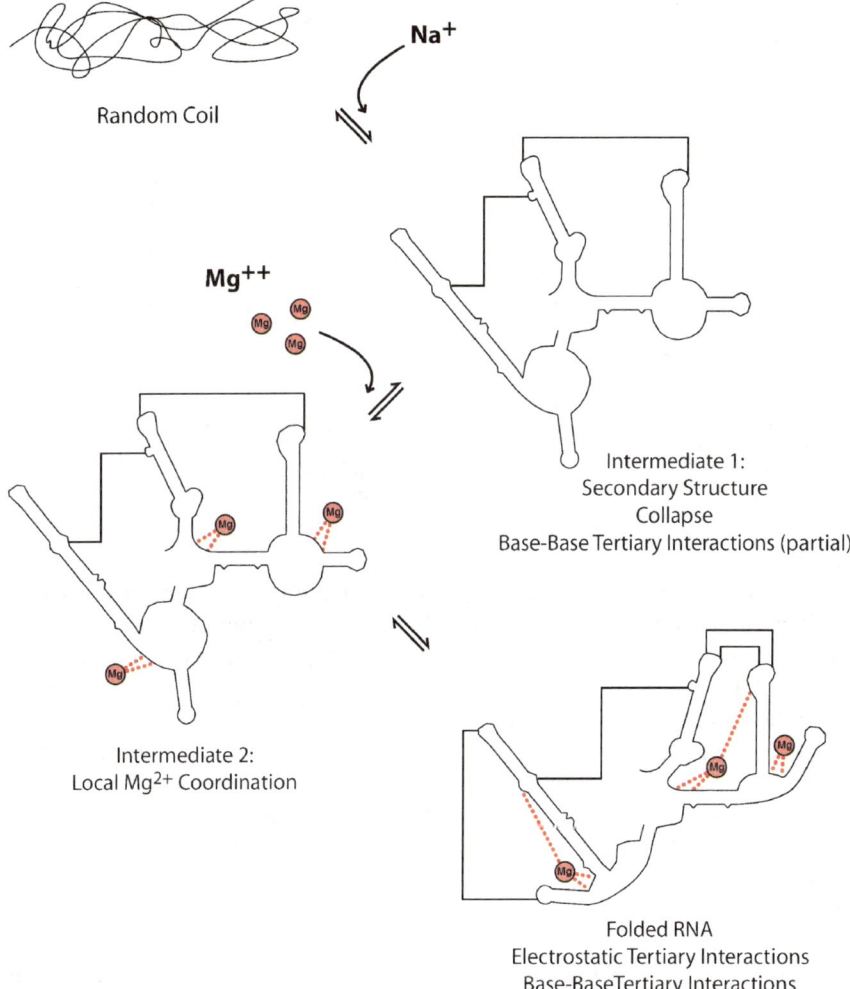

Figure 1.12 Simplified mechanism of RNA folding. The first step of RNA folding is Na$^+$- (or K$^+$-) dependent. In this step, secondary structure forms, along with some tertiary structure (solid lines). The second step is Mg^{2+}-dependent. In this step, local Mg^{2+}(OP)$_2$ chelation occurs (short dashed lines), preferentially at single-stranded regions. This Mg^{2+} binding dampens conformational fluctuations of the single-stranded regions. The third step forms electrostatic tertiary interactions (long dashed lines), which are Mg^{2+} mediated interactions between OPs that are remote in primary structure. At the same time, the final tertiary interactions are locked into place.

canonical conformations ($n_{10}^{obs} = 20$) and 13 of them bond to RNA in non-canonical conformations ($n_{01}^{obs} = 13$), etc.

To help determine if Mg^{2+} interacts preferentially with conformational motifs or with conformational deviants, we formulated a random binding

Table A1 Observed and calculated probabilities of Mg^{2+} interaction with conformational motifs and conformational deviants.

	State[a]		Observed n_{ij}^{obs}	Calculated		
				$n_{ij,ave}^{calc}$	σ_{ij}^{calc}	$\Delta_{ij}/\sigma_{ij}^{calc}$ [b]
Monodentate	$Mg^{2+}(RNA)_0$	(0,0)	25	25	4.4	0.0
	$Mg^{2+}(RNA)_1$	(1,0)	20	15	3.6	1.5
	$Mg^{2+}(RNA)_1$	(0,1)[c]	13	5	2.2	3.6
Bidentate	$Mg^{2+}(RNA)_2$	(2,0)	16	8	2.8	2.9
	$Mg^{2+}(RNA)_2$	(1,1)	10	6	2.4	1.7
	$Mg^{2+}(RNA)_2$	(0,2)[c]	14	1	1.0	12.7
Tridentate	$Mg^{2+}(RNA)_3$	(3,0)	0	5	2.2	−2.3
	$Mg^{2+}(RNA)_3$	(2,1)	3	5	2.2	−1.0
	$Mg^{2+}(RNA)_3$	(1,2)	4	2	1.3	1.6
	$Mg^{2+}(RNA)_3$	(0,3)[3]	5	0	0.45	10.4

[a] State (i,j) contains i bonds to conformational motifs and j bonds to conformational deviants.
[b] $\Delta_{ij}/\sigma_{ij}^{calc} = (n_{ij}^{obs} - n_{ij}^{calc})/\sigma_{ij}^{calc}$, which is the discrepancy between observation and calculation, normalized to the standard deviation of the calculated probability envelope.
[c] Most significant deviations between model and observation.

model that assumes no preference of Mg^{2+} binding to any given residue or conformation over another. We use this model to compute populations of states. We assume that the binding process occurs *via* a series of steps, where at each step a bond is formed with probability p_b. A total of n bonds requires n binding steps, followed by a terminal non-bonding step. The final result is essentially identical with that for a biased one-dimensional random walk. This approach allows us to compute $n_{ij,ave}^{calc}$, the average number of Mg^{2+} ions in state (i,j). In addition, we compute σ_{ij}^{calc}, the standard deviation of $n_{ij,ave}^{calc}$. Finally, we compute the statistical significance of the difference between the observed and calculated results $(n_{ij}^{obs} - n_{ij,ave}^{calc})/\sigma_{ij}^{calc}$. The results are tabulated in Table A1 and Figure 1.10.

The combined results indicate that the observations are not well described by the random model. In observed monodentate RNA complexes [*i.e.* $Mg^{2+}(RNA)_1$], Mg^{2+} ions are observed to form bonds with moderate preference for conformational deviants over conformational motifs. In observed multidentate complexes $[Mg^{2+}(RNA)_{n>1}]$, Mg^{2+} binds with strong preference for conformational deviants. Multidentate Mg^{2+} $[Mg^{2+}(RNA)_{n>1}]$ does not bind randomly, but binds with clear preference to conformational deviants. This preference is indicated by uniformly positive values of $(n_{ij}^{obs} - n_{ij,ave}^{calc})/\sigma_{ij}^{calc}$ for Mg^{2+} ions that form fewer bonds to conformational motifs than conformational deviants [*i.e.* for states (0,1), (0,2), (1,2) and (0,3)]. The (0,2) and (0,3) states show the greatest statistical deviations from the random model, with values of $(n_{ij}^{obs} - n_{ij,ave}^{calc})/\sigma_{ij}^{calc}$ of 12.7 and 10.4, respectively. For the (0,2) state, the random model predicts a population of 1 whereas the database actually contains 14. For the (0,3) state, the random model predicts a population of zero while 23S-rRNA[HM] contains 5. We excluded Mg^{2+} ions with more than three interactions, since there are very few of them and none of the corresponding states are occupied by more than a single Mg^{2+} ion.

Derivation of Functions Governing Random Mg^{2+} Ion Binding to RNA

We begin by first deriving the probability that a given Mg^{2+} has i canonical and j non-canonical interactions. We say that such an Mg^{2+} is in state (i,j) and we denote the probability of being in state (i,j) by p_{ij}.

Let us assume that the binding process occurs via a series of steps, where at each step a bond is formed with probability p_b. For a total of n bonds to form, there must be n binding steps, followed by a terminal nonbonding step. The total probability is then $p_b^n(1-p_b)$. Assuming that an Mg^{2+} is bonded n times, we can then ask what the probability is of it having i canonical and $j = n - i$ non-canonical interactions. If f denotes the fraction of canonical RNA, then the probability of binding to canonical RNA is f and the probability of binding to non-canonical RNA is $1-f$. In the $n = i+j$ binding steps, i of the steps must be canonical bonds and j must be non-canonical bonds. Hence there are $(i+j)!/i!j!$ distinct binding sequences and since each sequence has a probability of $f^i(1-f)^j$, we obtain a total probability of $(i+j)!/i!j! f^i (1-f)^j$. Multiplying by the probability of forming n bonds gives

$$p_{ij} = (i+j)!/i!j! f^i (1-f)^j p_b^{i+j}(1-p_b)$$

Consider the case of N Mg^{2+} atoms interacting with the RNA molecule. We may denote n_{ij} as the number of Mg^{2+} ions in state (i,j). Let $n = (n_{00}, n_{10}, n_{01}, n_{20}, n_{11}, n_{02}, \ldots)$ denote the vector of n_{ij} values. In principle, n can take on any set of values subject to the restriction that $\sum_{ij} n_{ij} = N$. We set p_b equal to the observed fraction of bonded Mg^{2+}s in order to normalize n_{00} to $n_{00,avg}$.

Let $P(n)$ denote the probability of obtaining the occupancy vector n. Note that there are $N!/\Pi_{ij} n_{ij}!$ ways of assigning states to the N Mg^{2+} ions consistent with the occupancy n. Since each such assignment has a probability $\Pi_{ij} p_{ij}^{n_{ij}}$, we obtain

$$p(n) = (N!/\Pi_{ij} n_{ij}!) \Pi_{ij} p_{ij}^{n_{ij}}$$

With this probability distribution in hand, we can now proceed to compute any statistical quantity of interest to us. Here, we are interested in the following quantities: (1) $P_{ij}(n)$, the probability of n Mg^{2+} ions being in state (i,j), (2) $n_{ave} = (n_{00,ave}, n_{10,ave}, n_{01,ave}, \ldots)$ and (3) $\sigma_{ij}^2 = n_{ij,ave}^2 - n_{ij,ave}^2$. To determine (2) and (3), we consider the general case of evaluating $f(n_{ij})_{ave}$ over the distribution $P(n)$, for some function f. Note that

$$f(n_{ij})_{ave} = \sum_n f(n_{ij}) P(n) = \sum_{n-ij=0,\ldots,N} f(n_{ij}) \sum_n P(n)$$

where the final summation is taken over all n consistent with the value n_{ij}. Note then that this sum is simply $P_{ij}(n)$ and thus,

$$f(n_{ij})_{avg} = \sum_{n-ij=0,\ldots,N} f(n_{ij}) P_{ij}(n_{ij})$$

To determine $P_{ij}(n_{ij})$, note that it is obtained by summing over all sequences consistent with having n_{ij} Mg^{2+} ions in state (i,j). We can readily evaluate such a sum as follows: There are $N!/n_{ij}!(N-n_{ij})!$ different ways of placing n_{ij} Mg^{2+} ions into state (i,j). The remaining $N - n_{ij}$ Mg^{2+} ions can each be in any state other than (i,j). Since the probability of being in state (i,j) is p_{ij}, the probability of being in any state other than (i,j) is $1 - p_{ij}$. Therefore, given a set of n_{ij} Mg^{2+} ions in state (i,j) and $N - n_{ij}$ Mg^{2+} ions in any other state, the probability of the sequence is $p_{ij}^{n_{ij}}(1-p_{ij})_{ij}^{N-n}$. Multiplying by the number of ways of choosing n_{ij} Mg^{2+} ions to be in state (i,j), we obtain

$$P_{ij}(n_{ij}) = [N!/n_{ij}!(N-n_{ij})!] p_{ij}^{n_{ij}} (1-p_{ij})^{N-n_{ij}}$$

From this equation, it is a standard result that $n_{ij,ave} = Np_{ij}$ and $\sigma_{ij}^2 = Np_{ij}(1-p_{ij})$. It may be noted that these results are identical with those for a biased one-dimensional random walk.

Acknowledgements

The authors thank Drs. J. Michael Schurr, Nicholas Hud and R. J.P. Williams for helpful discussions.

References

1. S. Basu, R. P. Rambo, J. Strauss-Soukup, J. H. Cate, A. R. Ferre-D'Amare, S. A. Strobel and J. A. Doudna, *Nat. Struct. Biol.*, 1998, **5**, 986.
2. J. H. Cate, R. L. Hanna and J. A. Doudna, *Nat. Struct. Biol.*, 1997, **4**, 553.
3. J. H. Cate, A. R. Gooding, E. Podell, K. Zhou, B. L. Golden, A. A. Szewczak, C. E. Kundrot, T. R. Cech and J. A. Doudna, *Science*, 1996, **273**, 1696.
4. J. H. Cate, A. R. Gooding, E. Podell, K. Zhou, B. L. Golden, C. E. Kundrot, T. R. Cech and J. A. Doudna, *Science*, 1996, **273**, 1678.
5. K. Juneau, E. Podell, D. J. Harrington and T. R. Cech, *Structure*, 2001, **9**, 221.
6. J. H. Cate, M. M. Yusupov, G. Z. Yusupova, T. N. Earnest and H. F. Noller, *Science*, 1999, **285**, 2095.
7. N. Ban, P. Nissen, J. Hansen, P. B. Moore and T. A. Steitz, *Science*, 2000, **289**, 905.

8. B. T. Wimberly, D. E. Brodersen, W. M. Clemons Jr., R. J. Morgan-Warren, A. P. Carter, C. Vonrhein, T. Hartsch and V. Ramakrishnan, *Nature*, 2000, **407**, 327.
9. J. Harms, F. Schluenzen, R. Zarivach, A. Bashan, S. Gat, I. Agmon, H. Bartels, F. Franceschi and A. Yonath, *Cell*, 2001, **107**, 679.
10. M. M. Yusupov, G. Z. Yusupova, A. Baucom, K. Lieberman, T. N. Earnest, J. H. Cate and H. F. Noller, *Science*, 2001, **292**, 883.
11. D. J. Klein, P. B. Moore and T. A. Steitz, *RNA*, 2004, **10**, 1366.
12. V. Berk, W. Zhang, R. D. Pai and J. H. Cate, *Proc. Natl. Acad. Sci. U.S.A.*, 2006, **103**, 15830.
13. M. Selmer, C. M. Dunham, F. V. Murphy, A. Weixlbaumer, S. Petry, A. C. Kelley, J. R. Weir and V. Ramakrishnan, *Science*, 2006, **313**, 1935.
14. N. R. Voss, M. Gerstein, T. A. Steitz and P. B. Moore, *J. Mol. Biol.*, 2006, **360**, 893.
15. R. J. P. Williams, *Adv. Chem. Ser.*, 1971, **100**, 155.
16. R. J. P. Williams, *J. Chem. Soc., Dalton Trans.*, 1991, 539.
17. C. B. Black, H. W. Huang and J. A. Cowan, *Coord. Chem. Rev.*, 1994, **135**, 165.
18. G. S. Manning, *Q. Rev. Biophys.*, 1978, **11**, 179.
19. M. T. Record, C. F. Anderson and T. M. Lohman, *Q. Rev. Biophys.*, 1978, **11**, 103.
20. K. A. Sharp and B. Honig, *Annu. Rev. Biophys. Biophys. Chem.*, 1990, **19**, 301.
21. C. F. Anderson and M. T. Record Jr., *Annu. Rev. Phys. Chem.*, 1995, **46**, 657.
22. J. M. Schurr, 2007.
23. V. Swaminathan and M. Sundaralingam, *CRC Crit. Rev. Biochem.*, 1979, **6**, 245.
24. L. McFail-Isom, C. Sines and L. D. Williams, *Current Op. Struct. Biol.*, 1999, **9**, 298.
25. L. D. Williams and L. J. Maher, *Annu. Rev. Biophys. Biomol. Struct.*, 2000, **29**, 497.
26. J. Sponer, J. Leszczynski and P. Hobza, *Biopolymers*, 2001, **61**, 3.
27. N. V. Hud and M. Polak, *Curr. Opin. Struct. Biol.*, 2001, **11**, 293.
28. J. A. Subirana and M. Soler-Lopez, *Annu. Rev. Biophys. Biomol. Struct.*, 2003, **32**, 27.
29. D. E. Draper and V. K. Misra, *Nat. Struct. Biol.*, 1998, **5**, 927.
30. D. E. Draper, *RNA*, 2004, **10**, 335.
31. B. Onoa and I. Tinoco Jr, *Curr. Opin. Struct. Biol.*, 2004, **14**, 374.
32. D. E. Draper, D. Grilley and A. M. Soto, *Annu. Rev. Biophys. Biomol. Struct.*, 2005, **34**, 221.
33. P. F. Agris, *Prog. Nucleic Acid Res. Mol. Biol.*, 1996, **53**, 79.
34. W. G. Scott and A. Klug, *Trends Biochem. Sci.*, 1996, **21**, 220.
35. A. M. Pyle, *J. Biol. Inorg. Chem.*, 2002, **7**, 679.
36. M. D. Been, *Curr. Top. Microbiol. Immunol.*, 2006, **307**, 47.

37. D. J. Klein, T. M. Schmeing, P. B. Moore and T. A. Steitz, *EMBO J.*, 2001, **20**, 4214.
38. J. S. Richardson, B. Schneider, L. W. Murray, G. J. Kapral, R. M. Immormino, J. J. Headd, D. C. Richardson, D. Ham, E. Hershkovits, L. D. Williams, K. S. Keating, A. M. Pyle, D. Micallef, J. Westbrook and H. M. Berman, *RNA*, 2008, **14**, 1.
39. D. W. Celander and T. R. Cech, *Science*, 1991, **251**, 401.
40. J. A. Latham and T. R. Cech, *Science*, 1989, **245**, 276.
41. J. R. Williamson, M. K. Raghuraman and T. R. Cech, *Cell*, 1989, **59**, 871.
42. N. V. Hud, P. Schultze, V. Sklenar and J. Feigon, *J. Mol. Biol.*, 1999, **285**, 233.
43. S. Neidle and G. N. Parkinson, *Curr. Opin. Struct. Biol.*, 2003, **13**, 275.
44. S. Burge, G. N. Parkinson, P. Hazel, A. K. Todd and S. Neidle, *Nucleic Acids Res.*, 2006, **34**, 5402.
45. M. L. Gill, S. A. Strobel and J. P. Loria, *Nucleic Acids Res.*, 2006, **34**, 4506.
46. P. Podbevsek, N. V. Hud and J. Plavec, *Nucleic Acids Res.*, 2007, **35**, 2554.
47. H. E. Moser and P. B. Dervan, *Science*, 1987, **238**, 645.
48. J. C. Francois, T. Saison-Behmoaras and C. Helene, *Nucleic Acids Res.*, 1988, **16**, 11431.
49. R. H. Shafer, *Prog. Nucleic Acid Res. Mol. Biol.*, 1998, **59**, 55.
50. J. L. Leroy, K. Gehring, A. Kettani and M. Gueron, *Biochemistry*, 1993, **32**, 6019.
51. D. E. Gilbert and J. Feigon, *Curr. Opin. Struct. Biol.*, 1999, **9**, 305.
52. W. C. Winkler and R. R. Breaker, *Annu. Rev. Microbiol.*, 2005, **59**, 487.
53. R. L. Coppins, K. B. Hall and E. A. Groisman, *Curr. Opin. Microbiol.*, 2007, **10**, 176.
54. S. H. Kim, F. L. Suddath, G. J. Quigley, A. McPherson, J. L. Sussman, A. H. Wang, N. C. Seeman and A. Rich, *Science*, 1974, **185**, 435.
55. A. Jack, J. E. Ladner and A. Klug, *J. Mol. Biol.*, 1976, **108**, 619.
56. P. Brion and E. Westhof, *Annu. Rev. Biophys. Biomol. Struct.*, 1997, **26**, 113.
57. I. Tinoco Jr. and C. Bustamante, *J. Mol. Biol.*, 1999, **293**, 271.
58. V. K. Misra and D. E. Draper, *J. Mol. Biol.*, 2000, **299**, 813.
59. P. E. Cole, S. K. Yang and D. M. Crothers, *Biochemistry*, 1972, **11**, 4358.
60. K. Takamoto, R. Das, Q. He, S. Doniach, M. Brenowitz, D. Herschlag and M. R. Chance, *J. Mol. Biol.*, 2004, **343**, 1195.
61. X. Shui, C. Sines, L. McFail-Isom, D. VanDerveer and L. D. Williams, *Biochemistry*, 1998, **37**, 16877.
62. N. V. Hud, P. Schultze and J. Feigon, *J. Am. Chem. Soc.*, 1998, **120**, 6403.
63. N. V. Hud, V. Sklenar and J. Feigon, *J. Mol. Biol.*, 1999, **286**, 651.

64. I. D. Brown, *Acta Crystallogr. Sect. B.*, 1992, **48**, 553.
65. I. D. Brown, *Acta Crystallogr. Sect. B.*, 1988, **44**, 545.
66. S. Komeda, T. Moulaei, K. K. Woods, M. Chikuma, N. P. Farrell and L. D. Williams, *J. Am. Chem. Soc.*, 2006, **128**, 16092.
67. D. A. Doyle, J. Morais Cabral, R. A. Pfuetzner, A. Kuo, J. M. Gulbis, S. L. Cohen, B. T. Chait and R. MacKinnon, *Science*, 1998, **280**, 69.
68. N. V. Hud, F. W. Smith, F. A. L. Anet and J. Feigon, *Biochemistry*, 1996, **35**, 15383.
69. F. H. Westheimer, *Science*, 1987, **235**, 1173.
70. C. W. Bock, A. K. Katz, G. D. Markham and J. P. Glusker, *J. Am. Chem. Soc.*, 1999, **121**, 7360.
71. C. C. Sines, L. McFail-Isom, S. B. Howerton, D. VanDerveer and L. D. Williams, *J. Am. Chem. Soc.*, 2000, **122**, 11048.
72. S. E. Rodriguez-Cruz, R. A. Jockusch and E. R. Williams, *J. Am. Chem. Soc.*, 1998, **120**, 5842.
73. C. W. Bock, A. Kaufman and J. P. Glusker, *Inorg. Chem.*, 1994, **33**, 419.
74. C. W. Bock, G. D. Markham, A. K. Katz and J. P. Glusker, *Theor. Chem. Acc.*, 2006, **115**, 100.
75. G. D. Markham, J. P. Glusker and C. W. Bock, *J. Phys. Chem. B*, 2002, **106**, 5118.
76. D. Bandyopadhyay and D. Bhattacharyya, *J. Biomol. Struct. Dyn.*, 2003, **21**, 447.
77. J. J. Cannone, S. Subramanian, M. N. Schnare, J. R. Collett, L. M. D'Souza, Y. Du, B. Feng, N. Lin, L. V. Madabusi, K. M. Muller, N. Pande, Z. Shang, N. Yu and R. R. Gutell, *BMC Bioinformatics*, 2002, **3**, 2.
78. J. A. Mears, M. R. Sharma, R. R. Gutell, A. S. McCook, P. E. Richardson, T. R. Caulfield, R. K. Agrawal and S. C. Harvey, *J. Mol. Biol.*, 2006, **358**, 193.
79. J. A. Mears, J. J. Cannone, S. M. Stagg, R. R. Gutell, R. K. Agrawal and S. C. Harvey, *J. Mol. Biol.*, 2002, **321**, 215.
80. P. B. Moore, *Annu. Rev. Biochem.*, 1999, **68**, 287.
81. N. B. Leontis and E. Westhof, *Curr. Opin. Struct. Biol.*, 2003, **13**, 300.
82. W. Saenger, *Principles of Nucleic Acid Structure*, Springer-Verlag, New York, 1984.
83. C. R. Woese, S. Winker and R. R. Gutell, *Proc. Natl. Acad. Sci. U. S. A.*, 1990, **87**, 8467.
84. C. Tuerk, P. Gauss, C. Thermes, D. R. Groebe, M. Gayle, N. Guild, G. Stormo, Y. d'Aubenton-Carafa, O. C. Uhlenbeck, I. Tinoco Jr., E. N. Brody and L. Gold, *Proc. Natl. Acad. Sci. U. S. A.*, 1988, **85**, 1364.
85. C. R. Woese and R. R. Gutell, *Proc. Natl. Acad. Sci. U. S. A.*, 1989, **86**, 3119.
86. C. C. Correll and K. Swinger, *RNA*, 2003, **9**, 355.
87. N. B. Leontis and E. Westhof, *J. Mol. Biol.*, 1998, **283**, 571.
88. P. Vallurupalli and P. B. Moore, *J. Mol. Biol.*, 2003, **325**, 843.

89. C. C. Correll, J. Beneken, M. J. Plantinga, M. Lubbers and Y. L. Chan, *Nucleic Acids Res.*, 2003, **31**, 6806.
90. A. A. Szewczak and P. B. Moore, *J. Mol. Biol.*, 1995, **247**, 81.
91. B. Wimberly, G. Varani and I. Tinoco Jr., *Biochemistry*, 1993, **32**, 1078.
92. T. A. Goody, S. E. Melcher, D. G. Norman and D. M. Lilley, *RNA*, 2004, **10**, 254.
93. S. Matsumura, Y. Ikawa and T. Inoue, *Nucleic Acids Res.*, 2003, **31**, 5544.
94. C. Hsiao, S. Mohan, E. Hershkovitz, A. Tannenbaum and L. D. Williams, *Nucleic Acids Res.*, 2006, **34**, 1481.
95. E. Hershkowitz, G. Sapiro, A. Tannenbaum and L. D. Williams, *IEEE/ACM Trans. Comp. Biol. Bioinformatics*, 2006, **3**, 33.
96. E. Hershkovitz, E. Tannenbaum, S. B. Howerton, A. Sheth, A. Tannenbaum and L. D. Williams, *Nucleic Acids Res.*, 2003, **31**, 6249.
97. A. S. Petrov, G. R. Pack and G. Lamm, *J. Phys. Chem. B*, 2004, **108**, 6072.
98. L. Rulisek and J. Šponer, *J. Phys. Chem. B*, 2003, **107**, 1913.
99. N. Gresh, J. E. Šponer, N. Spackova, J. Leszczynski and J. Šponer, *J. Phys. Chem. B*, 2003, **107**, 8669.
100. A. S. Petrov, G. Lamm and G. R. Pack, *J. Phys. Chem. B*, 2002, **106**, 3294.
101. M. M. Khan and A. E. Martell, *J. Am. Chem. Soc.*, 1966, **88**, 668.
102. L. D. Williams. Between Objectivity and Whim: Nucleic Acid Structural Biology. In *Top. Curr. Chem.*, J. B. Chaires, M. Waring, Eds., Springer: Heidelberg, 2005, **253**, 77.
103. X. Shui, L. McFail-Isom, G. G. Hu and L. D. Williams, *Biochemistry*, 1998, **37**, 8341.
104. S. B. Howerton, C. C. Sines, D. VanDerveer and L. D. Williams, *Biochemistry*, 2001, **40**, 10023.
105. J. P. Manners, K. G. Morallee and R. J. P. Williams, *J. Chem. Soc., Chem. Commun.*, 1970, 965.
106. F. J. Kayne, *Arch. Biochem. Biophys.*, 1971, **143**, 232.
107. R. L. Post, S. Kume, T. Tobin, B. Orcutt and A. K. Sen, *J. Gen. Physiol.*, 1969, **54**, 306.
108. C. E. Inturrissi, *Biochim. Biophys. Acta*, 1969, **178**, 630.
109. C. E. Inturrissi, *Biochim. Biophys. Acta*, 1969, **173**, 567.
110. J. S. Britten and M. Blank, *Biochim. Biophys. Acta*, 1968, **159**, 160.
111. J. Reuben and F. J. Kayne, *J. Biol. Chem.*, 1971, **246**, 6227.
112. P. A. Pedersen, J. M. Nielsen, J. H. Rasmussen and P. L. Jorgensen, *Biochemistry*, 1998, **37**, 17818.
113. V. Villeret, S. Huang, H. J. Fromm and W. N. Lipscomb, *Proc. Natl. Acad. Sci. U.S.A.*, 1995, **92**, 8916.
114. J. P. Loria and T. Nowak, *Biochemistry*, 1998, **37**, 6967.
115. S. Basu, A. A. Szewczak, M. Cocco and S. A. Strobel, *J. Am. Chem. Soc.*, 2000, **122**, 3240.

116. C. Caceres, G. Wright, C. Gouyette, G. Parkinson and J. A. Subirana, *Nucleic Acids Res.*, 2004, **32**, 1097.
117. T. Moulaei, T. Maehigashi, G. T. Lountos, S. Komeda, D. Watkins, M. P. Stone, L. A. Marky, J. S. Li, B. Gold and L. D. Williams, *Biochemistry*, 2005, **44**, 7458.
118. S. B. Howerton, A. Nagpal and L. D. Williams, *Biopolymers*, 2003, **69**, 87.
119. G. L. Conn, A. G. Gittis, E. E. Lattman, V. K. Misra and D. E. Draper, *J. Mol. Biol.*, 2002, **318**, 963.
120. J. Badger, Y. Li and D. L. Caspar, *Proc. Natl. Acad. Sci. U.S.A.*, 1994, **91**, 1224.
121. J. Badger, A. Kapulsky, O. Gursky, B. Bhyravbhatla and D. L. Caspar, *Biophys. J.*, 1994, **66**, 286.
122. H. S. Gill and D. Eisenberg, *Biochemistry*, 2001, **40**, 1903.
123. V. Tereshko, G. Minasov and M. Egli, *J. Am. Chem. Soc.*, 1999, **121**, 3590.
124. V. Tereshko, C. J. Wilds, G. Minasov, T. P. Prakash, M. A. Maier, A. Howard, Z. Wawrzak, M. Manoharan and M. Egli, *Nucleic Acids Res.*, 2001, **29**, 1208.
125. J. Eisinger, R. G. Shulman and B. M. Szymansk, *J. Chem. Phys.*, 1962, **36**, 1721.
126. J. Eisinger, F. Fawazest and R. G. Shulman, *J. Chem. Phys.*, 1965, **42**, 43.
127. J. Reuben and M. Cohn, *J. Biol. Chem.*, 1970, **245**, 6539.
128. A. L. Feig, *Met. Ions Biol. Sys.*, 2000, **37**, 157.
129. P. S. Salgado, M. A. Walsh, M. R. Laurila, D. I. Stuart and J. M. Grimes, *Acta Crystallogr., Sect D: Biol. Crystallogr.*, 2005, **61**, 108.
130. S. Rudisser and I. Tinoco, *J. Mol. Biol.*, 2000, **295**, 1211.
131. W. H. Braunlin, C. F. Anderson and M. T. Record Jr., *Biochemistry*, 1987, **26**, 7724.
132. D. Sen and D. M. Crothers, *Biochemistry*, 1986, **25**, 1495.
133. R. V. Gessner, G. J. Quigley, A. H.-J. Wang, G. A. van der Marel, J. H. van Boom and A. Rich, *Biochemistry*, 1985, **24**, 237.
134. M. Brannvall, N. E. Mikkelsen and L. A. Kirsebom, *Nucleic Acids Res.*, 2001, **29**, 1426.
135. J. A. Cowan, *J. Inorg. Biochem.*, 1993, **49**, 171.
136. D. Porschke, *Biophys. Chem.*, 1976, **4**, 383.
137. D. Porschke, *Nucleic Acids Res.*, 1979, **6**, 883.
138. P. R. Schimmel and A. G. Redfield, *Annu. Rev. Biophys. Bioeng.*, 1980, **9**, 181.
139. V. A. Buckin, B. I. Kankiya, D. Rentzeperis and L. A. Marky, *J. Am. Chem. Soc.*, 1994, **116**, 9423.
140. S. L. Heilman-Miller, J. Pan, D. Thirumalai and S. A. Woodson, *J. Mol. Biol.*, 2001, **309**, 57.
141. V. K. Misra and D. E. Draper, *Proc. Natl. Acad. Sci. U. S. A.*, 2001, **98**, 12456.

142. V. K. Misra and D. E. Draper, *J. Mol. Biol.*, 2002, **317**, 507.
143. F. Cesare Marincola, V. P. Denisov and B. Halle, *J. Am. Chem. Soc.*, 2004, **126**, 6739.
144. K. Woods, L. McFail-Isom, C. C. Sines, S. B. Howerton, R. K. Stephens and L. D. Williams, *J. Am. Chem. Soc.*, 2000, **122**, 1546.
145. F. H. Allen, S. E. Harris and R. Taylor, *J. Comput. Aided Mol. Des.*, 1996, **10**, 247.
146. R. Taylor, *Acta Crystallogr., Sect D: Biol. Crystallogr.*, 2002, **58**, 879.
147. R. Taylor, O. Kennard and W. Versichel, *J. Am. Chem. Soc.*, 1983, **105**, 5761.
148. R. Taylor, O. Kennard and W. Versichel, *J. Am. Chem. Soc.*, 1984, **106**, 244.
149. H. B. Burgi, *Inorg. Chem.*, 1973, **12**, 2321.
150. H. B. Burgi, J. D. Dunitz and E. Shefter, *J. Am. Chem. Soc.*, 1973, **95**, 5065.
151. F. H. Allen, R. Mondal, N. A. Pitchford and J. A. K. Howard, *Helv. Chim. Acta*, 2003, **86**, 1129.
152. M. Sundaralingam and Y. C. Sekharudu, *Science*, 1989, **244**, 1333.
153. F. A. Hays, A. Teegarden, Z. J. Jones, M. Harms, D. Raup, J. Watson, E. Cavaliere and P. S. Ho, *Proc. Natl. Acad. Sci. U. S. A.*, 2005, **102**, 7157.
154. H. B. Burgi, J. D. Dunitz, J. M. Lehn and G. Wipff, *Tetrahedron*, 1974, **30**, 1563.
155. J. M. Vargason, K. Henderson and P. S. Ho, *Proc. Natl. Acad. Sci. U.S.A.*, 2001, **98**, 7265.
156. P. J. Flory, *Principles of Polymer Chemistry*, Cornell University Press, Ithaca, NY, 1953.
157. E. Koculi, D. Thirumalai and S. A. Woodson, *J. Mol. Biol.*, 2006, **359**, 446.
158. B. I. Kankia, *Biophys. Chem.*, 2003, **104**, 643.
159. M. T. Record, W. T. Zhang and C. F. Anderson, Analysis of effects of salts and uncharged solutes on protein and nucleic acid equilibria and processes: A practical guide to recognizing and interpreting polyelectrolyte effects, Hofmeister effects, and osmotic effects of salts. *In Adv. Protein Chem.*, 1998; Vol. **51**; pp. 281.
160. A. A. Rashin and B. Honig, *J. Phys. Chem.*, 1985, **89**, 5588.

CHAPTER 2
Coordinative Bond Formation Between Metal Ions and Nucleic Acid Bases

BERNHARD LIPPERT

Fakultät für Chemie, Technische Universität Dortmund, Otto-Hahn-Strasse 6, D-44227 Dortmund, Germany

2.1 Introduction

2.1.1 Nucleic Acids and Metal Ions

Nucleic acids are excellent targets for metal ions and metal-containing compounds, in that they are negatively charged and provide a plethora of potential binding sites. These are the phosphate oxygen atoms, the various atoms of the heterocyclic nucleobases and to some extent even the hydroxyl groups of the sugars.[1-3] The natural intracellular counter ions of DNA and RNAs are K^+, Mg^{2+}, Na^+ and Ca^{2+}, in decreasing order as far as concentrations are concerned, with K^+ being present in a remarkably high concentration of almost 0.2 M. The other essential metal trace elements such as Fe, Zn or Cu appear not to be relevant in the context of nucleic acid binding because they are under the tight control of chaperone proteins, except in cases of disease states (*e.g.* excess of Cu in Wilson's disease or excess of Fe in thalassemia or hematochromatosis) or acute metal poisoning. However, metal–nucleic acid chemistry becomes highly relevant again when transition metal compounds are applied as chemical probes in nucleic acid biochemistry or in chemotherapy. In fact, the use of platinum drugs in cancer chemotherapy over the last 35 years has boosted research on transition metal–nucleic acid chemistry strongly.

2.1.2 Types of Interactions

The types of interactions between metal ions and nucleic acids can be roughly divided into two categories, namely non-covalent interactions and coordinative bonding. As discussed in several other chapters, the first type comprises electrostatic attraction, 'outer-sphere' binding *via* hydrogen bonds, π–π interactions between a ligand of the metal complex and the heterocyclic nucleobases (intercalation) or shape-selective binding to the grooves employing weak forces such as van der Waals interactions. Generally, this type of interaction refers to 'coordinatively saturated' metal species, irrespective of the nature of the metal ion and that of the ligands. The second type, 'inner-sphere' binding, the focus of this chapter, takes place directly between the metal and any of the donor sites mentioned above or combinations thereof. It involves the interaction between a filled orbital of the ligand atom (donor) and a suitable, empty orbital of the metal species (acceptor). Concerning relative binding energies, the non-covalent bonds are in general weaker than the coordinative bonds, which in turn are weaker than typical covalent bonds. Of course, several non-covalent bonds may eventually outweigh a coordinative bond and in certain cases coordinative bonds can be stronger than typical covalent bonds, *e.g.* that of Cl_2. A coordinative bond formation is, in principle, reversible. Hence strong nucleophiles are capable of displacing metal ions bonded to nucleobases. However, the kinetics of such displacement reactions may be very slow, up to the point that the coordinated metal remains bonded to its original target. For example, the antitumor agent cisplatin, when bound to DNA, cannot be completely removed by excess CN^-, despite the fact that the thermodynamic stability of $[Pt(CN)_4]^{2-}$ is higher than that of the most stable Pt^{II}–nucleobase adducts.[4] As demonstrated with isolated model nucleobase complexes of Pt^{II}, exocyclic nucleobase groups adjacent to the metal binding sites can effectively shield the Pt center from attack by CN^- and render such complexes virtually 'inert'.[5]

There is a general problem associated with the pattern of differentiation between non-covalent bonds on the one hand and coordinative bonds on the other, in that a particular metal ion, depending on conditions, can engage in both types of interactions. Thus, the sodium ion, as its hexaaqua complex $[Na(H_2O)_6]^+$, is a typical outer-sphere binder, but after partial or complete loss of its aqua ligands, it can directly coordinate to nucleobase donor atoms (*e.g.* in G-quartet stabilization, see Chapter 4). For K^+ and Mg^{2+} the same is true. The concentration-dependent thermal stabilization/destabilization of duplex DNA by a series of transition metal ions (Cu^{2+}, Cd^{2+}, Zn^{2+}) is likewise believed to reflect this dual property of these metals,[6] and even solvolysis products of cisplatin (*e.g. cis*-$[PtCl(H_2O)(NH_3)_2]^+$) are likely to undergo a non-covalent pre-association with DNA prior to coordinatively binding to target nucleobases.

2.1.3 Brief History

Since the discovery by Hammarsten in the 1920s that metal cations are the natural counter ions of (what is now called) DNA, the field was dominated for more than 30 years by physico-chemical studies in which essentially alkali and alkaline earth metal ions were employed. It was not until the 1950s that interest in metal–DNA interactions started to include transition metal ions, notably Hg^{2+}, CH_3Hg^+, Ag^+, Cu^{2+} and Zn^{2+}. The qualitative difference between alkali and alkaline earth metal ions on the one hand and the d^{10} transition metal ions on the other is that the latter, rather than forming only weak 'outer-sphere' complexes with DNA, become engaged in much stronger coordinative bond formation with the nucleobases. Consequently, coordinative bond formation has more profound effects on DNA structure and properties than just 'outer-sphere' binding of M^+ and M^{2+}, which provides an 'averaged' influence on DNA structure due to fast exchange kinetics. In contrast, 'inner-sphere' binding can lead to slow exchange kinetics, to displacement of protons from the nucleobases and to certain preferences of metal ions for particular bases. For example, Ag^+ is found to prefer GC-rich sequences, whereas Hg^{2+} shows a preference for AT-rich sequences (it may be added at this point that the structural basis of this difference is not fully understood even today).

The discovery of cis-$PtCl_2(NH_3)_2$ (cisplatin) as a powerful antitumor agent by Rosenberg in the late 1960s, and the realization that coordinative bond formation of the d^8 metal ion Pt^{II} with nucleobase donor atoms in DNA is most likely responsible for its mode of action,[7] led to an enormous upsurge in related research activities in the 1970s. This development is still continuing, with numerous other transition metal species, notably of Ru and Rh, now also in the focus (see also Chapter 10). This research has substantially extended our understanding on these interactions. The use of model systems (isolated nucleotides, nucleosides and model nucleobases) has greatly assisted this development, as have two main analytical methods, X-ray crystallography and NMR spectroscopy, which have become routine methods over the last three decades. Both methods were absolutely instrumental in firmly establishing 'inner-sphere' metal binding sites at the nucleic acid bases. Milestones in our understanding of molecular details of transition metal–DNA and metal–nucleobase coordinative bond formation, all based on important contributions of a community of researchers, were, among others, studies of Hg^{2+} and CH_3Hg^+ binding to nucleobases,[8,9] the first X-ray crystal structure of a metal–nucleobase complex,[10] detailed Raman and NMR spectroscopic studies on Pt^{II}–nucleobase interactions[11] and the X-ray crystal structure determination of the GG intrastrand adduct of cis-$(NH_3)_2Pt^{II}$ with a DNA dodecamer.[12] Of course, in more general terms, the X-ray crystal structure determinations of tRNAs in the 1970s, which revealed the importance of Mg^{2+} binding to nucleobases and phosphate groups in stabilizing the particular tertiary structure of these biomolecules, were extremely important in understanding metal–nucleic acid interactions.

The more recent developments in this field are characterized by the attempt to elucidate even very specific details of metal–nucleic acid interactions, such as the significance of hydrogen bonding interactions between co-ligands of the metal and the nucleic acid target (see Chapters 1 and 3), for example, and to understand fully the role of metal ions in catalytically active nucleic acids (ribozymes; DNAzymes) (see Chapter 8), to name a highly topical area of research. On the other hand, there appears to be a tendency in many cases to bypass molecular details of metal–nucleic acid interactions and to concentrate primarily on functional aspects of such complexes. Studies on 'metallized' DNA, to be used as conducting molecular wires, are an example where a picture of the molecular conditions within the wire has yet to be provided – where are metal ions attached and what are the consequences of metal ion reduction? – but the principle appears to work.

2.1.4 Scope

This chapter is not intended to cover exhaustively the field of metal–nucleic acid interactions employing coordinative bond formation. It will essentially concentrate on coordinative bond formation of transition metal ions with the heterocyclic part of the nucleobases as derived from isolated nucleotides, nucleosides and model nucleobases (with the N9 positions of purines and the N1 positions of pyrimidines blocked by alkyl groups). The parent nucleobases (proton at N9 and N1, respectively) will not be considered because they offer additional metal binding positions not available in the natural nucleic acids. Metal coordinative bond formation with phosphate oxygen atoms, albeit of importance with the physiologically important alkali metal ions and occasionally with Mg^{2+}, will be mentioned only briefly, as this topic is covered in other chapters. Metal binding to sugar oxygen atoms (*e.g.* in RNA mononucleosides and mononucleotides) will be ignored because it requires relatively high pH. Finally, interactions of coordinatively saturated metal complexes and nucleic acids, which take place in a strictly 'outer-sphere' fashion, will not be covered at all, despite their indisputable excitement and usefulness as chemical probes of nucleic acid structures.[13]

The emphasis of this chapter will be on basic features of metal binding, namely on binding patterns of and their consequences for the individual bases. In general, larger nucleic acids will not be treated, as their metal binding properties are dealt with in several other chapters.

2.2 Nucleobases as Ligands

2.2.1 Location of Donor Sites in DNA and RNAs

As much of the research on metal binding to nucleic acids involves double-stranded (ds) DNA, it is instructive to recall the relative location of the various binding sites in duplex DNA (Figure 2.1). Ignoring differences between the various polymorphic forms of DNA (A-, B-, Z-DNA and others), it is unquestionable that in particular the purine-N7 sites are well exposed and

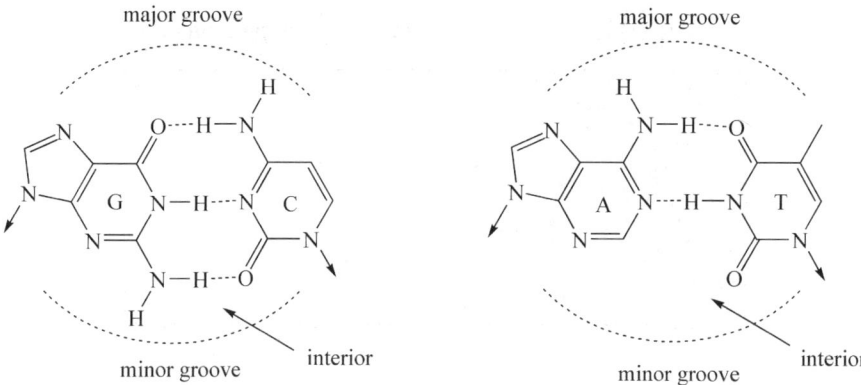

Figure 2.1 Watson–Crick base pairs (G·C; A·T) and relative dispositions of various potential metal binding sites in the major groove, the minor groove and the interior of DNA.

therefore excellent metal binding sites. As will be outlined in Section 2.3, the potential donor sites in the center of dsDNA which are normally involved in Watson–Crick base pairing, are by no means excluded from becoming metal binding sites. Binding to donor atoms in the minor groove, finally, puts some stereochemical restrictions on the positioning of co-ligands at the metal entity, because the metal has to insert into a relatively narrow groove generated by two antiparallel strings of sugar–phosphate backbones.

It seems much more difficult to predict preferred metal binding sites in RNA molecules, given their high structural variety. Folding and metal binding are reciprocally dependent, as discussed in other chapters. Metal binding can generate a fold, and a molecular scaffold of negative charges provide a pocket for metal binding.[14] X-ray crystal structures have characterized the binding patterns of Mg^{2+} with large molecules and have provided ample evidence for the multitude of possibilities (see Chapters 1 and 3). Moreover, the role of metal ions for RNA structure and catalysis in group II intron ribozymes is beginning to become clear.[15] It appears likely that the metal specificity of catalytically active single-stranded DNA molecules (DNAzymes) likewise depends on specific folding patterns.[16]

2.2.2 Acid–Base Equilibria of Nucleobases

N9-substituted purine (pu) and N1-substituted pyrimidine (pym) nucleobases can take part in acid–base chemistry and can therefore occur in their protonated, neutral and deprotonated forms.[17] In Figure 2.2, approximate existence ranges of the various forms of the four common nucleobases are shown, separated by average pK_a values of the bases. To give an example: Adenine nucleobases (N9 position blocked) become protonated below pH ≈ 4 and deprotonated above pH ≈ 17 (Figure 2.3). Accurate pK_a values depend on the

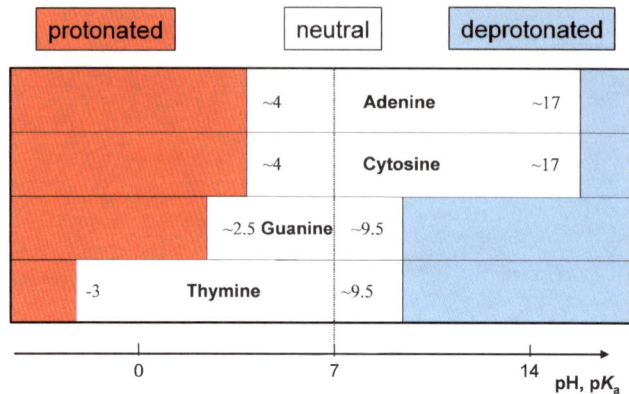

Figure 2.2 Approximate existence ranges of protonated (left), neutral (middle) and deprotonated forms of nucleobases (right). Note that the boundaries, as given by (average) pK_a values, represent 50% distributions of both adjacent forms.

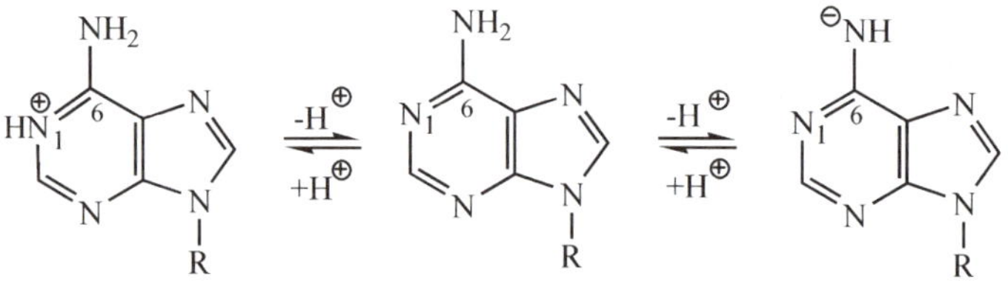

Figure 2.3 Acid–base equilibria of N9-blocked adenine. As in all Lewis structures shown, additional appropriate mesomeric structures have been omitted for simplicity.

substituents at the N1 (pym) and N9 (pu) positions and also, in a polynucleotide, the sequence of the particular nucleobase as a consequence of the electrostatic effects of neighboring phosphate groups. Moreover, pK_a shifts in either direction, sometimes substantial, can be achieved by neighboring group effects that stabilize either the protonated or the deprotonated form of a particular nucleobase. Multiple protonation (*e.g.* AH$_2^{2+}$) or multiple deprotonation processes [*e.g.* (G–2H)$^{2-}$] are not considered in Figure 2.2.

It may appear that deprotonation reactions of A and C (at the exocyclic amino groups) are irrelevant for aqueous chemistry because of the high pK_a values. Although true for the nucleobases in the presence of the physiological metal cations Na$^+$, K$^+$ and Mg^{2+}, in the presence of strongly polarizing transition metal ions such deprotonation reactions can occur readily (see below).

2.2.3 'Conventional' Metal Binding Sites in Neutral Nucleobases

The unprotonated endocyclic N atoms and the exocyclic carbonyl O atoms of nucleobases in their *preferred keto and amino tautomeric forms* are obvious

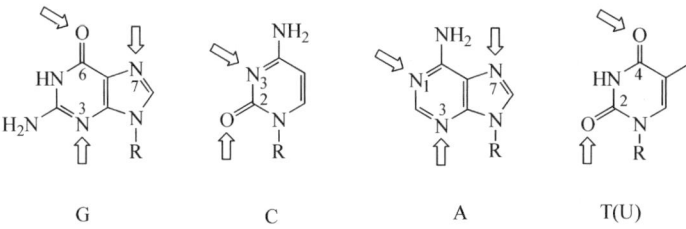

Figure 2.4 Potential metal binding sites of the 4 (5) common nucleobases in their neutral, preferred tautomeric structures.

metal binding sites. These include G-*N7*, G-*N3*, G-*O6*, C-*N3*, C-*O2*, A-*N7*, A-*N1*, A-*N3*, T(U)-*O4* and T(U)-*O2* (where U = uracil) (Figure 2.4). These principal binding sites can be combined in numerous ways. For example, one or more nucleobases can be bonded *via* identical binding sites to a single metal ion, *e.g.* M(G-*N7*)$_2$, or two nucleobases can be bonded via two different metal binding sites, *e.g.* M(C-*N3*)(C-*O2*). Moreover, a single metal may be chelated by a nucleobase, *e.g.* M(C-*N3*,*O2*), or a single nucleobase may serve as a bridge for two metal ions, *e.g.* M$_2$(C-*N3*,*O2*), a neutral nucleobase may act as a ligand for three metal ions, *e.g.* M$_3$(A-*N1*,*N3*,*N7*) with Ms being identical or different metal ions or a single metal ion may bind three different nucleobases via different binding sites, *e.g.* M(C-*N3*)(G-*N7*)(A-*N7*). All these examples have been structurally characterized by X-ray crystallography.[18]

In Figure 2.5, a mixed-nucleobase, mixed-metal complex is depicted which displays four different metal binding sites in a single compound: Pt^{II} is bonded (twice) to N7 of 9-methylguanine and (once) to N3 of 1-methylcytosine, whereas Na$^+$ uses O6 of the guanine base (twice) and O2 of cytosine as donor sites.[19]

2.2.4 Exocyclic Amino Groups: Not Used Unless Deprotonated

Unlike in primary amines RNH$_2$, which have available a lone electron pair at the N atom that consequently renders it a good metal binding site, the exocyclic amino groups of the nucleobases are not metal binding sites. This is because the lone electron pair is delocalized into the heterocyclic ring, as reflected by the very low basicity of these NH$_2$ groups (no protonation possible in aqueous medium), by the sp^2 hybridization of the N atom and the appreciable double bond character of the C–NH$_2$ bond. Consequently, there is no single X-ray crystal structure available of a metal bonded to an NH$_2$ group of A, G or C. The situation is different if either the amino group becomes deprotonated or the nucleobase adopts a rare imino tautomer structure (see below). In these cases, metal coordination to the exocyclic N atom is possible. The experimental findings on the absence of metal binding to exocyclic NH$_2$ groups are in contrast to results of theoretical calculations in the gas phase, which suggest that the exocyclic N atom may carry some electron density, which causes it to undergo some pyramidalization,[20] thereby allowing metal binding. Usually,

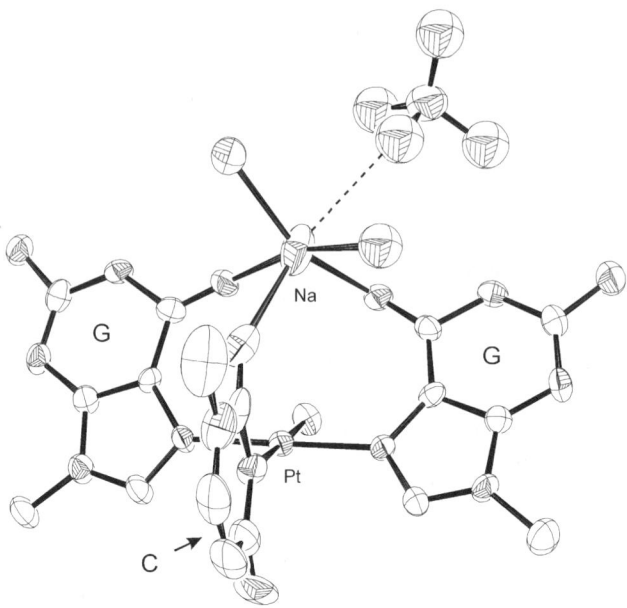

Figure 2.5 X-ray crystal structure of model nucleobase complex *trans*-[Pt(NH$_3$)(1-MeC)(9-MeGH)Na(H$_2$O)$_2$]ClO$_4$ (where 1-MeC = 1-methylcytosine; 9-MeGH = neutral 9-methylguanine), displaying the use of four different binding sites, G-*N7*, G-*O6*, C-*N3* and C-*O2*. Reproduced with permission from ref. 19.

this metal binding takes place in conjunction with a good donor atom, *e.g.* N7 of A (chelation), according to calculations.

2.2.5 Metal Coordination to Rare Nucleobase Tautomers

Proper Watson–Crick base pairing between complementary bases depends crucially on the correct tautomeric structures of the nucleobases, hence the presence of the keto and amino tautomeric forms given in Figure 2.1. The natural abundance of the second most stable tautomer of any of the four bases (enol tautomers of G and T; imino tautomers of A or C) has been estimated to be 10^{-5} or lower. Its occasional appearance during DNA replication has been associated with spontaneous mutations.[21] Given this relatively low value of the tautomer equilibrium constant K_T, it may seem unreasonable to consider direct complexation of a metal ion with a rare nucleobase tautomer at all (with respect to complexation of a rare tautomer in a large genome of 3 billion base pairs, such a possibility may indeed be reasonable). However, the occurrence of metal–nucleobase complexes with the nucleobases being present in rare tautomeric forms is possible and is considerably more relevant than has been anticipated for a long time. It has to be realized that a rare tautomer structure *can be generated* under the influence of a metal *coordinated to a nucleobase in its preferred tautomer structure*. Two scenarios are feasible.

In the first scenario, the chemical modification of the electronic structure of the nucleobase introduced by metal binding causes a shift of the tautomer equilibrium by partially moving a weakly acidic proton to another site of the nucleobase. *The metal ion remains at its initial binding site.* As a consequence, the tautomer equilibrium constant K_T of the nucleobase in its preferred (nb) and rare (nb*) tautomer form [eqn (2.1)] becomes larger [eqn (2.2)]:

$$nb \rightleftharpoons nb^* \qquad K_T = \frac{[nb^*]}{[nb]} \approx 10^{-5} \qquad (2.1)$$

$$M(nb) \rightleftharpoons M(nb^*) \qquad K_{T,M} = \frac{[M(nb^*)]}{[M(nb)]} \qquad K_{T,M} > 10^{-5} \qquad (2.2)$$

hence metal binding generates more of the rare tautomer than could be expected in the absence of a metal. According to a theoretical study,[22] (dien)Pt^{II} (dien=diethylenetriamine) binding to A-*N7* produces a shift from the amino tautomer form of A towards an energetically more favorable imino tautomer A* (Figure 2.6). Earlier attempts to demonstrate such a behavior in aqueous solution were inconclusive.[23] It needs to be mentioned that, at least in the gas phase, an analogous shift in tautomer structure of guanine, from oxo to enol form, is not favored.[24]

In the second scenario, the metal ion and a proton *switch places* or, in other words, the metal ion resides at a site normally occupied by the proton. This process may either be 'spontaneous', as seen for C and A (migration of M from N3 to N4; migration of M from N1 or N7 to N6), or it may be taking place in a stepwise manner that can formally be divided into three stages, (i) nucleobase deprotonation, (ii) metal binding to the deprotonated site and (iii) re-protonation of the anionic nucleobase at a site different from the one where the proton used to be and which is now occupied by the metal. The last possibility is realized for metals complexes of G and T(U), with M occupying N1 and N3 positions, respectively (Figure 2.7). These metal compounds can be viewed as 'metal-stabilized' forms of rare nucleobase tautomers and have been verified for all four bases.[25] Formation of these 'metal-stabilized' rare tautomers is frequently recognized by the unusual pK_a values of the complexed rare tautomers. To give an example for M–C* with M being Pt^{II}: Pt^{II} coordination to the

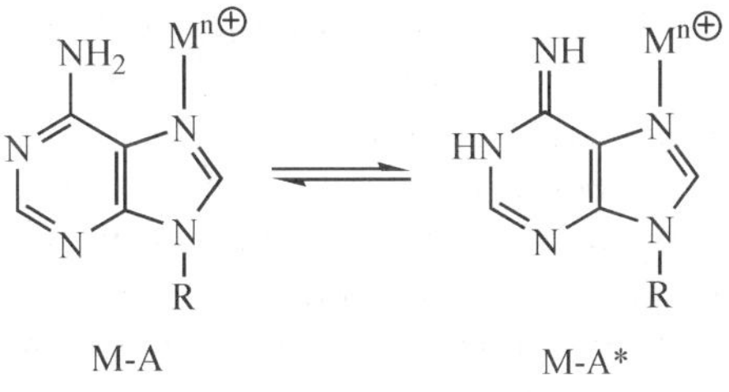

Figure 2.6 Alteration of adenine tautomer structure from the amino (A) to the imino (A*) form under the influence of a coordinated metal.

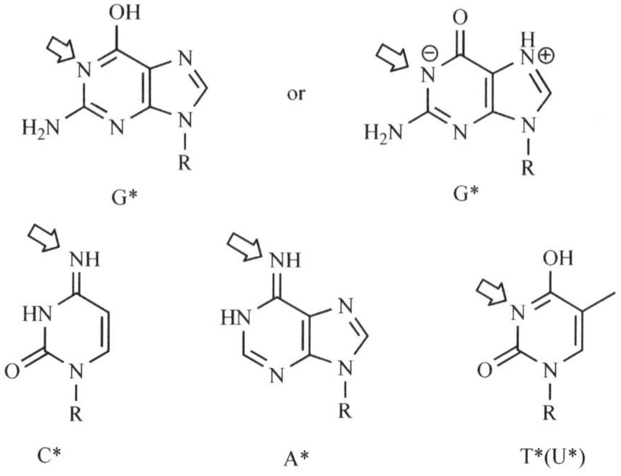

Figure 2.7 'Metal-stabilized' rare tautomer forms of the common nucleobases. Arrows indicate the locations of the metal entities.

'conventional' binding site N3 [M(C-*N3*)] causes an acidification of the exocyclic amino group by some 3–4 log units, to give a pK_a of ~13, as expected. However, in M(C*-*N4*) the cytosine deprotonates with a pK_a of ~7.5, with the deprotonation occurring at the N3 position. A more comprehensive coverage of this aspect can be found elsewhere.[25]

2.2.6 Deprotonated Nucleobases

On the basis of the pH-dependent nucleobase species distribution given in Figure 2.2 and considering physiologically relevant pH conditions (7–7.4), it may seem unnecessary at first glance even to discuss metal binding sites in deprotonated nucleobases. Although indeed true for normal intracellular conditions with K^+, Na^+ and Mg^{2+} as nucleic acid counter ions, it is not true if transition metal ions come into play. *It would be a serious misconception to assume that binding of transition metal ions such as Pt, Pd, Ru, Hg and Zn to deprotonated nucleobases requires strongly alkaline conditions!* Rather, many of these reactions occur at neutral or even at moderately acidic pH. For example, with inosine and Pd^{II}(dien), the 'crossover pH' is 6.1, meaning that inosine-*N1* becomes more favorable than N7 for this particular metal entity above this pH.[4] There are two major reasons for the existence of deprotonated nucleobases in their transition metal complexes. First, due to the polarizing power of the metal(s) bonded to neutral nucleobases (in their preferred or rare tautomeric structure), protons remote from the metal(s) undergo a profound acidification. In other words, the pK_a value shifts to lower values. Numerous factors have an influence on the extent of acidification, such as metal oxidation state, number of metals bonded, distance between metal(s) and proton under consideration or the microenvironment. Concerning the last factor, a co-ligand of the metal(s) capable of stabilizing the deprotonated nucleobase can further shift

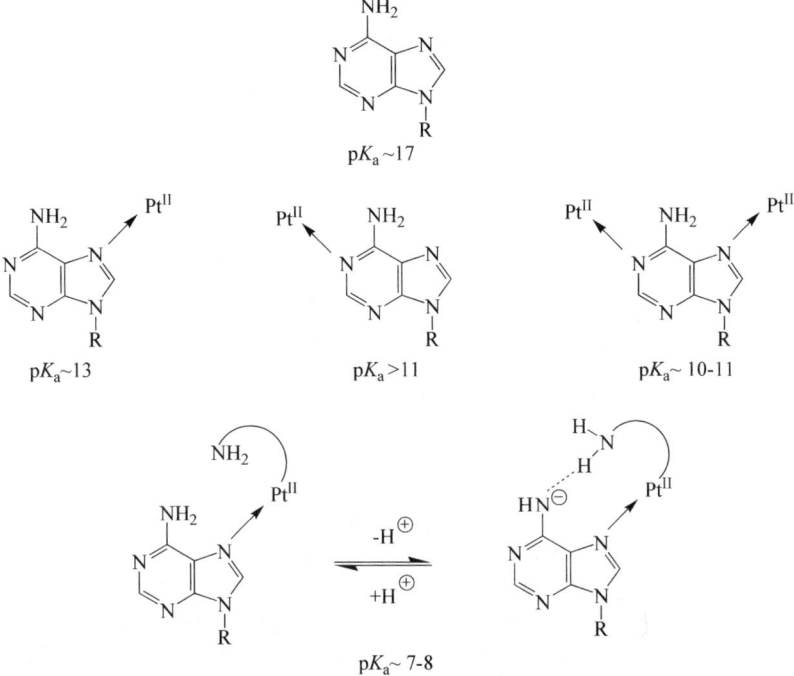

Figure 2.8 Effects of metal (Pt^{II}) coordination on acidity of exocyclic amino group of adenine.

the pK_a to lower values. To give an example (Figure 2.8): the pK_a of the N(6)H$_2$ group of 9-methyladenine, which is close to 17 in the free base, is successively lowered, if one or two Pt^{II} entities are bonded to either N7 or N1 or to both sides. If, in addition, a suitable hydrogen bonding donor function from a co-ligand is present, the pK_a is lowered to 7–8 and hence well within the physiological pH range.[26] Second, the metal ion is frequently present as M–OH and therefore carries its own 'base' and is, in a pH-independent fashion, capable of deprotonating a nucleobase site in a condensation-type reaction.

This phenomenon was reported by Simpson in 1964,[9] when studying the coordination behavior of CH$_3$HgOH toward the four common nucleobases. Deprotonation of endocyclic N(3)H (U, T) and N(1)H positions (G) are possible, as is deprotonation of exocyclic amino groups of A, C, G. Even twofold nucleobase deprotonation (C, A, G), accompanied by metal coordination, has been reported for complexes containing Hg^{II} or both Pt^{II} and Hg^{II}.[27,28] An incomplete list of patterns is presented in Figure 2.9 and an example of a cyclic mixed Pt^{II},Hg^{II} complex containing monoanionic 9-methyladenine and dianionic 9-methylguanine ligands is provided in Figure 2.10.[28]

It appears likely that a combination of the two scenarios described is particularly effective in generating anionic nucleobase ligands. Multiple metal binding to 'conventional' sites of a nucleobase, be it transient or permanent,

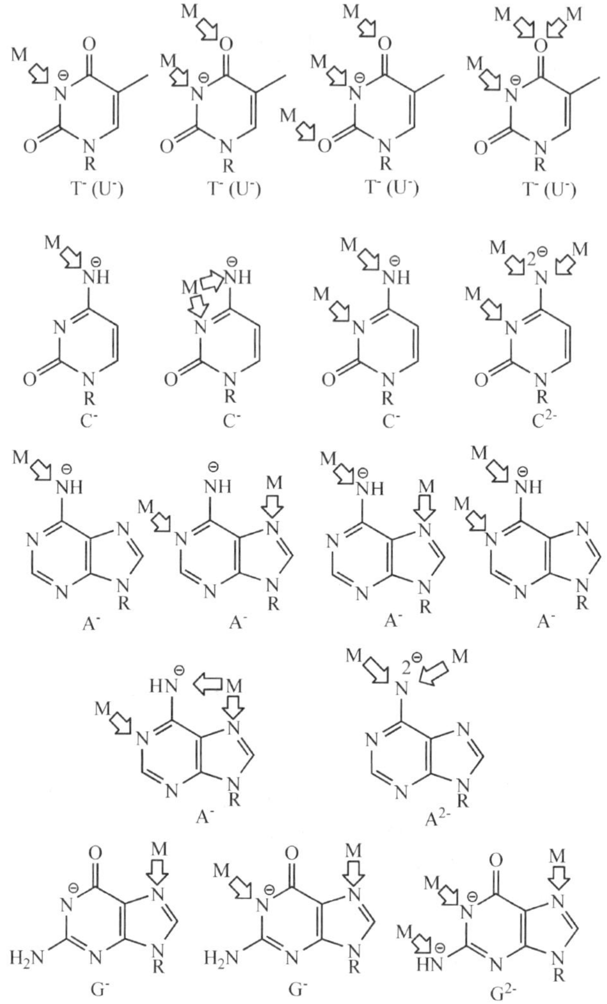

Figure 2.9 Examples of metal complexes containing anionic nucleobases. Note that charges of the deprotonated nucleobases have been put on the site of deprotonation and other mesomeric electronic structures are not considered.

may very well be a highly efficient way to acidify a particular proton that subsequently is replaced by reaction with an M–OH entity.[29]

2.2.7 Organometallic Nucleobase Complexes

There is a limited number of examples of organometallic nucleobase complexes, and hence of nucleobase compounds containing the metal (Hg^{2+}, Pt^{2+}, Pt^{3+},

G
A
Pt(2)
Pt(1)
Hg(3)
Pt(3)
Hg(2)
A
G
Hg(1)

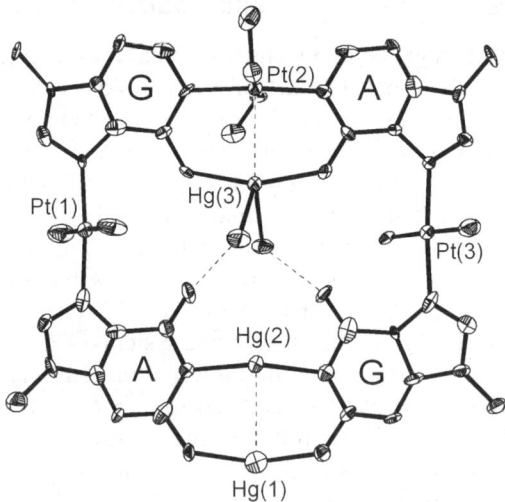

Figure 2.10 View of the cation of a cyclic hexanuclear purine complex comprised of monodeprotonated (at exocyclic amino group) 9-methyladenine and twofold deprotonated (at N1 and at exocyclic amino group) 9-methylguanine. Reproduced with permission from ref. 28.

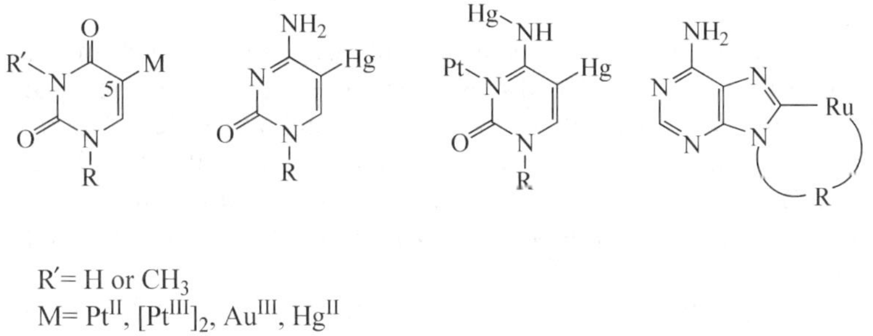

R′= H or CH3
M= Pt^II, [Pt^III]2, Au^III, Hg^II
Hg
NH
NH2
Pt
Ru
R

Figure 2.11 Examples of X-ray structurally characterized organometallic nucleobase complexes.

Au^{3+}) bonded through a ring-C atom of a nucleic acid base. These compounds form under mild conditions in water and are for this reason remarkable. Binding sites are typically C5 of cytosine and uracil, and also C8 of purine bases.[18,30] Formally these complexes also contain (mono- or di-) anionic nucleobases (Figure 2.11). Mercuration of C and U bases by means of $Hg(CH_3COO)_2$ takes place under very mild conditions and has been applied in various structural studies of polynucleotides,[31] and is in fact still in use today in multiple isomorphous replacement (MIR) methods for solving RNA crystal structures.[32]

2.2.8 Chelates, Macrochelates and 'Indirect' Chelates

Chelation of both main group and transition metal ions by nucleic acid bases is possible. If only the heterocyclic part of a nucleoside, nucleotide or a model nucleobase is involved, ring sizes are four (*e.g.* N3 and O2 of G; N3 and N4 of C$^-$; N1 and N6 of A$^-$) or five (*e.g.* N7 and N6 of A$^-$). It appears that metal ions adopting higher coordination numbers (≥ 6) more readily form four- and five-membered chelates than four-coordinate *cis*-square-planar metal ions, such as Pt^{II} and Pd^{II}, and that deprotonation of the nucleobase facilitates chelate formation. In contrast, so-called 'macrochelates' involve a donor site of the heterocyclic part (*e.g.* N3 of T$^-$ or C; N4 of C$^-$; N7 of pu) and a phosphate oxygen atom and hence form much larger ring systems. Terminal phosphates in mononucleotides, being more basic than internal ones in oligonucleotides, are essentially deprotonated at physiological pH (p$K_a \approx 6$) and particularly prone to macrochelate formation. Favorable stacking interactions between co-ligands at the metal and the nucleobase in ternary metal complexes likewise favor macrochelate formation, as extensively studied by Sigel.[33]

There has been a long discussion as to whether the uniqueness of cisplatin as an antitumor agent may in fact be related to its propensity for forming a chelate with the N7 and O6 positions of guanine. However, the existence of such a chelate has never been proven, *e.g.* by an X-ray structure determination. In contrast, with adenine a five-membered ring chelate involving the N7 and the deprotonated N6 position has been demonstrated to exist for *cis*-(PR$_3$)$_2$Pt^{II}.[34]

A third type of chelate, the so-called 'indirect chelate', which in fact is not a chelate in the strict sense, is discussed in the literature. This term describes a situation in which a small ligand, *e.g.* a water molecule or ammonia, is coordinated to the metal and is hydrogen bonded to an atom of the heterocyclic ring (*e.g.* O6 of G) or the oxygen atom of a phosphate of a coordinated nucleotide. Such interactions appear to be ubiquitous in metal–nucleic acid complexes.[35]

As a concluding note of caution, it should be added that equilibria frequently exist in solution between the various forms of 'chelates', especially if stability constants are not too high and if the complexes are kinetically labile.[36]

2.2.9 Coordination Behavior of Di- and Multinuclear Metal Complexes

Bridging rather than chelating binding modes are realized when dinuclear, metal–metal bonded species such as dirhodium, diruthenium or dirhenium carboxylates, acetamidates or formamidinates interact with isolated nucleobases, nucleotides or DNA fragments.[37] Interest in these reactions stems from early reports concerning their antitumor activity. Although these metal species can bind in a monofunctional way to nucleobases *via* the axial sites, hence with retention of the original paddle-wheel structure, it is obvious that substitution of one or of two of the bridging equatorial ligands by purine nucleobases

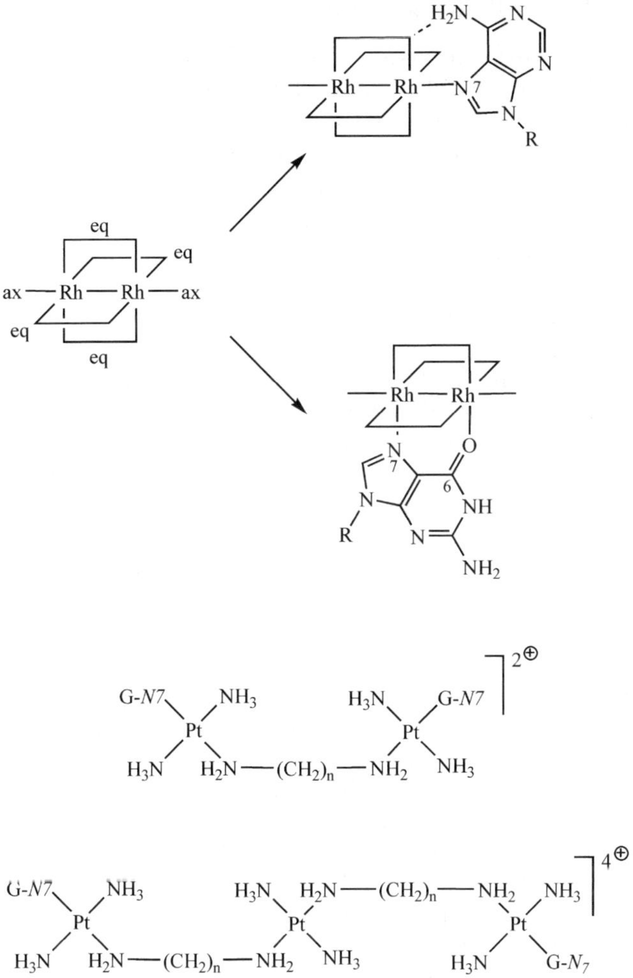

Figure 2.12 Examples of di- and trinuclear metal complexes interacting with nucleic acid bases.

(G-*N7,O6*; A-*N7,N6*) is particularly favorable for steric reasons (Figure 2.12, top). With the bis(purine) adducts both *head–head* and *head–tail* dinuclear structures are possible. Dinuclear Cu^{II} and Zn^{II} complexes have likewise been shown to bind to G in a bridging fashion, involving N7 and O6 sites.[38]

Flexible di- and trinuclear Pt^{II} complexes with monofunctional metal entities at both ends and with the ability to form DNA cross-links separated by several base pairs have been intensively studied by Farrell (Figure 2.12, bottom). Some of these compounds turned out to be distinctly different in their antitumor properties from cisplatin and its analogs, although binding patterns with nucleic acid bases are similar to those of other monofunctional Pt drugs.[39]

2.2.10 Protonated Purine Bases as Ligands

Most likely biologically irrelevant, there are cases known where purinium cations, *e.g.* 9-methyladeninium and 9-alkylguaninium (alkyl = ethyl or methyl), are complexed to Pt^{II}. The metals are coordinated to N7 and the protonation sites are N1 (adenine compound) and N3 (guanine compounds).[18] No analogous compounds with pyrimidine nucleobases are known.

2.2.11 Other Modes

For the sake of completeness it should be mentioned that there is one case reported in which an Ru^{II}-containing anion forms a π-complex with the C5–C6 double bond of cytidine and uridine.[40] Finally, oxidation of C5–C6 double bonds of the pyrimidine bases T and C to *cis*-diol structures can occur with the strong oxidants MnO_4^- and OsO_4. In particular, Os compounds[41] have been widely applied as chemical probes for detecting mis- or unpaired and also bulged-out thymine bases in duplex DNA or 5-methylcytosine, a modified nucleobase associated with modulating gene expression.

2.2.12 Metal Migration

It has never been questioned that with kinetically labile metal complexes dynamic exchange processes can easily take place. However, for some time there has been the belief that metal ions forming kinetically inert (robust) complexes remain bound to particular donor atoms of nucleobases once they have reached them. This view originally also applied to cisplatin adducts of DNA, but over the years it has changed. In fact, from model studies with isolated nucleobases,[26,42,43] and also studies with oligonucleotides in DNA[44–46] it has become evident that even Pt^{II} is capable of migrating along different donor atoms of an isolated nucleobase, namely from a kinetically to a thermodynamically preferred binding site, and that in DNA a favorable disposition of another donor atom may lead to a cross-linking adduct not expected on the basis of 'genuine' binding preferences. An example is the change of a 1,2-intrastrand *GG* cross-link of cisplatin to an interstrand *GG* cross-link.[46] It appears that different mechanisms exist for how such processes occur, with only a few being reasonably well understood at present.

Finally, even migration of Pt^{II} from a sulfur donor atom (amino acid methionine) to G-*N7* has been reported.[47,48] Theoretical calculations have provided some rationale for why there is a preference of N over S donor sites in this case, although the HSAB (*H*ard and *S*oft *A*cids and *B*ases) concept might have suggested otherwise.[49]

2.2.13 Metals as Chameleons in Nucleic Acid Coordination

It is tempting to categorize metal ions according to their nucleobase preference. However, such a simplistic view is not (always) valid and in fact may discredit other opportunities and lead to wrong conclusions. Rather, it depends on numerous factors such as electrostatic potential and hence sequence, accessibility of the particular donor atom and co-ligands providing an element of 'recognition' for which binding pattern is eventually realized. Although it is true that many Pt^{II} species favor G-$N7$ over other binding sites, other possibilities must not be overlooked. Thymine, although generally not considered a prime target of cisplatin, can become platinated at the N3 position, following its deprotonation, not only in model systems, but also in oligo-T stretches.[50]

In the following, the versatility of Zn^{II} in its reactions with DNA and its constituents will be briefly summarized to underline this aspect. In a way, the chemistry of Zn^{II} with nucleic acids is special. Not only are Zn ions integral constituents of the DNA binding zinc finger proteins and part of nucleic acid polymerases and hydrolases, but also they appear to be unique in their ability to rewind the duplex of melted DNA. In addition, Zn ions are capable of stabilizing particular triplex structures (see below) and Zn complexes have been demonstrated to induce B → Z transitions. Whereas in dsDNA a preference of Zn^{2+} for GC-rich sequences is observed,[51] in melted DNA other binding patterns are also likely to be possible, as judged from numerous model studies. With the variable coordination geometries (tetrahedral, five-coordinate, octahedral) that Zn^{II} is able to adopt, its balance in donor sites preference (N, O) may account for this propensity. To give examples: octahedral [Zn(H$_2$O)$_4$(IMP-$N7$)$_2$] geometries[52] (where IMP = inosine 5′-monophophate) and tetrahedral [ZnCl$_2$XY] (where X = H$_2$O or X, Y = nucleobases) are feasible, for example in cross-linking N7 of one G with O6 of another.[53] Surprisingly, Zn^{II} binding to A-$N7$ even in the presence of G is possible,[54] and when incorporated in a suitable macrocyclic polyamine ligand, Zn^{II} can be become a highly specific chemical probe for T in DNA or U in RNA.[55] Multiple Zn units are even capable of unzipping stretches of AT-rich regions of DNA, thereby inhibiting DNA transcription. Finally, a model compound containing three Zn^{II}, two μ-hydroxo ligands and eight 1-methylcytosine bases, six of which display π–π stacking, needs to be mentioned,[56] because some of its structural details may be relevant to the role of Zn^{II} during DNA rewinding or to the structure of M-DNA, a form of DNA in which metal ions have been proposed, but still remains to be proven, to replace complementary hydrogen bonds in the center of the double helix.[57] In Figure 2.13, various scenarios are sketched.

2.2.14 Complex Stabilities

Thermodynamic stability constants K of metal complexes with isolated nucleobases, nucleosides and nucleotides have been determined for a number of transition metal ions. Frequently, a linear relationship is observed between log$K$ and pK_a of a particular metal ion with related ligands that provide similar donor groups.[4] Stability constants for the slowly reacting Pt^{II} species are difficult to obtain and are frequently estimated by taking values of the (fast reacting) Pd^{II} analogues and by allowing some additional stabilization (factor of 10) in the case

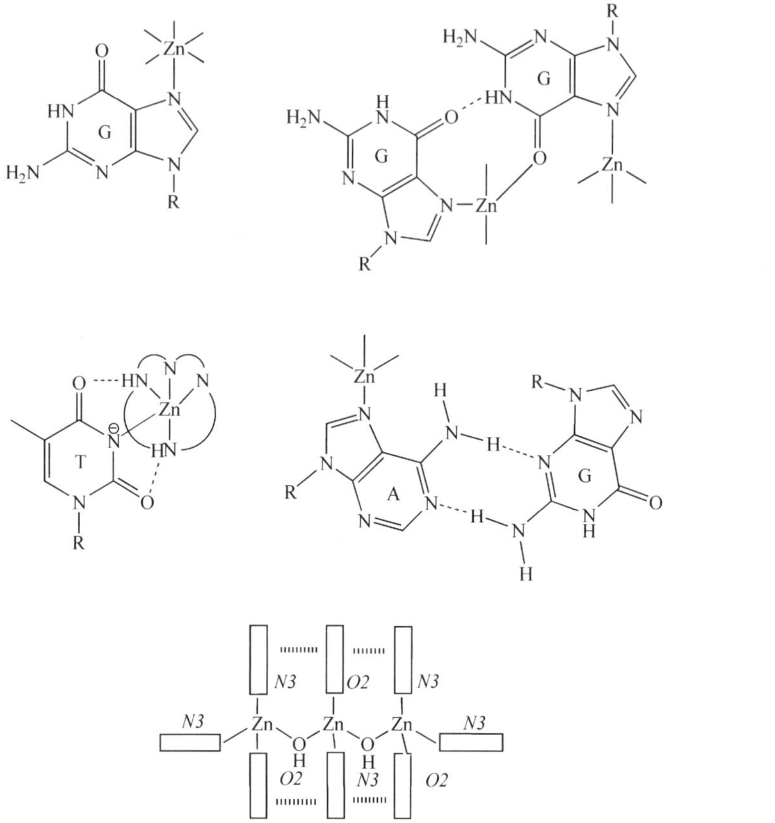

Figure 2.13 Different interactions of Zn^{2+} ions with nucleobases as outlined in the text.

of Pt^{II}. As compared with typical first row transition metal ions (Ni^{2+}, Cu^{2+}, Zn^{2+}), stability constants for Pd^{II}(dien) and Pt^{II}(dien) are substantially higher. As shown by Martin,[4] the stabilities of Pd^{II}(dien) complexes with the common nucleosides/nucleotides span a range of 7 log units between the highest (log$K$ up to 9 for T-$N3$, U-$N3$, G-$N1$) and lowest (A-$N7$). Clearly, the high stability constants of the compounds containing the deprotonated nucleobases correlate with the high basicity of these 'pyridine-like' sites. In addition, substituents in *ortho* position to the metal binding site have an influence on complex stability. If complex stabilities are considered at a particular pH, *e.g.* physiological pH = 7.4, competition of the metal binding sites with H^+ binding has to be taken into consideration, leading to a different sequence of the so-called 'stability values'. For Pd^{II}(dien) the sequence is G-$N7$ > T-$N3$,G-$N1$ > C-$N3$ > A-$N1$ > A-$N7$, now spanning a range of only 4 log units. Under these conditions, the N7 site of G is the strongest binding site for Pd^{II}(dien).[4] If binding of transition metal ions to oligonucleotides and nucleic acids is considered, adduct stabilities are also markedly affected by the ionic strength of the medium.

2.3 Consequences of Metal Coordination

2.3.1 Cross-link Formation in dsDNA

2.3.1.1 DNA Distortion by cis-$Pt^{II}(NH_3)_2$

It is obvious that the cross-linking of nucleic acid bases by a metal ion has a more profound effect on the properties of a nucleic acid than has monodentate binding to a single site, *e.g.* on the exterior of a DNA double helix. There has been particular interest in the structural and functional consequences of metal binding of *cis*-$Pt^{II}(NH_3)_2$ because of its suspected role in the mode of action as an antitumor agent. Specifically, the intrastrand *GG* cross-link, which is the most abundant one during the reactions of cisplatin with DNA, has been the subject of a large number of studies, both in solution and in the solid state. Despite differences in detail between the results of NMR studies[46,58] and the X-ray crystal structures of a platinated DNA fragment,[12] there is general agreement derived from both methods concerning the major features of DNA distortion. Namely, DNA bending towards the major groove as a consequence of tilting of the two adjacent G nucleobases (angle 30–50°), unwinding of the DNA duplex as a consequence of pulling together the N7 sites of the two bases to a distance of *ca.* 2.8 Å and a perturbation of hydrogen bonding of the G·C pair at the 5′-side. It now appears certain that *GG* adducts, like their models,[59] can occur in different conformations, which are in equilibrium, and for this reason difficult to study by 'slow' spectroscopic techniques. The cellular response to the resulting DNA distortion is binding of specific damage-recognition proteins that trigger a cascade of reactions leading eventually to necrosis and/or apoptosis.[60] Details of these processes are discussed further in Chapter 10.

Interstrand cross-linking of *cis*-$Pt^{II}(NH_3)_2$ with two G bases via the N7 positions, albeit representing a minor DNA adduct of cisplatin only, likewise causes severe distortions in DNA structure, namely DNA bending toward the minor groove and duplex unwinding. The two complementary C bases are rotated out of the original double helix and are now extrahelical.[61,62]

Numerous other possible cross-linking adducts of cisplatin with DNA bases have been prepared and studied, frequently at the level of model nucleobases,[63] but also at the level of dinucleotides.[64]

2.3.1.2 Interstrand Cross-linking

The degree of DNA distortion imposed by cross-linking two complementary bases in DNA by a *trans*-$Pt^{II}(NH_3)_2$ entity is, as expected, less dramatic. A 2D NMR study and gel electrophoresis studies revealed in the case of *GC* cross-linking only minor DNA unwinding (12°) and a moderate bending of the double helix of 26° toward the major groove, with the local distortion essentially concentrated on a few base pairs around the adduct.[65] Molecular details of the adduct are very similar to those of a model compound, the structure of which has been determined by X-ray crystallography.[66] According to the crystal structure, metal cross-linking takes place *via* G-*N7* and C-*N3*, with the guanine adopting a *syn* rather than an *anti* orientation of the sugar and the two nucleobases being hydrogen bonded *via* G-*O6* and C-*N(4)H$_2$*, with the two nucleobases remaining nearly coplanar. Formation of this adduct in dsDNA conceivably takes place *via* initial Pt^{II} binding to G-*N7* in the major groove, a subsequent swing of the metalated guanine around the glycosidic bond from *anti* to *syn* and eventually coordination to C-*N3* (Figure 2.14).

Interstrand cross-linking of *trans*-Pt^{II}(NH$_3$)$_2$ that involves two non-complementary bases, *e.g.* A and G, with coordination sites being *N1* and *N7*, respectively, leads to a larger distortion and causes T to adopt an extrahelical position.[67] In essence, formation of interstrand metal–DNA adducts can also be considered the consequences of DNA 'breathing' or of base-flipping mechanisms used by nature during nucleobase alkylation, for example.

Similar pathways are also feasible for long-known DNA metalation reactions with Hg^{2+} and Ag$^+$, which likewise lead to interstrand adducts. Models of possible adducts are available and have been discussed, *e.g.* the 'chain-slippage' scenario leading to T-*N3*, T-*N3* cross-linking. There is strong evidence that in DNA hairpin loops containing T bases and in dsDNA containing T–T mispairs, Hg^{II} binds with high preference to such sites. In the latter case, coordination thermally stabilizes DNA against denaturation.[68]

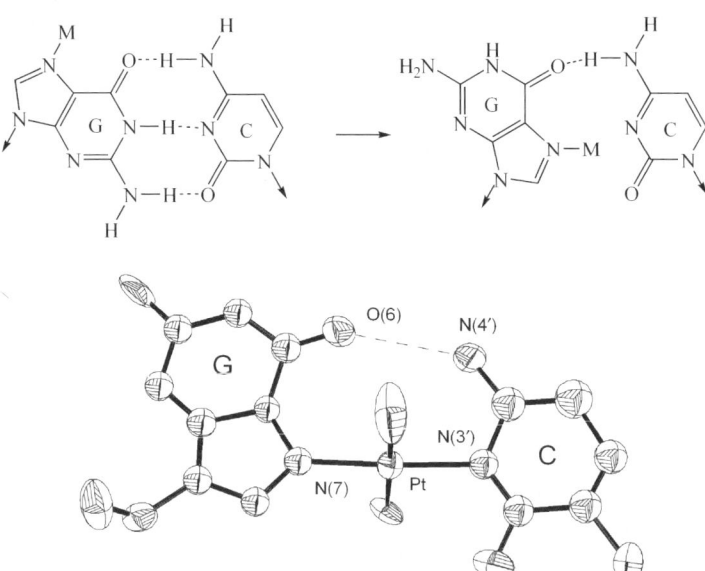

Figure 2.14 Proposed way of formation of an interstrand cross-link between the complementary bases G and C with M = *trans*-Pt^{II}(NH$_3$)$_2$. The model compound (bottom) is reproduced with permission from ref. 66.

2.3.1.3 DNA Condensation

DNA condensation and higher order DNA compaction, as characterized by a substantial decrease in volume, involve numerous ways which includes the actions of proteins (histones; condensins), protonated polyamines, coordinatively saturated metal cations (especially if highly charged), and also of divalent, partially dehydrated cations. Details of these processes are poorly understood as yet. Crystallographic studies on packing of dsDNA molecules (*e.g.* of B-DNA dodecamers) in the presence of divalent metal ions have revealed that successive dehydration of the crystals favors DNA compaction by direct metal cross-linking, either via phosphate-O or possibly G-*N7* sites.[69]

2.3.1.4 Inversion of DNA Helicity

The inversion of right-handed B-DNA into left-handed Z-DNA is among the most dramatic changes conceivable for a dsDNA. Accomplished in nature presumably mainly by specific Z-DNA binding proteins and accompanied by a base-pair flipping motion in the B–Z junction, Z-DNA has also been shown to be stabilized by the action of outer-sphere binding of $[Co(NH_3)_6]^{3+}$ or cyclic polyamine complexes of Zn^{II}. Multiple hydrogen bonding contacts between the NH_3 ligand and phosphate-O, which include also G-*N7* and G-*O6* acceptors, are responsible. However, more recent findings, obtained with flexible dinuclear transition metal complexes and bite distances equivalent to separations of N7 sites of adjacent G bases in Z-DNA, strongly suggest that direct coordination to nucleobases may be yet another mechanism for changing DNA helicity.[70]

2.3.2 Multistranded Nucleic Acids

2.3.2.1 Natural Systems

The important role of metal–nucleobase interactions becomes particularly obvious if complicated nucleic acid structures involving more than two strands are considered (*e.g.* tRNAs; RNAs in general; catalytically active nucleic acids). Nucleobase quartets and their extended forms with five, six or even eight nucleobases interacting *via* hydrogen bonds provide ample evidence for the significance of metal coordination to exocyclic nucleobase functions,[71,72] for example, the coordination of metal ions to the O6 carbonyl groups of guanine nucleosides in G-quartets (see Chapter 4). Screening of the repulsive negative charges of the phosphate groups of the nucleotides appears to be one of the main functions of the metal ions in multi-stranded nucleic acid structures.

For triple stranded nucleic acid structures, *e.g.* DNA triplexes, both proton binding (*e.g.* CH^+–G·C) or metal coordination to N7 of the purine base in the third strand (*e.g.* in (M–G)–G·C) are other ways to achieve screening of phosphate repulsions (Figure 2.15). Both Mg^{2+} and a number of transition

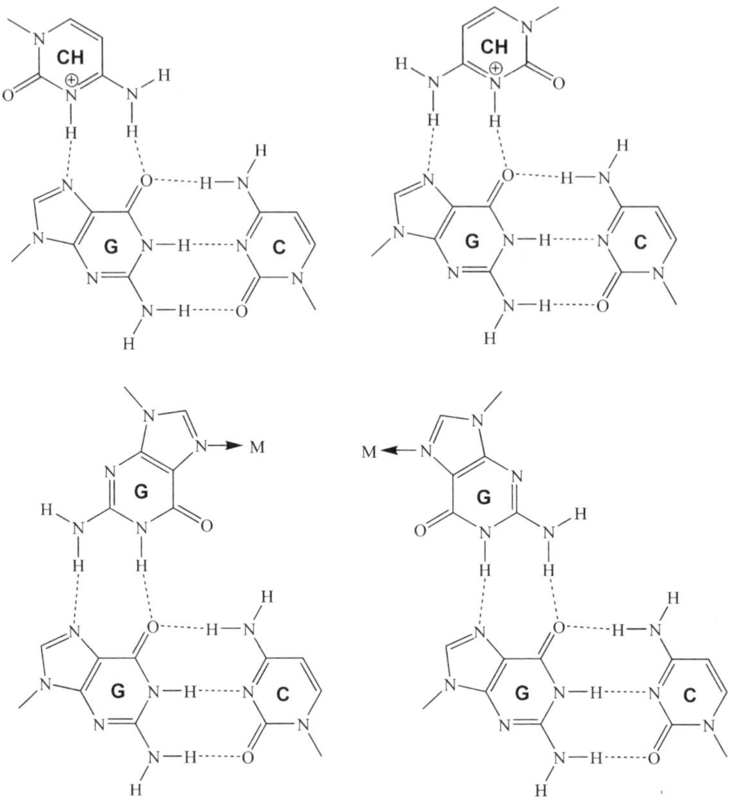

Figure 2.15 Examples of nucleobase triplets involving protonated or metalated nucleobases: CH^+–G·C (top) and (M–G)–G·C (bottom). The difference between individual triplets refers to the mutual orientations of the glycosidic bonds of CH^+ and G and of (M–G) and G. In the structures on the left they are *cis*, whereas on the right they are *trans*. This difference has an effect on the direction of the third strand (antiparallel or parallel).

metal ions (Mn^{2+}, Co^{2+}, Ni^{2+}, Zn^{2+}, Cd^{2+}) have been shown to exercise such an effect.[73] In numerous cases it has been shown that under the influence of these metal ions, a DNA duplex can be converted to an intramolecular triplex (and a single-stranded loop). Intermolecular triplexes, on the other hand, can form by combining a suitable dsDNA with a single stranded oligonucleotide plus the corresponding metal ions (or protons). Theoretical calculations have suggested that in the case of G–G·C triplexes, in addition to the electrostatic attraction between the metal ions and *both* guanine bases, a polarization mechanism is at work, which causes a strengthening of the G–G hydrogen bonding pairing scheme, irrespective of the relative orientation of the two Gs (Figure 2.15, bottom).[74] Interestingly, no such polarization effect is seen for the analogous A–A·T triplets.

2.3.2.2 Artificial M-Triplets and M-Quartets; Possible Uses

The concept of partially replacing protons in hydrogen bonds between nucleic acid base pairs by suitable transition metal ions (frequently having a linear coordination geometry) has been successfully applied to a series of pairs involving complementary bases in Watson–Crick or Hoogsteen arrangements, and also between mismatched base pairs. The concept can be extended to larger nucleobase aggregates such as triplets, quartets or sextets. A number of such compounds, in particular with *trans*-$Pt^{II}a_2$ (a = NH_3 or amine), Hg^{2+} and Ag^+, have been prepared and structurally characterized.[18,25] As outlined schematically in Figure 2.16 for nucleobase quartets, numerous variants are possible. A common feature of all quartets shown is that the linear metal has been moved from the center of the quartets to the periphery. The resulting compounds display interesting details that include, for example, unusual hydrogen bonding patterns (*e.g.* C–H···N hydrogen bonding; pairing between neutral and

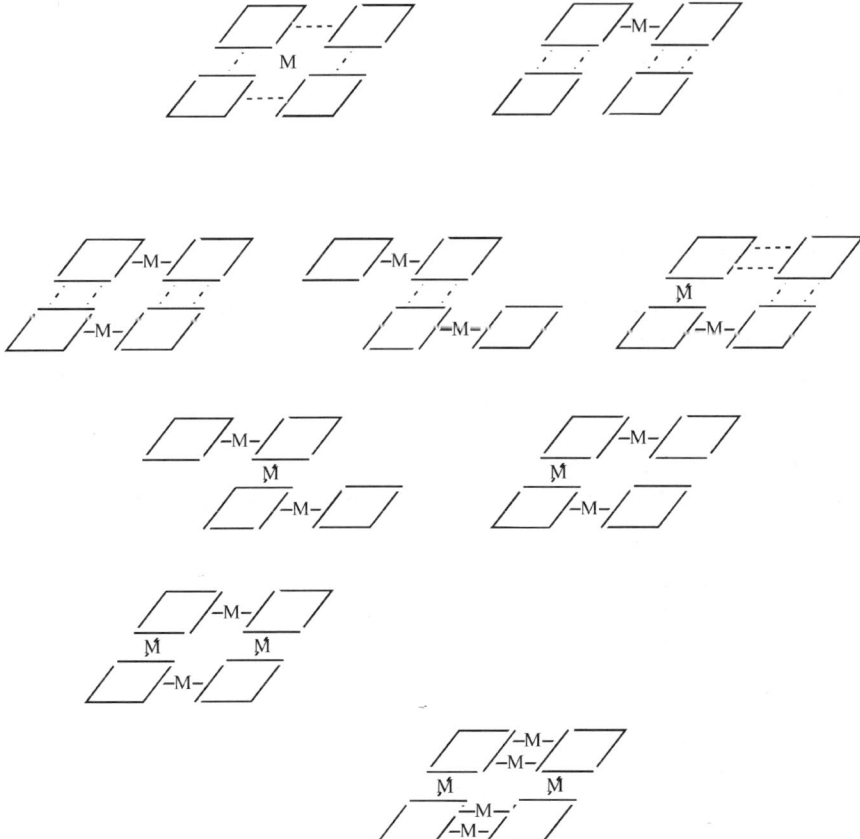

Figure 2.16 Schematic outline of X-ray structurally characterized metal–nucleobase quartets containing coordinated metal ions and hydrogen bonds.

anionic guanine bases),[75] or a way of stabilizing quartets formed by two Gs and two As,[76] to name only a few. The compounds prepared may be considered models or artificial analogues of nucleobase quartets that are possible or even realized by nature. As early as 1980, it was demonstrated that polyinosine is converted into a four-stranded, metal-cross-linked helix if exposed to Ag^+ ions.[77] More recently, it has been shown that an adenine model nucleobase forms quartet structures with Ag^+ ions.[78] Chemically robust metal–purine squares, because of their flat geometries, their size and their positive charge, may prove to be suitable agents to interact with natural guanine quartets present in the telomere ends of the chromosomes.[79] If successful, these compounds could be useful inhibitors of the enzyme telomerase, which is overexpressed in an estimated 80–90% of all cancers.[80]

Concerning nucleobase triplets, the natural ones (see above) can likewise become 'metal-modified' by either replacing an acidic nucleobase proton [e.g. N(3)H in $CH^+-G\cdot C$] by a metal of linear coordination geometry[81] or by substituting the 'natural' metal M' in a $G-G\cdot C$ triplet by a transition metal ion capable of cross-linking the two guanine bases (Figure 2.17).[82] This idea has also been pursued with oligonucleotides and using trans-$Pt^{II}a_2$ (a=NH_3 or amine), in an attempt to apply such platinated oligonucleotides as antisense[83] and antigene reagents, respectively.[84,85] Although the principle has been shown to work,[86] the slow kinetics of cross-link formation with the target sequence represent a major drawback toward this goal at present.

2.3.3 Subtle Effects of Metal Coordination

Metal coordination to the heterocyclic part of a nucleobase causes a redistribution of electrons in the heteroaromatic ring, up to the point that it has an effect on the tautomer equilibrium (see above). In nucleosides and nucleotides, this influence of the metal ion may even extend to the sugar entity, thereby having an effect on the sugar puckering.[87] Moreover, the stacking property of the nucleobase is affected,[88] although it is difficult to differentiate electronic effects from steric effects resulting from the presence of co-ligands of the metal.

2.3.3.1 Effects on Acid–Base Equilibria

The effect of a coordinated metal on the chemical properties of a nucleobase that is easiest to determine experimentally is that on its acid–base equilibria. As already mentioned, transition metal coordination to a ring-N atom of a nucleobase in its predominant tautomeric structure causes acidification of other protons of the nucleobase, hence leading to a decrease in pK_a. At the same time, the basicity of nucleobase groups (e.g. ring-N, exocyclic O) is reduced.[89] Similarly, metal coordination to a rare nucleobase tautomer can cause a dramatic shift in pK_a. If metal binding occurs to a ring-N atom of a deprotonated site, other sites of the nucleobase frequently undergo an apparent increase in

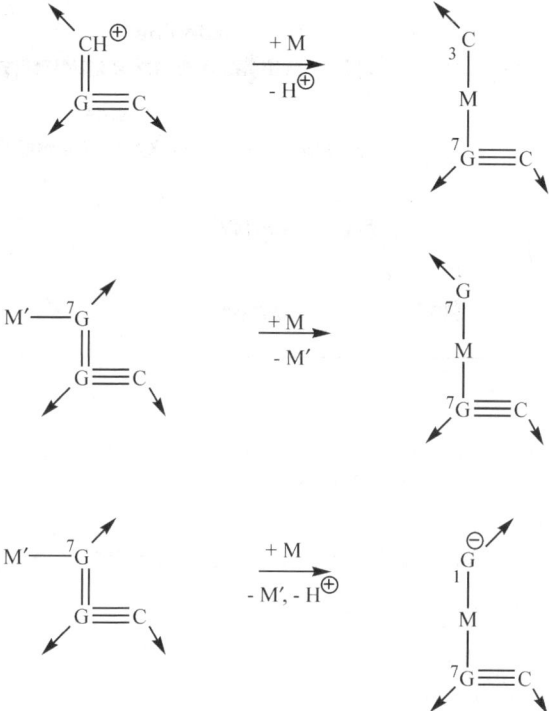

Figure 2.17 Schematic representations of metal(M)-modified nucleobase triplets (right) and of natural base triplets (left), with M being a transition metal ion with linear coordination geometry, cross-linking two nucleobases. G$^-$ denotes a guanine base deprotonated at N1 and carrying the transition metal at this position and M' is a hexacoordinate metal ion such as Mg^{2+}, Mn^{2+}, Co^{2+}, Ni^{2+}, Zn^{2+} or Cd^{2+}.

basicity, simply because the negative charge resulting from deprotonation is not relocalized in the M–N bond. Without going into detail (a more detailed discussion can be found elsewhere[25]), it suffices to keep in mind that metal coordination, particularly in combination with a favorable microenvironment that is capable of stabilizing a particular protonated or deprotonated nucleobase, can make available virtually the entire pH range of water for acid–base chemistry (Figure 2.18). As a consequence, the pH range given in Figure 2.2 for neutral nucleobases becomes irrelevant when metals and specifically transition metal ions are involved. This aspect may be of considerable significance when discussing the role of metal ions in catalytic reactions involving nucleic acids. It tentatively suggests that there may be yet another function of metal ions in such reactions in addition to polarizing bonds, stabilizing transition states or providing a hydroxo nucleophile, namely simply acting as an acid–base catalyst. An attractive aspect of such a possibility would be that the metal ion in question would not need to reside at the active site but rather could be remote from it and could represent just the origin of a relay for H$^+$ shuttling.[90] For a

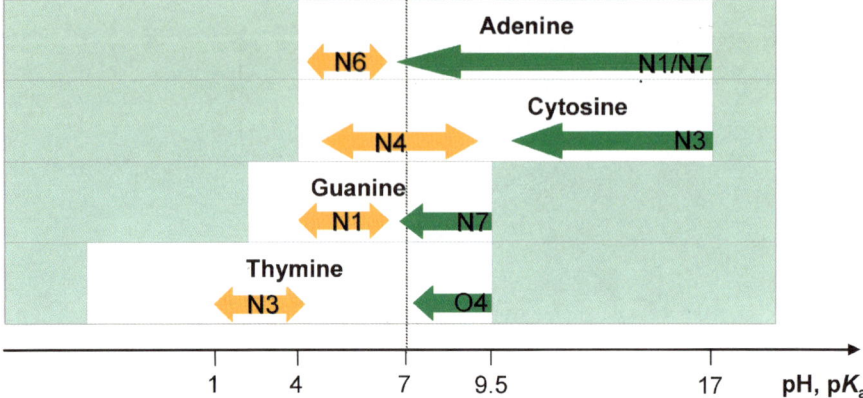

Figure 2.18 Shifted nucleobase pK_a values as a consequence of metal coordination to preferential and rare tautomers.

more in-depth discussion of the role of metal ions in RNA-catalyzed reactions, see Chapter 8.

2.3.3.2 Base Pair Stability

Closely related to the question of nucleobase acidity and basicity is the question of the stability of a hydrogen-bonded nucleobase association (*i.e.* base pairs, base triplets, *etc.*). As mentioned above, theoretical calculations (gas phase, 0 K) have provided evidence that polarization effects, as evidenced by metal coordination to G-*N7* (yet not to A-*N7*), lead to a slight to moderate increase in nucleobase interaction energies and hence to more stable base pairing. In other words, it effectively depends on the sum of ligand acidification (which causes an increase in H bonding strength) and loss of ligand basicity (which causes a loss in H bonding strength), whether base pairing is strengthened or weakened by metal ion coordination. In the case of guanine model nucleobase complexes carrying a Pt^{II} ion at the N7 position, the predicted perturbation of guanine–cytosine pairing has been confirmed experimentally.[91] What this finding suggests is that the overall thermal destabilization of dsDNA as a function of coordination of monofunctional or *cis*-bifunctional Pt^{II} to one strand is most likely not primarily a consequence of an inherent destabilization of the Watson–Crick pair(s) with cytosine, but rather must be due to other reasons, *e.g.* a steric disturbance of the dsDNA structure. Recent density functional theory calculations carried out with model guanine bases containing various metal entities bonded to O6 revealed that the pairing energy with cytosine is significantly reduced.[92]

A case where acid–base behavior and hydrogen bond formation appear to be particularly strongly interrelated, is that of base pairing between a neutral guanine base (carrying a Pt^{II} at *N7* or no metal at all) and an anionic guanine

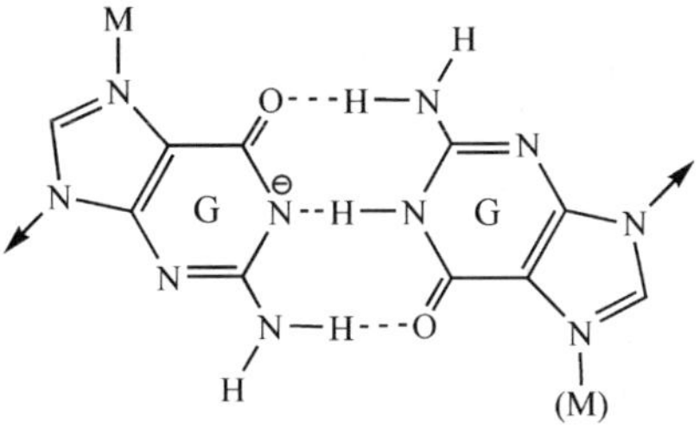

Figure 2.19 Schematic representation of hemideprotonated guanine base pair between *N1*-deprotonated guanine and neutral guanine. While the former needs to have a metal coordinated to N7, the latter may also carry a metal ion or represent a neutral base.

base, deprotonated at N1 and having a metal (Pt^{II}) at *N7* (Figure 2.19). The shift in pK_a of guanine toward lower values as a consequence of metal binding allows formation of such hemideprotonated guanine pairs at neutral or even weakly acidic pH and has produced numerous variants of this motif.[93,94] Association constants can be exceptionally high, exceeding that of the Watson–Crick G·C pair by a factor of 100.[93c]

2.3.3.3 Isotopic Exchange for Carbon-bonded H Atoms

Isotopic hydrogen exchange of the C(8)H atoms of the purine nucleobases A and G has long been known and the usefulness of both ^{2}D and ^{3}T analogues for the assignment of complicated nucleic acid structures has been demonstrated. Today, it is generally accepted that the mechanism of proton exchange involves an ylide intermediate, which results from an attack of OD$^-$ on C(8)H in an N7-protonated ligand.[95] As protonation of the purines occurs at low pH only, metal coordination to the N7 sites accelerates isotopic exchange at neutral or weakly acidic pH, because it places a positive charge close to the site of exchange ('metal activating factor'[96]). The effect of a coordinated metal ion is clearly smaller (by a factor of 10^4–10^6) than that of a proton, yet comparable to that of a methyl group such as present in 7-methylguanosine. As a consequence of this isotopic exchange, metal–(purine-*N7*) complexes show a partial or complete loss of H8 signal intensity in their ^{1}H NMR spectra in D$_2$O. The facile mercuration of purine nucleotides at the C8 position also appears to be a consequence of ylide formation.

Isotopic exchange reactions are also known for C(5)H of C and U bases and occasionally even for C(6)H. In all cases they proceed considerably more slowly than those of purines. They can be accelerated by agents with the ability to add reversibly to the C5–C6 double bond, such as HSO_3^- and 2-mercaptoethylamine.[97] Coordination of metal ions, *e.g.* of Pt^{II}, can likewise favor isotopic exchange of C(5)H, frequently under very mild conditions. This applies both to the parent nucleobase (*e.g.* unsubstituted uracil with Pt^{II} at N1, pD 2–3,[98] or

Pt^{II} simultaneously at N1 and N3[99]) and to model pyrimidine nucleobases (*e.g.* 1-methylcytosine with Pt^{II} at N3[100]). Mechanistic details of the metal-catalyzed isotopic C–H exchange reactions with pyrimidine bases are not well understood.[101]

2.3.4 Irreversible Alterations

2.3.4.1 Backbone Cleavage

Hydrolytic cleavage and reformation of sugar–phosphate bonds are central to many processes involving nucleic acids, such as synthesis, repair and degradation. These reactions are frequently carried out by metalloenzymes, hence proteins having two or three metal ions in their active site. The metal ions serve different purposes, such as polarizing bonds, generating the OH$^-$ nucleophile or stabilizing transition states, yet are not directly involved in coordination to the heterocyclic bases. The various approaches used today to generate artificial metallonucleases in a sense mimic nature, the difference being that the carrier ligands of the metal ion(s) in general are not proteins, but rather synthetic ligands with an affinity for nucleic acids.[102]

There is also the possibility of a non-enzymatic hydrolytic degradation of (at least) RNA and possibly even of DNA, with the metal ions directly coordinating to nucleobases. The Pb^{II}-catalyzed hydrolysis of yeast tRNA^{phe}, originally observed during X-ray crystal structure determinations of heavy atom (Pb^{II}) derivatives, revealed that in all three sites where Pb^{II} is located, extensive coordination to nucleobases occurs.[103] The actual cleavage process is brought about by the OH$^-$ ligand at the metal and is favored by the p$K_a \approx 7$ of the Pb–OH$_2$ species. This feature is not unique to the main group element lead, but is also realized by numerous other metal ions, including transition metal ions such as Zn^{II} and Cd^{II}.[104] In fact, the production of RNA nucleosides, mononucleotides and short oligonucleotides relied for some time on the degradation of natural RNA by treatment with various metal hydroxides.[105]

2.3.4.2 Sequencing DNA and RNA

DNA and RNA sequencing methods rely on irreversible, nucleobase-specific modifications of these bases and further treatment that leads to strand breakage. Metal-containing reagents are not among the classical sequencing agents applied in the Maxam–Gilbert method. However, a number of metal-containing compounds are in use as chemical probes to detect exposed and hence reactive nucleobases. Addition of OsO$_4$ across the 5,6-double bonds of un- or mispaired pyrimidine nucleobases has already been mentioned (see above) and MnO$_4^-$ causes oxidation of the 5,6-double bond of T to give 5,6-dihydroxythymine. Redox chemistry involving coordinated nickel ions will be dealt with below. The footprinting techniques that are based on the redox chemistry of transition metal ions (*e.g.* Cu$^+$, Fe^{2+}) and the generation of

2.3.4.3 Other Hydrolytic Processes

Here we describe two examples of irreversible nucleobase alterations that involve 'inner-sphere' binding of metal ions to the heterocyclic part of a nucleic acid base. Cleavage of the N-glycosidic bonds of nucleotides at low pH or following a chemical modification of the base, e.g. ring-N alkylation, is a well-known phenomenon. It is also utilized by natural glycosylases during DNA repair. Given the preference of many metal ions for the N7 positions of purine bases, the effect of metal ions on 'depuration' reactions has been of interest for some time. From studies employing a series of transition metal ions (Ag^+, Co^{2+}, Ni^{2+}, Cu^{2+}, Zn^{2+}, Cd^{2+}, Pd^{2+}, Pt^{2+}) and including also acyclic nucleoside analogues, it now seems clear that metal ion coordination to the purines (G-N7, A-N1,N7) retards the acid-catalyzed hydrolysis, because the metal ion competes with H^+ for binding sites.[106] However, in a pH range where no nucleobase protonation takes place, the metal entity accelerates spontaneous hydrolysis of the glycosidic bond. Thus, Pt^{II}(dien) retards depuration of G at pH < 4, but facilitates depurination at pH > 4 because of a pK_a of ~2–3 for GH^+.[107] Similar arguments apply to A, yet only when both N1 and N7 carry a metal ion.[108]

Another type of hydrolysis reaction takes place when an exocyclic amino group of a nucleobase is converted into an exocyclic carbonyl group. Such reactions, occurring under the influence of a coordinated metal ion, have been reported in recent years in two cases: Pt^{II} coordination to N3 of a cytosine model nucleobase dramatically accelerates the conversion into a uracilato ligand at alkaline pH.[109] If a similar deamination process were to occur with a 5-methylcytosine base in DNA, it would generate T, which, unlike U, would not be excised from DNA and therefore could give rise to a point mutation. It is interesting, in this context, to note that 5-methylcytosine residues are in fact hotspots for spontaneous mutations. Platinated adenine nucleobases can behave similarly, yielding hypoxanthine.[110] It is feasible that N1-metalated guanine might be converted into a xanthine base by a similar mechanism.

2.3.4.4 Redox Chemistry with Nucleobases

Oxidative DNA damage is implicated in various pathological states of organisms, such as inflammations and cancer, and is believed to be a contributing factor for aging. Of all nucleobases, the electron-rich G has the lowest ionization potential and consequently is most susceptible to oxidation. Depending on the conditions, one- and two-electron oxidation processes are possible, which can lead to a considerable number of different intermediates and end products.[111] There are numerous cases known today of metal complexes that generate highly reactive oxygen species, *e.g.* OH˙ radicals, in the vicinity of the nucleic acid without binding to them in an inner-sphere fashion. Fe^{II}–EDTA complexes, Cu^{I}–phenanthroline complexes and porphyrin compounds having high-valent $Mn^{V}=O$ entities in the center are typical examples. Qualitatively similar are coordinatively saturated polypyridyl complexes of Ru^{III}, generated by photochemical excitation from the corresponding Ru^{II} compounds, which can lead to one-electron oxidation of G. All these cases will not be dealt with here. Rather, examples will be summarized where the oxidation involves direct coordination of one or more metal ions to a nucleobase.

As far as reactions with G bases are concerned, a macrocyclic Ni^{II} complex has been demonstrated to cleave DNA and RNA backbones following coordination to G-*N7* and oxidation to Ni^{III} by peroxysulfate.[112] Accessibility of G is particularly important to achieve oxidation. A phenolate-bridged dicopper(II) complex, which has shown to bridge guanosine via N7 and O6, is capable of cleaving DNA at GG sites, following reduction to the $[Cu^{I}]_2$ species by 3-mercaptopropionic acid in the presence of oxygen.[38] Pt^{IV} complexes are, under certain conditions, capable of converting guanine to 8-oxoguanine.[113] This finding adds to the discussion of whether antitumor-active Pt^{IV} drugs can interact directly with DNA or whether they are only prodrugs that need to become reduced intracellularly to the Pt^{II} species. Finally, Au^{III} binding (as $AuCl_3$) to N7 of the model base 9-ethylguanine has been found to lead to sluggish degradation of the purine ring, in both the absence and presence of O_2; however, different products are formed.[114]

Compared with G oxidation, reports on redox transformations of the other nucleobases are scarce. This refers in particular to metal-mediated reactions. A rare case of such a reaction, namely the formation of a 5,5′-diuracilyl compound from uracil, brought about by $AuCl_4^-$, has been described.[115]

2.4 Summary and Outlook

This chapter has dealt with metal–nucleobase interactions from an inorganic chemistry perspective, with the focus on coordinative bond formation involving transition metal ions. As pointed out, the essential transition metal ions such as Fe^{n+} ($n=2$ or 3), Cu^{n+} ($n=1$ or 2) and Zn^{2+} are not supposed to interact directly with nucleic acids because of their potential for causing unwanted oxidative or hydrolytic damage. However, numerous scenarios are feasible where interactions between nucleic acids and transition metal ions are relevant and highly significant. Examples include severe disturbances of metal homeostasis in an organism, mutagenic events caused by toxic exogenous metal ions, treatment of diseases with metal-containing pharmaceuticals (see Chapter 10 for further discussion), the use of metal coordination compounds as chemical probes for the study of nucleic acid structure and function and novel analytical sensing techniques based on nucleic acid hybridization in combination with metal-catalyzed cleavage reactions. Moreover, numerous applications in the field of the material sciences involving metal–nucleobase or metal–nucleic acid interactions have been proposed and are being actively pursued at present, such as the generation of molecular wires, surface patterning and nanostructure design.[3b] Finally, the enormous potential of 'blue sky' research in the field of supramolecular chemistry needs to be mentioned. A profound understanding of the basic principles governing metal–nucleobase chemistry is essential in all these cases.

Some of the most important lessons learned during the last three or four decades may be summarized as follows:

- In a living organism, the interaction of a 'non-physiological' transition metal species with nucleic acids leads to a massive response up to the point that cells are irreversibly damaged and undergo apoptosis.
- The ways in which a particular transition metal ion reacts with nucleic acids are extremely versatile. There is nothing like a 'metal-specific' binding pattern.
- Once coordinated to a nucleic acid, the metal ion affects base pairing schemes (strength, pattern), nucleobase acid–base equilibria and tautomer structures. These features make transition metal ions distinctly different from the 'natural' counter ions of nucleic acids such as K^+, Na^+ and Mg^{2+}.

The field of metal–nucleic acid chemistry is a fascinating area at the crossroads of inorganic chemistry, biochemistry, medicine and lately also of material sciences. It is now developed to a point where novel challenges can be taken on.

References

1. See various volumes of the series A. Sigel and H. Sigel (eds), *Metal Ions in Biological Systems*, Marcel Dekker, New York, e.g. volumes 8 (1979), 32 (1996) and 33 (1996).
2. See various articles in (a) T. G. Spiro (ed.), *Nucleic Acid–Metal Ion Interactions*, Wiley-Interscience, New York, 1980; (b) G. L. Eichhorn and L. G. Marzilli (eds), *Metal Ions in Genetic Information Transfer*, Elsevier/North-Holland, New York, 1981.
3. (a) K. Aoki, in *Comprehensive Supramolecular Chemistry*, J. L. Atwood, J. E. D. Davies, D. D. Macnicol and F. Vögtle, ed., Elsevier, Oxford, 1996, Vol. 5, p. 249; (b) A. Houlton, *Adv. Inorg. Chem.*, 2002, **53**, 87.
4. (a) R. B. Martin, in *Cisplatin – Chemistry and Biochemistry of a Leading Anticancer Drug*, B. Lippert, ed., VHCA, Zürich, and Wiley-VCH, Weinheim, 1999, p. 183; (b) R. B. Martin, *Acc. Chem. Res.*, 1985, **18**, 32; (c) R. B. Martin, *Met. Ions Biol. Syst.*, 1996, **32**, 61.
5. G. Frommer and B. Lippert, *Inorg. Chem.*, 1990, **29**, 3259, and references cited therein.
6. G. L. Eichhorn and Y. A Shin, *J. Am. Chem. Soc.*, 1968, **90**, 7323.

7. B. Rosenberg, in *Cisplatin – Chemistry and Biochemistry of a Leading Anticancer Drug*, B. Lippert, ed., VHCA, Zürich, and Wiley-VCH, Weinheim, 1999, p. 3.
8. T. Yakane and N. Davidson, *J. Am. Chem. Soc.*, 1961, **83**, 2599.
9. R. S. Simpson, *J. Am. Chem. Soc.*, 1964, **86**, 2059.
10. E. Sletten, *J. Chem. Soc., Chem. Commun.*, 1967, 1119.
11. M. R. Moller, M. A. Bruck, T. O'Connor, F. J. Armatis Jr., E. A. Knolinski, N. Kottmair and R. S. Tobias, *J. Am. Chem. Soc.*, 1980, **102**, 4589, and references cited therein.
12. P. M. Takahara, A. C. Rosenzweig, C. A. Frederick and S. J. Lippard, *Nature*, 1995, **377**, 649.
13. See, e.g., (a) A. Oleksi, A. G. Blanco, R. Boer, I. Usón, J. Aymami, A. Rodger, M. J. Hannon and M. Coll, *Angew. Chem. Int. Ed.*, 2006, **45**, 1227; (b) A. M. Pyle and J. K. Barton, *Prog. Inorg. Chem.*, 1990, **38**, 413.
14. T. Hermann and E. Westhof, *Structure*, 1998, **6**, 1303.
15. R. K. O. Sigel and A. M. Pyle, *Chem. Rev.*, 2007, **107**, 97.
16. Y. Lu, *Chem. Eur. J.*, 2002, **8**, 4588.
17. See, e.g., V. A. Bloomfield, D. M. Crothers and I. Tinoco, Jr., *Nucleic Acids. Structures, Properties and Functions*, University Science Books, Sausalito, CA, 2000, p. 28.
18. B. Lippert, *Coord. Chem. Rev.*, 2000, **200–202**, 487.
19. E. Freisinger, A. Schreiber, M. Drumm, A. Hegmans, S. Meier and B. Lippert, *J. Chem. Soc., Dalton Trans.*, 2000, 3281.
20. B. Luisi, M. Orozco, J. Šponer, F. J. Luque and Z. Shakked, *J. Mol. Biol.*, 1998, **279**, 1123.
21. (a) M. D. Topal and J. R. Fresco, *Nature*, 1976, **263**, 285; (b) H. T. Miles, *Proc. Natl. Acad. Sci. USA*, 1961, **47**, 791.
22. J. V. Burda, J. Šponer and J. Leszczynski, *J. Biol. Inorg. Chem.*, 2000, **5**, 178.
23. B. Lippert, H. Schöllhorn and U. Thewalt, *Inorg. Chim. Acta*, 1992, **198**, 723.
24. I. L. Zilberberg, V. I. Avdeev and G. M. Zhidomirov, *J. Mol. Struct. (Theochem)*, 1997, **418**, 73.
25. B. Lippert, *Prog. Inorg. Chem.*, 2005, **54**, 385.
26. (a) M. Garijo Añorbe, M. S. Lüth, M. Roitzsch, M. Morell Cerdà, P. Lax, G. Kampf, H. Sigel and B. Lippert, *Chem. Eur. J.*, 2004, **10**, 1046, and references cited therein; (b) M. Roitzsch and B. Lippert, *J. Am. Chem. Soc.*, 2004, **126**, 2421.
27. J.-P. Charland, M. T. P. Viet, M. St-Jacques and A. L. Beauchamp, *J. Am. Chem. Soc.*, 1985, **107**, 8202.
28. M. S. Lüth, E. Freisinger, F. Glahé and B. Lippert, *Inorg. Chem.*, 1998, **37**, 5044.
29. M. Morell Cerdà, D. Amantia, B. Costisella, A. Houlton and B. Lippert, *Dalton Trans.*, 2006, 3894.

30. See, e.g., F. Zamora, M. Kunsman, M. Sabat and B. Lippert, *Inorg. Chem.*, 1997, **36**, 1583.
31. R. M. K. Dale, E. Martin, D. C. Livingston and D. C. Ward, *Biochemistry*, 1975, **14**, 2447.
32. C. C. Correll, B. Freeborn, P. B. Moore and T. A. Steitz, *J. Biomol. Struct. Dyn.*, 1997, **15**, 165.
33. H. Sigel, *ACS Symp. Ser.*, 1989, **402**, 159.
34. B. Longato, L. Pasquato, A. Mucci, L. Schenetti and E. Zangrando, *Inorg. Chem.*, 2003, **42**, 7861.
35. L. G. Marzilli and T. J. Kistenmacher, *Acc. Chem. Res.*, 1977, **10**, 146.
36. H. Sigel, S. S. Massoud and N. A. Corfù, *J. Am. Chem. Soc.*, 1994, **116**, 2958.
37. H. T. Chifotides and K. R. Dunbar, *Acc. Chem. Res.*, 2005, **38**, 146.
38. L. Li, N. N. Murthy, J. Telser, L. N. Zakharov, G. P. A. Yap, A. L. Rheingold, K. D. Karlin and S. E. Rokita, *Inorg. Chem.*, 2006, **45**, 7144.
39. N. Farrell, *Met. Ions Biol. Syst.*, 2004, **41**, 252.
40. S. Zhang, L. A. Holl and R. E. Sheperd, *Inorg. Chem.*, 1990, **29**, 1012.
41. (a) T. J. Kistenmacher, L. G. Marzilli and M. Rossi, *Bioinorg. Chem.*, 1976, **6**, 347; (b) A. Okamoto, K. Tainaka and T. Kamei, *Org. Biomol. Chem.*, 2006, **4**, 1638.
42. J. Arpalahti, in: *Cisplatin – Chemistry and Biochemistry of a Leading Anticancer Drug*, B. Lippert, ed., VHCA, Zürich, and Wiley-VCH, Weinheim, 1999, p. 207.
43. B. Lippert, H. Schöllhorn and U. Thewalt, *J. Am. Chem. Soc.*, 1986, **108**, 6616.
44. K. M. Comess, C. E. Castello and S. J. Lippard, *Biochemistry*, 1990, **29**, 2102.
45. R. Dalbiès, D. Payet and M. Leng, *Proc. Natl. Acad. Sci. USA*, 1994, **91**, 8147.
46. D. Yang, S. S. G. E. van Boom, J. H. van Boom and A. H.-J. Wang, *Biochemistry*, 1995, **34**, 12912.
47. J. Reedijk, *Chem. Rev.*, 1999, **99**, 2499.
48. K. J. Barnham, M. I. Djuran, P. del Socorro Murdoch and P. J. Sadler, *J. Chem. Soc., Chem. Commun.*, 1994, 721.
49. (a) D. V. Deubel, *J. Am. Chem. Soc.*, 2003, **125**, 15308; (b) D. V. Deubel, *J. Am. Chem. Soc.*, 2004, **126**, 5999.
50. J. Vinje, E. Sletten and J. Kozelka, *Chem. Eur. J.*, 2005, **11**, 3863.
51. (a) J. Vinje, J. A. Parkinson, P. J. Sadler, T. Brown and E. Sletten, *Chem. Eur. J.*, 2003, 9, 1620; (b) X. Jia, G. Zon and L. G. Marzilli, *Inorg. Chem.*, 1991, **30**, 228.
52. S. K. Miller, D. G. VanDerVeer and L. G. Marzilli, *J. Am. Chem. Soc.*, 1985, **107**, 1048.
53. F. Zamora and M. Sabat, *Inorg. Chem.*, 2002, **41**, 4976.

54. P. Amo Ochoa, O. Castillo, P. J. Sanz Miguel, M. Sabat, B. Lippert and F. Zamora, *J. Biol. Inorg. Chem.*, 2007, **12**, 543.
55. S. Aoki and E. Kimura, *Chem. Rev.*, 2004, **104**, 769.
56. E. C. Fusch and B. Lippert, *J. Am. Chem. Soc.*, 1994, **116**, 7204.
57. (a) J. S. Lee, L. J. P. Latimer and R. S. Reid, *Biochem. Cell Biol.*, 1993, **71**, 162; (b) M. Fuentes-Cabrera, B. G. Sumpter, J. E. Šponer, J. Šponer, L. Petit and J. C. Wells, *J. Phys. Chem. B*, 2007, **111**, 870.
58. L. G. Marzilli, J. S. Saad, Z. Kuklenyik, K. A. Keating and Y. Xu, *J. Am. Chem. Soc.*, 2001, **123**, 2764.
59. See, *e.g.*, (a) D. Over, G. Bertho, M.-A. Elizondo-Riojas and J. Kozelka, *J. Biol. Inorg. Chem.*, 2006, **11**, 139; (b) G. Natile and L. G. Marzilli, *Coord. Chem. Rev.*, 2006, **250**, 1315.
60. (a) D. Wang and S. J. Lippard, *Nat. Drug Discov.*, 2005, **4**, 307; (b) H. Zorbas and B. K. Keppler, *ChemBioChem.*, 2005, **6**, 1157.
61. F. Coste, J.-M. Malinge, L. Serre, W. Shepard, M. Roth, M. Leng and C. Zelwer, *Nucleic Acids Res.*, 1999, **27**, 1837.
62. H. Huang, L. Zhu, B. R. Reid, G. P. Drobny and P. B. Hopkins, *Science*, 1995, **270**, 1842.
63. See, *e.g.*, (a) B. Lippert, *Prog. Inorg. Chem.*, 1989, **37**, 1; (b) G. Schröder, J. Kozelka, M. Sabat, M.-H. Fouchet, R. Beyerle-Pfnür and B. Lippert, *Inorg. Chem.*, 1996, **35**, 1647; (c) R. Bau and M. Sabat, in *Cisplatin – Chemistry and Biochemistry of a Leading Anticancer Drug*, ed. B. Lippert, VHCA, Zürich, and Wiley-VCH, Weinheim, 1999, p. 319.
64. See, *e.g.*, M.-A. Elizondo-Riojas, F. Gonnet, J.-C. Chottard, J.-P. Girault and J. Kozelka, *J. Biol. Inorg. Chem.*, 1998, **3**, 30.
65. F. Paquet, M. Boudvillain, G. Lancelot and M. Leng, *Nucleic Acids Res.*, 1999, **27**, 4261.
66. A. Erxleben, S. Metzger, J. F. Britten, C. J. L. Lock, A. Albinati and B. Lippert, *Inorg. Chim. Acta*, 2002, **339**, 461.
67. B. Andersen, E. Bernal-Méndez, M. Leng and E. Sletten, *Eur. J. Inorg. Chem.*, 2000, 1201.
68. Y. Miyake, H. Togashi, M. Tashiro, H. Yamaguchi, S. Oda, M. Kudo, Y. Tanaka, Y. Kondo, R. Sawa, T. Fujimoto, T. Machinami and A. Ono, *J. Am. Chem. Soc.*, 2006, **128**, 2172.
69. G. R. Clark, C. J. Squire, L. J. Baker, R. F. Martin and J. White, *Nucleic Acids Res.*, 2000, **28**, 1259.
70. A. Medina-Molner and B. Spingler, personal communication.
71. D. J. Patel, S. Bouaziz, A. Kettani and Y. Wang, in *Oxford Handbook of Nucleic Acids Structure*, S. Neidle, ed., Oxford University Press, Oxford, 1999, p. 389.
72. J. Deng, Y. Xiong and M. Sundaralingam, *Proc. Natl. Acad. Sci. USA*, 2001, **98**, 13665.
73. V. N. Soyfer and V. N. Potaman, *Triple-Helical Nucleic Acids*, Springer, New York, 1996.
74. (a) J. Šponer, M. Sabat, J. V. Burda, A. M. Doody, J. Leszczynski and P. Hobza, *J. Biomol. Struct. Dyn.*, 1998, **161**, 139; (b) J. Muñoz,

J. L. Gelpi, M. Soler-López, J. A. Subirana, M. Orozco and F. J. Luque, *J. Phys. Chem. B*, 2002, **106**, 8849; (c) F. Moroni, A. Famulari, M. Raimondi and M. Sabat, *J. Phys. Chem. B*, 2003, **107**, 4196.
75. (a) G. Tamasi, F. Botta and R. Cini, *J. Mol. Struct. (Theochem)*, 2006, **766**, 61; (b) J. Gu, J. Wand and J. Leszczynski, *J. Am. Chem. Soc.*, 2004, **126**, 12651.
76. M. Roitzsch and B. Lippert, *Inorg. Chem.*, 2004, **43**, 5483.
77. Y. A. Shin and G. L. Eichhorn, *Biopolymers*, 1980, **19**, 539.
78. C. S. Purohit and S. Verma, *J. Am. Chem. Soc.*, 2006, **128**, 400.
79. E. Freisinger, I. B. Rother, M. S. Lüth and B. Lippert, *Proc. Natl. Acad. Sci. USA*, 2003, **100**, 3748.
80. J. E. Reed, A. Arola Arnal, S. Neidle and R. Villar, *J. Am. Chem. Soc.*, 2006, **128**, 5992.
81. I. Dieter-Wurm, M. Sabat and B. Lippert, *J. Am. Chem. Soc.*, 1992, **114**, 357.
82. R. K. O. Sigel, M. Sabat, E. Freisinger, A. Mower and B. Lippert, *Inorg. Chem.*, 1999, **38**, 1481.
83. M. B. L. Janik and B. Lippert, *J. Biol. Inorg. Chem.*, 1999, **4**, 645.
84. B. Lippert and M. Leng, *Top. Biol. Inorg. Chem.*, 1999, **1**, 117.
85. J. Müller, M. Drumm, M. Boudvillain, M. Leng, E. Sletten and B. Lippert, *J. Biol. Inorg. Chem.*, 2000, **5**, 603.
86. C. Colombier, B. Lippert and M. Leng, *Nucleic Acids. Res.*, 1996, **24**, 4519.
87. M. Polak and J. Plavec, *Eur. J. Inorg. Chem.*, 1999, 547.
88. A. Robertazzi and J. A. Platts, *Chem. Eur. J.*, 2006, **12**, 5747.
89. (a) R. Griesser, G. Kampf, L. E. Kapinos, S. Komeda, B. Lippert, J. Reedijk and H. Sigel, *Inorg. Chem.*, 2003, **42**, 32, and references cited therein; (b) H. Sigel, *Pure Appl. Chem.*, 2004, **76**, 1869.
90. M. Roitzsch, M. Garijo Añorbe, P. J. Sanz Miguel, B. Müller and B. Lippert, *J. Biol. Inorg. Chem.*, 2005, **10**, 800.
91. R. K. O. Sigel, E. Freisinger and B. Lippert, *J. Biol. Inorg. Chem.*, 2000, **5**, 287.
92. A. Robertazzi and J. A. Platts, *J. Biol. Inorg. Chem.*, 2005, **10**, 854.
93. (a) R. Faggiani, C. J. L. Lock and B. Lippert, *J. Am. Chem. Soc.*, 1980, **102**, 5418; (b) G. Schröder, B. Lippert, M. Sabat, C. J. L. Lock, R. Faggiani, B. Song and H. Sigel, *J. Chem. Soc., Dalton Trans.*, 1995, 3767; (c) R. K. O. Sigel, S. M. Thompson, E. Freisinger, F. Glahé and B. Lippert, *Chem. Eur. J.*, 2001, **7**, 1968.
94. G. McGowan, S. Parsons and P. J. Sadler, *Chem. Eur. J.*, 2005, **11**, 4396.
95. B. Noszál, V. Scheller-Krattiger and R. B. Martin, *J. Am. Chem. Soc.*, 1982, **104**, 1078.
96. J. R. Jones and S. E. Taylor, *Chem. Soc. Rev.*, 1981, **10**, 329.
97. Z. Shabarova and A. Bogdanov, *Advanced Organic Chemistry of Nucleic Acids*, VCH, Weinheim, 1994, 410.
98. B. Lippert, *Inorg. Chem.*, 1981, **20**, 4326.

99. H. Rauter, E. C. Hillgeris, A. Erxleben and B. Lippert, *J. Am. Chem. Soc.*, 1994, **116**, 616.
100. B. Lippert, C. J. L. Lock and R. A. Speranzini, *Inorg. Chem.*, 1981, **20**, 335.
101. M. Goodgame and D. A. Jakubovic, *Coord. Chem. Rev.*, 1987, **79**, 97.
102. See, *e.g.*, (a) F. Mancin, P. Scrimin, P. Tecilla and U. Tonellato, *Chem. Commun.*, 2005, 2540; (b) S. I. Kirin, R. Krämer and N. Metzler-Nolte, in *Concepts and Models in Bioinorganic Chemistry*, ed. H.-B. Kraatz and N. Metzler-Nolte, Wiley-VCH, Weinheim, 2006, p. 159; (c) A. O'Donoghue, S. Y. Pyun, M.-Y. Yang, J. R. Morrow and J. P. Richard, *J. Am. Chem. Soc.*, 2006, **128**, 1615.
103. R. S. Brown, J. C. Dewan and A. Klug, *Biochemistry*, 1985, **24**, 4785.
104. G. L. Eichhorn, in *Inorganic Biochemistry*, G. L. Eichhorn ed., Elsevier, New York, 1973, p. 1220.
105. See, *e.g.*: K. Dimroth and H. Witzel, *Justus Liebigs Ann. Chem.*, 1959, **620**, 109.
106. J. Arpalahti, R. Käppi, J. Hovinen, H. Lönnberg and J. Chattopadhyaya, *Tetrahedron*, 1989, **45**, 3945, and references cited therein.
107. (a) S. Kuusela and H. Lönnberg, *Met. Ions Biol. Syst.*, 1996, **32**, 271; (b) J. Arpalahti, A. Jokilammi, H. Hakala and H. Lönnberg, *J. Phys. Org. Chem.*, 1991, **4**, 301.
108. B. L. Iverson and P. B. Dervan, *Nucleic Acids Res.*, 1987, **15**, 7823.
109. J. E. Šponer, P. J. Sanz Miguel, L. Rodriguez-Santiago, A. Erxleben, M. Krumm, M. Sodupe, J. Šponer and B. Lippert, *Angew. Chem. Int. Ed.*, 2004, **43**, 5396.
110. K. D. Klika and J. Arpalahti, *Chem. Commun.*, 2004, **1**, 666.
111. G. Pratviel and B. Meunier, *Chem. Eur. J.*, 2006, **12**, 6018.
112. C. M. Burrows and J. G. Muller, *Chem. Rev.*, 1998, **98**, 1109.
113. S. Choi, R. B. Cooley, A. Voutchkova, C. H. Leung, L. Vastag and E. D. Knowles, *J. Am. Chem. Soc.*, 2005, **127**, 1773.
114. A. Schimanski, E. Freisinger, A. Erxleben and B. Lippert, *Inorg. Chim. Acta*, 1998, **283**, 223.
115. F. Zamora, P. Amo-Ochoa, B. Fischer, A. Schimanski and B. Lippert, *Angew. Chem. Int. Ed.*, 1999, **38**, 2274.

CHAPTER 3
Sequence-specific DNA–Metal Ion Interactions

NICHOLAS V. HUD AND AARON E. ENGELHART

School of Chemistry and Biochemistry, Parker H. Petit Institute of Bioengineering and Bioscience, Georgia Institute of Technology, Atlanta, GA 30332-0400, USA

3.1 Introduction

The common depiction of DNA as a homogeneous double helix, although aesthetically pleasing, is an inadequate model for understanding how DNA functions within the cell. High-resolution crystal structures and spectroscopic data have revealed considerable sequence-specific structural variations in the DNA double helix and associated sequence-dependent interactions with water and cations.[1–7] Such local variations in the structure and solvation of DNA are of obvious importance with regard to DNA sequence recognition by transcription factors, restriction enzymes and small molecule ligands (*e.g.* DNA-binding drugs).[8–16]

The helical structure of DNA has also been shown to vary as a function of hydration level and salt conditions.[17] For example, some of the earliest fiber diffraction studies of DNA demonstrated that the sodium salt of DNA undergoes a structural transition from the B-form to the A-form as the water content in a DNA fiber is decreased, a transition that is fully reversible if the atmosphere surrounding the fiber is increased back to 92–100% relative humidity.[18] In contrast, the lithium salt of DNA does not undergo the B- to A-form transition, but a transition to the C-form (an over-wound member of the B-form family) between 44 and 66% relative humidity.[19] Such observations from fiber diffraction studies provided the first evidence that the helical structure of DNA, a duplex of polyanions, depends on solvation in addition to the identity and concentration of its surrounding cations.

For many years, DNA–cation interactions were viewed as a more or less sequence-independent phenomenon, with cations being delocalized in a diffuse cloud around the DNA helix, as described by the polyelectrolyte models discussed by Schurr in Chapter 9. Polyelectrolyte models are certainly indispensable tools in cases where DNA can be modeled as a semi-rigid rod with a perfectly uniform or periodic charge density, such as in modeling DNA condensation (Chapter 9). However, these models do not capture the sequence-specific interactions between DNA and its associated cations that are now known to exist close to the surface of DNA.

In the present chapter, we discuss results from studies that have sought to characterize the nature of DNA–cation interactions at the ångstrom scale and, in particular, how specific bases and sequence elements modulate such interactions within the grooves of the double helix. We will also discuss how sequence-dependent localization of cations correlates with sequence-dependent variations in DNA structure. These correlations include the aforementioned B- to A-form transition[20–23] and less dramatic structural transitions, such as narrowing of the minor groove in A-tract DNA.[6,24] Several non-duplex secondary structures of DNA are known to exist, including G-quadruplexes (the topic of Chapter 4) and triplex structures (briefly discussed in Chapter 2), which exhibit their own particular propensities for cation coordination. In this chapter, we restrict our discussion to duplex DNA with Watson–Crick pairs. Compared with the G-quadruplex, duplex DNA exhibits comparatively subtle sequence-dependent cation-binding characteristics, but duplex DNA is, by far, the most abundant and important form of DNA in living cells.

Following a brief overview of some concepts relevant to understanding DNA–cation interactions, we will discuss sequence-specific DNA–cation interactions. We will address cation localization in the major and minor grooves separately. We will also address divalent and monovalent cation localization separately. We use this organizational format because the major and minor grooves exhibit different propensities for, and sequence dependence in, cation localization. Similarly, the observed modes of interaction are different for monovalent and divalent cations. A subset of DNA sequences that contain multiple base pairs of the same type, the so-called AT-rich 'A-tracts' and the GC-rich 'G-tracts', exhibit a strong correlation between cation-localization properties and structural deviations from canonical B-form (or 'generic') DNA. This correlation suggests a direct physical connection between DNA sequence, cation localization and helix structure. Therefore, we pay specific attention to the cation localization properties of A- and G-tract DNA in this chapter.

3.2 A Brief Overview of Cation Binding Modes to DNA

At physiological pH, each nucleotide residue contributes one negative charge to a DNA polymer. As one would predict, cation–phosphate interactions are an important part of DNA–cation interactions for all nucleotide sequences. Because the topic of the present chapter is sequence-specific cation interactions, we will focus on how sequence-directed placement of nucleotide functional groups within the grooves modulates sequence-specific cation localization therein. Structural aspects of cation association with the phosphate groups are covered in Chapter 1 and a review of hydrated metal ion interactions with the phosphate backbone of DNA has recently been presented by Subirana and Soler-López.[25]

Computational studies of DNA electrostatics have long indicated that the functional groups in the major and minor grooves create sequence-dependent electrostatic surface potentials, which were predicted to induce non-uniform cation localization.[26–29] The minor groove of B-form DNA is lined with electronegative groups: the deoxyribose O3′ atom that is esterified with the phosphate, the deoxyribose ring O4′, the O2 of the pyrimidine bases and the N3 of the purine bases (Figure 3.1). In the case of an A·T base pair, the O2 of T and the N3 of A are accessible for direct interaction with cations (*i.e.* upon partial dehydration) or indirect interaction *via* shared waters of hydration that bridge DNA and cations. In contrast, the minor groove placement of the N2 amino group of G interferes with cation localization near the O2 of C and the N3 of G in a G·C base pair, as the N2 amino protons present a positive electrostatic potential that both disfavors cation localization and depletes electron density from the O2 of C. Hence, even at the single base pair level, the A·T pair is expected to localize cations preferentially in the minor groove relative to the G·C pair.

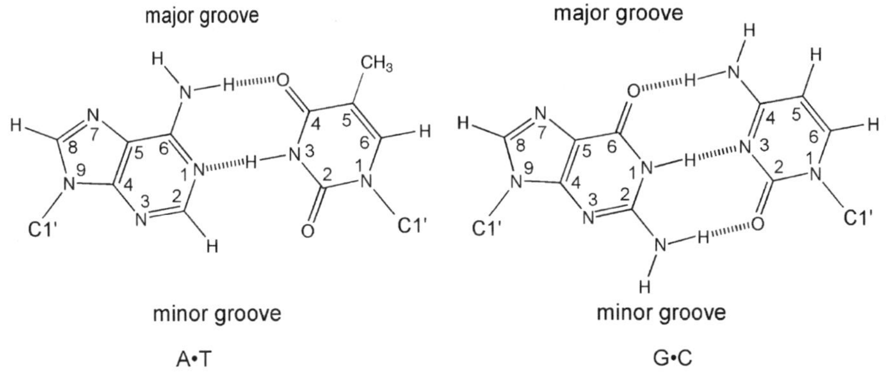

Figure 3.1 Chemical structures and atom numbering of the Watson–Crick base pairs.

With regard to the major groove, both A·T and G·C base pairs present the same polar groups in the groove: a purine N7, one exocyclic amine and one carbonyl oxygen (Figure 3.1). Therefore, predicting sequence-specific cation localization in the major groove is less intuitive than it is for the minor groove, where there are clear electrostatic differences between the two base pairs. However, the placement of the major groove carbonyl and amino groups of the two Watson–Crick base pairs is not equivalent with respect to the center of the major groove or relative to the purine N7. As discussed extensively below, cation localization in the major groove has been shown to be more favored at G·C base pairs, due to the N7 and O6 of guanine presenting a more favorable site for cation localization than the N7 of adenine, which has an adjacent N6 amino group (Figure 3.1).

Preferential cation localization becomes pronounced, particularly for divalent ions, at the dinucleotide level (*versus* the single base pair level). Above the dinucleotide level, certain sequence elements also contribute to the differential localization of cations in the grooves, with length-dependent variations being observed up to a length of at least eight base pairs.

In this chapter, we focus on the interaction of DNA with the alkali and alkaline earth metal ions Na^+, K^+, Mg^{2+} and Ca^{2+}, and also other cations (*e.g.* NH_4^+, Rb^+, Cs^+, Tl^+ and Mn^{2+}) that are spectroscopically useful mimics of these physiologically important cations. We will not discuss the direct coordination of transition metals to single nucleotide bases; this topic is addressed in Chapter 2. In the present chapter, a distinction is drawn between 'cation localization' and 'cation coordination' by DNA; cation localization does not necessarily imply direct contact between a cation and the atoms of DNA. The interactions discussed here between DNA and the alkali and alkaline earth metal ions at specific locations along the double helix frequently take place through waters of hydration that are shared by DNA and the cation. In the crystal state, where the DNA duplex might be expected to be somewhat dehydrated with respect to aqueous solution, cations are still primarily observed to interact with DNA through waters of hydration. For example, in an X-ray crystallographic analysis of Mg^{2+} localization in the major groove of DNA at the N7 position of G, only water-mediated contacts were observed between Mg^{2+} and the guanine base, except in the case of Z-form DNA.[25] Nevertheless, cations, particularly monovalent cations, can dehydrate when associated with DNA – in both the solid state and aqueous solution. The degree of dehydration upon binding to DNA in solution depends on the species of cation and the DNA sequence.[30,31] The analysis of the hydration state of alkali and alkaline earth metal ions when bound to nucleic acids in solution, in crystals and *in silico* continues to provide an ever more refined picture of these complex interactions.[25,30,32–34]

Finally, results from experimental studies are the primary focus of this chapter. Results from computational studies, such as molecular dynamics (MD) simulations, are discussed only when they are particularly illuminating with regard to experimental results. For an overview of MD simulations directed towards understanding the dynamics of DNA–cation interactions, we refer the reader to reviews and recent reports of long-duration simulations.[35–43]

3.3 Defining A-tracts, G-tracts and Generic DNA

In the following discussions, references are made to DNA sequence motifs known as A-tracts, G-tracts and generic DNA, hence it is necessary to define these terms as they are to be understood in this chapter. A-tract DNA refers to any sequence in duplex DNA that contains three or more d(ApA), d(ApT) or d(TpT) steps without a d(TpA) step. This definition includes sequences such as the d(AATT) portion of the Dickerson–Drew dodecamer (DDD) duplex, with sequence d(CGCGAATTCGCG), and also the poly(dA)·poly(dT) duplex. This definition of an A-tract excludes poly(dA–dT)$_2$, for example, because half of the steps are d(TpA) steps. As will be illustrated below, the cation localization properties and helical structure associated with a d(TpA) step are very different from those of d(ApA), d(ApT) and d(TpT) steps, and this step is therefore excluded from the definition of an A-tract. Sequence rules are not as well defined for G-tract DNA, but it can be stated that G-tracts are generally GC-rich DNA sequence elements with multiple d(GpG) steps. For example, the sequence elements d(GGGGGG) and d(GGGCCC) in duplex DNA are G-tracts, in addition to the homoduplex poly(dG)·poly(dC). Finally, all remaining DNA sequences are defined as 'generic DNA' by exclusion. Generic DNA sequences typically contain both A·T and G·C base pairs and comprise the bulk of genomic DNA. Due to the different placement of electronegative functional groups in A·T and G·C base pairs, the cation localization properties of A-tracts and G-tracts have been found to be profoundly different. The cation localization properties of generic DNA lie somewhere between those of the other two sequence motifs, as one might predict.

3.4 Divalent Cation Localization

3.4.1 Divalent Cations and the Major Groove

3.4.1.1 NMR Spectroscopic Studies

NMR studies of ^{25}Mg^{2+} and ^{43}Ca^{2+} lineshape and chemical shift in the presence of naturally derived DNA provided some of the earliest evidence for heterogeneous and sequence-specific binding of alkaline earth metal cations to DNA in solution, and also specific sites where DNA atoms appeared to provide inner-shell ligands for divalent cations (*i.e.* partial replacement of the first shell of hydration).[44–50] These studies demonstrated a positive correlation between tight cation binding sites and high GC content – clearly indicating the preferential involvement of G·C base pairs in divalent cation coordination.[45,48]

To identify the precise location of divalent cation binding sites on DNA, Sletten and co-workers used the paramagnetic cation Mn^{2+} as a probe in ^1H NMR studies of synthetic oligonucleotide duplexes.[51–56] In such experiments, resonance line broadening of a particular proton indicates the close proximity of that proton to a Mn^{2+} binding or localization site.[57] Although the binding of Mn^{2+} to duplex DNA is somewhat tighter than that of Mg^{2+} and Ca^{2+}, the ability of these physiologically relevant cations to displace Mn^{2+} from DNA provides support for the appropriateness of Mn^{2+} as a probe of Mg^{2+} and Ca^{2+} localization sites on DNA.[58]

The studies of Sletten and co-workers revealed a clear preference for divalent cation localization at guanine residues (Table 3.1).[55] Furthermore, the sequence specificity for Mn^{2+} localization was found to be defined at the dinucleotide level, with the relative propensity for localization being GpG ≥ GpA > GpT ≫ GpC. These investigators found no evidence of Mn^{2+} binding at GpC steps or at dinucleotide steps that lack a guanine residue. Mn^{2+} localization was determined to be near the 5′-G of d(GpN) steps and in the major groove, as the ^1H resonance most broadened by Mn^{2+} was the H8 of the 5′-G. These results were clearly consistent with earlier predictions that the O6 and N7 atoms of guanine would make the major groove edge of a G·C base pair a more favorable coordination site than that of an A·T base pair.

As a specific example of data resulting from Mn^{2+} resonance line broadening experiments, ^1H NMR spectra are shown in Figure 3.2 for the duplex of $d(CATG_3C_3ATG)$ in the absence and presence of 1 µM $MnCl_2$. These spectra illustrate the extreme line broadening of the H8 protons of G4 and G5 by Mn^{2+}. Both protons are at the 5′ side of d(GpG) steps. G6H8, which is the 3′ residue of a d(GpG) step and the 5′ residue of a d(GpC) step, is also significantly broadened but to a lesser extent than G4H8 or G5H8. These results are consistent with Mn^{2+} localization at d(GpG) over d(GpC) steps and localization at the 5′-G. Importantly, the spectra shown also illustrate the near-total lack of line broadening of A2H8 and A10H8 resonances and of the H6 resonances of all pyrimidine residues. We note that these resonances have essentially the same linewidth in the presence of 1 µM Mn^{2+} or 1 mM EDTA (which chelates even trace paramagnetic impurities), whereas the H8 resonances of G4 and G5 are broadened to baseline by 1 µM Mn^{2+} (Figure 3.2).

Mn^{2+} localization on $[(CATG_3C_3ATG)]_2$ was originally studied by Hud et al.[3] to test the hypothesis that cation localization in the G-tract element $d(G_3C_3)$ could be the origin of this sequence element's propensity to induce

Table 3.1 Sequence-specific paramagnetic broadening of guanine H8 resonances by M^{2+} (bold letters indicate bases with the most broadened aromatic ^1H resonances).[a]

1. CGCGAATTCGCG	3. GCCAGATCTGGC	5. GCCGTTAACGGC
2. GCCGATATCGGC	4. CGCGTATACGCG	6. CAT**GGG**CCCATG

[a]Sequences 1–5 were studied by Sletten and co-workers.[55] Data for sequence 6 are shown in Figure 3.2 (originally discussed by Hud et al.[3]).

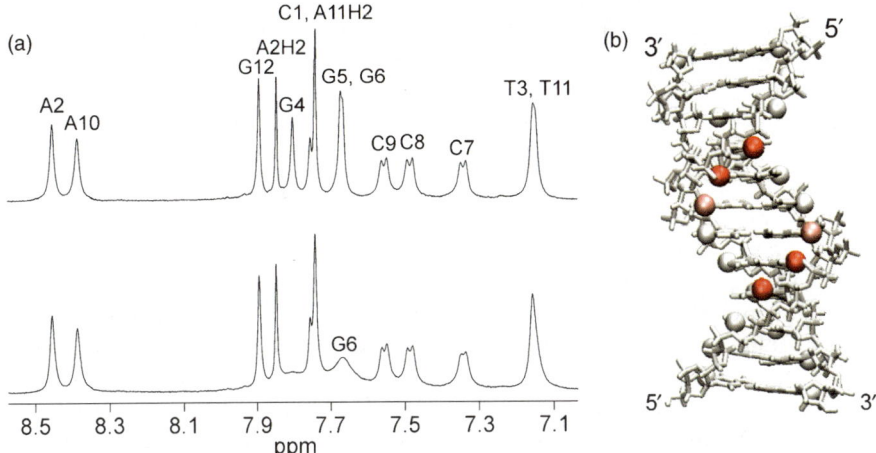

Figure 3.2 Resonance line broadening of the aromatic protons of guanine residues in a G-tract. (a) 500 MHz ^{1}H NMR spectra of the aromatic region of d(CATG$_3$C$_3$ATG) in the presence of 0.1 mM EDTA (top) and 1.0 μM MnCl$_2$ (bottom). Both samples were 2 mM in oligonucleotide strand in D$_2$O, 50 mM NaCl, pH 6.0, at 283 K. The residual peak in the bottom spectrum labeled G6 was assigned based on cross-peaks in a 2D NOESY spectrum. (b) A model duplex structure of [d(CATG$_3$C$_3$ATG)]$_2$ with the H6/H8 base aromatic protons shown as spheres. The aromatic protons of G4 and G5, which are most broadened by Mn^{2+}, are shown in red, the aromatic proton of G6, which exhibits an intermediate sensitivity to Mn^{2+}, is shown in pink. Note that the sequence element d(G$_3$C$_3$) localizes Mn^{2+} on one face of the double helix.

helical axis bending towards the major groove.[59] A model of this duplex is shown in Figure 3.2, with the most broadened resonances shown in color. This model illustrates that the sequence element d(G$_3$C$_3$) localizes divalent cations on one face of the DNA duplex, an observation that is consistent with the proposal that cation localization by d(G$_3$C$_3$) is responsible for major groove narrowing and associated helical axis bending (discussed below in Section 3.7).[3,60] Mn^{2+} localization by the d(G$_3$C$_3$) sequence element is also consistent with data obtained by a number of other methods, which also indicate preferential cation binding in the major groove of G-tract DNA.[1,2]

3.4.1.2 X-ray Crystallographic Studies

While NMR spectroscopy has provided valuable insights regarding sequence-specific localization of divalent cations on DNA in solution, high-resolution X-ray crystallography has provided a more precise determination of where these cations bind and how they are coordinated to DNA. As we shall see throughout this chapter, the combination of data from X-ray crystallography and NMR spectroscopy provides a more detailed picture of DNA–cation interactions than is possible with either technique alone. Divalent cations, particularly Mg^{2+}, can be distinguished from solvent water molecules in high-resolution crystal structures by their ordered inner-shell ligands. Most divalent metal ions observed in DNA crystal structures exist as the hexaaqua or octaaqua species, with octahedrally disposed water ligands, but some metal ions are partially coordinated by DNA base or phosphate atoms.

An analysis of high-resolution crystal structures by Subirana and Soler-López[25] resulted in several conclusions regarding Mg^{2+} binding in the major groove of B-form DNA that are, for the most part, consistent with the solution-state NMR spectroscopic studies discussed above. These investigators found that guanine is the preferred base for Mg^{2+} coordination in the major groove and that one inner-shell water of Mg^{2+} typically acts as an anchor between the cation and the N7/O6 atoms of the guanine base. In Figure 3.3, portions of DNA crystal structures are shown in which hexaaqua Mg^{2+} ions are bound at all four possible d(GpN) steps. As Subirana and Soler-López noted, the first-shell coordination waters of Mg^{2+} typically participate as well in hydrogen bonds with the 3′ base of a d(GpN) step (Figure 3.3). This observation is consistent with the finding of Sletten and co-workers that divalent cation localization is defined at the dinucleotide level, as opposed to the single base pair level.[55]

The results from X-ray crystal structures, shown in Figure 3.3, seem to suggest that hexaaqua Mg^{2+} binding at d(GpC) steps of duplex DNA is as favorable as its binding at the other three d(GpN) steps. However, such a conclusion would be inconsistent with the solution-state NMR line broadening data discussed above. We note that during the crystallization process, low-dielectric dehydrating agents [e.g. 2-methylpentane-2,4-diol (MPD)] are used to drive DNA out of solution (i.e. DNA is dehydrated). Such conditions lead to enhanced association of cations with DNA, undoubtedly resulting in binding of cations at sites that might not be preferential localization sites in aqueous solution. The d(GpC) step may be an example of this caveat, as hydrated divalent cations are almost always found in the major groove of the Dickerson–Drew dodecamer (DDD) within the CGCG sequence element, yet neither the d(CpG) nor the d(GpC) dinucleotide step appears to be a favorable site for Mn^{2+} localization in solution.

An X-ray crystallographic study by Chiu and Dickerson has proven very useful for appreciating the sequence-specific nature of Mg^{2+} and Ca^{2+} binding to DNA, and also the subtle differences between the binding of these two alkaline earth metal ions to DNA.[61] In this study, two DNA sequences, d(CCAACGTTGG) and d(CCAGCGCTGG), were both crystallized in the presence of Ca^{2+} and Mg^{2+}, respectively. The four resulting crystal structures were solved to 1 Å resolution. The Mg^{2+}-form structure of duplex d(CCAGCGCTGG) revealed a hexaaqua Mg^{2+} bound in the major groove at the d(GpG) step. All contacts with DNA are made through the inner shell waters of this cation, as noted above to be typical for Mg^{2+} bound in the major groove of B-form DNA. In the Ca^{2+}-form structure of the same DNA

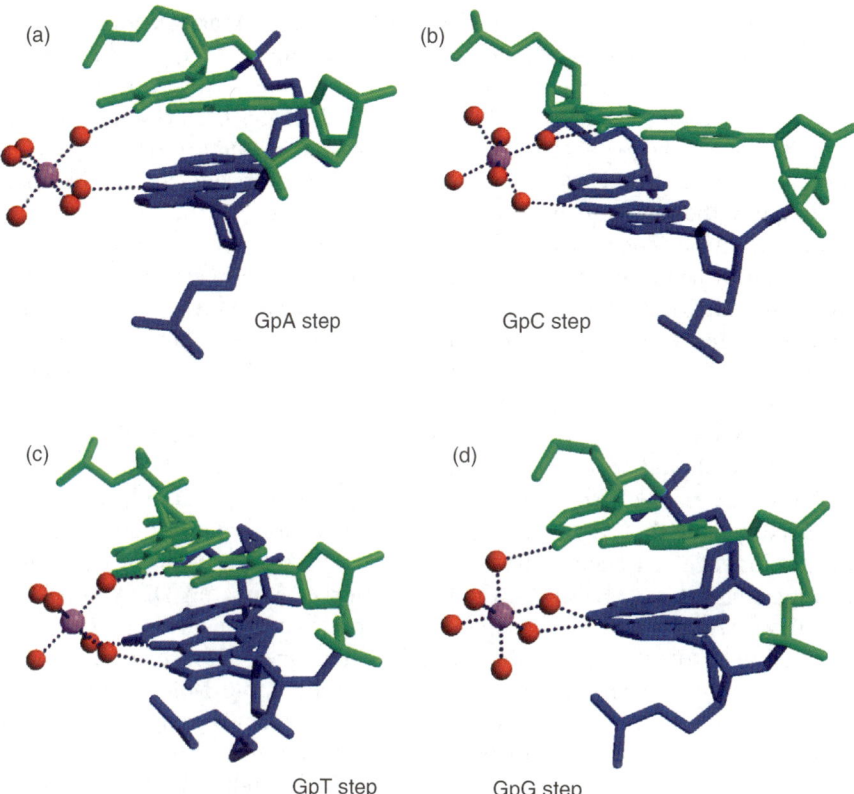

Figure 3.3 Portions of X-ray crystal structures of B-form DNA that illustrate the modes of Mg^{2+} binding to the four possible d(GpN) steps. Guanine is the front base of the lower base pair in each structure shown. The examples shown are from the crystal structures with NDB entry codes (a) bd0037, (b) bd0007, (c) bd0033 and (d) bd003. Copyright of Annual Reviews. Reprinted from ref. 25 with permission.

sequence, a Ca^{2+} is also bound in the major groove at the d(GpG) step. However, this Ca^{2+} makes direct contacts with guanine bases. This hepta-coordinate Ca^{2+} is pentaaquated; the remaining two ligands are G9N7 and G10O6. Ca^{2+} likely exhibits this difference in binding compared with Mg^{2+} because of its lower enthalpic penalty for dehydration, and also access to higher coordination numbers and geometries than Mg^{2+}.[62]

The study by Chiu and Dickerson provided additional insights regarding the difference between Mg^{2+} and Ca^{2+} binding to DNA, and also differences between solid-state and solution-state cation localization. The duplex with sequence d(CCAGCGCTGG) was found to bind fully hydrated Mg^{2+} and Ca^{2+} at the d(ApG) step,[61] which contradicts the rule derived by Sletten and co-workers for divalent cation localization in the major groove only at d(GpN) steps. For example, Sletten and co-workers reported Mn^{2+} localization at the d(GpG) steps of duplex d(CCAAGCTTGG), but did not report evidence for localization at the d(ApG) step for the same duplex.[55] Hence the d(ApG) step may only be a favorable site for divalent cation localization under the dehydrating conditions of crystallization. Interestingly, the crystal structures of d(CCAACGTTGG) also revealed Mg^{2+} bound at the d(GpT) step in the Mg^{2+}-form crystal, but not Ca^{2+} in the Ca^{2+}-form crystal. Such differences are certainly the result of the interplay that exists between the free energies of cation hydration, coordination and orbital geometry, all of which could be modulated by sequence-dependent interactions with the nucleotide bases. For either divalent cation, there is good support for accepting the consensus that guanine residues are essential for major groove divalent metal ion coordination [either d(GpN) or d(NpG)], with or without inner-shell metal–DNA contacts.

Although there is reasonable agreement between the Mn^{2+} localization studies by NMR and the divalent cation binding sites identified in X-ray crystal structures, it should be kept in mind that the precise mode of Mn^{2+} binding to DNA may not be the same as that of Mg^{2+} and Ca^{2+}. We have already discussed the fact that Mg^{2+} tends to bind in the major groove at d(GpN) steps without loss of inner-shell hydration, whereas DNA groups can provide inner shell ligands to Ca^{2+}. Based on the localization of Mn^{2+} on DNA in a 2.5 Å crystal structure of the nucleosome core particle, it appears that Mn^{2+} preferentially coordinates directly at the N7 of guanine.[63] Hence there appears to be a true difference between the nature of Mn^{2+} and Mg^{2+} coordination at d(GpN) steps, even though the sequence rules for major groove localization of these two cations appear to be the same. Computational studies suggest that the respective energetics of Mn^{2+} and Mg^{2+} binding to the nucleotide bases could contribute to these differences observed in coordination modes.[64]

X-ray crystal structures of A-form DNA have provided significant insights into the interplay between helix structure and divalent cation binding in the major groove. Wang and co-workers analyzed Mg^{2+} binding to the A-form crystal structure of duplex d(ACCGGCCGGT), along with three A-form RNA–DNA chimeras with the same nucleotide sequence.[65] Several hexaaqua Mg^{2+} ions were identified in the major grooves of these duplexes. Based on their structures and previously reported crystal structures, Wang and co-workers concluded that divalent cations have two distinct modes for binding within the major groove of A-form DNA (or RNA). The first type of binding site is deep in the groove at d(GpN) dinucleotide steps, similar to that observed for divalent cations bound in the major groove of B-form DNA (Figure 3.4). The second binding site is near the top of the major groove, at what is sometimes called the 'mouth' or 'lips' of the groove, where a fully hydrated divalent cation can interact with phosphate groups from both strands (Figure 3.4). This second mode is possible for an A-form helix, but not a B-form helix, because of the closer approach of the phosphate groups across in the major groove in the A-form helix (Figure 3.4). As will be discussed in the following section, an analogous mode of cation binding is observed across the narrow minor groove of the B-form helix (Figure 3.4).

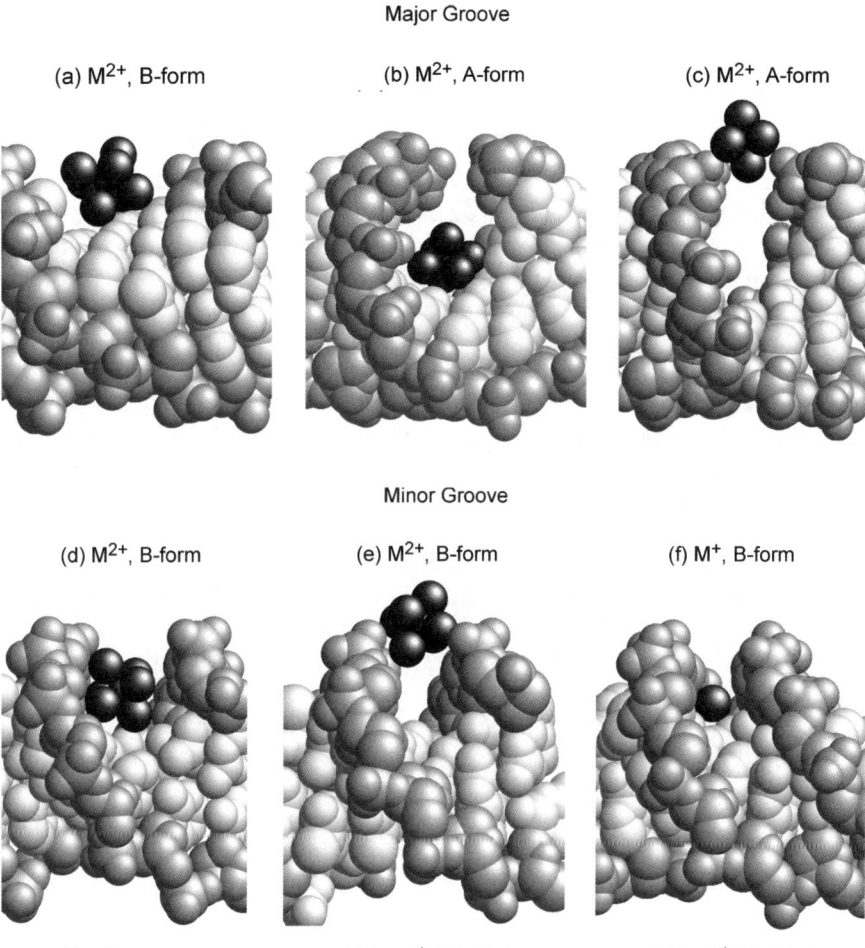

Figure 3.4 An illustration of the correlation that exists between DNA groove width and cation coordination using space-filling representations of X-ray crystal structures. (a) $Ca(H_2O)_8^{2+}$ in the major groove of $[d(CCAGTACTGG)]_2$ at the d(ApG) step (ref. 76; NDB entry bd0023). (b) $Mg(H_2O)_6^{2+}$ in the major groove at the d(GpG) step of $[d(ACCGGCCGGT)]_2$ in the A-form helix (ref. 65; NDB entry ad0006). (c) $Co(NH_2)_6^{3+}$ spanning the major groove at the d(CpG) step of $[d(ACCGGCCGGT)]_2$ in the A-form helix (ref. 65; NDB entry adj065). (d) $Mg(H_2O)_6^{2+}$ in the minor groove of $[d(CCAGCGCTGG)]_2$ at the d(CpT) step (ref. 61; NDB entry bd0035). (e) $Mg(H_2O)_6^{2+}$ spanning the minor groove of T7-fluorinated DDD at the d(GpA) step (ref. 136; NDB entry bd0007). (f) Rb^+ in the minor groove at the d(ApT) step of $[d(CGCGAA(TAF)TCGCG)]_2$ (T7-fluorinated DDD) (ref. 114; NDB entry bd0012). DNA base atoms are white and backbone atoms are light gray; cations and water oxygen atoms are dark gray. Adapted from Figure 1 in ref. 2.

3.4.2 Divalent Cations and the Minor Groove

3.4.2.1 *NMR Studies of Divalent Cations and the Minor Groove*

Divalent cation localization in the minor groove of B-form DNA has also been studied by NMR spectroscopy using Mn^{2+} in resonance line broadening experiments.[5,66] These studies demonstrated that Mn^{2+} localization is more favorable in the minor groove near A·T base pairs than near G·C base pairs, as predicted, due to the greater negative electrostatic potential of the minor groove at A·T base pairs. However, the rules that govern which sequences localize Mn^{2+} in the minor groove, and the mode of cation binding at these sequences, have proven more complicated than the d(GpN) rule discussed above for the major groove.

Hud and Feigon investigated Mn^{2+} interaction with a series of 12 bp DNA duplexes containing the sequence elements $d(T_nA_n)$ and $d(A_nT_n)$ (where $n = 2, 3$ or 4).[5,66] Entry of Mn^{2+} deep into the minor groove at specific locations along duplexes containing these sequence elements was most clearly indicated by the selective resonance line broadening of AH2 protons, as these protons are located on the floor of the minor groove (Figure 3.1). Of the duplexes studied, $[d(CGT_4A_4CG)]_2$ exhibited the most dramatic resonance line broadening by Mn^{2+} (Figure 3.5). Quantitative analysis of line broadening as a function of $MnCl_2$ revealed that the most favored site for Mn^{2+} entry deep into the minor groove of $[d(CGT_4A_4CG)]_2$ is at the d(ApA) steps [or, equivalently, the d(TpT) steps] immediately adjacent to the central d(TpA) step.[66] Although $[d(CGT_4A_4CG)]_2$ contains two symmetry-related d(ApA)·d(TpT) sites, their close proximity (separated by the single d(TpA) step) makes it unlikely that two Mn^{2+} ions simultaneously occupy both sites of a given duplex, as substantial electrostatic repulsion would exist between the two divalent cations. Consistent with this proposal, NMR studies also revealed that Mn^{2+} is in fast exchange between multiple binding sites on DNA and with bulk solvent (*i.e.* residence times <1 ms).[66] As a particular example, in the case of $[d(CGT_4A_4CG)]_2$ resonance line broadening data were consistent with Mn^{2+} being in exchange between the two symmetry related d(ApA)·d(TpT) sites and with less favored binding sites within the minor groove.[66]

A comparison of resonance line broadening by Mn^{2+} of three dodecamer duplexes containing the sequence elements $d(T_4A_4)$, $d(T_3A_3)$ and $d(T_2A_2)$ revealed that cation entry and localization sites in the minor groove are defined by sequence elements that are larger than a dinucleotide step. As illustrated in Figure 3.5, line broadening of AH2, H4' and H1' resonances reveal that Mn^{2+} localization at the 5'-most d(ApA) step [and the 3'-most d(TpT) step] of the sequence element $d(T_4A_4)$ is more pronounced than it is at the corresponding steps of $d(T_3A_3)$, which is, in turn, more pronounced than it is for $d(T_2A_2)$. Thus, base pair changes that are even two nucleotides away from the most apparent Mn^{2+} binding sites of $d(T_4A_4)$ are able to reduce the localization of Mn^{2+} at these sites.

The homo-AT sequence elements of the form $d(A_nT_n)$ also localize Mn^{2+} deep in the minor groove, but not as dramatically as $d(T_nA_n)$ sequence elements

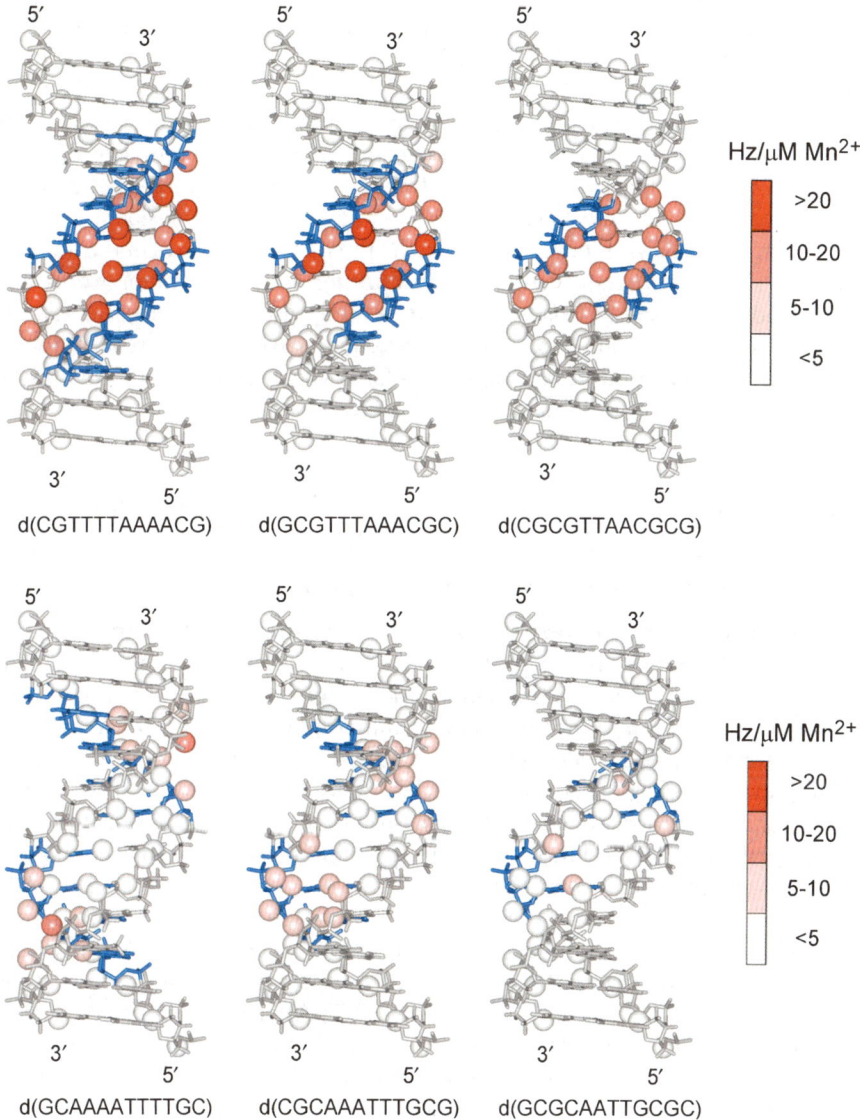

Figure 3.5 Graphical representation of the relative line broadening exhibited by AH2, H1′ and H4′ protons as a function of $MnCl_2$ concentration for a series of duplexes containing the $d(T_nA_n)$ and $d(A_nT_n)$ sequence element. AH2, H1′ and H4′ protons are represented with van der Waals radius spheres. Adenine residues are shown in blue. The DNA models shown are based on the canonical B-form helix and are orientated such that the reader is looking into the minor groove at the middle of each helix. Linewidth information was extracted from 1D and 2D NMR spectra as described in ref. 66. Adapted from Figures 3 and 7 in ref. 66.

(Figure 3.5). Mn^{2+} entry and localization in the minor groove of the $d(A_4T_4)$ sequence element are similar to those exhibited by $d(T_4A_4)$, in that localization is also most apparent at the 5'-most d(ApA) step [and equivalently the 3'-most d(TpT) step]. $d(A_4T_4)$ is, however, different from $d(T_4A_4)$, in that the cation localization sites at the 5'-most d(ApA) step of $d(A_4T_4)$ are at the outer edges of this sequence element, as opposed to being near the center of $d(T_4A_4)$ (Figure 3.5). Mn^{2+} localization was also found to decrease slightly as the $d(A_4T_4)$ sequence element was shortened to $d(A_3T_3)$ and to decrease considerably upon shortening to $d(A_2T_2)$ (Figure 3.5). Thus, the three DNA duplexes in the $d(A_nT_n)$ series also illustrate that divalent cation localization sites in the minor groove can be defined by sequence elements that are several base pairs in length. As discussed below, this phenomenon may be particular to A-tracts and related homo-AT sequences that exhibit sequence-dependent helical structures and hydration motifs that also depend on the length of the sequence element.

3.4.2.2 Comparison of Divalent Cation Localization Sites of A_nT_n and T_nA_n Sequence Elements as Defined by NMR and X-ray Crystallography

A comparison of results obtained from solution-state Mn^{2+} resonance line broadening of minor groove protons with X-ray crystal structures is enlightening, albeit not without apparent contradictions. The first comparison we present is between solution-state and crystallographic studies of DNA duplexes that contain sequence elements representative of the motif $d(T_nA_n)$. Three crystal structures deposited in the Nucleic Acids Data Base (NDB) are considered most appropriate and useful for this comparison. These structures are all better than 2.0 Å resolution and were crystallized from solutions containing either magnesium or calcium salts. The crystal structures of duplex d(CGAT-TAATCG), determined by Quintana et al. (NDB entry bdj031),[67] and duplex d(CCTTTAAAGG), determined by Mack et al. (NDB entry bd0051),[68] both contain a single hexaaqua Mg^{2+} that is located deep in the minor groove at the d(ApA) step that is adjacent to the central d(TpA) step. The X-ray structure of duplex d(CTTTTAAAAG) by Han et al. (NDB entry bd0071), which was determined from a crystal grown in the presence of calcium acetate, contains a single octaaqua Ca^{2+} deep in the minor groove within the $d(T_4A_4)$ sequence element. This hydrated cation is also located at the d(ApA) step that is adjacent to the central d(TpA) step. Hence X-ray crystal structures of DNA duplexes with sequence elements similar to or identical with the $d(T_nA_n)$ sequence elements used in the Mn^{2+} resonance line broadening studies reveal a bound divalent cation at a position that is perfectly consistent with the solution state data.

Correlating the solution-state and crystal-state results for cation localization at $d(A_nT_n)$ sequence elements is not as straightforward as for the $d(T_nA_n)$ sequence elements. To appreciate fully what both the NMR spectroscopic and X-ray crystallographic studies have revealed about cation localization in the minor groove of $d(A_nT_n)$ and $d(T_nA_n)$ sequences, it is necessary to consider the unique structural properties of A-tract DNA. As defined in Section 3.2, A-tract sequence elements contain four or more consecutive adenine or thymine residues without a d(TpA) step [*i.e.* only d(ApA), d(ApT) and d(TpT) steps]. A structural feature particular to A-tract sequences is an unusually narrow minor groove. As one measure, the minor groove of a typical B-form helix is about 5.9 Å, whereas the minor groove of an A-tract can be as narrow as 2.8 Å (groove widths are given as the distance between pairs of phosphorus atoms across the grooves from which 5.8 Å was subtracted to account for the van der Waals radii of the two phosphate groups).[1] Additionally, the minor groove of an A-tract of the form $d(A_n)$ (with $n \geq 4$) has been shown to narrow from the 5′ to the 3′ end.[3,69–72] That is, the width of the minor groove at the first adenine of $d(A_n)$ is similar to that of canonical B-form DNA, whereas the minor groove at the fourth and following adenine residues has a width that approaches the minimum observed for A-tracts. For symmetric A-tracts of the form $d(A_nT_n)$, the minor groove is of typical B-form width at the two ends and tapers to the narrow A-tract width near the central d(ApT) step. As noted above, d(TpA) steps are excluded from the definition of an A-tract. This exclusion is based on the observation that a single d(TpA) step disrupts the structural features that are particular to an A-tract, including the narrow minor groove.[5,69,70] Thus, the sequence element $d(T_4A_4)$ has a minor groove that is typical of B-form DNA width at the central d(TpA) step and narrows progressively from the center outwards to the outermost base pairs of the sequence element.[5,67]

Given what is known about the minor groove width of A-tract DNA and related sequence elements, Mn^{2+} entry deep into the minor groove appears to be regulated by both electrostatic potential and steric effects. Entry is most favored at homo-AT sequences, which have more electronegative minor grooves than GC-containing sequences and where the minor groove is relatively wide (*i.e.* typical of the B-form). In the case of $d(T_4A_4)$, Mn^{2+} entry deep into the minor groove is most favored near the two centrally located 5′-most d(ApA) steps, which have a minor groove more typical of the canonical B-form width. Entry deep into the minor groove is decreasingly favored for the d(ApA) steps that are further from the center of this sequence element where the groove is known to narrow. Considering the size of a fully hydrated divalent cation compared to the groove width of DNA (Figure 3.4), these observations suggest that the minor groove is of optimal size for entry of a fully hydrated divalent cation near the central d(TpA) step, but too narrow for entry at the more narrow regions of an A-tract. In the case of $d(A_4T_4)$, the minor groove is widest at the outside base pairs and narrowest around the central d(ApT) step.[5,69,70] Hence entry of Mn^{2+} into the minor groove at the ends of this sequence element corresponds to entry at the widest minor groove region of this homo-AT sequence.

One would expect that the narrow minor groove associated with A-tracts, which brings phosphate groups in particularly close proximity, would create a preferred site for cation binding. Thus, at first it seems paradoxical that Mn^{2+}

is not observed by solution-state NMR studies to be localized in the minor groove near the center of $d(A_nT_n)$ sequence elements where the phosphates have their closest approach across the minor groove. Entry of a divalent cation into the minor groove at the center of a $d(A_nT_n)$ sequence element would require at least partial removal of the first shell of hydration, which may be energetically prohibitive. The H1′, H4′ and AH2 protons have proven useful as indicators of Mn^{2+} localization *deep* in the minor groove at A·T base pairs (Figure 3.1). However, these protons may be too deep in the minor groove to report the binding of a Mn^{2+} at the top of the groove (*i.e.* along or between the closely-spaced phosphate backbones of an A-tract minor groove). The hypothesis that divalent cations bind atop the narrow minor grooves of A-tracts but are missed by Mn^{2+} resonance line broadening studies is fully supported by X-ray crystallography studies discussed in the following section. Furthermore, the ^{31}P relaxation studies by Redfield and co-workers have demonstrated that the backbone of $[d(GGAATTCC)]_2$ is most rigidified near the central d(ApT) step upon addition of Mg^{2+} to a DNA sample containing only monovalent cations, an observation that is also consistent with a Mg^{2+} binding across the minor groove of the d(AATT) sequence element in solution.[73]

3.4.2.3 Additional Divalent Cation Localization Sites Revealed in X-ray Crystal Structures

Williams and co-workers performed a systematic analysis of divalent cation binding sites within 10 previously determined high-resolution crystal structures (0.74–1.5 Å) of the Dickerson–Drew dodecamer, d(CGCGAATTCGCG).[34] These authors identified 44 divalent cations (Ca^{2+} and Mg^{2+}) in contact with DNA phosphate groups; 73% of these cations were found to be atop the minor groove (*i.e.* touching at least one backbone) or deep within the minor groove. In some cases, Mg^{2+} ions that were identified as bound in the major groove were also found, based on the structure of the asymmetric unit cell of the crystal, to be bound atop the minor groove of an adjacent duplex of the crystal lattice. An example structure of a hexaaqua Mg^{2+} bound atop the center of the d(AATT) sequence element of the DDD is shown in Figure 3.4e. The systematic analysis performed by Williams and co-workers, and also studies in other laboratories of individual crystal structures, have provided substantial evidence that divalent cation binding atop the narrow minor groove of A-tract sequences is common in crystal structures.[34,74] Divalent cations bound at the top of the minor groove can be fully hydrated or have inner-shell ligands provided by phosphate oxygen atoms.[34] As noted above, Mn^{2+} may also bind at similar sites in solution but reside too far from DNA protons to cause appreciable resonance line broadening and therefore go undetected in Mn^{2+} resonance line broadening studies.

Egli and co-workers also compared Ca^{2+} and Mg^{2+} binding sites in crystal structures of the Dickerson–Drew dodecamer.[74] In the Ca^{2+} form, these authors observed five hydrated ions transversing the minor groove and one bound deep in the groove near the start of the d(AATT) sequence element. In the Mg^{2+} structure, several hydrated ions were observed to bridge the groove, but none was found deep within the groove. Based on these observations, it appears that the lower penalty of dehydration and geometric flexibility for coordination afforded by Ca^{2+} may permit Ca^{2+} entry into the minor groove where it is impossible for Mg^{2+} (or Mn^{2+}) to enter with an unperturbed inner shell of hydration.[62] Subirana and co-workers also concluded, based on high-resolution structures of two DNA duplexes containing the AATT sequence element, that Ca^{2+} and Mg^{2+} do not enter the minor groove of this sequence element, as the structure of ordered water in this groove (the minor groove spine of hydration, discussed below) is identical for crystals grown in the presence of either divalent cation.[75]

As mentioned above, cation localization in the minor groove is not expected to be as favored at G·C base pairs due to the presence of the amino group at the C2 position of guanine. The amino group of guanine also presents an added challenge for observation of Mn^{2+} localization in solution, as the protons of this amino group rapidly exchange with solvent protons, whereas the non-exchangeable AH2 proton at the analogous position on adenine has proven extremely valuable for Mn^{2+} localization in the minor groove near A·T base pairs. Thus, Mn^{2+} localization within the minor groove at G·C base pairs by NMR spectroscopy has primarily focused on line broadening measurements of deoxyribose protons, particularly H1' and H4'.[66] A comprehensive investigation of H1' and H4' resonance line broadening by Mn^{2+} has not been reported for a wide variety of DNA sequences. Nevertheless, the lack of significant AH2 line broadening detected for A·T base pairs within mixed G·C/A·T sequences,[55] together with the minimal line broadening detected in homo-GC regions that flank homo-AT sequences,[5,66] suggest that Mn^{2+} localization is not favored in solution within the minor groove of B-form DNA at G·C base pairs.

The high-resolution crystal structures solved by Chiu and Dickerson of duplex d(CCAACGTTGG) and d(CCAGCGCTGG) for crystals grown with Mg^{2+} and Ca^{2+}, respectively, again illustrate cation binding in the crystal state where it not is indicated by Mn^{2+} localization studies in solution. These structures also provide excellent examples of the cation-specific nature of minor groove coordination. In the Mg^{2+}-form crystal structure of duplex d(CCAACGTTGG), a Mg^{2+} is observed in the minor groove near the center of the d(ApC) step. In the Ca^{2+}-form structure of the same sequence, a Ca^{2+} is located at the d(ApA) step, more than 3 Å away from the position at which Mg^{2+} is bound in the Mg^{2+}-form structure. The divalent cations are fully hydrated in both crystal structures and their bound waters form hydrogen bonds with base edges and phosphate groups. The authors suggested that these hydrogen bonds largely dictate the positions of these divalent cations in the minor groove.

The crystal structures of Chiu and Dickerson for duplex d(CCAGCGCTGG) and of Kielkopf et al. for duplex d(CCAGTACTGG) have provided important insights regarding divalent cation binding (as defined by crystal structures) in the minor groove of mixed AT/GC sequences.[61,76]

These high-resolution structures revealed that divalent cations in the minor groove of [d(CCAGCGCTGG)]$_2$ and [d(CCAGTACTGG)]$_2$ reside at more than one location in the minor groove, within a single crystal. In the case of d(CCAGTACTGG), Kielkopf *et al.* observed two locations for octaaquated Ca^{2+}; both sites are at the d(GpT) step, but they are separated by 1.0 Å.[76] In the case of d(CCAGCGCTGG), Chiu and Dickerson reported two alternative sites for Ca^{2+} binding in the minor groove, one at the d(ApG) step and one at the neighboring d(GpC) step.[61] Additionally, Chiu and Dickerson observed that the Mg^{2+} form of the d(CCAGCGCTGG) crystal structure contained three alternative site for Mg^{2+} binding which span the d(ApG) and d(GpC) steps identified as Ca^{2+} binding sites. These alternative coordination sites are discussed further in Section 3.8.

The combination of results from NMR spectroscopy and X-ray crystallography support, for the most part, the most favored and well-defined sites for cation binding in the minor groove being AT-rich sequence elements. Solution-state NMR studies have revealed a preference for AT-rich sequences and a minor groove of typical B-form width in order for divalent cations to enter *deep* into the minor groove. X-ray crystallographic studies have revealed that divalent cation binding is also favored above the narrow minor groove of A-tract sequences, with ^{31}P studies providing indirect support for this mode of binding in solution. The binding of divalent cations in the minor groove within GC-rich regions is expected to be less favorable than in AT-rich regions, as demonstrated by solution-state NMR.[5,55,56,66]

The divalent cations observed in the minor groove of [d(CCAGCGCTGG)]$_2$ in the crystal state may, again, be partially due to the action of the dehydration of crystallization.[2] The observation of alternative minor groove binding sites for divalent cations in this duplex may result from a poor definition of a single well-defined divalent binding site in the GC-rich minor groove. The result that X-ray crystal structures of [d(CCAACGTTGG)]$_2$, which is more AT-rich, do not exhibit such alternative sites[61] is consistent with the proposal that multiple A·T base pairs are required for a well-defined divalent cation binding site in the minor groove that is favorable in the absence of dehydration.

3.5 Monovalent Cation Localization

3.5.1 Monovalent Cations and the Major Groove

For several years, experimental characterization of monovalent cation localization in the DNA major groove lagged behind that of divalent cation localization, despite the availability of ultrahigh-resolution (<1 Å) crystal structures.[75] A major difficulty in discerning monovalent cation localization *via* X-ray crystallography is that Na$^+$, a common cation in DNA crystals, is isoelectronic with water and therefore extremely difficult to discern from solvent water molecules. Furthermore, both Na$^+$ and K$^+$, the two most abundant physiological monovalent cations, have highly variable and irregular coordination geometries. Therefore, even when solvent densities are well defined in a high-resolution DNA crystal structure, the inner-shell coordination geometry of Na^+ or K^+ cannot be used to identify these cations in crystal structures, whereas the regular coordination geometry of divalent cations such as Mg^{2+} (Figures 3.3 and 3.4) are often sufficient to identify them in crystal structures.

The use of the thallium ion by Williams' group in 2001 opened the door for the high-resolution analysis of monovalent cation binding in the major groove of DNA.[77] Tl^+ is an excellent mimic of the potassium ion. Tl^+ and K^+ have similar ionic radii (K^+, 1.33 Å; Tl^+, 1.49 Å), irregular coordination geometries[78] and almost identical energies of hydration (K^+, -77 kcal mol^{-1}; Tl^+, -78 kcal mol^{-1}).[79] Furthermore, Tl^+ has been shown to substitute for K^+ in the potassium binding sites of proteins, including sodium–potassium ion pumps,[80-84] in G-quartets[85,86] and in the P4–P6 domain of the *Tetrahymena* ribozyme.[87] From an X-ray crystallography perspective, Tl^+ is an excellent probe of monovalent cation binding sites because of its strong and anomalous scattering of X-rays.[88,89]

Williams and co-workers obtained an atomic resolution structure of the DDD duplex, [d(CGCGAATTCGCG)]$_2$, from a crystal grown in the presence of Tl^+, Mg^{2+} and spermine^{4+}.[77] The RMS deviation between their refined Tl^+-form structure and a previously refined K^+-form structure of the same DNA sequence was only 0.32 Å. With the two DNA structures being isomorphous and the above-mentioned similar coordination properties of Tl^+ and K^+, it was proposed that Tl^+ binding sites identified in the Tl^+-form structure are likely to be K^+ binding sites. Their final refined model of the Tl^+-form structure contained 13 partially occupied Tl^+ binding sites – the most monovalent cation binding sites identified to date on the DDD. The majority of Tl^+ identified were in the grooves and not in direct contact with phosphate groups (Figure 3.6). The majority of Tl^+ were also found to interact with a single DNA duplex and not to participate in lattice contacts between two duplexes, which is in contrast to divalent cations, which are often found to provide lattice contacts between duplexes. None of the monovalent cation binding sites identified were found to be fully occupied by Tl^+. Those sites with the highest occupancy, 20–35%, were located in the major groove.

The Tl^+-form structure of the Dickerson–Drew dodecamer provided, for the first time, direct experimental evidence for the sequence-specific localization of monovalent cations in the major groove. As discussed above, the major groove was predicted by calculations and shown by divalent cation localization to bind cations more favorably at G·C base pairs. Williams and co-workers found that all six O6 and five out of six N7 atoms of non-terminal G residues are within 3.4 Å of a Tl^+ binding site. That is, *all* non-terminal G residues participate in monovalent cation coordination. The participation of the GO6 and GN7 atoms in Tl^+ coordination is not surprising given the prior observation of divalent cations coordinated to the same sites of guanine bases. However, monovalent coordination in the major groove may only require the participation of a single base pair, whereas divalent cations are observed to span two base pairs. All Tl^+

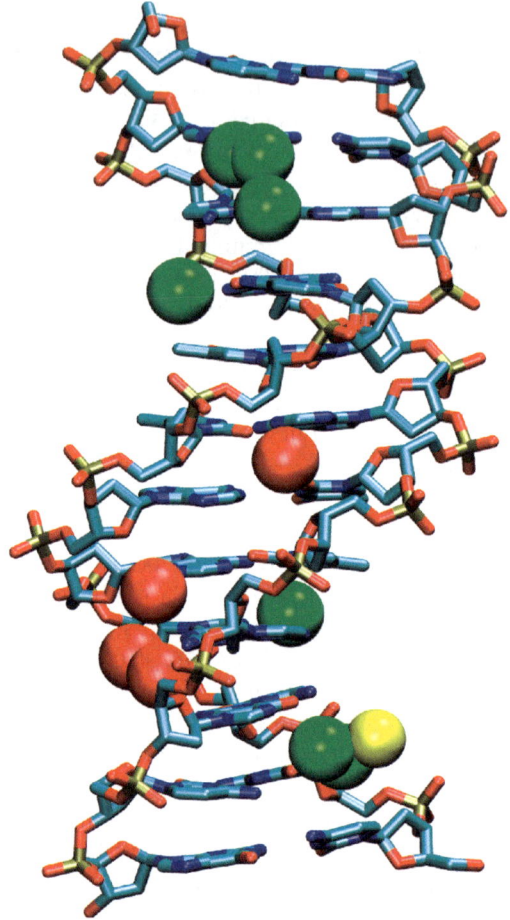

Figure 3.6 X-ray crystal structure of the DDD, d(CGCGAATTCGCG), from a crystal grown in the presence of Tl^+ (NDB entry bd0054). Tl^+ located in the minor groove are shown in red, Tl^+ located in the major groove are shown in green. A single Mg^{2+} is shown in yellow. For clarity, no water molecules are shown. All cation binding sites shown were determined by Williams and co-workers to be partially occupied,[77] and closely spaced cation binding sites would not be occupied in the assymetric unit cell of the crystal.

bound in the major groove were, with one exception, observed to lie within the plane of a guanine base (Figure 3.6). One Tl^+ binding site was located between the planes of two G bases of a d(GpC) step, with Tl^+ coordinated by the two GO6 atoms (*i.e.* between the two strands).

Of equal importance to the identification of Tl^+ binding sites in major groove at G · C base pairs was the complete absence of major groove Tl^+ binding sites in the A-tract region of the same duplex. Williams and co-workers concluded that the selective localization of monovalent cations at G·C base pairs in the major groove is dictated by several factors. Those factors deemed important include the appropriate geometry of functional groups for bidentate coordination, which can be provided by the O6 and N7 of a single G base or by two O6 atoms of a d(GpC) step, and also the avoidance of the electropositive amino groups of A·T base pairs. Additional factors considered important were the relatively low energetic cost for partial Tl^+ dehydration, which allows direct coordination to G bases, and the overall high electronegative potential of the major groove floor in a GC-rich sequence element. The particular Tl^+ binding sites identified in the crystal structure of the DDD are in agreement with prior predictions based on electrostatic surface potential calculations[26] and MD simulations.[90–93]

A less-expected result of the Tl^+-form DDD crystal structure was the observation that partially dehydrated Tl^+ and fully hydrated Mg^{2+} partially occupy a common binding site in the GC-rich major groove, apparently competing on a similar footing for DNA binding in the same crystal. At this particular site, partial occupancy by Tl^+ and hexaaqua Mg^{2+} sum to 100% occupancy. A similar competition between Rb^+ and hexaaqua Mg^{2+} in the GC-rich region of the DDD has also been observed.[77]

In this chapter, we primarily focus on the sequence-specific localization of cations by B-form DNA; however, investigations of cation binding to A-form DNA have provided important insights into the nature of cation localization in the major groove of DNA, particularly with respect to the B- to A-form transition. Egli and co-workers solved the structure of [d(GCGTA-TACGC)]$_2$ from crystals grown in the presence of several different cations. All structures were of the A-form. Using the anomalous X-ray scattering of K^+, Rb^+ and Cs^+, they were able to demonstrate the binding of each of these monovalent cations at the d(Gp1) step.[94] Cations were not observed at any other position within the major groove, suggesting that the d(GpT) is also a more favorable site for localization than d(GpC) in the A-form helix. A study by Fuller and co-workers that combined X-ray and neutron diffraction of DNA fibers revealed that monovalent cations are localized down the center of the major groove in what is effectively a water-lined channel.[95]

Given the comparatively meager experimental data regarding the localization of cations in the major groove of B-form DNA, for this class of interactions MD simulations have been particularly important for helping to provide an overall view of DNA–cation interactions. MD simulations by McConnell and Beveridge revealed cation localization in the major groove of duplex DDD near d(GpC),[96] which is consistent with the crystallographic studies discussed above for the Tl^+-form structure of the same duplex. The MD simulations also revealed that these major groove sites are occupied up to 10% by Na^+, and Na^+ binding near a particular d(CpG) site was found to cause a narrowing of the major groove.[96] A previous MD study by Feig and Pettitt reported simulations on the duplex $d(A_5G_5) \cdot d(C_5T_5)$ in the presence of high salt with both the CHARMM and AMBER force fields (preferentially stabilizing the A- and

B-DNA conformers, respectively). Na$^+$ was likewise observed to localize in the B-form major groove at guanine bases with 10–20% occupancy.[92] Simulations in the same study of the A-form DDD suggested greater ordering of G-tract major groove cations and a residence time as much as an order of magnitude higher than in B-form DNA (up to 160 ps).[92] MD simulations by Auffinger and Westhof on d(CG)$_{12}$ with K$^+$ showed transient contacts between the cation and all heteroatoms on DNA.[91] What they termed cation coordination, however, only occurred in the major groove, which is understandable given the lack of A·T base pairs in their sequence. K$^+$ was found to be correlated with guanine-O6 30% of the time, N7 10% of the time and phosphate oxygens 30% of the time. K$^+$ residence times in the major groove were 5–10 times longer than coordination times for K$^+$–phosphate contacts. They noted K$^+$ coordinated between sequential guanine-O6 in a d(GpC) step but not a d(CpG) step, which illustrates that monovalent cation localization can also be defined at the dinucleotide level. MD studies that compared Na$^+$ and K$^+$ interactions with the same DNA duplexes also indicated that these two monovalent cations have distinct preferences for interactions in the groove and with the phosphate backbone.[97,98]

3.5.2 Monovalent Cations and the Minor Groove

Monovalent localization sites in the minor groove are perhaps now the best characterized cation binding sites of DNA. Nevertheless, for more than a decade, the lack of identification of monovalent cations in high-resolution DNA crystal structures gave many investigators the impression that cations do not enter the minor groove of B-form DNA to replace, even transiently, water molecules bound in this groove. Undoubtedly, the aforementioned similar scattering of X-rays by Na$^+$ and water confounded the identification of what is often the most abundant cation present in DNA crystal structures. This impression was simultaneously challenged on three fronts in the mid-1990s, when researchers using MD simulations,[93] NMR spectroscopy[3,5] and X-ray crystallography[4,99] each provided compelling evidence that monovalent cations enter deep into the minor groove of DNA.[2] It is now widely appreciated that monovalent cation coordination in the minor groove is a sequence-specific phenomenon. At A·T base pairs, the AN3 and TO2 groups provide favorable sites for coordination, but this coordination still depends on the sequence context. Direct coordination of monovalent cations to the nucleotide bases within the minor groove is generally less favored at G·C base pairs, due to the electropositive amino group at the C2 position of G. In the following discussion, it should be kept in mind that water molecules and monovalent cations apparently share similar solvent sites within the minor groove of B-form DNA. In most cases, monovalent cations are found to occupy these sites less frequently than water molecules. However, in cases where relative water *versus* cation occupancy has been determined, water was present at a much higher concentration than cations.

3.5.2.1 NMR Spectroscopy of Monovalent Cations in the Minor Groove

The first direct localization of a monovalent cation bound to a nucleic acid in the solution state was, arguably, within a G-quadruplex using $^{15}NH_4^+$ as an NMR spectroscopic probe.[100] The appeal of $^{15}NH_4^+$ as a spectroscopic probe of cation coordination sites lies in the possibility of nuclear Overhauser effect (NOE) spin state transfer between protons of the ammonium ion and protons of a macromolecule, which could be used to determine directly the location of a cation binding site.[85,100,101] The ammonium ion has long been considered an excellent model of K^+, because it has nearly the same ionic radius[102] and can substitute for K^+ in a number of K^+-dependent enzymes.[103,104] Initial studies using $^{15}NH_4^+$ as a probe revealed the coordination of ammonium ions between the planes of stacked G-quartets of the bimolecular G-quadruplex $[d(G_4T_4G_4)]_2$ (see Chapter 4 for more details).[100,101] The subsequent observation of K^+ bound at identical sites in a crystal structure of the same quadruplex[105] provided strong validation for the utility of $^{15}NH_4^+$ as a probe of monovalent cation binding sites on nucleic acids.

NMR spectroscopic experiments of duplex DNA in the presence of $^{15}NH_4^+$ also provided the first direct experimental, solution-state evidence for the sequence-specific localization of monovalent cations in the minor groove of B-form DNA. Hud *et al.* studied $^{15}NH_4^+$ binding to three duplexes that contained A-tract sequence elements: $[d(GCA_4T_4GC)]_2$, $[d(CGT_4A_4CG)]_2$ and $d(GCA_5CG) \cdot d(CGT_5GC)$. For all three duplexes, $^{15}NH_4^+$ was observed to be in fast exchange between the surface of DNA and bulk solvent, as only one ^1H and one ^{15}N resonance were observed for $^{15}NH_4^+$. Nevertheless, NOE magnetization transfer was detected between the protons of $^{15}NH_4^+$ and DNA in 2D ^1H NOESY spectra and in 1D ^{15}N-filtered ^1H spectra.[3] As the 1D spectra shown in Figure 3.7 illustrate, NOEs were observed between $^{15}NH_4^+$ and AH2 resonances of A·T base pairs within in the A-tracts of all three DNA duplexes studied. 2D spectra reported in the original publication also revealed NOEs between $^{15}NH_4^+$ and the H4' sugar protons of both A and T bases.[3] Given that the AH2 and H4' sugar protons are in the minor groove, the observed NOEs confirmed the localization of ammonium ions deep within the minor groove and likely coordinated directly to the O2 carbonyl groups of the T bases. NOEs from these same DNA protons to the proton resonance of bulk water also confirmed that the minor groove solvent sites are shared by $^{15}NH_4^+$ ions and water molecules that are both in fast exchange with solution.

The binding of $^{15}NH_4^+$ in the $d(A_4T_4)$ sequence element was found to be distinct from that observed for the $d(T_4A_4)$ sequence element. In the case of $[d(GCA_4T_4GC)]_2$, $^{15}NH_4^+$ localization is most pronounced near the central d(ApT) step. In contrast, for $[d(CGT_4A_4CG)]_2$, $^{15}NH_4^+$ localization is most apparent near the outermost d(ApA) step [*i.e.* the symmetry-equivalent d(T3pT4) and d(A9pA10) steps]. For the non-self-complementary duplex $d(GCA_5CG) \cdot d(CGT_5GC)$, $^{15}NH_4^+$ localization was most apparent towards the 3' end of the $d(A_5)$ sequence element. As discussed above, the minor groove

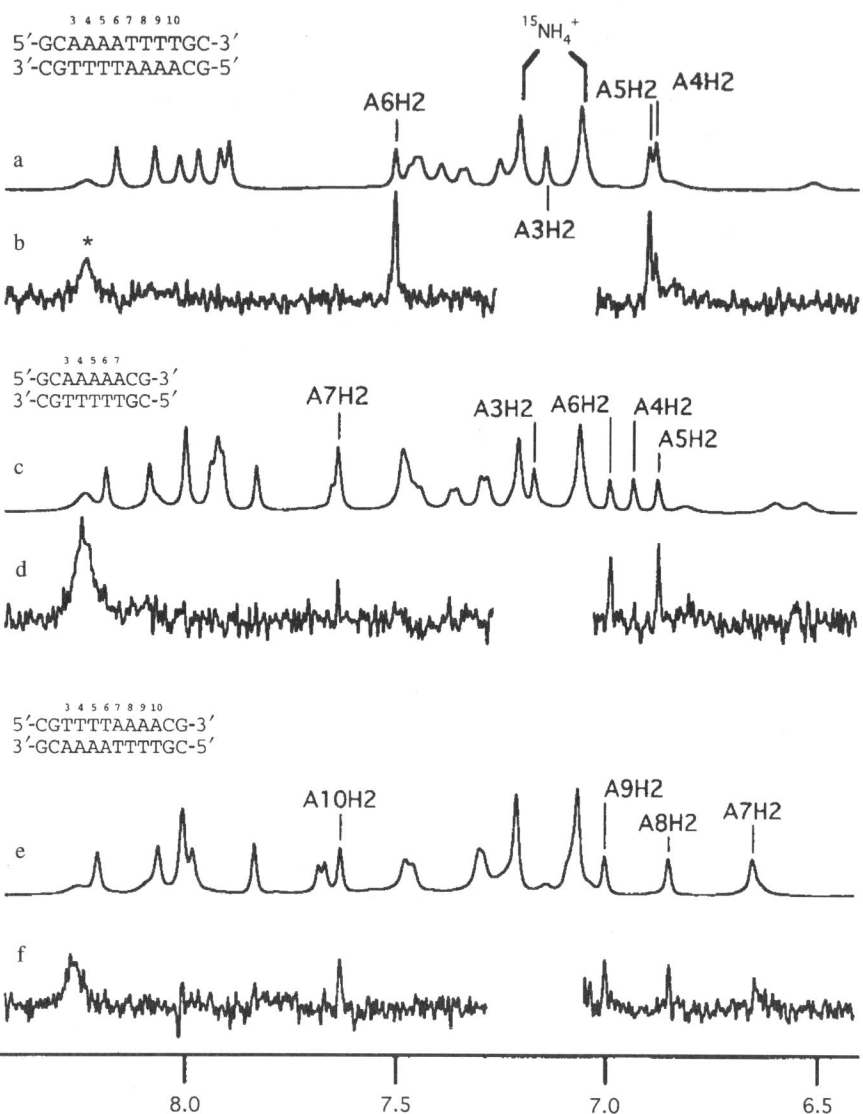

Figure 3.7 (a), (c), (e) ^{1}H and (b), (d), (f) 1D ^{15}N-filtered NOESY spectra, of (a), (b) 2.0 mM [d(GCA$_4$T$_4$GC)]$_2$, (c), (d) 3.5 mM d(GCA$_5$CG)·d(CGT$_5$GC) and (e), (f) 2.0 mM [d(CGT$_4$A$_4$CG)]$_2$, all with 4.0 mM ^{15}NH$_4$Cl at 273 K (pH 5.0). Missing regions of the 1D ^{15}N-filtered NOESY spectra correspond to the most intense portion of the ^{15}NH$_4^+$ doublet, which could not be adequately baseline corrected. The broad resonance observed near 8.2 ppm in all 1D ^{15}N-filtered NOESY spectra was attributed to proton exchange with the amino group of the terminal C residues (designated by *).[3] NMR samples contained 60 mM Li$^+$ (pH 5.5), 95% ^{1}H$_2$O, 5% ^{2}H$_2$O. Mixing time for the ^{15}N-filtered NOESY was 60 ms. Copyright Academic Press. Reprinted with permission from ref. 3.

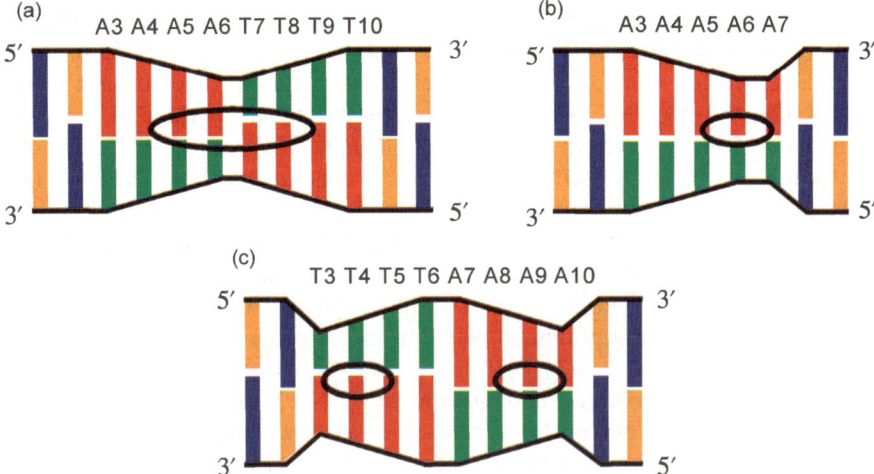

Figure 3.8 A schematic representation sites identified for $^{15}NH_4^+$ binding deep in the A-tract minor groove of (a) [d(GCA$_4$T$_4$GC)]$_2$, (b) d(GCA$_5$CG)·d(CGT$_5$GC) and (c) [d(CGT$_4$A$_4$CG)]$_2$. The known trend of the A-tract minor groove to narrow from the 5' to 3' end is illustrated to emphasize the correlation between minor groove width and $^{15}NH_4^+$ localization deep in the minor groove. Copyright Academic Press. Reprinted with permission from ref. 3.

of an A-tract sequence element is known to be particularly narrow relative to the B-form minor groove. Schematic representations of the $^{15}NH_4^+$ localization sites for the three duplexes studied and the known relative minor groove widths of the same sequences reveal a similar pattern of $^{15}NH_4^+$ localization for all three duplexes (Figure 3.8). That is, $^{15}NH_4^+$ coordination deep in the minor groove is most favored in an A-tract at the narrowest section of its minor groove and not particularly favored at its widest point (*i.e.* at the 5' end of the A-tract). We note that observed localization of $^{15}NH_4^+$ in the minor groove of A-tract DNA is essentially the opposite of what is observed for Mn^{2+} localization. It appears that the size of a hexaaqua divalent cation dictates localization in the widest region of an A-tract minor groove, whereas as the size of a partially dehydrated monovalent cation favors localization in the narrowest region. This suggestion that size, not charge *per se*, is responsible for the precise location of cation entry into the A-tract minor groove is addressed in Section 3.6.

The d(ApT) step appears to be particularly favorable for ammonium ion localization in the minor groove, as the AH2 protons of this step exhibit the largest NOE transfers from $^{15}NH_4^+$ (Figure 3.7). In addition to this being the narrowest point within the minor groove of the d(A$_4$T$_4$) sequence element, the O2 carbonyl oxygens of cross-strand thymine bases appear to be optimally positioned for direct coordination of a monovalent cation.

Qualitatively, ammonium ion binding at specific sites in an A-tract minor groove must be favorable compared with the binding of water molecules, as NOE cross-peaks in 2D NOESY spectra between $^{15}NH_4^+$ and AH2 protons were observed for some sites to be of similar intensity to those between water and AH2 protons, even though $^{15}NH_4^+$ was present at 4 mM and the concentration of neat water is a factor of $>10^4$ higher (*i.e. ca.* 56 M).[3] The concentration of $^{15}NH_4^+$ and DNA present in the experiments conducted by Hud *et al.* also indicated that ammonium ions are in fast exchange between multiple sites within the minor groove of the $d(A_4T_4)$ sequence element and that each ammonium binding site is, on a time average basis, only partially occupied.[3] This observation is consistent with results from MD simulations[41,93] and other experimental data presented below from both solution- and crystal-state studies.

No NOE transfers were observed between $^{15}NH_4^+$ and any of the major groove aromatic protons in the three A-tract sequences studied by Hud *et al.* (Figure 3.7) or to the non-A-tract duplex $[d(GCA_3CGT_3GC)]_2$.[3] These observations are consistent with crystallographic studies of Williams and co-workers, who only observed the localization of Tl^+ in the major groove at G · C base pairs. A comprehensive study of $^{15}NH_4^+$ interaction with non-A-tract sequences has not been reported. It therefore remains possible that monovalent cation binding sites could be detected in the major groove of G-tracts using $^{15}NH_4^+$ as a probe.

In 2000, Denisov and Halle provided important solution-state evidence for the sequence-specific binding of sodium in the minor groove of DNA.[106] Using magnetic relaxation dispersion, $^{23}Na^+$ relaxation was measured in the presence of three DNA dodecamer duplexes with sequences d(CGCGAATTCGCG), d(CGAAAATTTTCG) and d(CGCTCTAGAGCG). The first sequence is the DDD and the second is essentially the DDD with an extended A-tract and a shortened G-tract. The $d(A_2T_2)$ and $d(A_4T_4)$ sequence elements span the length commonly accepted for A-tracts (*i.e.* four to eight A or T residues without a TpA step). The last sequence can be considered generic DNA.

Denisov and Halle found that a small population of Na^+ binds to these duplexes with a correlation time of approximately 50 ns. This correlation time is a full order of magnitude longer than that previously estimated by linewidth measurements (from a study that assumed degenerate binding sites).[107] In their initial study, Denisov and Halle reported that $[d(CGAAAATTTTCG)]_2$, the duplex with the extended A-tract, exhibited the greatest number of slowly exchanging Na^+ ions, but that the occupancy of minor groove binding sites by Na^+ was only around 5%. In a more recent study, the same group concluded that the occupancy of minor groove cation binding sites by Na^+ is as great as 50%, being shared almost equally with water.[108] Saba and co-workers confirmed the binding of Na^+ in the minor groove by $^{23}Na^+$ NMR relaxation studies,[109] but they concluded that the number of Na^+ ions in the groove is considerably lower than the more recent estimates of Halle and co-workers.

An important component of the work by Halle and co-workers was their demonstration that when the DDD was studied in the presence of netropsin, which binds to the minor groove of the A-tract and is in slow exchange with solution,[110] slowly exchanging Na^+ ions were not observed.[106] Such experiments confirmed the location of bound Na^+ within the minor groove of the A-tract sequence element of the DDD. Additional competition experiments with K^+, Rb^+, Cs^+ and NH_4^+ demonstrated that these cations share the same binding sites as Na^+.[106,108] Halle and co-workers reported weak selectivity for Na^+ over Rb^+ in DDD. The occupancy time for groove-bound Na^+ was longer than that found in previous MD experiments – between 10 and 100 000 ns. The residence time for groove-bound Rb^+ was estimated at *ca.* 200 ns.[108] This work has done much to resolve the debate over whether Na^+ and K^+ share the minor groove with water in the solution state,[3,4,93,99] a proposal that was initially met with some scepticism.[111]

3.5.2.2 X-ray Crystallography of Monovalent Cations in the Minor Groove

Williams and co-workers provided some of the first crystallographic evidence for monovalent cation localization in the minor groove of DNA. Their initial studies on this topic involved the analysis of solvent geometries and electron difference maps for high-resolution structures of the DDD grown in the presence of Na^+ and K^+.[4,99] In the case of the K^+-form DDD crystal, their analysis suggested that the solvent site at the central d(ApT) step is shared by water and K^+, with about 30% occupancy by K^+.[34] Around this time, the validity of electron difference maps for Na^+- and K^+-form DNA crystal structures were questioned, with other groups reporting conflicting results.[111,112]

Working with the higher alkali metals (Rb^+ and Cs^+), the Williams and Egli groups subsequently provided irrefutable support for cation localization in the minor groove of DDD crystal structures.[113,114] Like Tl^+ discussed above, these late monovalent cations exhibit profound X-ray scattering compared with the lighter metals. Electron difference maps of the Cs^+ form of the DDD revealed the partial occupation by Cs^+ of four distinct sites within the inner layer of the so-called DDD 'spine of hydration' (Figure 3.9), the most highly occupied of these sites being the central ApT step.

More recent work by Williams' group with Tl^+-form DNA crystals has provided additional support for the co-occupation of the spine of hydration by monovalent cations.[11,77,115] Some positions identified as Tl^+ binding sites are in the A-tract minor groove and correspond to previously identified hybrid water/ion solvent binding sites, including a Tl^+ binding site at the central d(ApT) step. However, additional Tl^+ were identified in the minor groove at positions that did not correspond to previously identified hybrid solvent sites. Two Tl^+ are located at the junction between the central A-tract and the flanking GC region, outside the classic A-tract spine of hydration. In one case, Tl^+ interacts with the lone pair of a G amino group.[77] The occupancy of this site by Tl^+ appears to be approximately 10%. Recent extended MD

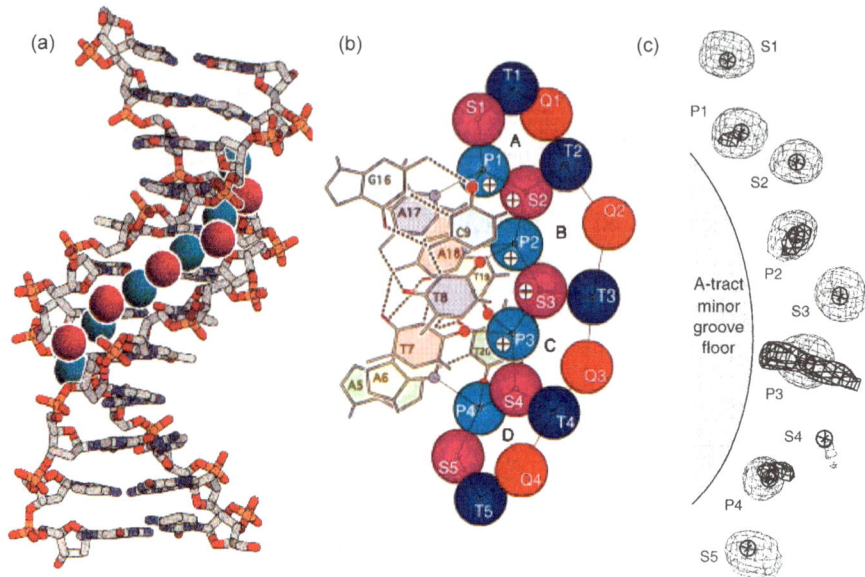

Figure 3.9 X-ray crystallographic evidence for co-occupation of the DDD minor groove by water and Cs^+ (a) The crystal structure of duplex DDD with the spine of hydration shown as solid spheres (NDB entry bdl084). (b) A schematic representation of the location of the two layers of spine solvent molecules within the minor groove (only a subsection of the minor groove is show). (c) Electron density surrounding the primary (P) and secondary (S) layers of solvent densities in the A-tract minor groove of the DDD for a crystal grown in the presence of Cs^+ (NDB entry bd0029). Sum of the Fourier electron density ($2F_o - F_c$) contoured at 1.0 σ is indicated by thin lines. The Fourier electron density ($F_o - F_c$) contoured at 2.5 σ is indicated by thick lines, which demonstrates the partial occupancy of certain solvent sites by Cs^+. Copyright American Chemical Society. (a) and (b) reprinted with permission from ref. 99, (c) reprinted with permission from ref. 113.

simulations also revealed the localization of Na^+ at the junction between the A-tract and GC-tract of the DDD.[41,43]

Finally, it is important to mention that possibly the first experimental evidence that monovalent cations are localized in the minor groove of DNA was presented by Bartenev *et al.*, who noted that fiber diffraction models of the Cs^+ salt of DNA were consistent with the binding of Cs^+ in the minor groove of B-form DNA.[116] It is also noteworthy that Seeman *et al.* assigned the solvent density in the proto-minor groove of the crystal structure of the Watson–Crick base paired RNA dinucleotide ApU as a Na^+ coordinated directly to the two UO2 carbonyls atoms of the base pair.[117] Although these two studies were particularly insightful, their relevance regarding the nature of DNA–cation interactions was not fully appreciated when high-resolution DNA crystal structures were eventually obtained, as all solvent electron density in the minor groove was assigned to ordered water.

3.6 Testing Cation Size Selection of the A-tract Minor Groove Using a Large Monovalent Cation

As discussed above, the localization of monovalent cations within the narrowest regions of the minor groove of A-tract DNA and the localization of divalent cations in the widest regions have been suggested to be governed by the relative size of partially dehydrated monovalent cations *versus* fully hydrated divalent cations.[3,5,66] As a means to test this proposal, we have investigated the binding of the tetramethylammonium cation to the DNA duplex [d(GCA$_4$T$_4$GC)]$_2$. The localization of this relatively large cation in the minor groove was revealed by NOE magnetization transfer between the cation methyl protons (2.97 ppm) and AH2 protons (Figure 3.10). NOE cross-peaks are observed between tetramethylammonium cations and all four AH2 protons within the d(A$_4$T$_4$) sequence element. However, localization of these cations is most pronounced near the outer 5'-most ApA step, as the NOE cross-peaks to A3H2 and A4H2 are approximately equal and more intense than the cross-peaks to the other, more central, AH2 protons. This position for the preferred localization of a large monovalent cation is the same as that which has been

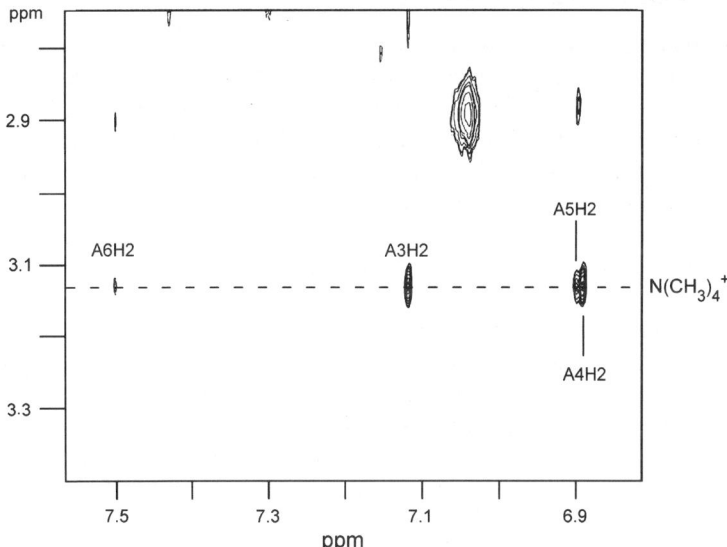

Figure 3.10 A region of a 2D NOESY spectrum of the DNA duplex [d(GCA$_4$T$_4$GC)]$_2$ in the presence of N(CH$_3$)$_4^+$. The region shown reveals NOE cross-peaks between the protons of N(CH$_3$)$_4^+$ and the AH2 protons of the DNA. Note that the H6/H8 major groove protons of the DNA also lie within the chemical shift region shown on the horizontal axis, but no cross-peaks are observed between the resonances of these protons and those of N(CH$_3$)$_4^+$. The cross-peaks observed indicate the preferential localization of N(CH$_3$)$_4^+$ near the 5'-most d(ApA) step of the duplex. Sample was 2 mM in DNA duplex, 22 mM N(CH$_3$)$_4^+$, pH 5.5, 283 K.

shown to be the preferred position of divalent cation localization deep in the minor groove and not the same as that revealed for monovalent cations.[3,5,66] Hence, these experimental results, and also other work discussed in this chapter, clearly indicate that cation localization deep in the minor groove of an A-tract is determined by a combination of electrostatic potential of the bases, effective cation size and nucleotide sequence.

3.7 Additional Experimental Evidence of Sequence-specific Binding of Cations to DNA

Thus far, our discussion of sequence-specific cation binding by DNA has focused primarily on experimental results from studies that utilized NMR spectroscopy or X-ray crystallography. Several other physical techniques have also provided substantial support for the sequence-dependent binding of cations to DNA. Here we briefly discuss the results from capillary electrophoresis and ultrasonic velocity measurements.

3.7.1 Capillary Electrophoresis

Stellwagen *et al.* used capillary electrophoresis to examine the relative mobility of 20 bp oligonucleotide duplexes with different nucleotide sequences but identical diffusion coefficients.[118] Four 20 bp duplexes were studied: one contained two repeats of the A-tract sequence element $d(A_3T_3)$; another contained two repeats of the A-tract sequence element $d(A_5)$; a third contained two repeats of the pseudo A-tract sequence $d(T_3A_3)$; and a fourth duplex, serving as a control, contained mixed AT/GC sequences in place of the aforementioned sequence elements. Their investigation revealed that the duplex containing the longest A-tract sequence, $d(A_3T_3)$, exhibited the slowest migration.[118] The duplexes containing the $d(A_5)$ and the $d(T_3A_3)$ sequences exhibited intermediate mobility, while the control sequence exhibited the fastest migration. Because the four duplexes studied have matched diffusion coefficients, observed differences in migration speed are essentially due to differences in the net charge of the duplexes. This investigation thereby provided evidence for a class of relatively tightly bound Na^+ ions to A-tract DNA and particularly to sequence elements of the form A_nT_n.[118]

3.7.2 Hydration Measurements

Buckin *et al.* used ultrasonic velocity measurements of solutions containing DNA to study the interaction of Mg^{2+} with DNA duplexes of different sequence composition.[31] In these experiments, changes in solute (*i.e.* cations and DNA) hydration were monitored as Mg^{2+} was added to solutions containing the Cs^+ salt of each DNA duplex. Observed changes in solute hydration are associated with Mg^{2+} replacing Cs^+ as the counterion of DNA. From these measurements, Buckin et al. determined apparent Mg^{2+}-DNA binding constants, K_{app}, of $150 M^{-1}$ for $A_8 \cdot T_8$ and $40 M^{-1}$ for $[d(GC)_4]_2$. The duplexes $[d(A_4T_4)]_2$ and $[d(AT)_4)]_2$ exhibited intermediate apparent binding constants of 100 and $70 M^{-1}$, respectively. In addition to their intrinsic value, these measurements are very important for understanding the NMR spectroscopic and X-ray crystallographic studies discussed above. First, the more favorable binding of Mg^{2+} to $d(A_8) \cdot d(T_8)$ and $[d(A_4T_4)]_2$ with respect to $[d(AT)_4]_2$ lends additional support to the proposal that Mg^{2+} is binding to the top of the minor groove of A-tracts in the solution state, even though this binding is not reported in ^1H resonance line broadening studies. Second, the greater affinity of Mg^{2+} for the three AT-rich sequence with respect to $[d(GC)_4]_2$ is consistent with Mn^{2+} resonance line broadening studies, which indicate that d(GpC) steps are not particularly favorable for divalent cation localization in solution, even though divalent cations are often observed at these steps in crystal structures.

The sequence-dependent changes in hydration volumes observed upon Mg^{2+} replacement of Cs^+ from the surface of duplex DNA are of great value for understanding DNA–cation interactions and fully consistent with the localization studies discussed above. Buckin et al. observed a change in the molar concentration increment of ultrasonic velocity per mole of bound Mg^{2+}, $\Delta A_{Mg^{2+}}$, of $-4.4 \, cm^3 \, mol^{-1}$ for $d(A_8) \cdot d(T_8)$, as compared with $-18 \, cm^3 \, mol^{-1}$ for $[d(GC)_4]_2$. These results indicate that replacement of Cs^+ by Mg^{2+} as the counterion of $d(A_8) \cdot d(T_8)$ is associated with a smaller net decrease in waters of hydration than the replacement of Cs^+ by Mg^{2+} around $[d(GC)_4]_2$. Buckin et al. interpreted these results as an indication that Mg^{2+} interacts with A-tract DNA through its solvent shell, whereas Mg^{2+} is partially dehydrated when associated with GC-rich DNA. Considering the very different modes of divalent cation interaction discussed above for A-tract and G-tract DNA, it is not surprising that Mg^{2+} apparently maintains a different number of waters of hydration when bound to these different DNA sequence elements. It is surprising, however, that Mg^{2+} association with $d(A_8) \cdot d(T_8)$ is more energetically favorable than its association with $[d(GC)_4]_2$, given that Mg^{2+} may directly coordinate to this GC-rich DNA and not to the A-tract DNA. It is important to note that the apparent association constants and different net hydration changes measured by Buckin et al. are affected by the intrinsic association constant of Cs^+ for DNA, and also net change in Cs^+ hydration upon displacement from DNA, two parameters that might also depend on DNA sequence.

Tikhomirova and Chalikian have more recently reported relative molar sound velocity increments, $[U]$, partial molar volumes and partial molar adiabatic compressibilities of various polynucleotides in the presence of the full alkali metal series (Li^+ through Cs^+), plus NH_4^+ and NR_4^+. cations.[30] They found that monovalent cations associated with the DNA duplex $[poly(dG-dC)]_2$ were not significantly dehydrated. In contrast, cations associated with $[poly(dA-dT)]_2$ are significantly dehydrated, retaining only 65

(± 18)% of their original hydration shell.[30] These results are fully consistent with the solution- and crystal-state studies which indicate that the d(ApT) step is a favorable binding site for dehydrated monovalent cations. Additionally, these results suggest that the observation of dehydrated Tl^+ bound in the major groove of the DDD at GpC steps may be promoted by crystallographic dehydration, as the monovalent cations studied by Tikhomirova and Chalikian do not dehydrate upon association with [poly(dG–dC)]$_2$.

3.8 Sequence-specific Cation Localization and DNA Fine Structure

As discussed in the Introduction, much of the interest concerning the sequence-specific binding of cations to DNA has resulted from the hypothesis that local variations in DNA helix are due, in part, to sequence-specific cation binding.[1–3,5,33,34,60,90,99,113,115,119,120] It has been proposed that the grooves of DNA sequences which preferentially localize cations are narrower than the grooves of generic DNA, as electrostatic attractions between cations and the backbone phosphate groups cause narrowing of the grooves. We have proposed a 'tug-of-war' model for DNA fine structure in which the *differential* localization of cations between the major and minor grooves is the most important cation-dependent modulator of helix structure.[1] For example, in the case of A-tract DNA, cations are more favorably localized in the minor groove (with respect to generic DNA) *and* less favorably localized in the major groove (with respect to generic DNA). This relative difference in cation localization for both grooves is suggested to contribute to the minor groove of A-tract DNA being narrower than that of generic DNA and the A-tract major groove being wider than that of generic DNA. Similarly, cations are more favorably localized in the major groove of G-tract DNA (with respect to generic DNA) *and* are less favorably localized in the minor groove of G-tract DNA (with respect to generic DNA). In this case, the differential localization of cations has been proposed to explain why G-tract DNA is more prone to undergo the B- to A-form transition than generic or A-tract DNA, as this transition involves narrowing of the major groove and widening of the minor groove.[1,121]

In earlier sections, we have discussed the correlation between cation localization in the minor groove of A-tract DNA and the anomalously narrow width of the minor groove in these sequence elements. Here we discuss specific examples in which changes in cation species or cation localization sites have been directly correlated with changes in DNA helix structure for the same DNA sequence in the same crystal lattice or in the solution state.

The ultrahigh-resolution (0.74 Å) structure of duplex d(CCAGTACTGG) determined by Kielkopf *et al.* included eight bound Ca^{2+} per DNA duplex.[76] Two of these cations are located in the minor groove and are octaaquated. The authors noted that one hydrated cation exhibits two alternative positions and a

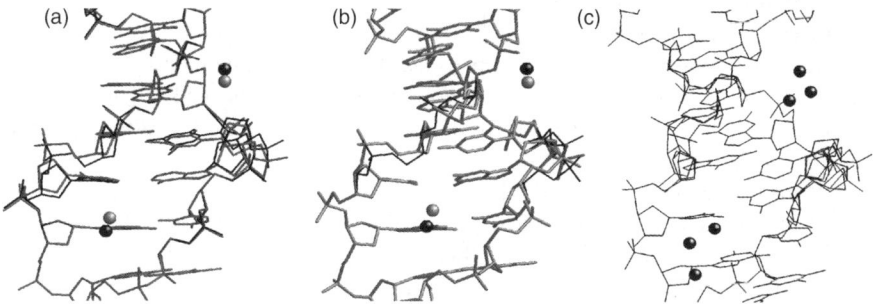

Figure 3.11 X-ray crystal structures of DNA duplexes showing alternative cation binding sites in the minor groove and corresponding alternative backbone conformations. The DNA structures are orientated such that the reader is looking into the minor groove. (a) Structure of the d(CCAG-TACTGG) duplex with two alternative backbone conformations and two Ca^{2+}-binding sites (NDB entry bd0023). (b) Structure of the d(CCAGCGCTGG) duplex with two alternative backbone conformations and two Ca^{2+}-binding sites (NDB entry bd0036). (c) Structure of the d(CCAGCGCTGG) duplex with three alternative backbone conformations and three Mg^{2+}-binding sites (NDB entry bd0035). Copyright Elsevier Science Ltd. Reprinted with permission from ref. 2.

polymorphism of the DNA duplex backbones. Specifically, at the d(GpT) step, the backbones of the alternative conformers vary by 1.0 Å. The minor groove Ca^{2+} binding sites are very near this location (Figure 3.11). The authors suggested that alternative DNA conformers (*e.g.* with variations in groove width) are caused by the binding of divalent cations at alternative positions in the minor groove.[76]

In Section 3.4.1, we discussed the set of crystal structures solved by Chiu and Dickerson to 1 Å resolution for which two DNA sequences were crystallized in the presence of both Mg^{2+} and Ca^{2+}.[61] These authors also observed two closely spaced divalent cation binding sites (each partially occupied) for the same species of cation bound in the minor groove of [d(CCAGCGCTGG)]$_2$. This sequence also exhibited alternative backbone conformations. For the Ca^{2+}-form structure, the difference in minor groove width between the two conformations is as great as 1.5 Å. Hud and Polak[2] noted that this crystal structure exhibits the same correlation reported by Kielkopf *et al.* for alternative cation positions and alternative DNA conformers. In the case of the Ca^{2+}-form crystal, the alternative backbone conformations and alternative cation binding sites are very similar to those observed by Kielkopf *et al.* (Figure 3.11), which is understandable, considering that the sequence d(CCAGCGCTGG) crystallized by Chiu and Dickerson only differs in the minor groove by the amino group G7N2. An even more intriguing correlation was recognized in the Mg^{2+}-form crystal of the same sequence.[2] Three alternative minor groove binding sites for Mg^{2+} were observed in the crystal structure of [d(CCAGCGCTGG)]$_2$, in addition to three alternative backbone conformations (*i.e.* three distinct minor groove widths). A visual inspection of the three DNA conformers and the three alternate and partially occupied Mg^{2+} binding sites again suggests that movement of the divalent cation along the minor groove results in changes in groove width (Figure 3.11). These observations are perfectly supportive of the proposal that minor groove width is sensitive to the exact location of bound cations.

The comparative analysis of Mg^{2+}- and Ca^{2+}-form crystal structures of the DDD by Egli and co-workers also led to the conclusion that divalent cations modulate minor groove width.[74] In their Ca^{2+}-form structure of the DDD, five hydrated Ca^{2+} traverse the minor groove and one is bound deep in this groove, near the junction of the A-tract (*i.e.* AATT) and the GC section. In their Mg^{2+} structure, hydrated Mg^{2+} ions also bridge phosphate groups across the minor groove, but no Mg^{2+} is found deep inside this groove. Based on groove width measurements and cation placement, Egli and co-workers concluded that Ca^{2+} penetration into the minor groove actually widens the groove, whereas Mg^{2+} binding at the top of the minor groove causes groove narrowing.

It is important to note that the Mg^{2+}- and Ca^{2+}-form crystal structures of the DDD compared by Egli and co-workers are from crystals with different crystal lattices.[74] Liu and Subirana also reported that the DDD crystallizes in different crystal lattices Mg^{2+} and Ca^{2+}.[122] These authors concluded that the structure of the central AATT sequence element of the DDD was 'practically identical' for the two crystal forms and therefore the structure is independent of the two crystallization conditions. Nevertheless, it is possible that some of the differences observed between the fine structure of the DDD could be due to differences in crystal lattice forces.

A recent ^{31}P NMR spectroscopic analysis of duplex DNA by Hartmann and co-workers has now provided evidence that even changing the cation associated with DNA from Na^+ to K^+ can affect DNA helical structure.[123] They demonstrated that the sequential distances $H2'_i$–$H6/8_{i+1}$, $H2''_i$–$H6/8_{i+1}$ and $H6/8_i$–$H6/8_{i+1}$, and also the conformational equilibrium between the backbone states BI and BII (which are defined by the backbone angles ε and ζ), were different for a DNA duplex (of generic sequence) depending on whether Na^+ or K^+ was present as the counterion. MD simulations have suggested that Na^+ and K^+ have distinct preferences for localization in the major and minor grooves of DNA, and also different degrees of dehydration upon binding to DNA.[97,124] These apparent differences in how Na^+ and K^+ interact with the same DNA sequence might be the origin of the small cation-dependent variations in helical structure detected by Hartmann and co-workers.[123]

A number of MD simulations have been conducted with the goal of determining the relative occupancy of the DNA grooves by cations and water and to determine if a correlation exists between the entry of cations in the major and minor grooves and the narrowing of these grooves. There is a general consensus that cations and water share the minor groove of the DDD. However, the relative occupancy levels of cations and water are still being investigated, as are the possible correlations between cation localization and groove narrowing.[43,90,92,96,125]

Here we discuss two extended MD simulations of the DDD that have provided recent insights into the nature of monovalent localization in the minor groove cations, and also the difficulty of modeling these interactions. The extended MD simulations by the Beveridge and Orozco groups have both confirmed earlier MD results that indicate the localization of Na^+ in the minor groove of the DDD at the central d(ApT) step (Figure 3.12b).[41,43] However,

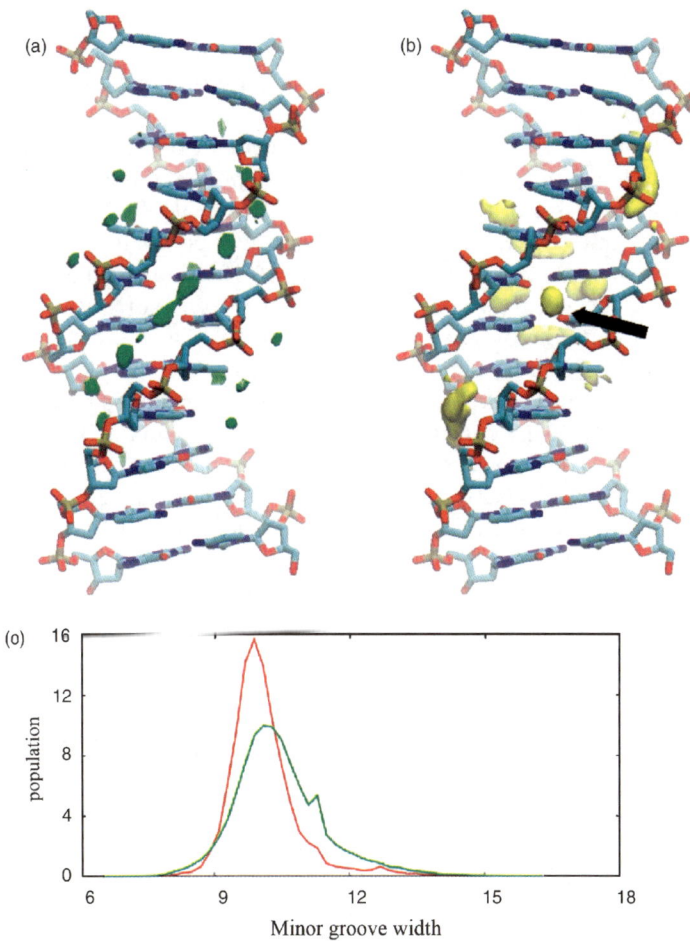

Figure 3.12 Results of water and Na^+ localization analysis over the course of a 1 ms MD simulation of the DDD by Orozco and co-workers.[43] Density contours for (a) water in green and (b) Na^+ in yellow. Contours represented correspond to 2.5 (water) and 2.0 (Na^+) times the background densities. The position of a long residence Na^+ is marked with an arrow. (c) Population (in %) of structures with AATT minor grooves of different widths (in Å) considering all the snapshots (green) or only those where the central position is occupied by a Na^+ (red). Copyright American Chemical Society. Reproduced with permission from ref. 43.

Na$^+$ is more often localized in the minor groove at the junction between the AT and GC regions of the DDD (sites at which Tl$^+$ is observed to be bound in the crystal structures of DDD) (Figures 3.6 and 3.12b). Although these results appear to be contrary to the NMR studies discussed above, it is important to note that both the ^{15}NH$_4^+$ localization studies and the ^{23}Na$^+$ relaxation studies are almost certainly only sensing partially dehydrated cations that are directly chelated to the electronegative DNA atoms within the minor groove, whereas MD studies could be revealing that cation localization in the minor (and major) groove by fully hydrated monovalent cations is much more common. Thus, NMR spectroscopic studies that show the sequence-selective localization of monovalent cations may, more precisely, be reporting the sequence-selective chelation/dehydration of these cations.

Confirmation or negation of the proposed connection between Na$^+$ localization and minor groove narrowing by MD has proven more challenging. On the one hand, analyses of the 60 ns simulation of the DDD duplex performed by Beveridge and co-workers indicated negligible correlation between ion–phosphate proximity and groove width ($R = 0.10$ for the minor groove).[41] However, over a smaller window of 10 ns, a much larger correlation ($R = 0.73$) was noted for one minor groove width (T7–G22). On the other hand, the analysis of a 1000 ns MD simulation by Orozco and co-workers revealed a small (*ca.* 0.5 Å) but positive correlation between narrowing of the minor groove of the d(AATT) sequence element of the DDD and the binding of Na$^+$ at the central d(ApT) step (Figure 3.12c).[43] This simulation also revealed that Na$^+$ only occupied this particular binding site for 4% of the MD trajectory (*i.e.* it was occupied 25-fold more often by a water molecule), with average Na$^+$ entry times of 400 ns and residence times of 10–15 ns for this binding site. These two extended MD studies underscore the difficulty of performing statistics on MD data with ions, as such large sampling times are required.

3.9 Investigations of the Role of Cations in DNA Fine Structure Using Non-natural Bases

A powerful tool has emerged to investigate the influence of localized cations on DNA structure: non-natural nucleotides that either eliminate cation localization sites or introduce covalently attached (*i.e.* tethered) cations along a DNA duplex. Such investigations are possible *via* the synthesis of non-natural nucleotide phosphoramidites that can be coupled along with natural nucleotides during automated oligonucleotide synthesis. Results from such studies have provided support for the proposed role of localized cations in the local modulation of DNA structure.

Modified DNA has now been prepared with a variety of cationic side-chains covalently tethered to the bases.[126–130] For example, L-aminoalkyl chemical groups, similar to the side-chain of lysine, have been attached to the C5 position of deoxypyrimidines. These modifications provide a means to investigate the effects of groove-localized cations on DNA structure. Most of the cationic base modifications investigated have been shown to cause axial bending in duplex DNA. In particular, it appears that charge modifications of DNA generally cause local distortions in DNA helical structure away from the canonical B-form,[126] and in some cases this distortion is towards the A-form.[131]

In 2005, Williams and co-workers reported crystal structures of the DDD persubstituted with dT residues having propylamino moieties replacing the methyl moieties. The crystals had been soaked in Tl^+-containing solution prior to structure determination. Surprisingly, only one of the four covalently attached amines interacted directly with the DNA (associating with a phosphate); the other three projected out from the helix.[115] The radially directed tethered cations did not appear to induce structural changes in the helix with respect to the Tl^+-form crystal structure of the unmodified DDD. However, one of the tethered cations was directed toward a phosphate group and appears to interact electrostatically. This interaction is associated with local changes in duplex helical parameters and by a displacement of the backbone with respect to the unmodified DDD structure. In addition, these interactions appear to be associated with the displacement of counterions from the major groove of the DNA.

Several studies have also been reported in which non-natural nucleotides have been used to investigate the effects of eliminating cation binding sites from the minor groove of A-tracts by removing the N3 of adenine and/or the O2 of thymine. For example, Kool and co-workers investigated how A-tract-induced helical axis bending is effected by difluorotoluene and 3-deazaadenine substitutions within a $dA_5 \cdot dT_5$ sequence element. This substitution of TO2 with fluorine and AN3 with carbon reduced the helical axis bend that normally results from the five base pair A-tract. Interestingly, these nucleotide substitutions were most effective in reducing DNA curvature when the adenine substitutions were made at the $3'$ end of the A-tract and the thymine substitutions were made at the $5'$ end;[132,133] these data are in agreement with the proposal that the localization of cations in the narrowest region of the A-tract minor groove contribute most to helical axis bending through local changes in helix structure.[1]

McLaughlin and co-workers have likewise demonstrated that single-atom substitutions in A-tracts (replacing the O2 of C-nucleoside analogues of dT with H, F or Me at the 2-position) also results in a loss of A-tract-induced helix bending.[134] The dual nucleoside substitution destabilized the duplex structure to a sufficient extent that the modified DDD exists only as an intramolecular hairpin. These investigators attributed the destabilizing effects of the mutation to a combined loss of ordered waters and minor groove cation localization.[135] It is important to bear in mind that the nucleotide base modifications discussed here could alter DNA helix structure by a mechanism that does not necessarily involve a change in cation localization. This possibility is again supported by the extended MD simulations discussed above that did not reveal a strong correlation between monovalent cation binding in the minor groove and groove width.[41,43]

3.10 Conclusions and Perspectives

Over the past decade, the accepted view of DNA–cation interactions has changed dramatically. In 1997, it was considered novel to suggest that cations enter the minor groove of A-tract DNA. Ten years later, it is surprising to think that such a suggestion was ever considered controversial. As demonstrated in this chapter, there is now substantial evidence for the sequence-specific interaction of cations with DNA. The salt dependence of DNA duplex stability illustrates that the double helix cannot exist without associated cations. Several studies have also provided compelling evidence that the specific location of cations along a DNA duplex alter the local structure of the helix. Nevertheless, the relative contributions of cation binding, hydration, base stacking and other non-covalent interactions to defining the fine structure of the double helix remain open to debate.

One of the most difficult problems to address regarding the relationship between cation binding and DNA structures is the issue of cause and effect. Do the grooves of DNA narrow in response to the binding of cations or do cations preferentially bind within grooves that are intrinsically narrow? It is likely that computer simulations will ultimately provide detailed answers to such questions. Many studies have already been reported on this topic and recent extended simulations indicate a link between cation binding and the changes in helical structure that are associated with DNA bending, but support a model that is more nuanced than simple groove narrowing. On the other hand, the direct involvement of cations in narrowing of the A-form minor groove and in driving the B- to A-form transition of G-tract DNA appears to be supported by various simulations. Assuming continuing increases in accessible computational power and improvements in nucleic acid potential energy functions, these questions will likely be resolved in the not too distant future.

Acknowledgements

The authors thank Professors Loren Williams and Juan Subirana for helpful suggestions. This work was supported by the NIH (GM62873).

References

1. N. V. Hud and J. Plavec, *Biopolymers*, 2003, **69**, 144.
2. N. V. Hud and M. Polak, *Curr. Opin. Struct. Biol.*, 2001, **11**, 293.
3. N. V. Hud, V. Sklenar and J. Feigon, *J. Mol. Biol.*, 1999, **286**, 651.
4. X. Shui, L. McFail-Isom, G. G. Hu and L. D. Williams, *Biochemistry*, 1998, **37**, 8341.
5. N. V. Hud and J. Feigon, *J. Am. Chem. Soc.*, 1997, **119**, 5756.
6. R. E. Dickerson and H. R. Drew, *J. Mol. Biol.*, 1981, **149**, 761.
7. H. Drew and R. Dickerson, *J. Mol. Biol.*, 1981, **151**, 535.

8. G. B. Koudelka, S. A. Mauro and M. Ciubotaru, *Prog. Nucleic Acid Res. Mol. Biol.*, 2006, **81**, 143.
9. S. A. Mauro and G. B. Koudelka, *J. Mol. Biol.*, 2004, **340**, 445.
10. S. Bergqvist, M. A. Williams, R. O'Brien and J. E. Ladbury, *J. Mol. Biol.*, 2004, **336**, 829.
11. S. B. Howerton, A. Nagpal and L. D. Williams, *Biopolymers*, 2003, **69**, 87.
12. B. Jayaram, K. J. McConnell, S. B. Dixit and D. L. Beveridge, *J. Comput. Phys.*, 1999, **151**, 333.
13. J. B. Chaires, *Biopolymers*, 1997, **44**, 201.
14. M. Record, W. Zhang and C. Anderson, *Adv. Protein Chem.*, 1998, **51**, 281.
15. T. M. Lohman and D. P. Mascotti, *Methods Enzymol.*, 1992, **212**, 400.
16. M. T. J. Record and R. S. Spolar, in *The Biology of Nonspecific DNA–Protein Interactions*, A. Rezvin ed., CRC Press, Boca Raton, FL, 1990, p. 33.
17. W. Saenger, *Principles of Nucleic Acid Structure*, Springer, New York, 1984, p. 556.
18. R. E. Franklin and R. G. Gosling, *Nature*, 1953, **171**, 740.
19. D. A. Marvin, L. D. Hamilton, M. Spencer and M. H. F. Wilkins, *J. Mol. Biol.*, 1961, **3**, 547.
20. L. E. Minchenkova, A. K. Schyokina, B. K. Chernov and V. I. Ivanov, *J. Biomol. Struct. Dyn.*, 1986, **44**, 463.
21. W. L. Peticolas, Y. Wang and G. A. Thomas, *Proc. Natl. Acad. Sci. USA*, 1988, **85**, 2579.
22. B. Basham, G. P. Schroth and P. S. Ho, *Proc. Natl. Acad. Sci. USA*, 1995, **92**, 6464.
23. V. I. Ivanov and L. E. Minchenkova, *Mol. Biol.*, 1995, **28**, 780.
24. M. Shatzky-Schwartz, N. D. Arbuckle, M. Eisenstein, D. Rabinovich, A. Bareket-Samish, T. E. Haran, B. F. Luisi and Z. Shakked, *J. Mol. Biol.*, 1997, **267**, 595.
25. J. A. Subirana and M. Soler-López, *Annu. Rev. Biophys. Biomol. Struct.*, 2003, **32**, 27.
26. R. Lavery and B. Pullman, *J. Biomol. Struct. Dyn.*, 1985, **2**, 1021.
27. R. Lavery and B. Pullman, *Nucleic Acids Res.*, 1981, **9**, 3765.
28. A. Pullman and B. Pullman, *Q. Rev. Biophys.*, 1981, **14**, 289.
29. R. Lavery and B. Pullman, *Int. J. Quantum Chem.*, 1981, **20**, 259.
30. A. Tikhomirova and T. V. Chalikian, *J. Mol. Biol.*, 2004, **341**, 551.
31. V. A. Buckin, B. I. Kankiya, D. Rentzeperis and L. A. Marky, *J. Am. Chem. Soc.*, 1994, **116**, 9423.
32. D. Bandyopadhyay and D. Bhattacharyya, *J. Biomol. Struct. Dyn.*, 2003, **21**, 447.
33. D. Hamelberg, L. D. Williams and W. D. Wilson, *Nucleic Acids Res.*, 2002, **30**, 3615.
34. C. C. Sines, L. McFail-Isom, S. B. Howerton, D. VanDerveer and L. D. Williams, *J. Am. Chem. Soc.*, 2000, **122**, 11048.
35. T. E. Cheatham and M. A. Young, *Biopolymers*, 2000, **56**, 232.

36. D. L. Beveridge and K. J. McConnell, *Curr. Opin. Struct. Biol.*, 2000, **10**, 182.
37. E. Giudice and R. Lavery, *Acc. Chem. Res.*, 2002, **35**, 350.
38. M. Orozco, A. Perez, A. Noy and F. J. Luque, *Chem. Soc. Rev.*, 2003, **32**, 350.
39. D. L. Beveridge, G. Barreiro, K. S. Byun, D. A. Case, T. E. Cheatham, S. B. Dixit, E. Giudice, F. Lankas, R. Lavery, J. H. Maddocks, R. Osman, E. Seibert, H. Sklenar, G. Stoll, K. M. Thayer, P. Varnai and M. A. Young, *Biophys. J.*, 2004, **87**, 3799.
40. T. E. Cheatham, *Curr. Opin. Struct. Biol.*, 2004, **14**, 360.
41. S. Y. Ponomarev, K. M. Thayer and D. L. Beveridge, *Proc. Natl. Acad. Sci. USA*, 2004, **101**, 14771.
42. J. Sponer and F. Lankas, *Computational Studies of RNA and DNA*, Springer, Dordrecht, 2006, Vol. **2**, p. 636.
43. A. Perez, F. J. Luque and M. Orozco, *J. Am. Chem. Soc.*, 2007, **129**, 14739.
44. W. H. Braunlin, *Adv. Biophys. Chem.*, 1995, **5**, 89.
45. W. H. Braunlin, T. Drakenberg and L. Nordenskiold, *J. Biomol. Struct. Dyn.*, 1992, **10**, 333.
46. W. H. Braunlin, T. Drakenberg and L. Nordenskiold, *Biopolymers*, 1987, **26**, 1047.
47. W. H. Braunlin, L. Nordenskiold and T. Drakenberg, *Biopolymers*, 1989, **28**, 1339.
48. W. H. Braunlin, L. Nordenskiold and T. Drakenberg, *Biopolymers*, 1991, **31**, 1343.
49. D. M. Rose, C. F. Polnaszek and R. G. Bryant, *Biopolymers*, 1982, **21**, 653.
50. D. M. Rose, M. L. Bleam, M. T. Record and R. G. Bryant, *Proc. Natl. Acad. Sci. USA*, 1980, **77**, 6289.
51. M. Montrel, V. P. Chuprina, V. I. Poltev, W. Nerdal and E. Sletten, *J. Biomol. Struct. Dyn.*, 1998, **16**, 631.
52. E. Sletten and N. A. Frøystein, *Met. Ions Biol. Syst.*, 1996, **32**, 397.
53. E. Sletten and N. A. Frøystein, in: *Metal Ions in Biological Systems*, A. Sigel and H. Sigel, ed., Marcel Dekker, New York, 1996, p. 397.
54. S. Steinkopf and E. Sletten, *Acta Chem. Scand.*, 1994, **48**, 388.
55. N. A. Frøystein, J. T. Davis, B. R. Reid and E. Sletten, *Acta Chem. Scand.*, 1993, **47**, 649.
56. N. A. Frøystein and E. Sletten, *Acta Chem. Scand.*, 1991, **45**, 219.
57. R. E. Hurd, Azhderian and B. R. Reid, *Biochemistry*, 1979, **8**, 4012.
58. S. D. Kennedy and R. G. Bryant, *Biophys. J.*, 1986, **50**, 669.
59. I. Brukner, S. Susic, M. Dlakic, A. Savic and S. Pongor, *J. Mol. Biol.*, 1994, **236**, 26.
60. I. Rouzina and V. Bloomfield, *Biophys. J*, 1998, **74**, 3152.
61. T. K. Chiu and R. E. Dickerson, *J. Mol. Biol.*, 2000, **301**, 915.
62. A. K. Katz, J. P. Glusker, S. A. Beebe and C. W. Bock, *J. Am. Chem. Soc.*, 1996, **118**, 5752.

63. J. Harp, B. Hanson, D. Timm and G. Bunick, *Acta Crystallogr. Sect. D*, 2000, **56**, 1513.
64. I. Solt, I. Simon, A. G. Csaszar and M. Fuxreiter, *J. Phys. Chem. B.*, 2007, **111**, 6272.
65. H. Robinson, Y. G. Gao, R. Sanishvili, A. Joachimiak and A. H. J. Wang, *Nucleic Acids Res.*, 2000, **28**, 1760.
66. N. Hud and F. Feigon, *Biochemistry*, 2002, **41**, 9900.
67. J. R. Quintana, K. Grzeskowiak, K. Yanagi and R. E. Dickerson, *J. Mol. Biol.*, 1992, **225**, 379.
68. D. R. Mack, T. K. Chiu and R. E. Dickerson, *J. Mol. Biol.*, 2001, **312**, 1037.
69. V. P. Chuprina, A. A. Lipanov, O. Y. Federoff, S.-G. Kim, A. Kintanar and B. R. Reid, *Proc. Natl. Acad. Sci. USA*, 1991, **88**, 9087.
70. M. Katahira, H. Sueta and Y. Kyogoku, *Nucleic Acids Res.*, 1990, **18**, 613.
71. M. Katahira, H. Sugeta, Y. Kyogoku, S. Fujii, R. Fujisawa and K. Tomita, *Nucleic Acids Res.*, 1988, **16**, 8619.
72. A. M. Burkhoff and T. D. Tullius, *Cell*, 1987, **48**, 935.
73. M. F. Roberts, Q. Cui, C. J. Turner, D. A. Case and A. G. Redfield, *Biochemistry*, 2004, **43**, 3637.
74. G. Minasov, V. Tereshko and M. Egli, *J. Mol. Biol.*, 1999, **291**, 83.
75. M. Soler-López, L. Malinina, J. Liu, T. Huynh-Dinh and J. Subirana, *A. J. Biol. Chem.*, 1999, **274**, 23683.
76. C. L. Kielkopf, S. Ding, P. Kuhn and D. C. Rees, *J. Mol. Biol.*, 2000, **296**, 787.
77. S. B. Howerton, C. C. Sines, D. VanDerveer and L. D. Williams, *Biochemistry*, 2001, **40**, 10023.
78. I. D. Brown, *Acta Crystallogr., Sect. B*, 1988, **44**, 545.
79. G. Wulfsberg, *Principles of Descriptive Inorganic Chemistry*, University Science Books, Sausalito, CA, 1991.
80. J. F. Hinton, W. L. Whaley, D. Shungu, R. E. D. Koeppe and F. S. Millett, *Biophys. J.*, 1986, **50**, 539.
81. K. A. Hill, S. A. Steiner and F. J. Castellino, *J. Biol. Chem.*, 1987, **262**, 7098.
82. G. D. Markham, *J. Biol. Chem.*, 1986, **261**, 1507.
83. J. Reuben and F. J. Kayne, *J. Biol. Chem.*, 1971, **246**, 6227.
84. P. A. Pedersen, J. M. Nielsen, J. H. Rasmussen and P. L. Jorgensen, *Biochemistry*, 1998, **37**, 17818.
85. J. Feigon, S. E. Butcher, L. D. Finger and N. V. Hud, *Methods Enzymol.*, 2001, **338**, 400.
86. S. Basu, A. A. Szewczak, M. Cocco and S. A. Strobel, *J. Am. Chem. Soc.*, 2000, **122**, 3240.
87. S. Basu, R. P. Rambo, J. Strauss-Soukup, J. H. Cate, A. R. Ferre-D'Amare, S. A. Strobel and J. A. Doudna, *Nat. Struct. Biol.*, 1998, **5**, 986.
88. J. Badger, Y. Li and D. L. Caspar, *Proc. Natl. Acad. Sci. USA*, 1994, **91**, 1224.

89. O. Gursky, Y. Li, J. Badger and D. L. Caspar, *Biophys. J.*, 1992, **61**, 604.
90. D. Hamelberg, L. McFail-Isom, L. D. Williams and W. D. Wilson, *J. Am. Chem. Soc.*, 2000, **122**, 10513.
91. P. Auffinger and E. Westhof, *J. Mol. Biol.*, 2000, **300**, 1113.
92. M. Feig and B. M. Pettitt, *Biophys. J.*, 1999, **77**, 1769.
93. M. A. Young, B. Jayaram and D. L. Beveridge, *J. Am. Chem. Soc.*, 1997, **119**, 59.
94. V. Tereshko, C. J. Wilds, G. Minasov, T. P. Prakash, M. A. Maier, A. Howard, Z. Wawrzak, M. Manoharan and M. Egli, *Nucleic Acids Res.*, 2001, **29**, 1208.
95. P. Langan, V. Forsyth, A. Mahendrasingam, D. Alexeev, S. Mason, W. Fuller, in *Water–Biomolecule Interactions, Proceedings of the EBSA International Workshop*, Vol. 43, *S. Flavia (Palermo)*, ed. M. Palma, M. Palma-Vittorelli and F. Parak, Società Italiana di Fisica, Bologna, 1992, p. 235.
96. K. J. McConnell and D. L. Beveridge, *J. Mol. Biol.*, 2000, **304**, 803.
97. P. Varnai and K. Zakrzewska, *Nucleic Acids Res.*, 2004, **32**, 4269.
98. Y. H. Cheng, N. Korolev and L. Nordenskiold, *Nucleic Acids Res.*, 2006, **34**, 686.
99. X. Q. Shui, C. C. Sines, L. McFail-Isom, D. VanDerveer and L. D. Williams, *Biochemistry*, 1998, **37**, 16877.
100. N. V. Hud, P. Schultze, V. Sklenar and J. Feigon, *J. Mol. Biol.*, 1999, **285**, 233.
101. N. V. Hud, P. Schultze and J. Feigon, *J. Am. Chem. Soc.*, 1998, **120**, 6403.
102. A. A. Rashin and B. Honig, *J. Phys. Chem.*, 1985, **89**, 5588.
103. C. H. Suelter, *Science*, 1970, **168**, 789.
104. H. J. Evans and G. J. Sorger, *Annu. Rev. Plant Physiol.*, 1966, **17**, 47.
105. S. Haider, G. N. Parkinson and S. Neidle, *J. Mol. Biol.*, 2002, **320**, 189.
106. V. P. Denisov and B. Halle, *Proc. Natl. Acad. Sci. USA*, 2000, **97**, 629.
107. L. Nordenskiöld, D. Chang, C. Anderson and T. J. Record, *Biochemistry*, 1984, **23**, 4309.
108. F. C. Marincola, V. P. Denisov and B. Halle, *J. Am. Chem. Soc.*, 2004, **126**, 6739.
109. F. Mocci, A. Laaksonen, A. Lyubartsev and G. Saba, *J. Phys. Chem. B*, 2004, **108**, 16295.
110. D. Patel, *Proc. Natl. Acad. Sci. USA*, 1982, **79**, 6424.
111. T. K. Chiu, M. Kaczor-Grzeskowiak and R. E. Dickerson, *J. Mol. Biol.*, 1999, **292**, 589.
112. E. Johansson, G. Parkinson and S. Neidle, *J. Mol. Biol.*, 2000, **300**, 551.
113. K. K. Woods, L. McFail-Isom, C. C. Sines, S. B. Howerton, R. K. Stephens and L. D. Williams, *J. Am. Chem. Soc.*, 2000, **122**, 1546.
114. V. Tereshko, G. Minasov and M. Egli, *J. Am. Chem. Soc.*, 1999, **121**, 3590.
115. T. Moulaei, T. Maehigashi, G. T. Lountos, S. Komeda, D. Watkins, M. P. Stone, L. A. Marky, J. S. Li, B. Gold and L. D. Williams, *Biochemistry*, 2005, **44**, 7458.

116. V. N. Bartenev, I. Golovamov Eu, K. A. Kapitonova, M. A. Mokulskii, L. I. Volkova and I. Y. Skuratovskii, *J. Mol. Biol.*, 1983, **169**, 217.
117. N. C. Seeman, J. M. Rosenberg, F. L. Suddath, J. J. Kim and A. Rich, *J. Mol. Biol.*, 1976, **104**, 109.
118. N. C. Stellwagen, S. Magnusdottir, C. Gelfi and P. G. Righetti, *J. Mol. Biol.*, 2001, **305**, 1025.
119. L. McFail-Isom, C. C. Sines and L. D. Williams, *Curr. Opin. Struct. Biol.*, 1999, **9**, 298.
120. L. D. Williams and L. J. Maher, *Annu. Rev. Biophys. Biomol. Struct.*, 2000, **29**, 497.
121. A. K. Mazur, *J. Chem. Theory Comput.*, 2005, **1**, 325.
122. J. Liu and J. A. Subirana, *J. Biol. Chem.*, 1999, **274**, 24749.
123. B. Heddi, N. Foloppe, E. Hantz and B. Hartmann, *J. Mol. Biol.*, 2007, **368**, 1403.
124. A. Savelyev and G. A. Papoian, *J. Am. Chem. Soc.*, 2006, **128**, 14506.
125. T. E. Cheatham and P. A. Kollman, *Structure*, 1997, **5**, 1297.
126. B. Gold, *Biopolymers*, 2002, **65**, 173.
127. P. Hardwidge, D. Lee, T. Prakash, B. Iglesias, R. Den, C. Switzer and L. R. Maher, *Chem. Biol.*, 2001, **8**, 967.
128. J. K. Strauss, T. P. Prakash, C. Roberts, C. Switzer and L. J. Maher, *Chem. Biol.*, 1996, **3**, 671.
129. J. K. Strauss, C. Roberts, M. G. Nelson, C. Switzer and L. J. Maher, *Proc. Natl. Acad. Sci. USA*, 1996, **93**, 9515.
130. G. N. Liang, L. Encell, M. G. Nelson, C. Switzer, D. E. G. Shuker and B. Gold, *J. Am. Chem. Soc.*, 1995, **117**, 10135.
131. R. Soliva, V. Monaco, I. Gomez-Pinto, N. J. Meeuwenoord, G. A. Van der Marel, J. H. Van Boom, C. Gonzalez and M. Orozco, *Nucleic Acids Res.*, 2001, **29**, 2973.
132. A. Maki, F. E. Brownewell, D. Liu and E. T. Kool, *Nucleic Acids Res.*, 2003, **31**, 1059.
133. A. S. Maki, T. W. Kim and E. T. Kool, *Biochemistry*, 2004, **43**, 1102.
134. Meena, Z. H. Sun, C. Mulligan and L. W. McLaughlin, *J. Am. Chem. Soc.*, 2006, **128**, 11756.
135. K. K. Woods, T. Lan, L. W. McLaughlin and L. D. Williams, *Nucleic Acids Res.*, 2003, **31**, 1536.
136. V. Tereshko, G. Minasov and M. Egli, *J. Am. Chem. Soc.*, 1999, **121**, 470.

CHAPTER 4
Metal Ion Interactions with G-Quadruplex Structures

AARON E. ENGELHART,[a] JANEZ PLAVEC,[b] ÖZGÜL PERSIL[a] AND NICHOLAS V. HUD[a]

[a] School of Chemistry and Biochemistry, Georgia Institute of Technology, Atlanta, GA 30332, USA; [b] Slovenian NMR Center, National Institute of Chemistry, SI–1001, Ljubljana, Slovenia

4.1 Introduction

A number of DNA secondary structures are known in addition to the Watson–Crick double helix. From the perspective of nucleic acid–metal ion interactions, the G-quadruplex is certainly the most unique and interesting DNA secondary structure. Additionally, G-quadruplexes are currently of significant medicinal interest due to their potential relevance in gene regulation and carcinogenesis.[1–34] In this chapter, we discuss several aspects of metal ion interactions with G-quadruplexes, a key element governing the stability and topology of these structures.

The basic subunit of a G-quadruplex structure is the G-quartet, comprised of a planar cyclic array of four Hoogsteen-paired guanine residues (Figure 4.1a). In 1962, Davies and co-workers proposed that guanine could form such structures upon observing ordered assemblies in X-ray diffraction patterns of fibers pulled from GMP gels; other nucleotides did not give ordered fibers.[35] The spatial proximity of the GO6 atoms in a quartet is unique, with their lone pairs in close approach. This arrangement would be expected to produce significant electrostatic repulsion between the guanine bases. However, as would

RSC Biomolecular Sciences
Nucleic Acid–Metal Ion Interactions
Edited by Nicholas V. Hud
© Royal Society of Chemistry 2009
Published by the Royal Society of Chemistry, www.rsc.org

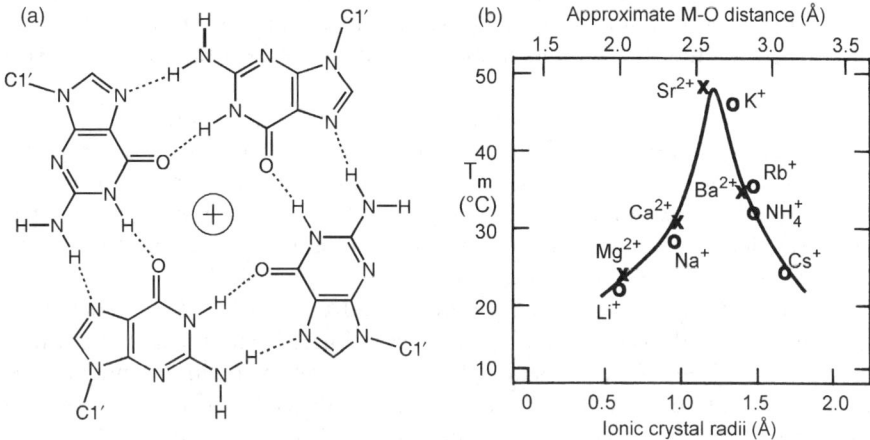

Figure 4.1 (a) Chemical structure of a G-quartet. (b) Plot of the melting temperature of 8-bromoguanosine gels as a function of cation size. Reprinted with permission from ref. 38; data originally published in ref. 39.

be appreciated some time later, the symmetric arrangement of multiple oxygen atoms also represents an excellent location for the coordination of dehydrated metal ions.[36,37] During the past few decades, researchers have performed numerous investigations of G-quadruplex structures, including studies that have verified and significantly characterized the nature of direct cation coordination by G-quartets.

The direct coordination of dehydrated cations lies at the crux of the well-documented cation-dependent stability of G-quadruplexes. Although X-ray crystallographic studies have now revealed numerous structures of G-quadruplexes with coordinated metal ions (discussed below), experimental evidence suggesting direct cation coordination existed long before crystal structures of G-quadruplexes were available. One such experimental result that is now explained by direct metal ion coordination was the observation that the melting temperature (T_m) of guanosine gels depends on the species of cation present and that this thermostability correlates strongly with ionic radius (Figure 4.1b).[38,39] For monovalent cations, guanosine gel stability follows the trend $K^+ \gg Rb^+ > NH_4^+ > Na^+ \gg Li^+$.

In the last two decades, much work has been performed with G-rich oligonucleotides that form quadruplexes. These structures exhibit similar cation selectivity to that reported earlier for guanosine gel stabilization.[40–47] Structural studies of oligonucleotides that form G-quadruplexes were first inspired by the observation that the G-rich sequences found at the ends of chromosomes (*i.e.* the single-stranded overhang of telomeres) and in immunoglobulin genes tend to form four-stranded structures *in vitro*.[48,49] Support for the participation of G-quadruplexes in telomere maintenance looks increasingly solid and continues to grow. Recent studies indicate that DNA sequences present throughout the genome may be involved in regulation of many genes through G-quadruplex formation.[50,51] Such observations continue to deepen interest in the structure and thermodynamics of G-quadruplexes formed by G-rich oligonucleotides.

The majority of physical studies reported for G-quadruplexes have been conducted in solutions containing Na^+ or K^+, as these are the most abundant monovalent cations in cells. However, G-quadruplexes are known to be stabilized by a number of different cations. Thus far, G-quadruplex formation has also been demonstrated in the presence of the monovalent cations NH_4^+, Rb^+ and Tl^+.[52-54] The divalent cations Sr^{2+}, Ba^{2+} and Pb^{2+} have likewise been shown to be directly coordinated within G-quadruplexes.[52,55-58] For some time there was no evidence that Ca^{2+} is ever coordinated within these structures. However, a recent X-ray crystal structure has demonstrated the presence of Ca^{2+} within a G-quadruplex.[59] Spectroscopic evidence has also been presented for the coordination of Ca^{2+} by modified guanine nucleosides in $CDCl_3$.[60] Mg^{2+}, on the other hand, has not been shown to be coordinated within any G-quadruplex to date, although low concentrations of Mg^{2+} ions will stabilize a G-quadruplex, presumably by providing electrostatic screening of phosphate repulsions, rather than participation in direct coordination with the G-quartets. Most recently, G-quartets have been shown to coordinate trivalent lanthanide metal ions,[61] but these studies are at a much earlier stage than those of monovalent and divalent cation coordination.

The space between the four GO6 carbonyl groups of a G-quartet is sufficiently large to accommodate a Na^+ coordinated in the plane of the guanine bases (Figure 4.2). The ionic radius of Na^+ is 0.95 Å, and, as discussed in this chapter, experimental studies indicate that this is around the maximum size of a cation that can be coordinated in the plane of a G-quartet. Larger cations, such as K^+, with an ionic radius of 1.33 Å, are too large to be coordinated within the plane of a G-quartet (Figure 4.2), but they are coordinated between the planes of two stacked G-quartets. Multiple cation coordination geometries and occupancies are possible and observed in G-quadruplexes. Coordinated cations screen electrostatic repulsion between the GO6 lone pair electrons of the G-quartets, but these same cations repel one another within G-quadruplexes that contain more than one coordinated cation. Thus, the details of cation localization within a G-quadruplex are the result of both cation-lone pair attractions and cation-cation repulsions.

In this chapter, we focus on the nature of cation coordination within G-quadruplex structures. Solid-state and solution-phase structural studies have informed our understanding of the localization of cations of varied size and charge within quadruplexes, in addition to the effects of such ions on quadruplex structure in general.[53,54,62-71] Solution-state studies have provided direct evidence for the dynamic exchange of cations between coordination sites and bulk solvent.[67,72] Various techniques have provided thermodynamic insight into the origin of cation selectivity phenomena.[70,73] Some discussion is also provided regarding the relationship between the species of coordinated cation and the topology of a G-quadruplex assumed by a particular DNA sequence.

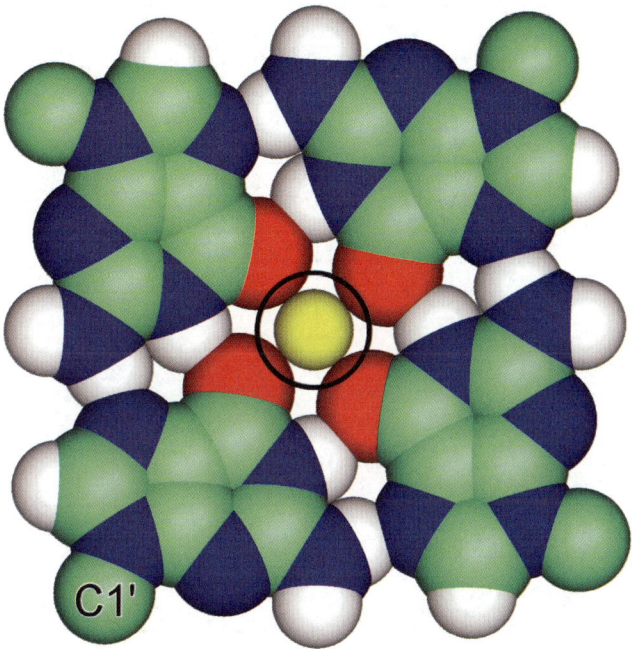

Figure 4.2 A space filling model of a G-quartet. A Na^+ is shown coordinated in the plane of the quartet. The black circle represents the ionic radius of K^+, which is too large for in-plane coordination.

For a more general and extensive overview of G-quadruplexes, including their medical and biological relevance, the reader is referred to a previous volume in this series that was dedicated specifically to G-quadruplex structures.[1]

4.2 A Brief Overview of G-quadruplex Structures and Folding Topologies

The polymorphism displayed by G-quadruplex structures is astounding in comparison to the relatively minor sequence-dependent variations associated with duplex DNA. Quadruplex polymorphisms include different strand orientations (parallel and antiparallel), molecularities (tetramolecular, bimolecular and unimolecular), glycosidic conformations (*syn* and *anti*) and topologies (lateral, diagonal, V-shaped and double chain reversal loops) (Figure 4.3).[74–76] G-quadruplex folding topology is strongly dependent on oligonucleotide sequence and some predictive trends have been recognized. In particular, oligonucleotides with one G-tract typically form parallel tetramolecular structures; those with two typically form antiparallel bimolecular structures; and those with four G-tracts typically form intramolecular

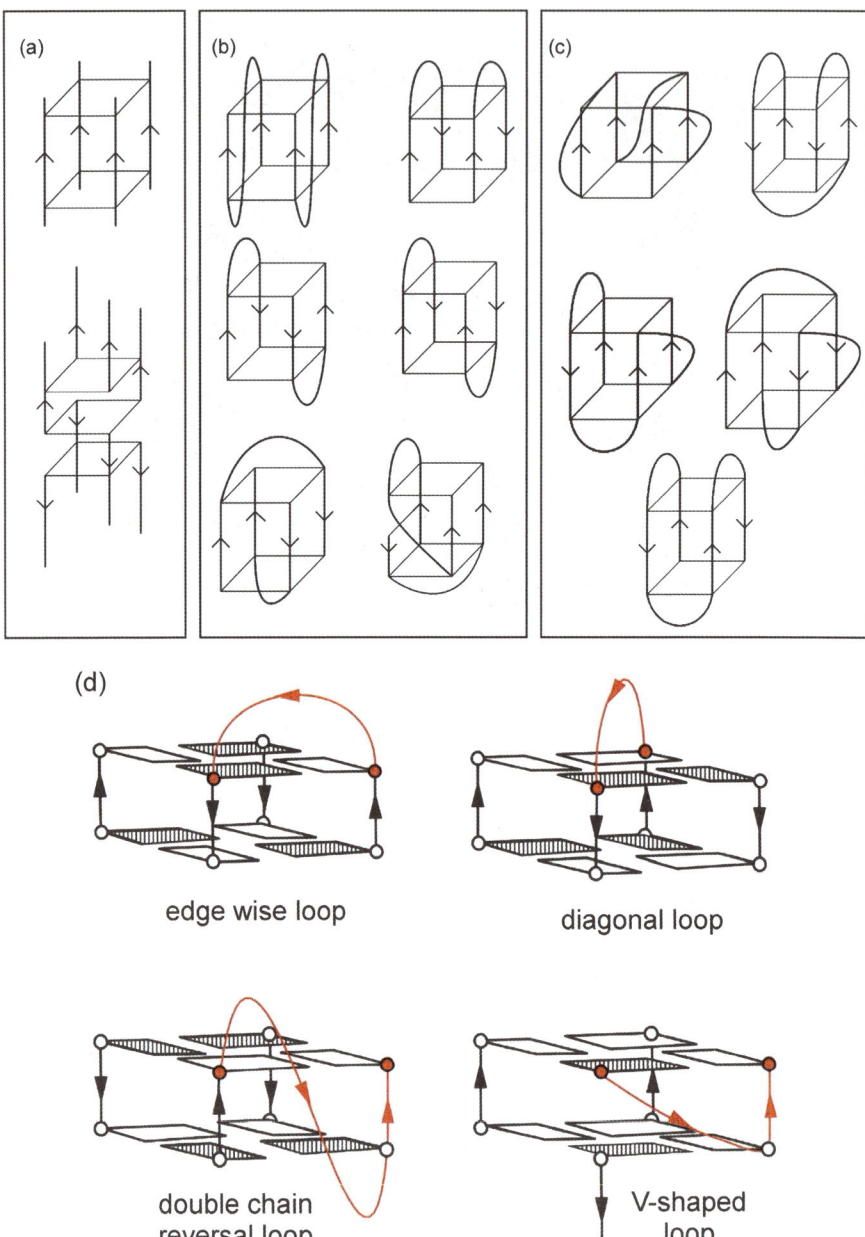

Figure 4.3 Molecular topologies associated with G-quadruplexes. (a) Strand orientations observed for tetramolecular G-quadruplexes. (b) Strand orientations and loop topologies observed for bimolecular G-quadruplexes. (c) Topologies observed for unimolecular G-quadruplexes. (d) Examples and definitions of four distinct loop topologies. Guanine bases in the *syn* conformation are shaded. Loop residues are not shown for clarity. (a)–(c) Copyright Oxford University Press, reprinted with permission from ref. 75; (d) Copyright Academic Press, reprinted with permission from ref. 76.

structures, with a variety of topologies. In many cases, the fold formed by a particular oligonucleotide depends strongly on the number of bases that separate guanine tracts, as these intervening bases become the loops of bimolecular and unimolecular quadruplexes. These loops can span all six sides of a quadruplex and their topologies are associated with orientating the guanine runs in parallel or antiparallel orientations (Figure 4.3). In a number of cases, as discussed below, the G-quadruplex fold adopted by a particular oligonucleotide sequence can vary with the species of coordinated cation.[40,45,77] The difficulty inherent in predicting the fold that will be adopted by an intramolecular G-quadruplex can be appreciated given the fact that the most stable fold is determined by numerous energetic contributions, including stacking interactions, hydrogen bonding, cation coordination and solvent interactions.

4.3 X-ray Crystallographic Studies of Cation Coordination by G-quadruplexes

4.3.1 X-ray Studies of Monovalent Cation Coordination

4.3.1.1 X-ray Studies of Na^+ Coordination

Crystal structures of the tetramolecular G-quadruplex formed by the oligonucleotide d(TG$_4$T) provided unprecedented details regarding cation coordination by stacked G-quartets. For crystals grown in the presence of Na^+ and Ca^{2+}, the parallel-stranded G-quadruplex [d(TG$_4$T)]$_4$ was solved to a resolution of 0.95 Å.[69] Although each oligonucleotide contains only four guanine residues, two of the tetramolecular G-quadruplexes stack head-to-head so that eight G-quartets are observed in an uninterrupted stack (Figure 4.4a). The continuous stack of G-quartets also provided additional insight into Na^+ coordination that may not have be obtained if only one G-quadruplex was present per unit cell. Seven Na^+ ions are coordinated within the channel that is created by the eight stacked G-quartets (Figure 4.4a and b). The two outermost G-quartets have Na^+ coordinated in the plane of the quartets, which, as discussed above, is possible for a cation the size of Na^+. Progressing towards the central G-quartets, at the junction between the two G-quadruplexes, the coordinated Na^+ ions are progressively further out of the planes of the G-quartets. At the center of the stack G-quartet, one Na^+ is coordinated half way between two G-quartets.

The Na^+-form crystal structure of [d(TG$_4$T)]$_4$ revealed the substantial diversity in the coordination geometry that is allowed by Na^+ within a G-quadruplex. The Na^+ ions at the ends are bound in a four-coordinate geometry to the G-quartet, the Na^+ at the center of the continuous G-quartet stack is bound in a bipyramidal coordination and all cations in between are bound by what is perhaps best described as an intermediate coordination geometry. Water molecules provide additional coordination to the two outermost Na^+ ions. This variation in Na^+ coordination geometry appears to be a combination

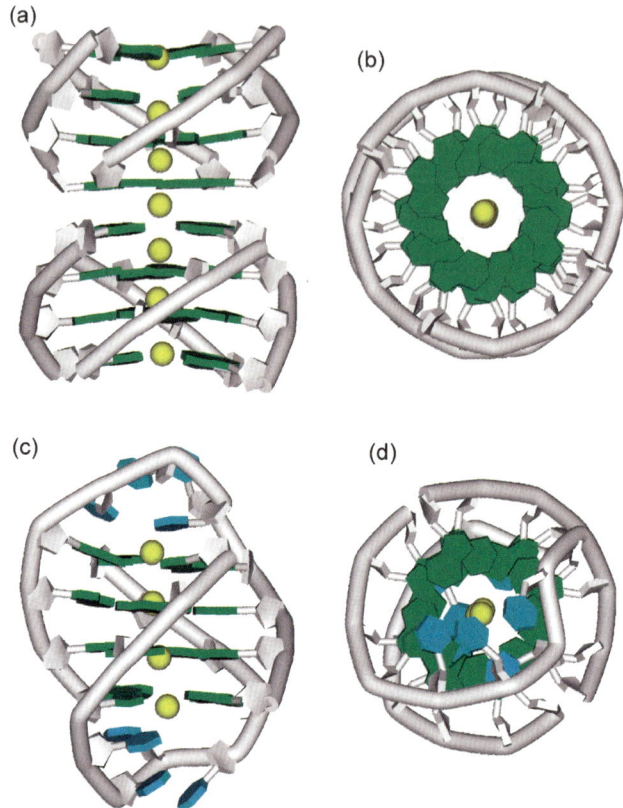

Figure 4.4 X-ray crystal structures of G-quadruplexes with coordinated Na^+ ions. (a) Side view of the Na^+ form of $d(TG_4T)$: PDB ID 352D. (b) Axial view of (a). (c) Side view of the Na^+ form of $d(G_4T_4G_4)$: PDB ID 1JB7. (d) Axial view of (c). Guanine bases are shown in green, and thymine bases in cyan. Na^+ ions are shown in yellow.

of the fact that only seven Na^+ ions are coordinated within the channel of a stack of eight G-quartets and the tendency for the Na^+ ions to move away from each other to minimize electrostatic repulsion. This repulsive energy has also been invoked to explain that seven Na^+ ions are coordinated within the eight stacked G-quartets, as opposed to eight ions, which one might expect (*i.e.* one Na^+ within the plane of each G-quartet). This hypothesis is consistent with the additional observation that the terminal G-quartets of the stack are bent slightly outward, presumably to increase further the distance between neighboring Na^+ ions. Interestingly, in the crystal structure of $[d(TG_4T)]_4$ with the drug molecule daunomycin bound to one end, the Na^+ ions coordinated within the central channel of the G-quadruplex are coordinated between the planes of the G-quartets.[24]

As mentioned above, the crystal that resulted in the high-resolution crystal structure of [d(TG$_4$T)]$_4$ was grown in the presence of both Na$^+$ and Ca^{2+}. It is well known that divalent cations aid crystallization by providing lattice contacts as bridges between the phosphates of nucleic acid molecules and Ca^{2+} serves this role in the [d(TG$_4$T)]$_4$ crystal. While there was no evidence that Ca^{2+} occupied any sites within the G-quartets in place of Na$^+$ in this structure, a more recent structure discussed below has now shown that Ca^{2+} can substitute for monovalent cations within a G-quadruplex.

The first high-resolution G-quadruplex for which any structure was determined (NMR or X-ray) was that formed by the oligonucleotide d(G$_4$T$_4$G$_4$), a sequence derived from the telomere DNA of the organism *Oxytricha nova*, which has a single-stranded 3' overhang with the sequence repeat d(G$_4$T$_4$)$_n$.[78,79] This DNA sequence was shown by 2D NMR spectroscopy to form a symmetric bimolecular G-quadruplex with diagonal T$_4$ loops. The structure of the G-quadruplex [d(G$_4$T$_4$G$_4$)]$_2$, co-crystallized with the telomere end binding protein of *Oxytricha nova*, was solved to a resolution of 1.86 Å (Figure 4.4c and d) and revealed the same structure as that solved earlier by NMR.[64] For this quadruplex, four Na$^+$ ions were observed within the four G-quartets. The two Na$^+$ ions near the center of the quadruplex are each almost co-planar with the two central G-quartets. The other two Na$^+$ ions are slightly out of the planes of the outside G-quartets, towards the T$_4$ loops. These outer Na$^+$ ions are also coordinated by TO2 carbonyl groups of the first and last thymine bases of the T$_4$ loops (Figure 4.4c and d). The occupancy of these outer sites by Na$^+$ is less than 100%.[64]

4.3.1.2 X-ray Studies of K$^+$ and Tl$^+$ Coordination

Because of the higher concentration of K$^+$ than Na$^+$ within cells and the selective binding of K$^+$ over Na$^+$ by G-quadruplexes (discussed below), there has been particular interest in determining the structure of G-quadruplexes in the presence of K$^+$. It is now apparent that some G-quadruplex sequences adopt alternative topologies in the presence of K$^+$ and Na$^+$. However, despite initial disagreement between NMR and X-ray structures,[80] the DNA sequence d(G$_4$T$_4$G$_4$) has been shown to adopt the same symmetric, biomolecular structure in Na$^+$- and K$^+$-form crystals. The first evidence for this similarity of structures was provided by NMR spectroscopy, which revealed the same fold in the presence of K$^+$ and Na$^+$, with only slight variations in loop structures.[78,79,81] The subsequent determination of the K$^+$-form crystal structure of [d(G$_4$T$_4$G$_4$)]$_2$, solved to a resolution of 1.49 Å (Figure 4.5a and b), provided definitive proof that the Na$^+$ and K$^+$ forms of this G-quadruplex are essentially the same.[63] However, the X-ray crystal structure of the K$^+$-form quadruplex revealed the coordination of cations at positions different from those in the Na$^+$-form structure. In the K$^+$-form crystal, five K$^+$ ions are coordinated by the four G-quartets (Figure 4.5a). The average interplane spacing of G-quartets in this structure is 3.3 Å.[63] The five equidistant K$^+$ ions lie along the axis of the central channel of all three quadruplexes in the asymmetric unit cell of the crystal. All of

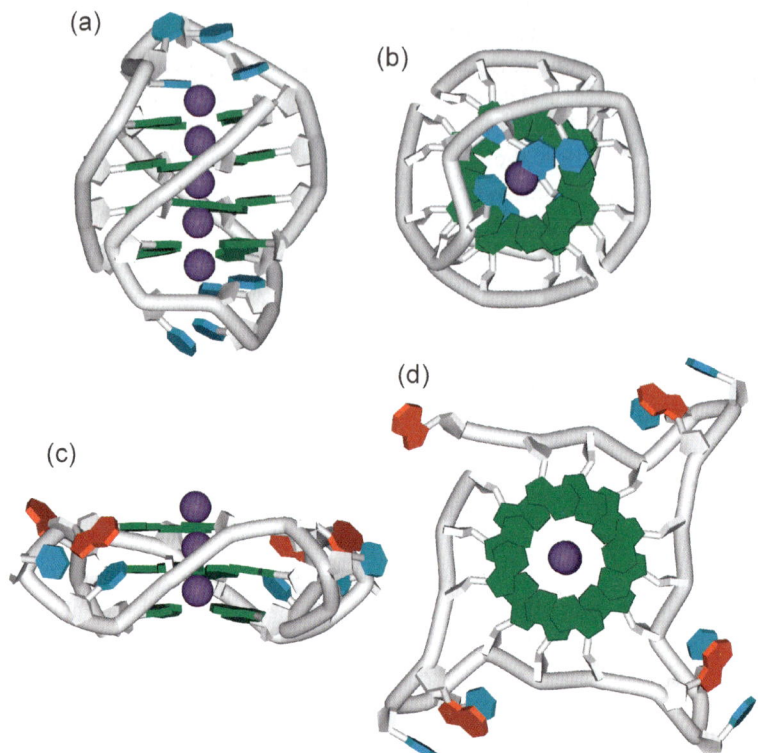

Figure 4.5 X-ray crystal structures of G-quadruplexes with coordinated K^+ ions. (a) Side view of the K^+ form of $d(G_4T_4G_4)$: PDB ID 1JRN. (b) Axial view of (a). (c) Side view of the K^+ form of $d[AGGG(TTAGGG)_3]$, PDB ID 1KF1. (d) Axial view of (c). Guanine bases are shown in green, adenine bases in red and thymine bases in cyan. K^+ ions are shown in purple.

these channel ion coordination sites exhibit full K^+ occupancy. The average inter-K^+ distance is 3.38 Å. All three quadruplexes in the asymmetric unit cell have three cations between two adjacent quartets, coordinated in square antiprismatic fashion. The remaining two K^+ ions are located within the loops, coordinated by the outer quartet, TO2 atoms from the loops and two water molecules. These ions are only slightly more mobile in the crystal state than the central three K^+ ions.[63] The presence of these cations in the loops is potentially responsible for the differences observed between the loop structures of the Na^+- and K^+-forms of $[d(G_4T_4G_4)]_2$.

An X-ray crystal structure has been reported at 2.1 Å for the monomolecular human telomere quadruplex formed by $d[AGGG(TTAGGG)_3]$ in the presence of K^+ (PDB ID 1KF1) (Figure 4.5c and d).[62] This quadruplex structure was found to adopt a 'propeller' shape (Figure 4.5d) with parallel GGG segments and with the TTA segments forming so-called double-chain reversal loops (Figure 4.3d). K^+ ions were again observed to be coordinated between

G-quartets. The distance between K^+ ions was, on average, 3.4 Å. The K^+-form crystal structure of the bimolecular quadruplex formed by the two-repeat telomere sequence d(TAGGGTTAGGGT) has also been solved (PDB ID 1K8P). This quadruplex structure exhibits a fold similar to the monomolecular quadruplex derived from the human telomere sequence, and likewise contains K^+ ions coordinated between adjacent G-quartets.[63]

The superior X-ray scattering potential of Tl^+ and its ionic radius of 1.44 Å have made this cation an excellent probe of K^+ binding sites in crystal structures. The first G-quadruplex crystal structure for which Tl^+ binding sites were characterized was the tetramolecular quadruplex formed by $d(TG_4T)$. From a mixed Na^+/Tl^+ solution, the oligonucleotide $d(TG_4T)$ crystallizes in two distinct forms.[82] These crystals provided lower resolution data (2.2 and 2.5 Å) than the Na^+-form crystals, but because Tl^+ exhibits a markedly greater scattering cross-section than Na^+, bound Tl^+ ions could be identified at lower resolution. One crystal form contained two quadruplexes in the asymmetric unit (PDB ID 1S45). The other contained three (PDB ID 1S47). In each case, quadruplexes were stacked in head-to-head fashion with 5'- to 5'- orientation, as in the Na^+-only structure.[69] The main difference in the Na^+/Tl^+ structures compared to the reported Na^+ structure relates, predictably, to the position of ions within the quadruplex. Tl^+ ions coordinated within the quadruplex are located between G-quartets, as expected due to the large ionic radius of Tl^+. Furthermore, incomplete occupancy of inner cation coordination sites by Tl^+ was observed (15–70%). The authors attributed the lower occupancy of Tl^+ to the high concentration of Na^+ present during crystallization.[82] However, as discussed below, solution-state studies of Na^+ and NH_4^+ (with a similar ionic radius to Tl^+) competition for binding within a G-quadruplex have revealed that the most energetically favored configuration of cations coordinated within a G-quadruplex structure can be a mixture of small and large cations, even though the energetics associated with the replacement of Na^+ by NH_4^+ are favorable at any one given coordination site within an Na^+-only G-quadruplex.

More recently, the Tl^+-form crystal structure of the G-quadruplex $[d(G_4T_4G_4)]_2$ has been solved to a resolution of 1.55 Å.[83] To obtain Tl^+-form crystals of this G-quadruplex, Gill et al.[83] first grew K^+-form crystals of $d(G_4T_4G_4)$ using a slightly modified protocol of that used by Neidle and co-workers.[63] The K^+-form crystals were then converted to the Tl^+ form by soaking the crystals in 50 mM thallium acetate. The resulting Tl^+-form crystals exhibited unit cell dimensions that differed from those of the K^+-form crystals by only 0.3–3.2% in all dimensions. Like the K^+-form structure of $[d(G_4T_4G_4)]_2$ discussed above, the Tl^+-form structure was found to have five cations coordinated along the central channel, with three Tl^+ ions coordinated between the four G-quartets and two coordinated between the terminal G-quartets and the T_4 loops. No Tl^+ ions were identified on the surface or in the grooves of the Tl^+-form $d(G_4T_4G_4)$ crystal structure. The average spacing of the five Tl^+ ions is 3.6 Å, which is close to the average spacing of 3.4 Å of the five K^+ ions in the analogous structure. Cation–O6 distances were also found to be similar between the two structures, with these distances ranging from 2.5 to 3.3 Å for Tl^+, compared with distances ranging from 2.6 to 3.1 Å for K^+.

Overall, the Tl^+- and K^+-form crystal structures are extremely similar, with an average RMSD of only 0.26 Å.

As illustrated by the example structures discussed here, X-ray crystal structures support the general conclusion that cations larger than Na^+, such as K^+ and Tl^+, are coordinated between stacked G-quartets because they are too large for in-plane coordination. In contrast, Na^+ can coordinate GO8 in three fashions: in-plane, between quartets and any intermediate configuration. Intercation distances in the Na^+-form crystals of $d(TG_4T)$ and $d(G_4T_4G_4)$ described above are greater than the average spacing between the planes of stacked G-quartets. The lower steric constraints on Na^+ allow it to occupy a wider array of positions, which can minimize repulsive electrostatic interactions between its neighboring cations.

4.3.2 X-ray Studies of Divalent Cation Coordination

It has been recognized for some time that Sr^{2+} can stabilize G-quadruplex structures to a similar extent as K^+.[57] However, the underlying reasons for the similar stabilizing abilities of K^+ and Sr^{2+} are not obvious. The ionic radius of Sr^{2+} (1.13 Å) is intermediate to Na^+ and K^+, but the repulsive interactions between neighboring Sr^{2+} ions would be four-fold higher if Sr^{2+} were coordinated as K^+ is within a G-quadruplex. A high-resolution X-ray crystal structure of $r(UG_4U)$ from a crystal grown in the presence of Sr^{2+} (PDB ID 1J8G) has shed some light on this enigma. Four strands form a parallel quadruplex, as seen for the analogous $d(TG_4T)$, but Sr^{2+} occupies *every other* inter-G-quartet site, with no cations coordinated in the intervening sites (Figure 4.6a).[65] It appears that the high stability of Sr^{2+} quadruplexes is due to the high coordination energy

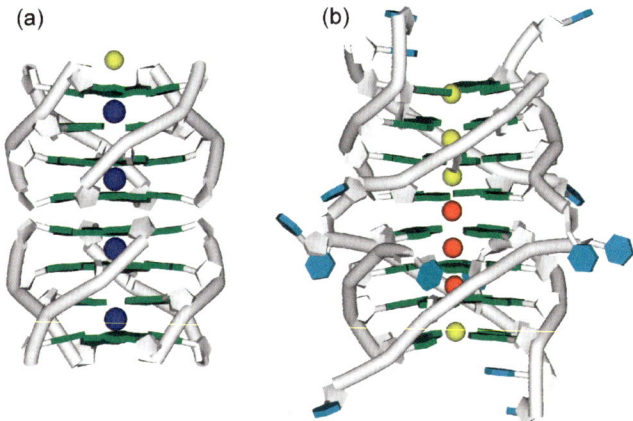

Figure 4.6 X-ray crystal structures of G-quadruplexes with coordinated divalent cations. (a) Sr^{2+} form of $r(UG_4U)$: PDB ID 1J8G. (b) Na^+/Ca^{2+} form of $d(TG_4T)$: PDB ID 2GW0. Guanine bases are shown in green and thymine bases in cyan. Na^+ ions are shown in yellow, Sr^{2+} ions in blue and Ca^{2+} ions in red.

associated with a divalent cation, while the repulsive electrostatic interactions are minimized by the spacing of cations.

Recently, Neidle and co-workers discovered the coexistence of Na^+ and Ca^{2+} ions along the central ion channel of the tetramolecular quadruplex of d(TG$_4$T) in a crystal grown in the presence of both cations (PDB ID 2GW0).[59] This crystal structure was refined to 1.55 Å and the asymmetric unit cell contained two co-axially stacked G-quadruplexes related by a pseudo-twofold axis (Figure 4.6b), similar to the crystal structure discussed above for the same oligonucleotide with only Na^+ ions identified within the G-quartets. The identification of Ca^{2+} by Neidle and co-workers was possible using an anomalous difference Fourier map. Ca^{2+} has a significant anomalous signal at the wavelength used by these authors (*i.e.* 1.54178 Å), whereas Na^+ does not. Hence significant peaks in the anomalous Fourier difference map could be assigned to Ca^{2+}. By this method, three high-density peaks were identified along the central channel of the two stacked G-quadruplexes of the asymmetric unit cell, and these peaks were assigned as fully occupied by Ca^{2+} in the refined structure. The positions of these Ca^{2+} are surprisingly asymmetric with respect to the pseudosymmetric G-quadruplex dimer (Figure 4.6b). One Ca^{2+} is located at the center of the stack of eight G-quartets, but the other two Ca^{2+} are located to one side of the central Ca^{2+}. Similarly to the Na^+-only structure discussed above, the Na^+ and Ca^{2+} ions within the central ion channel are more displaced from the midpoint between adjacent G-quartets the further cations are from the center of the eight G-quartet stack. This movement of Ca^{2+} to almost in-plane coordination by a G-quartet is possible due to the small ionic radius of Ca^{2+} (0.99 Å). Interestingly, Neidle and co-workers reported that attempts to model the G-quadruplex with the mixed Na^+ and Ca^{2+} coordination by molecular dynamics simulations were unsuccessful, as the Ca^{2+} ions rapidly diffused out of the central channel. HF and DFT calculations by the same group revealed that some of the charge on Ca^{2+} is delocalized within the G-quartets, a result that was consistent with their observation that molecular dynamics simulations were stable for the mixed Na^+/Ca^{2+} quadruplex if the charge placed on the calcium ions was reduced to +1.5.

4.4 NMR Studies of Cation Coordination by G-quadruplexes

4.4.1 Solution-state NMR Studies of Monovalent Cation Coordination

4.4.1.1 Quadrupolar NMR Studies of Cation Coordination

The first spectroscopic evidence for the coordination of dehydrated cations by G-quartets was presented by Laszlo and co-workers from their solution-state NMR studies of 5'-GMP.[84–86] These studied revealed changes in resonance lineshape of $^{23}Na^+$, $^{39}K^+$ and $^{87}Rb^+$ in the presence of 5'-GMP, consistent with G-quartet self-assembly and cation coordination. Subsequent studies by

Braunlin and co-workers, also using ^{23}Na NMR, demonstrated the coordination of dehydrated cations within the biomolecular and tetramolecular G-quadruplexes formed by the oligonucleotides d($G_4T_4G_4$) and d(T_2G_4T).[87,88] These studies also provided the first demonstration that Na$^+$ is in exchange between the coordination sites within a G-quadruplex and Na$^+$ free in solution and that the addition of K$^+$ to a Na$^+$-form G-quadruplex in solution results in the displacement of Na$^+$ ions from the inner coordination sites.

More recently, Wu and co-workers have made significant advances in the use of quadrupolar NMR to study alkali ion interactions with G-quartets.[89,90] Historically, the rapid relaxation of quadrupolar nuclei has resulted in broad resonance line widths that made it impossible to detect separate resonances for cations coordinated by a G-quartet and cations free in solution.[84–86] The application of higher magnetic fields by Wu and co-workers has demonstrated the tight binding of ^{23}Na$^+$, ^{39}K$^+$ and ^{87}Rb$^+$ to G-quartets in solution.[90] These authors also used *ab initio* calculations to relate Na$^+$ coordination geometry within G-quartets to observed ^{23}Na chemical shifts.[89] Results from these calculations were found to be consistent with a model of four stacked G-quartets that have three Na$^+$ ions coordinated within the central channel of [d($G_4T_4G_4$)]$_2$. Their model had the additional constraints that each Na$^+$ is located between two stacked G-quartets.[91] Chemical shift calculations for ^{23}Na by the same authors for models with Na$^+$ coordinated only in the plane of the G-quartets were not found to be in agreement with ^{23}Na NMR data.[89] This result is surprising, given that X-ray crystal structures (discussed above) have demonstrated that Na$^+$ is often coordinated within the plane of a G-quartet.

The combined ^{23}Na NMR and chemical shift prediction studies of Wu and co-workers have also raised new questions about previous assignments of ^{23}Na resonances associated with the G-quadruplex [d($G_4T_4G_4$)]$_2$.[92] Braunlin and co-workers had previously assigned ^{23}Na resonances for this quadruplex with exchange times of *ca.* 200 μs to ions coordinated within the central channel. Wu and co-workers have now presented evidence that these previously identified ions are actually Na$^+$ associated with the T_4 loops and that Na$^+$ ions in the central channel of [d($G_4T_4G_4$)]$_2$ have residence lifetimes of several milliseconds.[92] Wu and co-workers also contend that all Na$^+$ ions coordinated in the central channel of the G-quadruplexes [d($G_4T_4G_4$)]$_2$ and [d(TG_4T)]$_4$ are coordinated between the planes of stacked G-quartets. This assignment is, again, surprising, given its opposition to the observation of Na$^+$ coordinated both in-plane and in between the planes of G-quartets in the crystal structure of d(TG_4T). However, Wu and co-workers propose that the in-plane coordination of Na$^+$ observed in the crystal structure is due to crystal packing forces that are associated with the head-to-head stacking of two [d(TG_4T)]$_4$ G-quadruplexes in the crystal state.

4.4.1.2 *NMR Studies of Cation Coordination Using ^1H NMR of DNA Protons*

By monitoring the change in DNA ^{1}H resonances during KCl titrations, Hud *et al.* were able to monitor the competition between K$^+$ and Na$^+$ for coordination within the G-quadruplex formed by the oligonucleotide $d(G_3T_4G_3)$.[70] The bimolecular Na^+- and K^+-form G-quadruplexes of this oligonucleotide assume a foldback structure with three quartets and diagonal loops, a topology that is similar to that seen for $d(G_4T_4G_4)$ (Figure 4.4c). ^1H spectra taken over the course of a titration of KCl into an Na^+-form sample of $d(G_3T_4G_3)$ demonstrated that the equilibrium state of the G-quadruplex changes with increasing additions of KCl from the Na^+-only to the K^+-only form. Although this oligonucleotide adopts the same basic fold in the presence of either cation, local differences in structure or cation coordination geometry produce observable changes in the ^1H NMR spectra of $[d(G_3T_4G_3)]_2$. No additional resonances were observed in samples containing mixtures of Na^+ and K^+, illustrating that these two cations are in fast exchange on the NMR time-scale from the center channel of this particular G-quadruplex at 25 °C. Spectra acquired during the KCl titration were intermediate between the Na^+-only and the K^+-only spectra. Changes in DNA proton chemical shifts as a function of KCl concentration were fit perfectly by a model that included the Na^+-only form, the K^+-only form and two intermediate species of the asymmetric $[d(G_3T_4G_3)]_2$ that each have one coordinated Na^+ and one coordinated K^+.[70]

The quantitative analyses of chemical shift changes during KCl titrations also provided important thermodynamic insights regarding the origin of preferential cation coordination by G-quadruplexes. To a first approximation, hydration energy for the alkali cations varies inversely with ionic radius. The free energy of dehydration for K^+ is therefore less than that of Na^+. Since the net energy realized by coordination of a cation within a G-quadruplex is diminished by the penalty for removal from bulk solvent, the combination of the negative free energy of coordination and the positive free energy of dehydration determines cation selectivity in G-quadruplexes. Na^+ can actually realize greater negative coordination energy within a G-quadruplex,[93] but its reference state (*i.e.* in bulk solvent) is lower in energy than K^+. Therefore, K^+ realizes a more negative net energy change upon coordination. These competition studies revealed an energy change of $-1.7\,kcal/mol$ for the conversion of $[d(G_3T_4G_3)]_2$ from its Na^+- to its K^+-form.[70] Subsequent calculations by other groups have supported the assertion that the hydration term dominates the relative energy of Na^+- and K^+-form G-quadruplexes.[73]

4.4.1.3 Localization of Cation Binding Sites Using the Ammonium Ion

The ammonium ion, NH_4^+, has been shown to stabilize G-quadruplexes to an extent similar to that observed for Na^+.[38,58] Hud *et al.* used $^{15}NH_4^+$ with 2D ^1H–^1H and ^1H–^{15}N cross-relaxation experiments, and also ^{15}N-filtered ^1H experiments, to study cation localization within $[d(G_4T_4G_4)]_2$.[54,68] As discussed above, this oligonucleotide sequence forms a bimolecular quadruplex with four stacked G-quartets and diagonal T_4 loops on each end of the stack. Analyses of dipolar interactions between $^{15}NH_4^+$ and G-NH protons revealed coordination of three ammonium ions within this quadruplex; two in symmetry-related 'outer sites' (O) and one in a central 'inner site' (I) (Figure 4.7a) NOESY

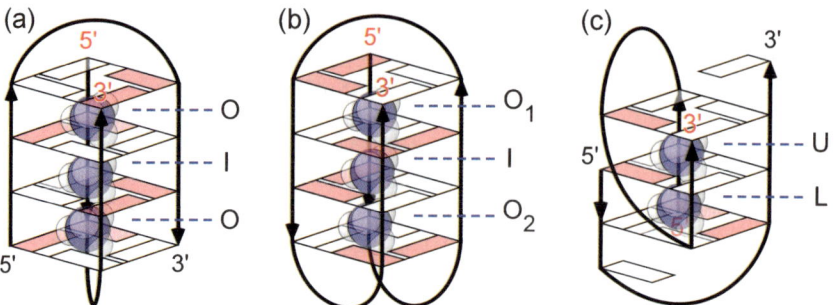

Figure 4.7 A schematic representation of ammonium binding sites within three G-quadruplexes: (a) [d(G$_4$T$_4$G$_4$)]$_2$, (b) d[G$_4$(T$_4$G$_4$)$_3$] and (c) [d(G$_3$T$_4$G$_4$)]$_2$. For clarity, thymine bases in loops are not shown. Guanine bases in the *anti* conformation are depicted as white rectangles and those in the *syn* conformation as pink rectangles.

cross-peaks revealed these cations are localized between adjacent stacked G-quartets, similar to the arrangement of the central three K$^+$ that were observed in the subsequent X-ray crystal structures of the same DNA sequence.[63] This technique has now been used to confirm the binding sites of ^{15}NH$_4^+$ within several other G-quadruplexes. For example, the localization of ^{15}NH$_4^+$ within the G-quadruplexes d[G$_4$(T$_4$G$_4$)$_3$] and [d(G$_3$T$_4$G$_4$)]$_2$ has also been reported (Figure 4.7).[94–96] The very different dynamics observed for the movement of ^{15}NH$_4^+$ within these G-quadruplexes is discussed below.

4.4.1.4 Cation Localization by ^{205}Tl NMR

As mentioned above, Tl$^+$ is a K$^+$ mimic and binds G-quadruplexes with similar affinity.[97] ^{205}Tl also has excellent spectral properties for NMR, being spin-1/2, of 70% natural abundance and exhibiting a sensitivity of 0.13 relative to ^{1}H. In 2000, Strobel and co-workers reported ^{205}Tl NMR studies of the quadruplex formed by d(T$_2$G$_4$T$_2$).[53] The ^{205}Tl spectrum obtained exhibited excellent dispersion and sensitivity to coordination, as three clearly resolved Tl$^+$ resonances and a bulk Tl$^+$ resonance were observed. The three resonances not associated with bulk ions were attributed to three distinct coordination sites within [d(T$_2$G$_4$T$_2$)]$_4$ (presumably the three interquartet sites). The linewidths of these peaks suggest motion of these ions on the NMR time-scale. One peak was narrower, which was attributed to a longer-lived coordination site, and the other two were assumed to be in free exchange with bulk solution. The bound lifetime of Tl$^+$ ions within this quadruplex was estimated to be at least 3 μs.

^{205}Tl$^+$ has also been used to study cation coordination within the bimolecular G-quadruplex [d(G$_4$T$_4$G$_4$)]$_2$. As discussed above, the Tl$^+$-form structure of this G-quadruplex is very similar to its K$^+$-form structure, as confirmed by both X-ray crystallography and NMR spectroscopy.[83,98] ^{1}H NMR experiments have also illustrated that Tl$^+$ and Na$^+$ freely exchange from coordination sites within the G-quadruplexes $[d(G_4T_4G_4)]_2$.[97] Gill et al. we able to localize individual $^{205}Tl^+$ coordinated between the G-quartets of $[d(G_4T_4G_4)]_2$ in solution. Four $^{205}Tl^+$ resonances, in addition to the bulk $^{205}Tl^+$ signal, were observed.[98] ^1H–^{205}Tl spin-echo difference spectra, between guanine protons and bound Tl^+, revealed that two of the $^{205}Tl^+$ resonances are associated with the three cations coordinated within the G-quadruplex (i.e. one cation at the inner coordination site, I in Figure 4.7a, and two bound at the symmetry related outer sites, O in Figure 4.7a, which have the same chemical shift). Although the origin of two additional ^{205}Tl resonances associated with $[d(G_4T_4G_4)]_2$ was not clear from their initial NMR studies, Gill et al. subsequently assigned these resonances to loop-bound $^{205}Tl^+$ following their observation of Tl^+ bound in the T_4 loops of $[d(G_4T_4G_4)]_2$ in their crystal structure.[83] Although the T_4 loops are symmetric in the solution-state structure, Gill et al. propose that two separate $^{205}Tl^+$ resonances are observed because these loops exist in two conformational states that are in sufficiently slow exchange for two $^{205}Tl^+$ resonances to be observed.

4.4.2 Solid-state NMR Studies of Cation Coordination

Wu and co-workers have made extensive use of solid-state NMR to characterize the coordination of cations by G-quartets.[91,99–101] In one particular study, solid-state ^{23}Na NMR was used to investigate the competition of monovalent cations for coordination within G-quartets formed by 5′-GMP.[100] From a series of ion titration experiments, the following thermodynamic parameters were derived for the coordination of selected cations relative to Na^+: K^+ (−1.9 kcal mol^{-1}), NH_4^+ (−1.8 kcal mol^{-1}), Rb^+ (−0.3 kcal mol^{-1}) and Cs^+ (1.8 kcal mol^{-1}). For the system studied, the authors reported G-quartet coordination preference as $K^+ > NH_4^+ > Rb^+ > Na^+ > Cs^+ > Li^+$, an ordering of cations that reflects the same trend as observed for the cation-dependent thermal stability of GMP gels (Figure 4.1b). These competition studies again illustrate that K^+ balances favorable coordination with the energetic penalty of dehydration. These same two energetic terms can also explain why Cs^+ and Li^+ are unknown within G-quadruplexes. Cs^+ dehydrates easily, but it is too large to coordinate well by G-quartets within a G-quadruplex. Li^+ could conceivably be coordinated within the plane of G-quartets, but its penalty for dehydration is too large.

Wu and co-workers have also used solid-state ^{23}Na NMR to study the coordination of cations by oligonucleotides that form G-quadruplexes. They have shown that three Na^+ ions reside inside the bimolecular G-quadruplex $[d(G_4T_4G_4)]_2$.[91] As mentioned above, chemical shift predictions by Wu and co-workers support each $^{23}Na^+$ in the central channel of this G-quadruplex being coordinated in between adjacent G-quartets, as opposed to being coordinated within the plane of a G-quartet. Their original solid-state ^{23}Na study did not provide evidence of Na^+ ions coordinated in the T_4 loops,[91] as is observed in the crystal structure. However, a re-evaluation of the solid-state spectrum (after completion of an ^{23}Na solution-state study) has resulted in their assignment of a resonance to loop-bound Na^+ in both the solution and solid-state ^{23}Na spectra.[92]

Finally, Wu and co-workers have begun to explore the use of solid-state ^{17}O NMR as a means to gain information about cation coordination with G-quadruplexes.[102] This work has thus far demonstrated that the effects of cation coordination and hydrogen bonding on the ^{17}O chemical shift tensor and quadrupole coupling tensor of the O6 carbonyl of guanine are quite distinct. Hence ^{17}O NMR will likely prove useful in the future for studying the ligand side of G-quartet-cation interactions.

4.5 Direct Observation of Movement and Exchange of Cations Within G-quadruplexes

NMR spectroscopy has proven to be a valuable technique for measuring the dynamics of coordinated cations within G-quadruplexes. Early studies by Braunlin and co-workers examined ^{23}Na NMR spectra of solutions containing the tetramolecular quadruplex $[d(T_2G_4T)]_4$ and the bimolecular quadruplex $[d(G_4T_4G_4)]_2$, from which they estimated bound lifetimes of coordinated cations.[87,88] Their results indicated that ^{23}Na$^+$ coordinated by a G-quadruplex exhibits a relaxation time that reflects the tumbling of the quadruplex, which occurs on the nanosecond time-scale. Thus, the coordinated cations have rotation times and bound lifetimes that are on the order of a few nanoseconds or longer. This observation was in stark contrast to the high-mobility nature of monovalent cations that are loosely associated with the surface of nucleic acid structures.[103,104]

The ^{23}Na relaxation experiments by Braunlin and co-workers suggested a bound lifetime of around 250 μs (at 10 °C) for Na$^+$ ions within $[d(G_4T_4G_4)]_2$ and substantially shorter lifetimes for Na$^+$ coordinated within $[d(T_2G_4T)]_4$.[87,88] The lifetime of Na$^+$ within a quadruplex is orders of magnitude shorter than that of quartet 'breathing' motions – some imino protons can persist for days to weeks in D$_2$O-exchanged quadruplex NMR samples.[105] These data suggest the dominant pathway for the exchange of bound cations with bulk solution is the central channel only (*i.e.* not *via* transient 'breathing' structures). The faster exchange of cations in the tetramolecular quadruplex supports this hypothesis. That is, the diagonal loops of $[d(G_4T_4G_4)]_2$ (Figure 4.4d) impede cation exchange, whereas diagonal loops are absent from $[d(TG_4T)]_4$ (Figure 4.4b). As mentioned above, more recent analyses by Wu and co-workers indicate that the bound lifetime for Na$^+$ within the G-quadruplex $[d(G_4T_4G_4)]_2$ appears to be much longer (*i.e.* on the millisecond time-scale) and the exchange rates reported by Braunlin and co-workers may actually reflect Na$^+$ bound in the T$_4$ loops of $[d(G_4T_4G_4)]_2$.[92]

^{15}NH$_4^+$ has also proven useful for studying the movement and dynamics of coordinated cations within G-quadruplexes.[54,67,60] As mentioned above, NMR spectroscopy was able to identify three ammonium coordination sites within the four stacked G-quartets of $[d(G_4T_4G_4)]_2$ (O and I; Figure 4.7a). These sites are apparently fully occupied by ^{15}NH$_4^+$ when this is the only counterion present.[54] ^{1}H–^{15}N 2D correlation spectra have revealed that ^{15}NH$_4^+$ move along the central axis of this G-quadruplex. Analysis of cross-peaks that correspond to the exchange of ammonium ion between the inner and outer sites (IO and OI in Figure 4.8) revealed an inner-site resonance time of 250 ms (at 10 °C).[54] An additional cross-peak, labeled OB in Figure 4.8b, revealed exchange between the outer sites and bulk solution. Analysis of cross-peak intensities demonstrated that an equal number of cations move from outer to inner coordination sites, inner to outer coordination sites and outer coordination sites to bulk solution. There was no spectral evidence to suggest direct exchange between the inner site and bulk solution occurred. These results provided further evidence for the in-channel mechanism of cation exchange within a G-quadruplex. The substantially longer residence time of $^{15}NH_4^+$ compared with Na^+ is likely due to the greater steric bulk of $^{15}NH_4^+$ compared with Na^+, as $^{15}NH_4^+$ would require some opening of a G-quartet to allow passage of the cation along the central channel.

More recently, Plavec and co-workers have shown that the residence time of $^{15}NH_4^+$ within a G-quadruplex is perturbed by the presence of Na^+. For example, NMR experiments with $[d(G_4T_4G_4)]_2$ have revealed that $^{15}NH_4^+$ transit from the inner to an outer site (i.e. I to O) is accelerated ca. 7.5-fold in the presence of Na^+.[72] The residence time of $^{15}NH_4^+$ in the inner site of $[d(G_4T_4G_4)]_2$ is reduced from 270 ms in the presence of $^{15}NH_4^+$ alone, to 36 ms in the presence of both cations (at 25 °C). This accelerated exchange appears to be due to the partial occupancy of coordination sites by Na^+ and not bulk ion concentration. Even 5 mM sodium decreases the residence time of $^{15}NH_4^+$ to its minimum value. Further addition of Na^+ does not result in further rate enhancements. Because Na^+ movement is not rate limiting to $^{15}NH_4^+$ movement, it appears that the 36 ms residence time represents the limiting rate for $^{15}NH_4^+$ movement from the inner to an outer site. The slower movement of $^{15}NH_4^+$ in the absence of Na^+ appears to be due to the requirement that concerted movement of two bulky ions (i.e. $^{15}NH_4^+$) occur in this regime.

Analysis of $^{15}NH_4^+$ within the unimolecular G-quadruplex $d[G_4(T_4G_4)_3]$ has revealed that the rate of $^{15}NH_4^+$ movement along the central channel of a G-quadruplex, in addition to movement out through the ends, depends greatly on G-quadruplex topology and the structure of the loops at each end.[96] In the presence of $^{15}NH_4^+$, $d[G_4(T_4G_4)_3]$ was shown to adopt a fold with two lateral loops and one diagonal loop (Figure 4.7b), the same fold as that was determined previously for this DNA sequence in the presence of Na^+.[106,107] This unimolecular G-quadruplex has three $^{15}NH_4^+$ binding sites within its four stacked G-quartets, designated O_1, I and O_2 in Figure 4.7b. ^1H–^{15}N 2D correlation spectra reveal that $^{15}NH_4^+$ ions coordinated within this G-quadruplex move between the internal binding sites and exchange with solution (Figure 4.8c and d). However, there was no evidence that $^{15}NH_4^+$ moves unidirectionally along the central channel.[96] Movement between the O_1 and O_2 sites and bulk solution is faster than movement between these sites and the central site I. The asymmetric loop structure of this unimolecular G-quadruplex exhibits a substantial effect on $^{15}NH_4^+$ movement. In particular, movement between site O_2 and bulk solution was five times faster than movement between site O_1 and bulk solution (cross-peak O_2B > cross-peak O_1B, Figure 4.8d). This observation is consistent with the two lateral loops at the end near site O_2 allowing more flexibility for G-quartet breathing and access to solution than the single diagonal loop near site O_1 (Figure 4.7b).

The G-quadruplex formed by the asymmetric DNA sequence $d(G_3T_4G_4)$ has further illustrated how G-quadruplex topology affects both the thermodynamics of cation binding and the kinetics of cation movement within the

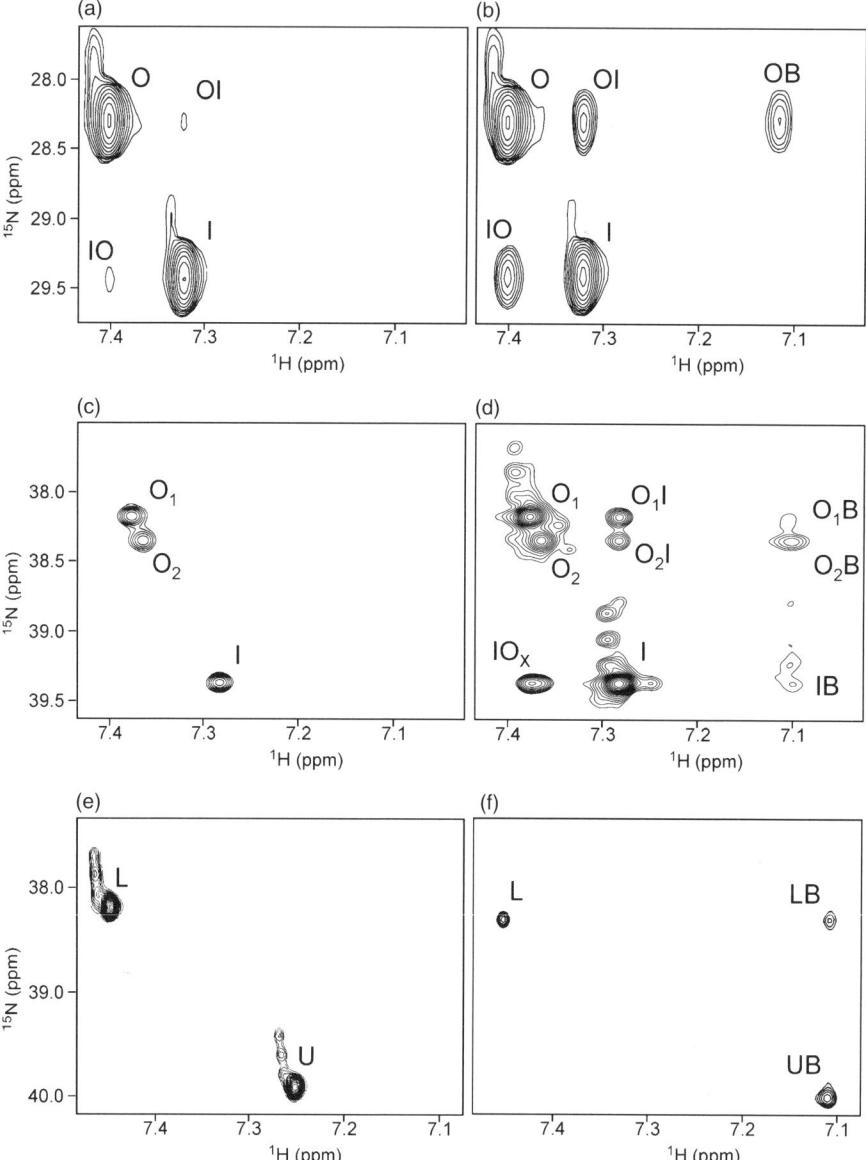

inner channel. Plavec and co-workers have shown that the bimolecular G-quadruplex [d($G_3T_4G_4$)]$_2$ folds into the same asymmetric bimolecular structure with three G-quartets in the presence of NH_4^+, K^+ and Na^+ (Figure 4.7c).[108] NMR studies have demonstrated that the three G-quartets of [d($G_3T_4G_4$)]$_2$ create two $^{15}NH_4^+$ coordination sites (designated U and L).[109] Titration of KCl into a solution of [d($G_3T_4G_4$)$_2$] folded in the presence of $^{15}NH_4^+$ ions revealed a mixed mono-K^+/mono-$^{15}NH_4^+$ form that represents an intermediate in the conversion of the di-$^{15}NH_4^+$ form into the di-K^+ form. $^{15}NH_4^+$ ions were similarly found to replace Na^+ ions inside the quadruplex. The preference for $^{15}NH_4^+$ over Na^+ ions for the two internal coordination sites is considerably smaller than the preference for K^+ over $^{15}NH_4^+$ ions. The two coordination sites within the G-quadruplex (Figure 4.7c) differ to such an extent that $^{15}NH_4^+$ ions bound to the site that is closer to the lateral-type loop (site U) are always replaced first during titration by KCl. That is, the second binding site (L) is not

---

Figure 4.8 2D ^{15}N–^{1}H NzExHSQC spectra of the three G-quadruplexes shown in Figure 4.7. For each G-quadruplex, two spectra are shown that illustrate the movement of $^{15}NH_4^+$ ions through the G-quadruplex that has occurred after a particular length of time. (a) NzExHSQC spectra of $^{15}NH_4^+$ ions in the presence of [d($G_4T_4G_4$)]$_2$. Total mixing time (τ_m) between chemical shift labeling of ^{15}N resonances and acquisition of ^{1}H free induction decay (FID) was 13 ms. (b) Same as (a), except with a τ_m of 63 ms. Spectra shown in (a) and (b) were acquired at 283 K for a sample 2.5 mM in G-quadruplex, 55 mM in $^{15}NH_4^+$ at pH 5.0. The $^{15}NH_4^+$ autocorrelation peaks are labeled inner (I) and outer (O), as shown in Figure 4.7. The cross-peak labeled IO corresponds to $^{15}NH_4^+$ ions that have moved from site I site to site O during τ_m, OI corresponds to $^{15}NH_4^+$ ions that have moved from O to I during τ_m and OB corresponds to $^{15}NH_4^+$ ions that have moved from O to bulk solution (B) during τ_m. (c) NzExHSQC spectra of $^{15}NH_4^+$ ions in the presence of d[$G_4(T_4G_4)_3$]. Intrinsic mixing time of experiment was 13 ms. (d) Same as (c), expect with a τ_m of 500 ms. Spectra shown in (c) and (d) were acquired at 293 K for a sample 1.8 mM in G-quadruplex, 130 mM in $^{15}NH_4^+$ at pH 6.0. Autocorrelation peaks, labeled O_1, I and O_2, correspond to the three unique inter-G-tetrad coordination sites of d[$G_4(T_4G_4)_3$], as defined in Figure 4.7b. Cross-peaks indicative of $^{15}NH_4^+$ movement during the τ_m of the pulse sequence labeled according to the initial and final position of the cation, as defined for panels (a) and (b). (e) NzExHSQC spectra of $^{15}NH_4^+$ ions in the presence of [d($G_3T_4G_4$)]$_2$. Intrinsic mixing time of experiment was 13 ms. (f) Same as (e), expect with a τ_m of 1.6 s. Spectra shown in (e) and (f) were acquired at 298 K for a sample 1.5 mM in G-quadruplex, 50 mM in $^{15}NH_4^+$ at pH 4.5. Autocorrelation peaks, labeled L and U, correspond to the two unique inter-G-tetrad coordination sites of [d($G_3T_4G_4$)]$_2$, as defined in Figure 4.7c. The cross-peaks LB and UB illustrate the movement of $^{15}NH_4^+$ from sites L and U to bulk solution, respectively. The lack of LU or UL cross-peaks indicates no movement of $^{15}NH_4^+$ between the two internal coordination sites of the G-quadruplex. (a) and (b) adapted from ref. 54; (c) and (d) adapted from ref. 96; (e) and (f) adapted from ref. 94.

occupied by a K^+ ion until a K^+ ion already resides at the first binding site. Quantitative analysis of the relative concentrations of the three di-cation forms (*i.e.* two $^{15}NH_4^+$, $^{15}NH_4^+/K^+$, two K^+) at equilibrium, which are in slow exchange on the NMR time-scale, showed that difference in standard Gibbs free energy between the di-$^{15}NH_4^+$ form and the $^{15}NH_4^+/K^+$ form is $-5.7\,kcal\,mol^{-1}$ and that between the $^{15}NH_4^+/K^+$ form and the di-K^+ form is $-4.3\,kcal\,mol^{-1}$.

The bimolecular G-quadruplex $[d(G_3T_4G_4)]_2$ also exhibits kinetics of $^{15}NH_4^+$ movement that were very different from those of the bimolecular G-quadruplex $[d(G_4T_4G_4)]_2$ and the unimolecular G-quadruplex $d[G_4(T_4G_4)_3]$.[94] First, there is no apparent movement of $^{15}NH_4^+$ within this G-quadruplex between sites U and L (no UL or LU cross-peak is observed in ^1H–^{15}N 2D spectra, Figure 4.8f). Second, the lifetime of $^{15}NH_4^+$ bound at site U (with respect to exchange with solution) is 139 ms, whereas the lifetime of $^{15}NH_4^+$ bound at site L is 1.7 s at 298 K. The 12-fold longer lifetime of $^{15}NH_4^+$ at site L, which is longer than that measured for the outer sites of the two G-quadruplexes discussed above, illustrates how $^{15}NH_4^+$ is also useful as a probe of G-quadruplex flexibility.

Finally, Plavec and co-workers have shown that analysis of 2D ^1H–^{15}N NMR spectra of $^{15}NH_4^+$ acquired during the folding of a G-quadruplex by titration of $^{15}NH_4Cl$ into the solution of an unfolded G-quadruplex-forming sequence can be used as a means to measure the affinity of individual sites within a G-quadruplex for the binding of $^{15}NH_4^+$.[95] The application of this method has shown tremendous differences between the binding affinities of different folds for $^{15}NH_4^+$. In particular, the U and L sites of $[d(G_3T_4G_4)]_2$ (Figure 4.7c) exhibit association constants of approximately 2700 and 2900 M^{-1}, respectively. In contrast, the O_1, I and O_2 site of $d[G_4(T_4G_4)_3]$ (Figure 4.7b) exhibit association constants of approximately 90, 30 and 45 M^{-1}, respectively. Thus, association constants of $^{15}NH_4^+$ for the two sites within $[d(G_3T_4G_4)]_2$ are nearly identical even though the flanking loops are very different, whereas the association constants for the three sites within $[d(G_3T_4G_4)]_2$ vary by a factor of three. It is possible that both loop topology and relative number of *syn* and *anti* guanine nucleotides modulate $^{15}NH_4^+$ binding, but it is so far not clear why the association constants for individual sites between these two G-quadruplexes can vary by almost a factor of ten.

4.6 G-quadruplex Cation-dependent Stability

As mentioned above, G-quadruplex stability depends strongly on the identity of the coordinated cation. In this section, we discuss some of the observations associated with the cation-dependent stability of G-quadruplexes. Most work to date on this topic has focused on measuring the difference between the stability of the K^+- and Na^+-forms of a G-quadruplex. Generally, G-quadruplexes exhibit greater thermostability in the presence of K^+ than Na^+, as measured by melting temperature (T_m). Both cation dehydration energy and sterically dictated differences in the allowed G-quartet coordination geometries of K^+ and Na^+ contribute to these stability trends.[46,70,110,111] The difference between the thermal stability of the K^+- and Na^+-forms of a G-quadruplex varies significantly with sequence; melting temperatures can differ by as little as 2 °C to as much as 30 °C for the two forms (Table 4.1).[52,111–114] This wide range of differential stabilities is likely associated with some G-quadruplexes assuming different global structural in the presence of the two ions. For example, G-quadruplexes formed by the human telomere repeat, such as d[A(GGGTTA)$_4$GGG], exhibit different conformations in the presence of Na^+ and K^+ (discussed below).[115] The relative stabilizing properties of monovalent cations have been examined systematically in the DNA oligonucleotide d[(TTAGGG)$_4$], comprised of four copies of the human telomere repeat. This work indicated the following order of quadruplex stabilization: $K^+ > Na^+ \approx Rb^+ > Li^+ > Cs^+$, consistent with the direct measurements of monovalent cation selectivity discussed above.[100]

G-quadruplexes stabilized by divalent cations, such as Sr^{2+}, Ba^{2+} and Pb^{2+}, also show cation-dependent stabilization.[55,58,116] For example, the melting temperatures of 8-bromoguanosine gels indicate that the relative order of

Table 4.1 Melting temperatures of G-quadruplexes.

Sequence	K^+		Na^+		Ref.
	$T_m/°C^a$	Salt concentration/mM	$T_m(Na^+)/°C^a$	Salt concentration/mM	
Monomolecular G-quadruplexes					
d[GGTTGGTGTGGTTGG]	48	100	~20	100	111
d[AGG(TTAGG)$_3$]	42	100	40	100	111
d[AGGG(TTAGGG)$_3$]	62	100	55	100	111
d[TTAGGG]$_4$	50.2	49	42.4	50	52
d[TTAGGG]$_4$	63	70	49	70	115
d[G$_4$(T$_4$G$_4$)$_3$]	84	100	67	100	114
d[TG$_3$(TTAG$_3$)$_3$]	51.1	10	37.7	10	176
d[TG$_3$(TTAG$_3$)$_3$]	81.8	100	62.8	100	176
d[G$_3$(TG$_3$)$_3$]	Too stable	1	86.2	10	176
d[G$_3$(T$_2$G$_3$)$_3$]	80.0	1	60.3	10	176
d[G$_4$(T$_2$G$_4$)$_3$]	79.1	10	51.9	10	176
d[G$_4$TTG$_4$TGTG$_4$TTG$_4$]	–		86.0	200	177
Bimolecular G-quadruplexes					
d[G$_4$T$_3$G$_4$]	–		44	70	130
d[G$_4$T$_4$G$_4$]	n.r.^{b}		53	100	111
d[G$_4$T$_4$G$_4$]	–		47.0^c/66.6	200	177
d[G$_3$TTAG$_3$]	42	70	31	70	115
Tetramolecular G-quadruplexes					
d[TGGT]	48	110	16	110	112
d[TGGGT]	>90	100	55	100	111
d[TG$_4$T]	–		74.6	200	177
d[TTAGGG]	50	110	17	110	112
d[TTAGGGT]	55	110	24	110	112
d[TTAGGG]	50	110	17	110	112

$^a T_m$ values refer to both reversible transitions and to apparent melting temperatures ($T_{1/2}$) for the non-reversible processes.
^{b} Non-reversible.
^{c} Premelting process.

stabilization of G-quartets by divalent cations is $Sr^{2+} \gg Ba^{2+} > Ca^{2+} > Mg^{2+}$.[38] This same ordering is generally observed in quadruplexes formed by DNA oligonucleotides. However, the effects of divalent cations are more complex and it is difficult to compare directly the stabilization by divalent and monovalent cations because, as discussed above, a particular G-rich sequence may not coordinate the same number of monovalent and divalent cations. Nevertheless, thermal denaturation studies of d[(TTAGGG)$_4$] indicate that Sr^{2+} is even more stabilizing than K^+.[52] Similarly, Pb^{2+} was shown to bind more tightly than K^+ to thrombin-binding aptamer, d(GGTTGGTGTGGTTGG), which adopts a similar fold in the Pb^{2+} and K^+-forms.[55] Interestingly, similar stabilization was observed for other intramolecular quadruplexes and Pb^{2+}, but not bimolecular quadruplexes. Davis and colleagues have used the quadruplex–Pb^{2+} interaction to induce the formation of quadruplex-type structures with lipophilic guanosine analogues in organic solvents.[117] Their work suggests that Pb^{2+} induces a lower energy and more compact structure than K^+, as seen in metal–oxygen bond lengths, intercarbonyl separation and interquartet separation.

Earlier work by Hardin and co-workers involved stability studies of G-quadruplexes formed by d(CGCG$_3$GCG) and d(TATG$_3$ATA) in the presence of Mg^{2+}, Ca^{2+}, Li^+ and Na^+. Their work illustrated that divalent cations stabilize the tetramolecular quadruplexes formed by these sequences more effectively than Li^+ and Na^+.[118,119] The effects are not so clear-cut, however. CD data from Sugimoto and colleagues suggest that the Na^+ form of the [d(G$_4$T$_4$G$_4$)]$_2$ quadruplex is destabilized by just 1 mM Mg^{2+}, Ca^{2+}, Mn^{2+}, Co^{2+} or Zn^{2+}.[120–123] The order of propensity for destabilization is $Zn^{2+} > Co^{2+} > Mn^{2+} > Mg^{2+} > Ca^{2+}$. Additionally, higher Ca^{2+} concentrations induce the quadruplex to undergo a transition to a parallel, tetramolecular quadruplex. This transition is complete at 20 mM Ca^{2+}. In the light of the recent X-ray structure with Ca^{2+}, it is possible that Ca^{2+} is being coordinated within the G-quadruplex.

In contrast, Ca^{2+} and Mg^{2+} fail to induce any ordered structure in [d(TTAGGG)]$_4$.[52] Nevertheless, all divalent cations can screen interphosphate Coulombic repulsion – and much more effectively than monovalent cations. This enhanced screening potential can make accessible alternate conformers when divalent cations are present.[58,124] However, only certain divalent cations appear to bind the quadruplex ion channel (e.g. Sr^{2+}, in some cases, Ca^{2+}, but never Mg^{2+}).

Investigations reported by Bolton and co-workers illustrate that site-selective binding of divalent cations to the outside of G-quadruplexes could also contribute to enhanced thermal stability.[125,126] The interaction of Mn^{2+} with the thrombin-binding aptamer, d(GGTTGGTGTGGTTGG) and the dimeric G-quadruplex [d(G$_4$T$_4$G$_4$)]$_2$ was studied by ESR and NMR spectroscopy. The ESR spectrum of this paramagnetic cation is quenched on addition of either quadruplex, indicating binding of the free cation to the quadruplex. Additionally, Mn^{2+} induces relaxation of the resonances for protons with which it makes close contacts, causing line broadening in the ^{1}H NMR spectrum. Based on the r^{-6} distance dependence for induced line resonance broadening, the authors treated the NMR-derived line broadening data as one would NOE constraints, mapping the Mn^{2+} onto the quadruplexes. The authors demonstrated that electrostatic potential maps accurately predicted the Mn^{2+} localization data.

Interestingly, Bolton and co-workers also showed that the free Mn^{2+} EPR signal is partially recovered on addition of thrombin to the TBA–Mn^{2+} complex, but not for the other quadruplex. This observation demonstrated both the selectivity of the TBA–thrombin interaction and that one or more of the divalent-binding sites in TBA made contact in the TBA–thrombin complex.[125,126] For additional discussion of the use of EPR to map divalent cation binding sites on the surface of a G-quadruplex, see Chapter 5.

4.7 Cations and G-quadruplex Polymorphism

Although G-quartets were first proposed in 1962,[35] it was more than 25 years before a specific biological function was proposed for the G-quadruplex. In the late 1980s, two laboratories independently proposed that G-rich sequences in the chromosomes of eukaryotic cells, including the single-stranded 3'-overhang of telomeres, could be involved in the physical pairing of genes through the formation of G-quadruplexes.[48,49,127] The observation that some G-rich DNA sequences adopted a parallel G-quadruplex structure in the presence of Na^+ and an antiparallel G-quadruplex structure in the presence of K^+ promoted further speculation that living organisms might use cation-driven G-quadruplex structural transitions as a means to regulate the formation of higher order DNA structures.[48] Historically, the G-rich DNA sequences associated with telomeres have attracted the most attention with regard to a possible function of G-quadruplexes *in vivo*. As a brief overview of cation-specific G-quadruplex structures, we present selected observations relating to the cation-dependent polymorphisms of G-quadruplexes that are primarily based on telomere-derived sequences. It should be noted, however, that many G-rich oligonucleotides show cation-dependent structures, but for reasons of space we have focused primarily on telomere-derived sequences.

Based on gel electrophoresis mobility studies, in 1990 Sen and Gilbert reported that DNA sequences associated with telomeres exhibit different structures in the presence of Na^+ and K^+.[48] In a subsequent study by Hardin *et al.*, a combination of gel electrophoresis, NMR and circular dichroism (CD) experiments confirmed that the *Tetrahymena* telomeric repeat $[d(T_2G_4)]_4$ existed as an intramolecular quadruplex in the presence of Na^+ and as a tetramolecular quadruplex in the presence of K^+.[128] A similar observation was also reported for the related sequences $d(G_3T_4G_3)$ and $d(G_4T_4G_4)$, which form bimolecular, diagonally looped structures in Na^+ and both bi- and tetramolecular structures in K^+.[129,130]

In a systematic study of cation-dependent G-quadruplex morphology, Thomas and co-workers created a phase diagram for the structures assumed by the *Oxytricha* repeat $d(T_4G_4)_4$ as a function of NaCl and KCl concentrations. At low salt concentrations, both cations promote the formation of an antiparallel, diagonally looped structure. At higher cation concentrations, this oligonucleotide assumes an extended, parallel, tetramolecular structure.

Although both cations can induce a transition from the antiparallel to the parallel structure, K^+ is more effective than Na^+. The midpoint of the intramolecular–tetramolecular transition was observed at 65 mM for KCl and 225 mM for NaCl.[131] The related sequence $d(G_3T_4G_3)$ exhibits a similar cation-dependent polymorphism. As mentioned earlier, this oligonucleotide adopts a bimolecular fold that is essentially the same in K^+ and Na^+.[78,79,81] However, several days after the addition of KCl to an Na^+-form sample of $[d(G_3T_4G_3)]_2$, broad resonances appear in the ^1H NMR spectra in addition to changes in the CD spectrum.[129] These changes indicate the formation of a larger structure that is presumably a linear, tetramolecular quadruplex. Without interfering loop residues, the parallel G-quadruplex formed by $d(G_3T_4G_3)$ can stack efficiently at the relatively high DNA concentrations normally required for NMR studies, giving rise to a long tumbling time and broad resonance peaks. No linear species is observed in the Na^+-only solution.[129]

The structure of the G-quadruplex formed by the oligonucleotide sequence $d(G_4T_4G_3)$, which has one less guanine residue than $d(G_4T_4G_4)$ and one more guanine residue than $d(G_3T_4G_3)$, exhibits a more pronounced sensitivity to cation species than either $d(G_4T_4G_4)$ or $d(G_3T_4G_3)$. In the presence of Na^+, $d(G_4T_4G_3)$ forms a bimolecular G-quadruplex with three stacked G-quartets, similar to $d(G_3T_4G_3)$, but two non-G-quartet guanine residues stacked between one terminal G-quartet and the diagonal T_4 loop.[132] In contrast, multiple structures are observed in samples containing K^+ or NH_4^+.[132,133] Thus, single base changes in a G-rich sequence can result in both changes in G-quadruplex structure and enhanced structural sensitivity to cation species. As discussed below, even changes in non-guanine residues can alter the cation-specific fold of a G-quadruplex.

DNA sequences derived from the human telomere have certainly generated the greatest interest in recent years regarding the cation-dependent morphologies of G-quadruplexes. The solution state structure of the human telomere-derived sequence $d[AGGG(TTAGGG)_3]$ was solved in 1993 by Wang and Patel.[134] This Na^+-form structure contains three stacked G-quartets with one diagonal and two lateral TTA loops that result in the four GGG segments being orientated with alternating anti-parallel and parallel strand orientations around the G-quadruplex core (Figure 4.9a). The glycosidic bond angles of the guanine residues exhibit a *syn–anti–anti–syn* pattern within each G-quartet (Figure 4.9a). An X-ray crystal structure of the K^+-form G-quadruplex of the same DNA sequence was determined almost 10 years later by Neidle and co-workers.[62] The Neidle crystal structure is surprisingly different from the solution-state structure of Wang and Patel. In the crystal structure, all four GGG segments have parallel strands, and each of the three TTA segments forms a double chain reversal loop (Figure 4.9b). All guanine residues in this structure are in the *anti* conformation. The overall shape of the Neidle crystal structure is much more planar compared to a G-quadruplex with loops that cross over the faces of the G-quartet stack (Figure 4.5).

The greater concentration of K^+ in the cell makes the K^+-form structure potentially more relevant than the Na^+-form structure. However, it is still not

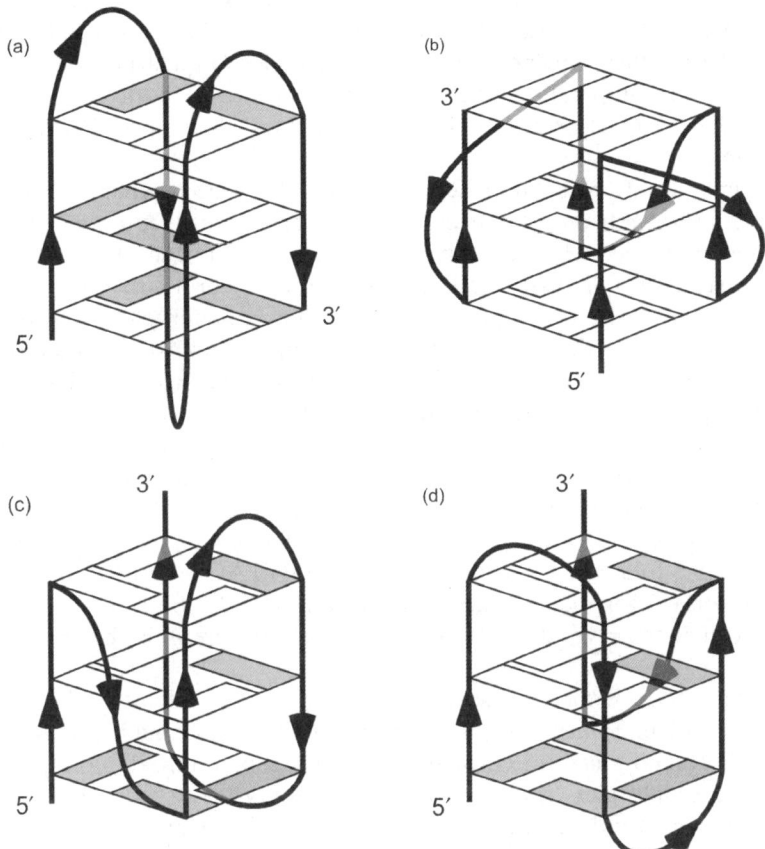

Figure 4.9 Schematic diagrams of human telomeric quadruplex folding topologies. (a) The fold determined by NMR spectroscopy of the Na^+-form G-quadruplex of $d[AG_3(T_2AG_3)_3]$. (b) The G-quadruplex fold, with parallel GGG elements and double chain reversal loops, as determined by X-ray crystallography for the same DNA sequence crystallized in the presence of K^+. (c) The Hybrid-1 solution-state structure determined for human telomere-derived sequences in the presence of K^+. (d) The Hybrid-2 solution-state structure determined for human telomere-derived sequences in the presence of K^+. Grey rectangles represent nucleosides with *syn* glycosidic torsion angles. For clarity, only G bases are represented. See text for references for corresponding NMR and crystal structures.

clear if the K^+-form crystal structure represents the structure of the 3′-overhang of the human telomere in living cells. Soon after the K^+-form crystal structure of the human telomere DNA was published, a number of studies appeared that questioned whether this structure was even representative of the K^+-form solution-state structure.[135–140] Initial attempts to determine the solution-state structure of human telomere-derived sequence $d[AGGG(TTAGGG)_3]$ in the presence of K^+ were unsuccessful due to the coexistence of multiple conformations.[141] Nevertheless, the results from other solution-based techniques indicated that this sequence forms an antiparallel G-quadruplex[136–140] or a mixed parallel/antiparallel structure, in the presence of both Na^+ and K^+, results that were clearly at odds with the parallel-stranded X-ray crystal structure.

Recently, K^+-form solution-state structures have been reported for DNA sequences derived from the human telomere. The two distinct topological folds observed are named 'Hybrid-1' and 'Hybrid-2'; both are considered hybrids of the two structures discussed above. That is, these hybrid structures exhibit a mixture of parallel and antiparallel strands that are connected by two loops that are similar to the loops of the Na^+-form solution-state structure and one that is similar to the loops of the K^+-form crystal structure (Figure 4.9c and d). The Hybrid-1 fold was determined the same year by three different groups: by the Yang laboratory for the oligonucleotide sequence d[AAAGGG(TTAGGG)$_3$AA],[142,143] by the Sugiyama laboratory for the oligonucleotide sequence d[TAGGG(TTAGGG)$_3$],[135] and by the Patel laboratory for the sequence d[TTGGG(TTAGGG)$_3$A].[144] Phan et al. soon after reported the Hybrid-2 fold for the sequence d[TAGGG(TTAGGG)$_3$TT][145] and Yang and co-workers for the sequence d[(TTAGGG)$_4$TT].[146] Both of these studies reported that the Hybrid-1 and Hybrid-2 structures are in exchange in a K^+ solution.

Shorter DNA sequences derived from the human telomere also exhibit structural polymorphism and cation-specific folds. For example, the sequence d(TAG$_3$TTAG$_3$T), comprised of two human telomere repeats, forms both a parallel G-quadruplex (similar to the K^+-form crystal structure of the four-repeat sequence) and an antiparallel G-quadruplex in the presence of K^+. Both structures are dimeric, with three quartets and interconvert freely. Each structure exhibits unique thermodynamics and folding kinetics. The antiparallel structure, with lateral loops, is favorable below 50 °C. The parallel structure, with double-chain reversals, is favored at higher temperatures.[141]

Overall, the NMR solution-state structures and the X-ray crystal structure of unimolecular G-quadruplexes formed by DNA sequences derived from the human telomere illustrate the dramatic sensitivity of G-quadruplex folds to the identity of the coordinated cation. Earlier studies also provided evidence for a complex interplay between G-quadruplex nucleotide sequence and cation-specific folds. For example, a study of oligonucleotides containing runs of guanine residues of various lengths and varying numbers of intervening residues revealed a cation-dependent pattern for the G-quadruplex folds adopted by these sequences. It was found that K^+ ions are required to stabilize 'chair'-type folds, that is, G-quadruplexes with lateral loops (e.g. structures in the middle of Figure 4.2b) rather than diagonal loops (e.g. structure at the bottom left of Figure 4.2b).[147] In general, G-quadruplexes with diagonal loops and tetramolecular parallel quadruplexes are observed in the presence of either K^+ or Na^+ ions, but intramolecular G-quadruplexes that have only lateral loops can form in the presence of K^+ and Na^+ ions, but not Na^+ alone.[147]

4.8 Non-biological Applications of G-quartet–Cation Interactions

In 1967, Pedersen described the crown ethers – compounds which, like G-quartets, form supramolecular coordination complexes.[148,149] From his Nobel lecture:

> "It seemed clear to me now that the sodium ion had fallen into the hole in the center of the molecule and was held there by the electrostatic attraction between its positive charge and the negative dipolar charge on the six oxygen atoms symmetrically arranged around it in the polyether ring. Tests showed that other alkali metal ions and ammonium ion behaved like the sodium ion so that, at long last, a neutral compound had been synthesized which formed stable complexes with alkali metal ions. Up to that point, no one had ever found a synthetic compound that formed stable complexes with sodium and potassium."[150]

In fact, such a compound did exist – guanosine. However, the coordination of metal ions by G-quartets was not yet appreciated. Reports of the propensity of guanosine derivatives to self-associate date back to Bang's 1910 description of GMP gels.[151] Five years prior to Pedersen's report of crown ethers, Davies and co-workers had described the hydrogen bonding motif found in G-quartets on the basis of fiber diffraction studies.[35] They suggested, however, that the central cavity of G-quadruplexes might contain water. Metal ion coordination by G-quartets was not appreciated until a decade later, in 1978, when Pinnavaia *et al.* reported NMR studies of GMP in the presence of various alkali metal ions.[36]

Today, G-quartet–metal interactions are as important a supramolecular tool as crown ether–metal interactions. Here, we describe some recent work on the non-biological uses of G-quartets, including work from the laboratory of Jean-Marie Lehn – a pioneer in supramolecular chemistry who shared the 1987 Nobel with Pedersen. Lehn and co-workers have shown that guanosine hydrazide yields stable gels in the presence of metal ions, as many guanosine analogues do. In the presence of aldehydes, concomitant hydrazone formation occurs on gelation. This process is described as "self-assembled reversible component amplification" that is driven by G-quartet formation.[152] The same authors have investigated such gels as biocompatible tools for the controlled release of small molecules.[153]

The self-assembly of certain modified guanine nucleosides can create ionophores for the extraction of specific ions of interest for the purpose of environmental remediation. Twelve guanosines or ten isoguanosines assemble to form a selective $^{226}Ra^{2+}$-binding channel – even in the presence of Na^+, K^+, Rb^+, Cs^+, Mg^{2+}, Ca^{2+}, Sr^{2+} and Ba^{2+} and even over pH 3–11. The guanosine-containing assembly requires picrate counterions to neutralize the assembly, whereas the isoguanosine ionophore does not. The ionophores were effective even in the presence of 10 mM of the above competing cations at extracting only 29 nM ^{226}Ra^{2+} – selectivity of at least 3.4×10^5. The isoguanosine ionophore appeared to be slightly more selective. ^{226}Ra^{2+} extraction also occurred at a Sr^{2+}:^{226}Ra^{2+} ratio of 10^6, but ceased at a 10^6 excess of Ba^{2+}. The guanosine assembly ceased to extract ^{226}Ra^{2+} in the presence of at 10^4 excess of Ba^{2+}. The isoguanosine-based device has been demonstrated to work efficiently and selectively on industrially produced ^{226}Ra-containing water.[154,155]

Investigators have demonstrated that the robust G-quartet architecture can form in organic solvents as well as in water. Gottarelli *et al.* demonstrated that a lipophilic guanosine analogue can extract potassium picrate from water, forming a stack of two quartets coordinating a potassium ion in organic solvents.[156] The same group also collaborated with the Davis group to illustrate that a lipophilic guanosine derivative could diastereoselectively form a hexadecameric supramolecular structure of over 8.5 kDa.[66,157]

Some investigators have hypothesized that G-quadruplexes' unique ability to bind monovalent cations makes them potential synthetic ion channel analogues. The Davis group recently showed that a dialkenyl-substituted guanosine analogue, when treated with Grubbs' catalyst, could undergo olefin metathesis, forming a covalently linked, lipophilic ion channel.[158] They demonstrated that the resulting macromolecule could transport sodium ions across phospholipid bilayers. Similarly, the Matile and Kato groups have demonstrated that a non-covalent complex, formed by a substituted folate analogue (which forms similar quartets to guanosine), can act as an ion channel.[159] For more information on the extensive supramolecular chemistry of G-quartets, see the review by Davis and Spada.[160]

Even the extended G-quadruplexes formed by poly-dG, termed G-wires, have attracted attention as possible structures to exploit for nanotechnology.[161,162] Guanosine's redox potential, the lowest of all nucleobases,[163–165] has led some to consider the use of these structures as literal conductive 'nanowires,' and it is appealing to modulate further the electronic properties of such wires by varying the coordinated cation species.[166] In a similar application, the duplex–'wire' transition has been driven by the addition of cations. Also in the realm of DNA-based nanotechnology, G·G mismatches incorporated within a DNA duplex have been used to drive the dimerization of two duplexes *via* quartet formation. The assembly of such 'synapsable' DNA is dramatically cation-dependent.[167,168] It has been suggested that such assemblies can drive 'nanomotors' *via* extension–contraction cycles, as duplex–quadruplex, duplex–I-motif and B–Z equilibria have been used in the past.[169–175]

4.9 Conclusion

We are now a century past what we now acknowledge, with hindsight, as the first recorded observation of G-quartets: the observation in 1910 by Bang that concentrated solutions of GMP form gels.[151] Far from showing its age, this supramolecular motif has proven to be as remarkably unpredictable as it is versatile. Although the biological functions of G-quadruplex DNA have been hard to pin down, we will not be surprised if the current speculations prove true that G-rich sequences throughout the human genome participate in gene regulation and chromosome maintenance through the formation of G-quartets. Guanine's unique, robust ability to self-direct the formation of a supramolecular helix – even in high-dielectric, hydrogen-bonding solvents – has encouraged exploration of its use in applications as diverse as drug delivery, environmental remediation and template-directed synthesis. Thus, G-quartets are likely also to find increasing applications outside biology. Regardless of the application identified for a G-quadruplex, natural or unnatural, cation coordination will play a 'central' role.

Acknowledgements

Financial support from the NATO Collaborative Programs Section (CLG grant 979520) is gratefully acknowledged. We thank Professor Gang Wu for comments on the manuscript.

References

1. S. Neidle and S. Balasubramanian, *Quadruplex Nucleic Acids* RSC Publishing, Cambridge, 2006, p. 301.
2. J. L. Huppert and S. Balasubramanian, *Nucleic Acids Res.*, 2007, **35**, 406.
3. G. N. Parkinson, R. Ghosh and S. Neidle, *Biochemistry*, 2007, **46**, 2390.
4. E. W. White, F. Tanious, M. A. Ismail, A. P. Reszka, S. Neidle, D. W. Boykin and W. D. Wilson, *Biophys. Chem.*, 2007, **126**, 140.
5. T. M. Ou, Y. J. Lu, C. Zhang, Z. S. Huang, X. D. Wang, J. H. Tan, Y. Chen, D. L. Ma, K. Y. Wong, J. C. O. Tang, A. S. C. Chan and L. Q. Gu, *J. Med. Chem.*, 2007, **50**, 1465.
6. J. Eddy and N. Maizels, *Nucleic Acids Res.*, 2006, **34**, 3887.
7. R. Kostadinov, N. Malhotra, M. Viotti, R. Shine, L. D'Antonio and P. Bagga, *Nucleic Acids Res.*, 2006, **34**, D119.
8. C. Le Sann, A. Baron, J. Mann, H. van den Berg, M. Gunaratnam and S. Neidle, *Org. Biomol. Chem.*, 2006, **4**, 1305.
9. N. Maizels, *Nat. Struct. Mol. Biol.*, 2006, **13**, 1055.
10. P. Rawal, V. B. R. Kummarasetti, J. Ravindran, N. Kumar, K. Halder, R. Sharma, M. Mukerji, S. K. Das and S. Chowdhury, *Genome Res.*, 2006, **16**, 644.
11. M. Franceschin, L. Rossetti, A. D'Ambrosio, S. Schirripa, A. Bianco, G. Ortaggi, M. Savino, C. Schultes and S. Neidle, *Bioorg. Med. Chem. Lett.*, 2006, **16**, 1707.
12. M. J. B. Moore, F. Cuenca, M. Searcey and S. Neidle, *Org. Biomol. Chem.*, 2006, **4**, 3479.
13. M. J. B. Moore, C. M. Schultes, J. Cuesta, F. Cuenca, M. Gunaratnam, F. A. Tanious, W. D. Wilson and S. Neidle, *J. Med. Chem.*, 2006, **49**, 582.

14. J. E. Reed, A. A. Arnal, S. Neidle and R. Vilar, *J. Am. Chem. Soc.*, 2006, **128**, 5992.
15. E. M. Rezler, J. Seenisamy, S. Bashyam, M. Y. Kim, E. White, W. D. Wilson and L. H. Hurley, *J. Am. Chem. Soc.*, 2005, **127**, 9439.
16. L. R. Kelland, *Eur. J. Cancer*, 2005, **41**, 971.
17. J. Seenisamy, S. Bashyam, V. Gokhale, H. Vankayalapati, D. Sun, A. Siddiqui-Jain, N. Streiner, K. Shin-ya, E. White, W. D. Wilson and L. H. Hurley, *J. Am. Chem. Soc.*, 2005, **127**, 2944.
18. K. Paeschke, T. Simonsson, J. Postberg, D. Rhodes and H. J. Lipps, *Nat. Struct. Mol. Biol.*, 2005, **12**, 847.
19. A. K. Todd, M. Johnston and S. Neidle, *Nucleic Acids Res.*, 2005, **33**, 2901.
20. J. L. Huppert and S. Balasubramanian, *Nucleic Acids Res.*, 2005, **33**, 2908.
21. A. Yafe, S. Etzioni, P. Weisman-Shomer and M. Fry, *Nucleic Acids Res.*, 2005, **33**, 2887.
22. J. Seenisamy, E. M. Rezler, T. J. Powell, D. Tye, V. Gokhale, C. S. Joshi, A. Siddiqui-Jain and L. H. Hurley, *J. Am. Chem. Soc.*, 2004, **126**, 8702.
23. C. M. Incles, C. M. Schultes, H. Kempski, H. Koehler, L. R. Kelland and S. Neidle, *Mol. Cancer Ther.*, 2004, **3**, 1201.
24. G. R. Clark, P. D. Pytel, C. J. Squire and S. Neidle, *J. Am. Chem. Soc.*, 2003, **125**, 4066.
25. S. M. Haider, G. N. Parkinson and S. Neidle, *J. Mol. Biol.*, 2003, **326**, 117.
26. R. J. Harrison, J. Cuesta, G. Chessari, M. A. Read, S. K. Basra, A. P. Reszka, J. Morrell, S. M. Gowan, C. M. Incles, F. A. Tanious, W. D. Wilson, L. R. Kelland and S. Neidle, *J. Med. Chem.*, 2003, **46**, 4463.
27. M. Y. Kim, M. Gleason-Guzman, E. Izbicka, D. Nishioka and L. H. Hurley, *Cancer Res.*, 2003, **63**, 3247.
28. J. L. Mergny, J. F. Riou, P. Mailliet, M. P. Teulade-Fichou and E. Gilson, *Nucleic Acids Res.*, 2002, **30**, 839.
29. S. Neidle and G. Parkinson, *Nat. Rev. Drug Discov.*, 2002, **1**, 383.
30. A. Siddiqui-Jain, C. L. Grand, D. J. Bearss and L. H. Hurley, *Proc. Natl. Acad. Sci. USA*, 2002, **99**, 11593.
31. M. Read, R. J. Harrison, B. Romagnoli, F. A. Tanious, S. H. Gowan, A. P. Reszka, W. D. Wilson, L. R. Kelland and S. Neidle, *Proc. Natl. Acad. Sci. USA*, 2001, **98**, 4844.
32. M. A. Read and S. Neidle, *Biochemistry*, 2000, **39**, 13422.
33. V. Caprio, B. Guyen, Y. OpokuBoahen, J. Mann, S. M. Gowan, L. M. Kelland, M. A. Read and S. Neidle, *Bioorg. Med. Chem. Lett.*, 2000, **10**, 2063.
34. H. Y. Han and L. H. Hurley, *Trends Pharmacol. Sci.*, 2000, **21**, 136.
35. M. Gellert, M. N. Lipsett and D. R. Davies, *Proc. Natl. Acad. Sci. USA*, 1962, **48**, 2013.
36. T. J. Pinnavaia, C. L. Marshall, C. M. Mettler, C. I. Fisk, H. T. Miles and E. D. Becker, *J. Am. Chem. Soc.*, 1978, **100**, 3625.
37. S. Arnott, R. Chandras and C. M. Marttila, *Biochem. J.*, 1974, **141**, 537.

38. W. Guschlbauer, J. F. Chantot and D. Thiele, *J. Biomol. Struct. Dyn.*, 1990, **8**, 491.
39. J. F. Chantot and W. Guschlbauer, *FEBS Lett.*, 1969, **4**, 173.
40. J. T. Davis, *Angew. Chem. Int. Ed.*, 2004, **43**, 668.
41. S. Neidle and G. N. Parkinson, *Curr. Opin. Struct. Biol.*, 2003, **13**, 275.
42. C. C. Hardin, A. G. Perry and K. White, *Biopolymers*, 2001, **56**, 147.
43. M. A. Keniry, *Biopolymers*, 2001, **56**, 123.
44. T. Simonsson, *Biol. Chem.*, 2001, **382**, 621.
45. D. E. Gilbert and J. Feigon, *Curr. Opinion Struct. Biol.*, 1999, **9**, 305.
46. J. R. Williamson, *Annu. Rev. Biophys. Biomol. Struct.*, 1994, **23**, 703.
47. D. Sen and W. Gilbert, *Methods Enzymol.*, 1992, **211**, 191.
48. D. Sen and W. Gilbert, *Nature*, 1990, **344**, 410.
49. J. R. Williamson, M. K. Raghuraman and T. R. Cech, *Cell*, 1989, **59**, 871.
50. V. K. Yadav, J. K. Abraham, P. Mani, R. Kulshrestha and S. Chowdhury, *Nucleic Acids Res.*, 2008, **36**, D381.
51. A. K. Todd, *Methods*, 2007, **43**, 246.
52. A. Wlodarczyk, P. Grzybowski, A. Patkowski and A. Dobek, *J. Phys. Chem. B*, 2005, **109**, 3594.
53. S. Basu, A. Szewczak, M. Cocco and S. A. Strobel, *J. Am. Chem. Soc.*, 2000, **122**, 3240.
54. N. V. Hud, P. Schultze, V. Sklenar and J. Feigon, *J. Mol. Biol.*, 1999, **285**, 233.
55. I. Smirnov and R. H. Shafer, *J. Mol. Biol.*, 2000, **296**, 1.
56. E. A. Venczel and D. Sen, *Biochemistry*, 1993, **32**, 6220.
57. F. M. Chen, *Biochemistry*, 1992, **31**, 3769.
58. J. S. Lee, *Nucleic Acids Res.*, 1990, **18**, 6057.
59. M. P. H. Lee, G. N. Parkinson, P. Hazel and S. Neidle, *J. Am. Chem. Soc.*, 2007, **129**, 10106.
60. I. C. M. Kwan, A. Wong, Y. M. She, M. E. Smith and G. Wu, *Chem. Commun.*, 2008, **682**.
61. I. C. M. Kwan, Y.-M. She and G. Wu, *Chem. Commun.*, 2007, 4286.
62. G. N. Parkinson, M. P. H. Lee and S. Neidle, *Nature*, 2002, **417**, 876.
63. S. Haider, G. N. Parkinson and S. Neidle, *J. Mol. Biol.*, 2002, **320**, 189.
64. M. P. Horvath and S. C. Schultz, *J. Mol. Biol.*, 2001, **310**, 367.
65. J. P. Deng, Y. Xiong and M. Sundaralingam, *Proc. Natl. Acad. Sci. USA*, 2001, **98**, 13665.
66. S. L. Forman, J. C. Fettinger, S. Pieraccini, G. Gottareli and J. T. Davis, *J. Am. Chem. Soc.*, 2000, **122**, 4060.
67. N. V. Hud, V. Sklenar and J. Feigon, *J. Mol. Biol.*, 1999, **286**, 651.
68. N. V. Hud, P. Schultze and J. Feigon, *J. Am. Chem. Soc.*, 1998, **120**, 6403.
69. K. Phillips, Z. Dauter, A. I. H. Murchie, D. M. J. Lilley and B. Luisi, *J. Mol. Biol.*, 1997, **273**, 171.
70. N. V. Hud, F. W. Smith, F. A. L. Anet and J. Feigon, *Biochemistry*, 1996, **35**, 15383.
71. G. Laughlan, A. I. H. Murchie, D. G. Norman, M. H. Moore, P. C. E. Moody, D. M. J. Lilley and B. Luisi, *Science*, 1994, **265**, 520.

72. P. Sket, M. Crnugelj, W. Kozminski and J. Plavec, *Org. Biomol. Chem.*, 2004, **2**, 1970.
73. J. D. Gu and J. Leszczynski, *J. Phys. Chem. A*, 2002, **106**, 529.
74. M. Webba da Silva, *Chem. Eur. J.*, 2007, **13**, 9738.
75. S. Burge, G. N. Parkinson, P. Hazel, A. K. Todd and S. Neidle, *Nucleic Acids Res.*, 2006, **34**, 5402.
76. N. Zhang, A. Gorin, A. Majumdar, A. Kettani, N. Chernichenko, E. Skripkin and D. J. Patel, *J. Mol. Biol.*, 2001, **311**, 1063.
77. D. Sen and W. Gilbert, *Curr. Opin. Struct. Biol.*, 1991, **1**, 435.
78. P. Schultze, F. W. Smith and J. Feigon, *Structure*, 1994, **2**, 221.
79. F. W. Smith and J. Feigon, *Nature*, 1992, **356**, 164.
80. M. Frankkamenetskii, *Nature*, 1992, **356**, 105.
81. P. Schultze, N. V. Hud, F. W. Smith and J. Feigon, *Nucleic Acids Res.*, 1999, **27**, 3018.
82. A. Caceres, G. Wright, C. Gouyette, G. Parkinson and J. A. Subirana, *Nucleic Acids Res.*, 2004, **32**, 1097.
83. M. L. Gill, S. A. Strobel and J. P. Loria, *Nucleic Acids Res.*, 2006, **34**, 4506.
84. A. Delville, C. Detellier and P. Laszlo, *J. Magn. Reson.*, 1979, **34**, 301.
85. M. Borzo, C. Detellier, P. Laszlo and A. Paris, *J. Am. Chem. Soc.*, 1980, **102**, 1124.
86. C. Detellier and P. Laszlo, *J. Am. Chem. Soc.*, 1980, **102**, 1135.
87. H. Deng and W. H. Braunlin, *J. Mol. Biol.*, 1996, **255**, 476.
88. Q. W. Xu, H. Deng and W. H. Braunlin, *Biochemistry*, 1993, **32**, 13130.
89. R. Ida, I. C. M. Kwan and G. Wu, *Chem. Commun.*, 2007, 795.
90. A. Wong, R. Ida and G. Wu, *Biochem. Biophys. Res. Commun.*, 2005, **337**, 363.
91. G. Wu and A. Wong, *Biochem. Biophys. Res. Commun.*, 2004, **323**, 1139.
92. R. Ida and G. Wu, *J. Am. Chem. Soc.*, 2008, **130**, 3590.
93. I. Mukerji, L. Sokolov and M. R. Mihailescu, *Biopolymers*, 1998, **46**, 475.
94. P. Sket and J. Plavec, *J. Am. Chem. Soc.*, 2007, **129**, 8794.
95. P. Podbevšek, P. Sket and J. Plavec, *Nucleos. Nucleot. Nucl. Acids*, 2007, **26**, 1547.
96. P. Podbevšek, N. V. Hud and J. Plavec, *Nucleic Acids Res.*, 2007, **35**, 2554.
97. J. Feigon, S. E. Butcher, L. D. Finger and N. V. Hud, *Methods Enzymol.*, 2001, **338**, 400.
98. M. L. Gill, S. A. Strobel and J. P. Loria, *J. Am. Chem. Soc.*, 2005, **127**, 16723.
99. G. Wu, A. Wong, Z. H. Gan and J. T. Davis, *J. Am. Chem. Soc.*, 2003, **125**, 7182.
100. A. Wong and G. Wu, *J. Am. Chem. Soc.*, 2003, **125**, 13895.
101. A. Wong, J. C. Fettinger, S. L. Forman, J. T. Davis and G. Wu, *J. Am. Chem. Soc.*, 2002, **124**, 742.
102. I. C. M. Kwan, X. Mo and G. Wu, *J. Am. Chem. Soc.*, 2007, **129**, 2398.
103. V. M. Stein, C. F. Anderson and M. T. Record, *Biophys. J.*, 1993, **64**, A267.

104. W. H. Braunlin, C. F. Anderson and M. T. Record, *Biopolymers*, 1986, **25**, 205.
105. F. W. Smith and J. Feigon, *Biochemistry*, 1993, **32**, 8682.
106. F. W. Smith, P. Schultze and J. Feigon, *Structure*, 1995, **3**, 997.
107. Y. Wang and D. J. Patel, *J. Mol. Biol.*, 1995, **251**, 76.
108. P. Sket, M. Cmugelj and J. Plavec, *Bioorg. Med. Chem.*, 2004, **12**, 5735.
109. P. Sket, M. Crnugelj and J. Plavec, *Nucleic Acids Res.*, 2005, **33**, 3691.
110. W. S. Ross and C. C. Hardin, *J. Am. Chem. Soc.*, 1994, **116**, 6070.
111. B. Sacca, L. Lacroix and J. L. Mergny, *Nucleic Acids Res.*, 2005, **33**, 1182.
112. J. L. Mergny, A. De Cian, A. Ghelab, B. Sacca and L. Lacroix, *Nucleic Acids Res.*, 2005, **33**, 81.
113. A. Risitano and K. R. Fox, *Nucleic Acids Res.*, 2004, **32**, 2598.
114. J. L. Mergny, A. T. Phan and L. Lacroix, *FEBS Lett.*, 1998, **435**, 74.
115. P. Balagurumoorthy and S. K. Brahmachari, *J. Biol. Chem.*, 1994, **269**, 21858.
116. I. V. Smirnov, F. W. Kotch, I. J. Pickering, J. T. Davis and R. H. Shafer, *Biochemistry*, 2002, **41**, 12133.
117. F. W. Kotch, J. C. Fettinger and J. T. Davis, *Org. Lett.*, 2000, **2**, 3277.
118. C. C. Hardin, M. Corregan, B. A. Brown and L. N. Frederick, *Biochemistry*, 1993, **32**, 5870.
119. C. C. Hardin, T. Watson, M. Corregan and C. Bailey, *Biochemistry*, 1992, **31**, 833.
120. W. Li, D. Miyoshi, S. Nakano and N. Sugimoto, *Biochemistry*, 2003, **42**, 11736.
121. D. Miyoshi, S. Matsumura, W. Li and N. Sugimoto, *Nucleos. Nucleot. Nucl. Acids*, 2003, **22**, 203.
122. D. Miyoshi, A. Nakao and N. Sugimoto, *Nucleic Acids Res.*, 2003, **31**, 1156.
123. D. Miyoshi, A. Nakao, T. Toda and N. Sugimoto, *FEBS Lett.*, 2001, **496**, 128.
124. S. W. Blume, V. Guarcello, W. Zacharias and D. M. Miller, *Nucleic Acids Res.*, 1997, **25**, 617.
125. V. M. Marathias, K. Y. Wang, S. Kumar, T. Q. Pham, S. Swaminathan and P. H. Bolton, *J. Mol. Biol.*, 1996, **260**, 378.
126. K. Y. Wang, L. Gerena, S. Swaminathan and P. H. Bolton, *Nucleic Acids Res.*, 1995, **23**, 844.
127. D. Sen and W. Gilbert, *Nature*, 1988, **334**, 364.
128. C. C. Hardin, E. Henderson, T. Watson and J. K. Prosser, *Biochemistry*, 1991, **30**, 4460.
129. G. D. Strahan, M. A. Keniry and R. H. Shafer, *Biophys. J.*, 1998, **75**, 968.
130. P. Balagurumoorthy, S. K. Brahmachari, D. Mohanty, M. Bansal and V. Sasisekharan, *Nucleic Acids Res.*, 1992, **20**, 4061.
131. T. Miura, J. M. Benevides and G. J. Thomas Jr., *J. Mol. Biol.*, 1995, **248**, 233.
132. M. Crnugelj, N. V. Hud and J. Plavec, *J. Mol. Biol.*, 2002, **320**, 911.
133. M. Crnugelj, P. Sket and J. Plavec, *J. Am. Chem. Soc.*, 2003, **125**, 7866.
134. Y. Wang and D. J. Patel, *Structure*, 1993, **1**, 263.
135. Y. Xu, Y. Noguchi and H. Sugiyama, *Bioorg. Med. Chem.*, 2006, **14**, 5584.

136. J. Y. Qi and R. H. Shafer, *Nucleic Acids Res.*, 2005, **33**, 3185.
137. J. Li, J. J. Correia, L. Wang, J. O. Trent and J. B. Chaires, *Nucleic Acids Res.*, 2005, **33**, 4649.
138. Y. J. He, R. D. Neumann and I. G. Panyutin, *Nucleic Acids Res.*, 2004, **32**, 5359.
139. S. Redon, S. Bombard, M. A. Elizondo-Riojas and J. C. Chottard, *Nucleic Acids Res.*, 2003, **31**, 1605.
140. L. M. Ying, J. J. Green, H. T. Li, D. Klenerman and S. Balasubramanian, *Proc. Natl. Acad. Sci. USA*, 2003, **100**, 14629.
141. A. T. Phan and D. J. Patel, *J. Am. Chem. Soc.*, 2003, **125**, 15021.
142. J. X. Dai, C. Punchihewa, A. Ambrus, D. Chen, R. A. Jones and D. Z. Yang, *Nucleic Acids Res.*, 2007, **35**, 2440.
143. A. Ambrus, D. Chen, J. X. Dai, T. Bialis, R. A. Jones and D. Z. Yang, *Nucleic Acids Res.*, 2006, **34**, 2723.
144. K. N. Luu, A. T. Phan, V. Kuryavyi, L. Lacroix and D. J. Patel, *J. Am. Chem. Soc.*, 2006, **128**, 9963.
145. A. T. Phan, K. N. Luu and D. J. Patel, *Nucleic Acids Res.*, 2006, **34**, 5715.
146. J. X. Dai, M. Carver, C. Punchihewa, R. A. Jones and D. Z. Yang, *Nucleic Acids Res.*, 2007, **35**, 4927.
147. V. M. Marathias and P. H. Bolton, *Biochemistry*, 1999, **38**, 4355.
148. C. J. Pedersen, *J. Am. Chem. Soc.*, 1967, **89**, 2495.
149. C. J. Pedersen, *J. Am. Chem. Soc.*, 1967, **89**, 7017.
150. C. J. Pedersen, in *Nobel Lectures, Chemistry 1981–1990*, B. G. Malmström ed., World Scientific, Singapore, 1992, p. 495.
151. I. Bang, *Biochem. Z.*, 1910, **26**, 293.
152. N. Sreenivasachary and J. M. Lehn, *Proc. Natl. Acad. Sci. USA*, 2005, **102**, 5938.
153. N. Sreenivasachary and J. M. Lehn, *Chem. Asian J.*, 2008, **3**, 134.
154. F. W. B. Van Leeuwen, C. J. H. Miermans, H. Beijleveld, T. Tomasberger, J. T. Davis, W. Verboom and D. N. Reinhoudt, *Environ. Sci. Technol.*, 2005, **39**, 5455.
155. F. W. B. van Leeuwen, W. Verboom, X. D. Shi, J. T. Davis and D. N. Reinhoudt, *J. Am. Chem. Soc.*, 2004, **126**, 16575.
156. G. Gottarelli, S. Masiero and G. P. Spada, *J. Chem. Soc. Chemical Commun.*, 1995, 2555.
157. M. S. Kaucher, Y. F. Lam, S. Pieraccini, G. Gottarelli and J. T. Davis, *Chem. Eur. J.*, 2005, **11**, 164.
158. M. S. Kaucher, W. A. Harrell and J. T. Davis, *J. Am. Chem. Soc.*, 2006, **128**, 38.
159. N. Sakai, Y. Kamikawa, M. Nishii, T. Matsuoka, T. Kato and S. Matile, *J. Am. Chem. Soc.*, 2006, **128**, 2218.
160. J. T. Davis and G. P. Spada, *Chem. Soc. Rev.*, 2007, **36**, 296.
161. E. Protozanova and R. B. Macgregor, *Biochemistry*, 1996, **35**, 16638.
162. T. C. Marsh and E. Henderson, *Biochemistry*, 1994, **33**, 10718.
163. R. N. Barnett, C. L. Cleveland, A. Joy, U. Landman and G. B. Schuster, *Science*, 2001, **294**, 567.

164. S. Steenken and S. V. Jovanovic, *J. Am. Chem. Soc.*, 1997, **119**, 617.
165. N. S. Hush and A. S. Cheung, *Chem. Phys. Lett.*, 1975, **34**, 11.
166. A. Calzolari, R. Di Felice, E. Molinari and A. Garbesi, *J. Phys. Chem. B*, 2004, **108**, 2509.
167. R. P. Fahlman and D. Sen, *J. Mol. Biol.*, 1998, **280**, 237.
168. E. A. Venczel and D. Sen, *J. Mol. Biol.*, 1996, **257**, 219.
169. P. Alberti and J. L. Mergny, *Proc. Natl. Acad. Sci. USA*, 2003, **100**, 1569.
170. P. Alberti and J. L. Mergny, *Cell. Mol. Biol.*, 2004, **50**, 241.
171. D. S. Liu and S. Balasubramanian, *Angew. Chem. Int. Ed.*, 2003, **42**, 5734.
172. C. D. Mao, W. Q. Sun, Z. Y. Shen and N. C. Seeman, *Nature*, 1999, **397**, 144.
173. Y. F. Wang, X. M. Li, X. Q. Liu and T. H. Li, *Chem. Commun.*, 2007, 4369.
174. B. Yurke, A. J. Turberfield, A. P. Mills, F. C. Simmel and J. L. Neumann, *Nature*, 2000, **406**, 605.
175. J. W. J. Li and W. H. Tan, *Nano Lett.*, 2002, **2**, 315.
176. A. Risitano and K. R. Fox, *Biochemistry*, 2003, **42**, 6507.
177. L. Petraccone, E. Erra, V. Esposito, A. Randazzo, L. Mayol, L. Nasti, G. Barone and C. Giancola, *Biochemistry*, 2004, **43**, 4877.

CHAPTER 5
Characterization of Nucleic Acid–Metal Ion Binding by Spectroscopic Techniques

VICTORIA J. DeROSE

Department of Chemistry, University of Oregon, Eugene, OR 97403-1253, USA

5.1 Introduction

The importance of metal ions in the structure and function of nucleic acids is discussed throughout this volume. In addition to obligate counterion condensation about the nucleic acid polyelectrolyte, there is ample evidence that biologically relevant structures are influenced by cations that associate with more specific properties.[1-6] Determining the nature of these metal ion–nucleic acid interactions, including structure, thermodynamics and dynamics of association, has been an active area of research for decades. This chapter reviews spectroscopic methods that are available for exploring the interactions of metals (inorganic cations) with nucleic acids. Structural investigations by X-ray crystallography, thermodynamics of metal association and methods of monitoring metal-induced nucleic acid folding are all described in other chapters. Here, the focus will be on direct spectroscopic probes that either monitor the metal ions themselves or provide local structural information about sites of metal ion association.

As described in several other chapters, metal ion 'association' with nucleic acids can be broadly categorized into three types: non-specific electrostatic association of hydrated cations in the counterion atmosphere of the biopolymer; localized association of fully hydrated cations to sites on the nucleic acid; and specific site binding that includes inner-sphere coordination between the cation and the nucleic acid (Figure 5.1).[1,2,5] Many spectroscopic techniques have been applied in efforts to analyze these categories of metal–nucleic acid interactions, which of course are all in continuous dynamic exchange. Spectroscopic methods that monitor cations will observe all ion populations present in solution,[9] whereas methods that monitor the nucleic acid will only detect interactions that cause structural changes or that can be observed due to distance-dependent phenomena connecting both metal and nucleic acid.

With at least three categories of interaction and significant dynamic interplay between them, it is challenging to carry out a meaningful experiment that directly measures the different metal ion populations associating with a nucleic acid in solution. Spectroscopic methods that are quantitative and provide high resolution and sensitivity to different environments would be ideal. Many techniques with these properties have been developed to investigate metal ions bound to proteins.[7] Unfortunately, the most relevant and abundant divalent

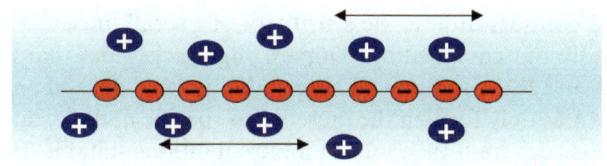

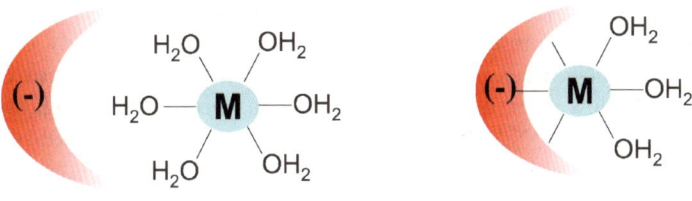

Figure 5.1 Three general classes of cation interactions with nucleic acids (after refs. 1 and 2). A mobile, hydrated layer of cations (top) is attracted to the nucleic acid polyelectrolyte. Hydrated cations may localize in particular sites or regions of a nucleic acid (bottom left) through 'outer-sphere' interactions. In the most specific type of interaction, cations may exchange aqua ligands for ligands from the nucleic acid to create 'inner-sphere' coordination. Typical inner sphere ligands include phosphodiester oxygen atoms and keto oxygen and imino nitrogen atoms from nucleobases.

cation for nucleic acids, Mg^{2+}, can be described as 'spectroscopically challenged', if not completely silent. For this reason, spectroscopic methods have relied strongly on replacing Mg^{2+} with metal ions having different properties that open up useful windows of observation.[6] Although these substitutions provide significant information, it is always important to note the caveats involved in changing one metal ion for another.

Other aspects can be noted that are unique to solution spectroscopic studies of nucleic acids in comparison with studies of metal ions in many metalloproteins. In nucleic acids, the apparent cation affinities tend to be low. As described below, apparent affinities are difficult to correlate to true thermodynamic binding constants, but phenomenologically it is the case that apparent K_d or $K_{1/2}$ values for localized cation–nucleic acid interactions are generally in the range of micromolar to millimolar, orders of magnitude weaker than those for many specific metal ion sites in proteins. Lower affinities lead to difficulties in reaching saturation of a type of site and generally are associated with relatively fast, microsecond to millisecond, exchange kinetics. Moreover, because of the polyanionic nature of nucleic acids, there are many possible cation interactions, corresponding to a possible range of apparent affinities. Poising a sample such that one type of site is dominantly occupied can require a judicious choice of conditions, including background ionic strength and nucleic acid and cation concentrations.

A related consideration is the sensitivity of overall nucleic acid structure to ionic strength and composition. Duplex oligonucleotides, hairpins and structured DNA and RNA samples all have stabilities that depend on ionic strength. Some structures only form in the presence of specific multivalent cations. These considerations can limit the range of ionic conditions suitable for spectroscopic studies. An independent measure of structural integrity is important for spectroscopic studies that narrowly focus on the cation or local changes in nucleic acid structure. Structure/function assays such as thermal denaturation, circular dichroism, footprinting, NMR features and, in the case of ribozymes, chemical activity can all be useful measurements of structural integrity.

Metal–nucleic acid interactions have been studied in a variety of different types of oligonucleotides ranging from calf thymus DNA and shorter duplex samples to highly structured ribozymes. A majority of investigations have focused on duplex DNA (see Chapter 3). In addition, the well-studied G-quartets provide highly structured molecules with fascinating and unique ion interactions (see Chapter 4). As described in the following sections, many recent investigations of cation–RNA interactions have focused on RNA subdomains that have intricate tertiary interactions. A favorite RNA model has been short hairpins containing the 5'GAAA tetraloop, a common hairpin-capping structure that mediates tertiary interactions in larger RNAs. The hammerhead ribozyme, one of the smallest of the nucleolytic ribozymes (~ 50 nucleotides), has provided a model system for developing spectroscopic studies (Figure 5.2), as has a metal ion site in the anticodon loop of tRNA. Metal-induced folding of the larger group I, group II and RNAseP ribozymes (~ 200–400 nucleotides) have all been explored by small-angle X-ray scattering but have not yet been

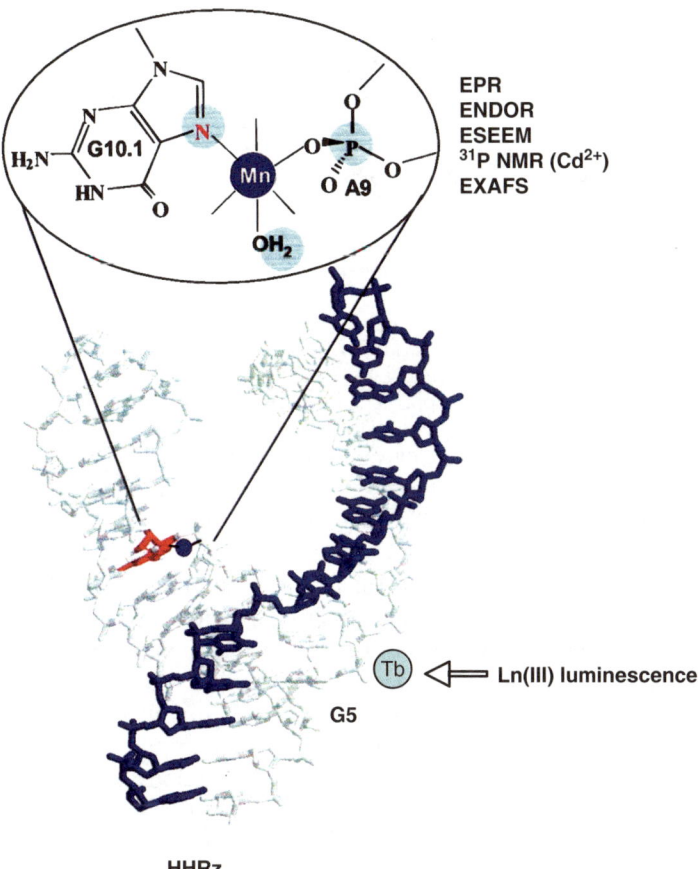

Figure 5.2 The hammerhead ribozyme has been used as a model RNA for probing metal–nucleic acid interactions by spectroscopic methods. A well-characterized divalent cation site coordinated by the phosphodiester 5′ to A9 and the N7 nitrogen from G10.1 has been explored using ^{31}P NMR with phosphorothioate substitutions and Cd^{2+},[22] Mn^{2+} EPR spectroscopy and related ESEEM and ENDOR methods,[65,67,68,71,73] and Mn^{2+} and Cd^{2+} EXAFS spectroscopy.[86,87] A different metal ion site near G5 that binds Ln(III) ions has been investigated using Tb(III) luminescence.[95] Details are described in the text.

probed intact with more specific metal-dependent spectroscopic methods. Smaller RNA subdomains from the ribosome, the spliceosome and the group II intron are, however, amenable to solution measurements.[3,6]

From the results of detailed RNA footprinting and global conformational assays, a general picture for cation-dependent RNA folding has emerged (Figure 5.3).[8,10] This overall picture predicts an initial global collapse to a condensed intermediate structure, followed by a conformational search for specific tertiary contacts. Global collapse, initiated by electrostatic screening,

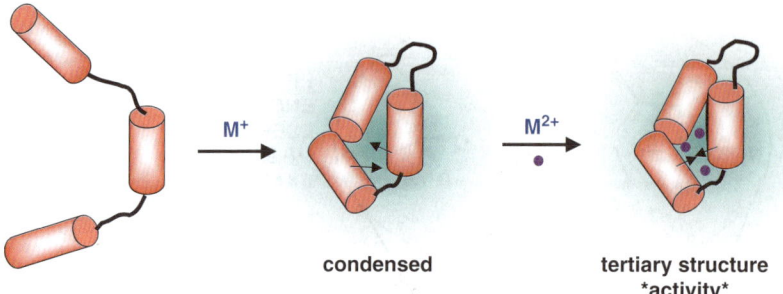

Figure 5.3 A general model for cation-supported folding of structured RNA molecules (adapted from ref. 8). Electrostatic collapse and formation of tertiary contacts can be supported by non-specific interactions from either monovalent or divalent cations. In some cases, stabilization of local 'interior' structures requires addition of specific divalent cations such as Mg^{2+}, presumably to localized sites.

may be supported by either monovalent or divalent ions and formation of tertiary contacts can also occur without specific cation interactions. However, stabilization of fine structure within the folded RNA may require binding of Mg^{2+} or other specific, usually divalent, cations.

This picture of specific cation-linked tertiary structure formation complicates the concept of an 'affinity' for a unique type of site in nucleic acids that have complex structures (see Chapter 6 for further discussion). Apparent affinities derived from metal ion titrations that are monitored by nucleic acid folding, NMR chemical shifts or ribozyme activities may be linked functions of both cation association and RNA structure changes and are also in the context of a condensed counterion atmosphere. Some apparent affinities or $K_{1/2}$ values of metal-induced structural changes can depend strongly on the background of monovalent cations, among other conditions. In order to clarify this situation completely, it is suggested that all 'K_d' values for metal ion association to oligonucleotides be reported as $K_{1/2, X \text{ M NaCl}}(M)$ or $K_{d,app\, X \text{ M NaCl}}(M)$. Thus, for example, the apparent affinity derived from an Mg^{2+} titration detected by NMR chemical shifts, in a background of 100 mM NaCl, might be reported as $K_{1/2, 0.1 \text{ M NaCl}}(Mg^{2+}) = X$. It has been pointed out that careful interpretation of these sorts of studies should also take into account the activity coefficients of the ions in solution.[11]

In the ensuing sections of this chapter, an attempt has been made to introduce the major types of spectroscopic methods that have been used (or, in a few cases, that are very close to being used) to probe metal–nucleic acid interactions. In each case, the basis for the method and interpretation are described on a relatively practical level, along with some information about sample preparation and examples from the literature. A good-faith effort has been made to mention both recent and seminal examples of the application of each method, but it is noted that many fine experiments could not be included in this particular review.

5.2 NMR Spectroscopy

NMR spectroscopy presents the powerful opportunity to detect nearly every nucleus – proton, nitrogen, carbon, phosphorus – in a nucleic acid. NMR techniques for oligonucleotide structure determination[12,13] continue to evolve, in particular with respect to longer range structure determinants. With respect to cation association with nucleic acids, this high-resolution structural tool provides the ability to observe structural and dynamic consequences of added metal ions.[14,15] These consequences of metal site occupation are generally obtained by monitoring changes in chemical shifts of different nuclei upon addition of metal ions. As described below, such NMR-observed titrations can allow the region of metal association to be located and also show longer range effects of metal interactions. Although powerful, observing the influence of metal ions is necessarily an indirect method of detecting metal ion association. Methods that directly probe the metal ion require an NMR signal from the metal itself, as can be the case with NMR-active metals such as ^{25}Mg, ^{113}Cd or ^{105}Tl or from the metal ion ligands, such as can be available from the protons of $Co(NH_3)_6^{3+}$. A hybrid method that also connects metals to the interaction sites involves addition of paramagnetic Mn^{2+}, which broadens features from nearby nuclei in a distance-dependent fashion.

5.2.1 (Indirect) Detection of Metal Interactions Using NMR Spectroscopy

Pinpointing the 'site' of metal ion association using NMR spectroscopy generally starts with assigning the proton or heteronuclear NMR signals from the oligonucleotide and is most powerful when an NMR-derived structure can be obtained. Once a structure or at least a set of resonance assignments is available, chemical shift perturbations that occur following addition of metals can be mapped to regions of the molecule; several recent examples of this approach are available for RNA[16,18–20,25] and DNA (see Chapter 3). Association of a metal ion can perturb the chemical shifts of surrounding nuclei due to electronic effects (electron withdrawing or electrostatic) *or* due to structural changes. It is important to note the latter possibility, since it can influence interpretation of these data in terms of localizing the site of metal ion association. Chemical shift effects that are opposite to those expected from direct metal ion coordination (*i.e.* downfield ^{31}P chemical shifts are expected due to deshielding upon cation association with phosphates) may indicate a metal-induced structural perturbation rather than the exact site of the metal interaction. In general, however, perturbation of a group of nuclei can help localize the region influenced by metal ion association.

An important characteristic of metal ion association with nucleic acids is that the ions are often relatively weakly associated. The 'on'–'off' rates can be fast on the NMR time-scale, in which case the 1-D spectra will reflect a weighted average of populated and unpopulated sites. Even at saturating concentrations of ions, the active association and loss of an ion from a site can result in significantly broadened lines. In and of itself, line broadening indicates an exchange process that is intermediate on the NMR time-scale, but does not indicate whether the source of line broadening is due to ion interaction at that site or conformational perturbations due to ion association elsewhere on the molecule.

One technique that can aid in localizing metal ion coordination to specific phosphodiester groups in an oligonucleotide involves the use of site-specific phosphorothioate substitutions.[21–25] Sulfur substitution results in a 50–60 ppm downfield shift of the phosphorothioate ^{31}P resonance relative to an unsubstituted phosphate group, allowing the site to be unambiguously observed in 1D ^{31}P spectra. Coordination of Cd^{2+} to a phosphorothioate results in a distinctive *upfield* chemical shift that can help separate effects of metal ion coordination from structural perturbations. Metal coordination to phosphodiester positions in structured RNA molecules has been monitored using this technique on the spliceosome U6 intramolecular stem-loop,[25] and on the hammerhead ribozyme,[22,24] with these studies confirming metal ion binding sites that had been predicted based on biochemical metal-rescue studies, including sites in a GAAA tetraloop.[23] Although Cd^{2+} is thiophilic, models indicate that single phosphorothioate substitutions do not recruit Cd^{2+} to otherwise unstructured oligonucleotides.[22,25] Importantly, the phosphorothioate substitution is not always structurally innocent. NMR solution-state structural data obtained so far indicate that while phosphorothioate substitutions at many positions cause little detectable difference,[23,26–28] at some sites this substitution can cause significant local perturbations.[23,27]

An interesting method that provides a more direct link between line broadening and the site of interaction involves the use of Mn^{2+}, whose unpaired electron spin dramatically shortens relaxation times of nuclei in a distance-dependent manner. Mn^{2+}-induced line broadening was used in early studies to monitor cation association with nucleic acid polymers.[29] Addition of very low, substoichiometric levels of Mn^{2+} can be used to locate areas populated by this ion.[17,19,29] Because of fast exchange around the polyanionic backbone, further addition of Mn^{2+} tends to broaden all features from an oligonucleotide. See Chapter 3 for a more detailed discussion of localization of cation binding sites using Mn^{2+} as a spin relaxation probe, including sample spectra.

5.2.2 Spectroscopy of NMR-active Cation Probes

Metal ions that have non-zero nuclear spin can be directly observed using NMR spectroscopy. Examples of some NMR-active metals relevant to nucleic acid interactions include ^{25}Mg, ^{113}Cd, ^{59}Co, ^{105}Tl, ^{23}Na and ^{195}Pt. Many of these nuclei have $I > 1/2$ and therefore have nuclear quadrupole interactions, which complicates direct detection due to effects on nuclear relaxation. On the other hand, relaxation effects from the nuclear quadrupole interaction can be sensitive to the symmetry of ligands around the metal ion, allowing different types of ion binding sites to be identified when the relaxation effects can be analyzed.[30,31] In some cases, NMR-active metal ion probes have broad ranges of characteristic chemical shifts that can confirm association with an oligonucleotide and, in favorable cases, also allow for identification of the metal coordination sphere based solely on the chemical shift of the bound nucleus. When observable, NOE interactions between metal and oligonucleotide nuclei can allow these metal binding sites to then be located within the nucleic acid.

The most ubiquitous divalent cation that is relevant to nucleic acids, Mg^{2+}, is NMR active as a low-sensitivity ^{25}Mg isotope. ^{25}Mg has a nuclear spin $I = 5/2$, a nuclear quadrupole moment and a chemical shift that is unfortunately relatively insensitive to its environment. The quadrupole interaction, however, has been used to differentiate populations of $^{25}Mg^{2+}$ bound to different ligands. In solution, this effect is manifested as a change in relaxation properties that can be directly analyzed in lifetime measurements and more qualitatively quantified from the resulting heterogeneous lineshape. Early solution studies measured bulk $^{25}Mg^{2+}$ binding to DNA and tRNA,[32–34] leading to estimates of fractional Mg^{2+} population on the nucleic acids and residence times of tens of milliseconds at room temperature. More recent solid-state experiments have differentiated ^{25}Mg bound to ATP phosphate ligands from unbound $Mg(OH_2)_6^{2+}$, assessed based on both quadrupole parameters derived from the lineshapes and on ^{25}Mg–^1H dipolar interactions.[35] Of note, solid-state experiments require that the sample be in a precipitated or crystallized form. If precipitated, partial rehydration of the nucleic acid is important to ensure a functional form.

The trivalent ion $Co(NH_3)_6^{3+}$ has long been utilized as a substitute for hydrated Mg^{2+} because the NH_3 (ammine) ligands are non-labile.[36] Cobalt hexammine is roughly the same size as hydrated Mg^{2+}, but may only fill nucleic acid sites that do not require inner-sphere coordination. The higher charge density of the ion, and slightly different hydrogen-bonding characteristics of the ammine ligands, may influence the exact binding characteristics of cobalt hexammine, but this ion has been extremely useful for locating metal interactions on oligonucleotides. Spectroscopically, both the ammine ligands and the Co^{3+} ion itself provide potential markers for describing $Co(NH_3)_6^{3+}$ association with oligonucleotides. With respect to the former experiment, ^1H–^1H NOESY spectroscopy can detect close interactions between the ammine protons and nearby nucleic acid nuclei (Figure 5.4). This method, used to localize $Co(NH_3)_6^{3+}$ within the GAAA tetraloop of an RNA hairpin,[37] has been applied to several other RNA structures.[17,37–39]

NMR signals from ^{59}Co, 100% natural abundance with $I = 7/2$, can be observed in the non-paramagnetic low-spin Co(III) form that is present in $Co(NH_3)_6^{3+}$. The chemical shift of $^{59}Co(III)$ is more sensitive to environment than that of ^{25}Mg and ^{59}Co has a relatively low nuclear quadrupole moment. The ^{59}Co chemical shift provides some information about bulk ion association with oligomers.[40,41] Again, higher resolution solid-state ^{59}Co NMR methods can differentiate between different binding sites, as exemplified in a recent study with tRNA.[42]

Monovalent cations in association with oligonucleotides have also been accessed using the NMR-active nuclei ^{23}Na ($I = 3/2$, 100% natural abundance)

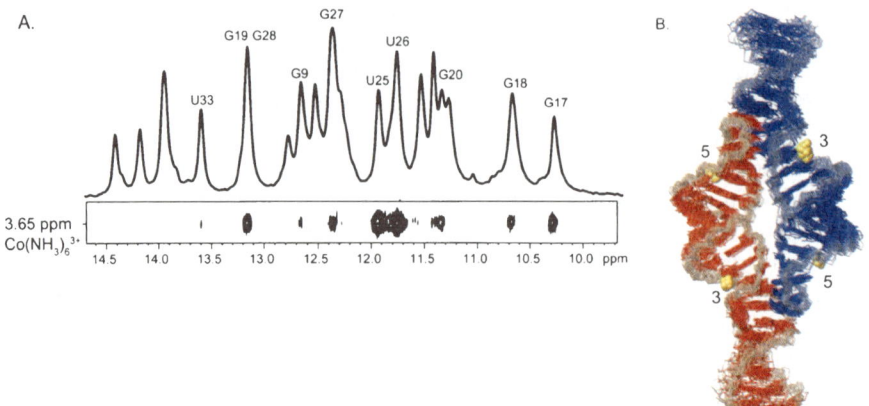

Figure 5.4 Example of metal ion association probed using intermolecular NOEs between protons of site-bound $Co(NH_3)_6^{3+}$ ions and a bimolecular GAAA tetraloop–receptor RNA model. 2D NOESY spectroscopy identifies cross-peaks between hexammine protons (lower strip) and imino protons on the RNA (upper spectrum). The $Co(NH_3)_6^{3+}$ sites identified in this way are modeled on the solution NMR-derived structure of the tetraloop–receptor complex. An accompanying experiment identified these metal sites (numbered on the RNA structure) and also three additional regions of metal interaction per monomer using Mn^{2+}-induced paramagnetic broadening. Reproduced with permission from ref. 17.

and its fairly sensitive surrogate thallium (^{105}Tl, $I = 1/2$, 70% natural abundance). In the case of ^{23}Na, as with ^{25}Mg, quadrupolar line-broadening and relaxation measurements allow the population and residence time of Na^+ ions to be established.[43–45] Although these experiments do not identify the location of Na^+, they can provide a source of information when combined with sequence-dependent studies and competitors that have known binding sites. For example, in a study of three different DNA sequences, Denisov and Halle found evidence for minor groove localization of Na^+ and Rb^+ with low, temperature-dependent occupancies and short residence times.[43,44] These authors provide a fairly specific description of the dynamics of Na^+ occupancy as having a short, ~ 50 ns, residence time, followed by ~ 1000 water binding events, each with an average residence time of ~ 1 ns.[44]

Monovalent cation association with nucleic acids has also been explored using solid-state NMR spectroscopy.[46–48] A series of studies on G-quartet models has detected distinct association of ^{39}K and ^{23}Na, and also mixtures of these and other ions within the DNA channel of G-quadruplexes (see Chapter 4). $^{15}NH_4^+$ has also proven very useful for defining the location and movement of cations within G-quadruplexes (see Chapter 4).[15]

By contrast with ^{23}Na, the chemical shift of ^{105}Tl ($I = 1/2$) is sensitive to the environment and thus able to report directly on different coordination sites.

^{105}Tl NMR was recently introduced as a spectroscopic tool for probing monovalent cation sites in nucleic acids through a series of studies on G-quartets, with emphasis on the *Oxytricha* telomeric G-quadruplex.[49–51] Initial ^{205}Tl spectra suggested three different Tl$^+$ environments, with fast exchange between two of them,[51] and subsequent experiments provided evidence for a fourth type of interaction.[49,50] ^{1}H–^{105}Tl NOE information could be obtained for two of the four ^{105}Tl features, identifying them as arising from Tl$^+$ bound inside of the G-quartet channel.[49] Structures of the Tl$^+$-bound G-quartet have now been obtained from both solution NMR and X-ray crystallography.[50] Comparison of the X-ray crystallographic data, which locate five different Tl$^+$ ions, with the ^{105}Tl NMR spectra[50] confirmed NMR assignments for the two interior ^{105}Tl cations and suggested assignments for the two other peaks as being due to Tl$^+$ bound to the quartet loop regions, with influence of medium-scale dynamic motions of the loops. The authors noted that the unusually large chemical shift dispersion of ^{105}Tl extends the dynamic range that is reflected in 1D ^{105}Tl NMR spectra.[50]

5.3 EPR Spectroscopy

Electron paramagnetic resonance (EPR) spectroscopy monitors signals from unpaired *electron* spins, found in organic radicals or paramagnetic metal ions. EPR and related techniques have been applied extensively in the field of bioinorganic chemistry to obtain detailed information about the coordination environments of paramagnetic metal ions in proteins.[7,52–54] In the case of nucleic acids, the paramagnetic ion Mn^{2+} has been used as a surrogate for Mg^{2+}, which is diamagnetic and not EPR active.[55,56] Some studies, mainly on nucleotide complexes, have also utilized the divalent, paramagnetic vanadyl ion (VO^{2+}).[57,58] Building on the EPR spectrum, a set of higher resolution experiments (ENDOR and ESEEM and the 2D HYSCORE sequence that are akin to heteronuclear NOESY/COSY NMR methods) can map out the immediate ligands of a nucleic acid-bound paramagnetic ion. For oligonucleotides, in principle all nearby (*i.e.* within 6–7 Å) NMR-active nuclei can be detected through their hyperfine interaction with the unpaired electron spin. Nuclei of interest include ^{31}P from coordinated phosphodiester groups, natural-abundance ^{14}N or site-specifically labeled ^{15}N or ^{13}C from nucleobases and ^{1}H from exchangeable or non-exchangeable sites. These higher-resolution ENDOR and ESEEM methods require the sample to be frozen, often with an added cryoprotectant such as sucrose or ethylene glycol.

Another source of paramagnetic labels in a nucleic acid can be available through site-specific conjugation to a nitroxide moiety. Protected in a ring system, the NO· radical in these 'spin labels' is a stable, $S = 1/2$ marker whose spectrum can be sensitive to molecular motion.[55] Moreover, two paramagnets will interact via a distance-dependent dipolar coupling, which can be a useful structural tool to monitor metal-induced folding of oligonucleotides labeled with two nitroxide radicals.[59] This dipolar interaction also has potential for distance measurements between one nitroxide and a site-bound paramagnetic metal ion. To date, site-directed spin labeling (SDSL) has been use to monitor metal-dependent conformational changes in several oligonucleotides.[55,59–61] Because it is currently an indirect probe of metal-induced structural changes, this topic will not be further discussed here and recent reviews are noted.[55,59]

The EPR time-scale of nanoseconds–microseconds is shorter than that of NMR and so EPR spectra are sensitive to motions on a faster time-scale. For EPR spectroscopy, the 'slow-exchange' limit, in which independent spectra are observed from each type of species in equilibrium, is reached for species exchanging on the time-scale of microseconds. These time-scales mean that metal site population or conformational changes that give rise to 'fast-exchange,' averaged NMR spectra can result in superimposed EPR spectra from each individual state that is present in solution.

5.3.1 EPR Spectroscopy of Oligonucleotide-bound Paramagnetic Ions

The somewhat dramatic EPR spectrum of Mn^{2+}, usually split into six main lines, arises from transitions between the $S=5/2$ sublevels that are further separated by a hyperfine interaction with the ^{55}Mn $I=5/2$ metal ion nucleus.[62] In experiments performed at room temperature, this EPR signal has been used to quantify ion association with oligonucleotides.[63–66] Whereas the freely rotating $Mn(OH_2)_6^{2+}$ ion gives rise to relatively sharp lines, the signal from Mn^{2+} ions associating with a slowly tumbling molecule is significantly broadened. Mn^{2+} association with tRNA,[63] G-quadruplex DNA[64] and the hammerhead ribozyme[65,66] has been quantified using this technique (Figure 5.5). In all cases, high-affinity classes of ions having $K_{d,app} \approx 1\,\mu M$ have been detected.

The vanadyl ion has the interesting solution property of forming EPR-silent oligomers at moderate pH. However, when isolated in a biomolecule or coordination complex the VO^{2+} EPR signal appears as an eight-line spectrum from the $S=1/2$, $I=7/2$ V(IV) ion. With the oxo ligand occupying one coordination position, the vanadyl ion can take on up to five additional ligands from its biomolecular site.[57,58]

5.3.2 Mn^{2+} and VO^{2+} Environments Probed by ENDOR and ESEEM Spectroscopies

Significant structural information can be obtained by evaluating the hyperfine interaction between the unpaired electron spin on a paramagnetic metal ion and NMR-active nuclei in the ion's vicinity. These hyperfine couplings can be relatively small and are normally not resolved in the 1D EPR spectral lineshape. 'Advanced' EPR techniques recover hyperfine interactions by one of two main methods. Electron spin-echo envelope modulation (ESEEM) spectroscopy monitors the modulation of an electron spin-echo response that occurs due to nuclear hyperfine and quadrupole interactions. HYSCORE is a 2D ESEEM variant that can aid in establishing the number of types of nuclei

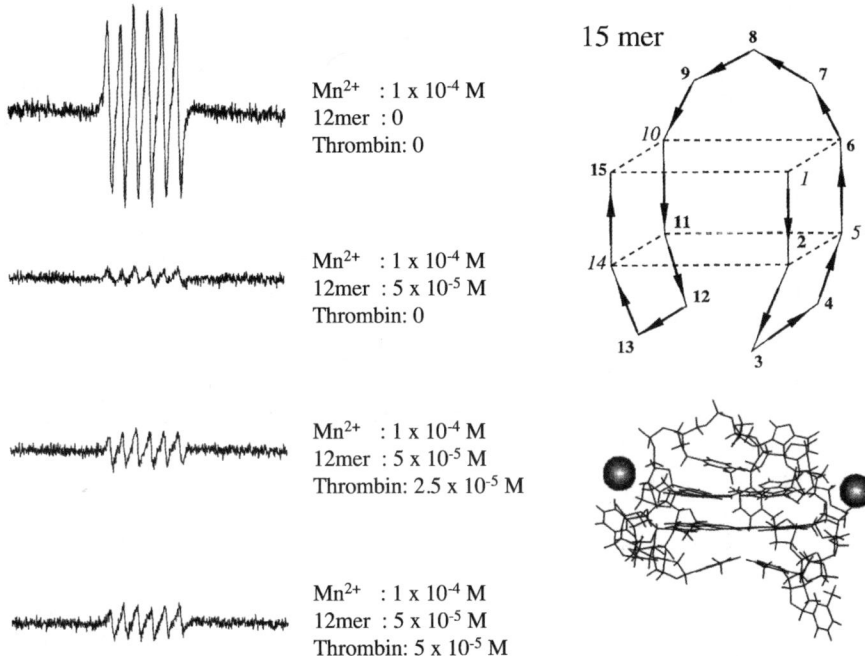

Figure 5.5 Metal ion binding to a 15-mer G-quartet observed using room temperature Mn^{2+} EPR spectroscopy. Mn^{2+} in solution has a strong EPR signal (top left). Addition of 0.5 equivalents of G-quartet DNA causes the Mn^{2+} EPR signal to disappear, consistent with binding of two Mn^{2+} ions per nucleic acid monomer. When thrombin is added in 0.5 and 1 equivalents per DNA monomer, the Mn^{2+} signal is increased as expected for release of Mn^{2+} into solution. Reprinted from ref. 64, with permission from Elsevier.

coupled to the electron spin. Electron nuclear double resonance (ENDOR) excites nuclear transitions directly using a separate radiofrequency and monitors the response of the EPR signal. These techniques rely on detecting the EPR signal of the nucleic acid-bound metal ion and the signals are additive. Thus, if multiple EPR-active metal ions are bound with overlapping EPR spectra, the ENDOR or ESEEM signal will be a superposition of signals reporting on hyperfine couplings to all ions. It is best to poise the system with a single bound metal ion, using a judicious choice of metal, nucleic acid and background ionic strength. For example, substoichiometric addition of metal:RNA, in the presence of high concentrations of monovalent ions to compete with non-specific cation sites, has been used in several studies of a metal binding site in the hammerhead ribozyme.[65,67,68]

ENDOR and ESEEM methods have been used to determine the exact manner in which Mn^{2+} and VO^{2+} coordinate to nucleotides in complexes formed in solution, addressing questions of number of coordinated phosphates, base nitrogens and aqua ligands.[57,58,67–73] The choice of technique can depend on aspects of the hyperfine coupling and, practically, on available instrumentation. Phenomenologically, for nucleic acids ESEEM has been particularly useful for detecting ^{14}N and ^{15}N from coordinated nucleobases[67,69,73] and for quantifying ^{2}H in ^{2}H$_2$O-exchanged samples.[67,70] ENDOR spectroscopy has provided clear ^{31}P spectra that allow distinction between directly coordinated phosphodiester ligands and 'outer-sphere' metal–aqua–phosphodiester interactions.[68,71,72] The former are distinguished by stronger hyperfine couplings between the metal ion electron spin and the ^{31}P nucleus. In the case of VO^{2+}, both ENDOR and ESEEM have been applied to the question of phosphate coordination.[57,58,74]

A particularly powerful application of ^{2}H ESEEM has been the ability to quantify the number of Mn^{2+}-bound water ligands through a comparison of D_2O- and H_2O-exchanged samples.[67,70] Based on results from X-ray crystallography, hydration levels of ions that are site-bound in nucleic acids appear to vary significantly for different sites.[3,5,6] ESEEM and ENDOR provide methods to confirm these predictions and further have the potential to probe for hydrogen bonding interactions involving metal-bound water ligands. In structured RNA molecules, ENDOR and ESEEM methods have been employed to describe Mn^{2+} sites in tRNA[70] and the hammerhead ribozyme.[56,67–69,71,73] In the case of a hammerhead ribozyme, incorporation of specific ^{15}N-labeled guanine allowed the identification of the dominantly-populated Mn^{2+} site (Figure 5.6).[67]

5.4 X-ray Absorption Spectroscopy and Small-angle X-ray Scattering

The use and interpretation of X-ray crystallography to explore metal–nucleic acid interactions are described in Chapters 1 and 2. Here, two additional applications of X-rays are presented. X-ray energies can be tuned for selective absorption by electrons in the s- or p-shell orbitals of different metal ions; this is the basis for anomalous scattering methods that aid in identifying metal ion occupancies by X-ray crystallography. When used as the basis for spectroscopic methods, selective metal ion excitation by X-rays in X-ray absorption spectroscopy (XAS) provides significant structural information.[75,76] XAS-based techniques are generally applied to first-row transition metals and later elements. The lower absorption energies of Mg are not amenable to solution-based X-ray spectroscopy, but the method is otherwise available for all transition metal ions.

Small-angle X-ray scattering (SAXS) is not a spectroscopic method *per se*, but rather provides a measurement of the overall size and shape of the biomolecule.[8,77] This technique allows observation of large structural changes, such as 'electrostatic collapse' of oligonucleotides upon addition of cations. Interestingly, the counterion atmosphere in differently sized cations can also be measured in careful SAXS experiments.[78,79] X-ray-induced processes occur on a very fast time-scale (femtoseconds), meaning that the time-averaged X-ray

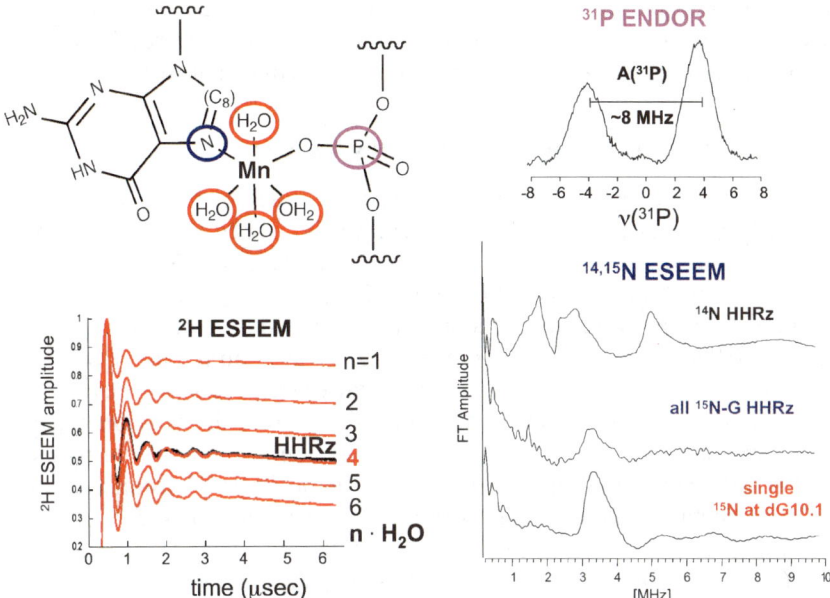

Figure 5.6 Summary of results from ESEEM and ENDOR studies of Mn^{2+} bound to the hammerhead ribozyme (HHRz). Top right: ^{31}P ENDOR spectroscopy shows two peaks centered at the ^{31}P nuclear Larmor frequency and separated by a Mn^{2+}–^{31}P hyperfine interaction of ~ 8 MHz, a value consistent with inner-sphere coordination to a phosphodiester oxygen ligand. Bottom right: ^{14,15}N ESEEM spectroscopy identifies modulations due to G10.1 nitrogen coordinated to the Mn^{2+} ion. Data shown are the Fourier transforms of time-domain traces. Modulation from ^{14}N ($I=1$) has features due to the Mn^{2+}–^{14}N hyperfine coupling and the ^{14}N nuclear quadrupole interaction. Substitution of all guanines in the HHRz enzyme strand with ^{15}N-labeled G results in an altered ESEEM spectrum that is consistent with Mn^{2+} coordinated to an $I=1/2$ ^{15}N ligand. Identical features are observed following site-specific substitution of guanine G10.1 with ^{15}dG, identifying the nitrogen ligand as G10.1. Bottom left: the number of water ligands on the bound Mn^{2+} ion can be determined using deuterium ESEEM spectroscopy. HHRz samples were exchanged into D_2O in order to label coordinated water ligands. The amplitude of the deuterium ESEEM modulation, shown as the time-domain signal, is directly proportional to the number of bound D_2O ligands. Quantitative comparison with a model of known hydration level shows that this HHRz-bound Mn^{2+} ion has four coordinated D_2O. Samples consist of $\sim 50\,\mu l$ of 200–400 μM RNA and 1:1 Mn^{2+}:RNA, 1 M NaCl, pH 7.5 and 20% (v/v) ethylene glycol or sucrose, loaded into capillary tubes and fast frozen in liquid nitrogen. Adapted from refs. 67 and 86.

spectra will be the sum of spectra from all populations present in solution. The intensity and tunability of X-rays created in synchrotron sources are generally required for XAS and SAXS studies of biomolecules. The spectroscopic techniques of XAS are usually performed on frozen solutions to minimize radical damage to the sample during signal collection. Scattering techniques such as SAXS are done with samples in solution, sometimes in a time-resolved manner that monitors oligonucleotide folding on a millisecond time-scale.

5.4.1 X-ray Absorption Spectroscopy (XAS)

The 1s electron (K-edge) absorption energies of different metal ions are well-separated, meaning that metals can be selectively probed through their X-ray absorption properties. The information available from X-ray absorption spectroscopy includes the oxidation state (X-ray near-edge structure, XANES) and immediate environment of the metal ion (extended X-ray absorption fine structure, EXAFS). With the latter technique, the structural information comes in the form of 'scattering shells,' or backscattering effects from electrons of neighboring atoms. Scattering shells appear as radial distribution functions that are modeled with the distance to, type and number of neighboring atoms. The EXAFS scattering function falls off dramatically with distance from the absorbing metal ion and has an average practical limit of 4–5 Å or less. A typical metal–nucleotide complex has first-shell metal–ligand bond lengths of 2.2–2.5 Å and distances of ~ 3 Å between the metal ion and phosphorus atom of a directly coordinated phosphate.

EXAFS has been used to examine metal ion interactions with a selection of oligonucleotides.[80–87] One of the earliest applications of EXAFS to metal–biomolecule complexes was an investigation of the Pt coordination environment in cisplatin-treated DNA.[80] It is difficult to discriminate between different low-Z elements, meaning that O- vs N-based ligands are challenging to discern. However, arguments based on (a) coordination number and (b) an observation of longer distance scattering from the semi-rigid nucleobase ring systems can aid in supporting models for different types of coordination sites. The higher electron density of phosphorus makes it relatively easy to detect, allowing discrimination between phosphodiester-bound metal ion sites and sites involving only the nucleobases.[81–84] For example, an EXAFS study of Pb^{2+} in G-quartets confirmed the expected coordination sphere for dehydrated Pb^{2+} of eight guanine oxo ligands,[81] whereas early EXAFS studies of Ca^{2+} association with DNA oligomers detected binding only to the phosphodiester backbone.[82]

XAS methods have been applied to questions regarding metal ion coordination in the hammerhead ribozyme (Figure 5.7).[85–87] An interesting study made use of the sulfur substitution in phosphorothioates, which substitutes a 'light' atom oxygen ligand with a more electron-dense sulfur. Since backscattering from sulfur should be distinguishable from oxygen, EXAFS was used to determine if Mn^{2+} is bound directly to the sulfur atom in a hammerhead ribozyme phosphorothioate-substituted site. Interestingly, the ligand shell with a higher backscattering amplitude was at a longer distance than the $< \sim 2.5$ Å expected for a Mn–sulfur bond. These results are interpreted to mean that Mn^{2+} coordinates to the oxygen of the phosphorothioate and the higher scattering sulfur is the non-coordinated atom at > 2.5 Å from the metal ion.

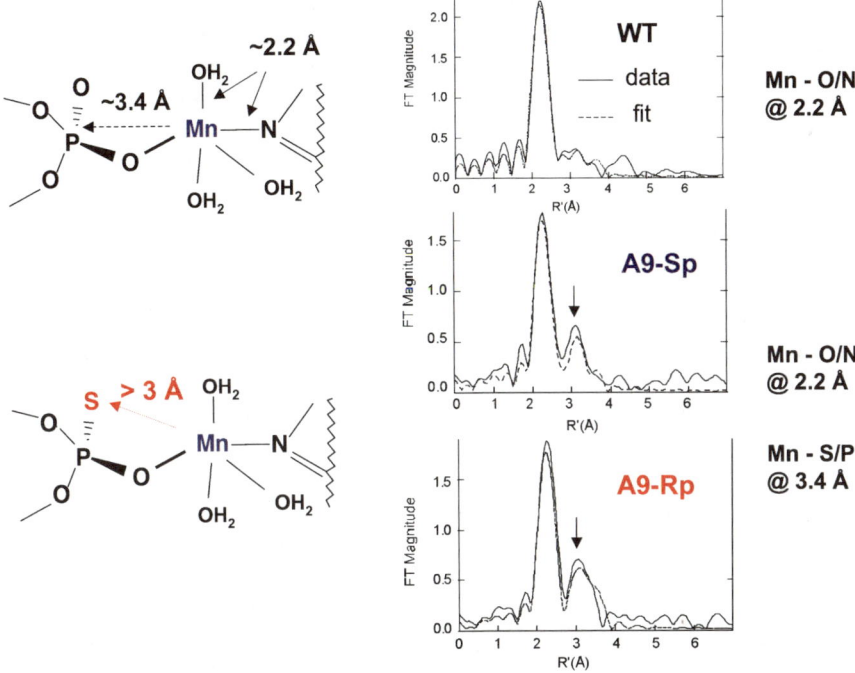

Figure 5.7 K-edge EXAFS spectroscopy of Mn^{2+} bound to the hammerhead ribozyme. The Fourier transform data display a radial distribution function with peaks corresponding to scattering from atoms at a given distance from the Mn^{2+} ion. A main peak from atoms ~ 2.2 Å from the Mn^{2+} is due to coordinated O and N atoms. The smaller peak at >3 Å arises from the phosphodiester P atom and also has contributions from atoms in the nucleobase ring. When the coordinated phosphodiester is substituted with an R_p or S_p phosphorothioate, the EXAFS peak at >3 Å increases in intensity due to the greater scattering power of S in comparison with native O. An Mn^{2+}–S distance of >3 Å demonstrates that the Mn^{2+} ion is coordinated to the O and not directly coordinated to the S atom of the phosphorothioate. Samples were prepared as described in Figure 5.6, except that they were frozen into EXAFS sample holders (~ 75 μl). Adapted from ref. 86.

5.4.2 Small-angle X-ray Scattering (SAXS) and ASAX

X-ray scattering from biomolecules results in a profile that depends on the biomolecule size, shape and distribution of electron density.[8,77] Recent applications of SAXS have used this size/shape dependency to monitor ion-dependent folding of large RNA molecules, both as a function of ionic strength and also in a time-resolved fashion.[8,78,88–90] At this time, models have been derived that are based on SAXS, in combination with hydroxyl radical footprinting, for ion-supported global RNA folding in large ribozymes such as the ~ 200–400-nucleotide RNaseP and group I and II introns. The major elements of these models include the formation of secondary structure in low concentrations of cations, followed by large-scale electrostatic collapse supported by non-specific counterion screening to a more compact structure. Evidence has been presented for a 'globule' form that precedes the formation of specific tertiary interactions. In the group I intron, formation of some critical internal structures such as those around the ribozyme active site appears to be mediated by site-bound cations. (More detailed discussions of the thermodynamics and kinetics associated with the folding of these RNA systems are presented in Chapters 6 and 7.)

The ionic atmosphere around an oligonucleotide also contributes to the X-ray scattering profile. The spatial distribution of monovalent and divalent cations including Na^+, Mg^{2+} and heavier substitutes Rb^+ and Sr^{2+} around a 25-nucleotide DNA duplex have been measured.[78,79] The data and analysis provide quantitative agreement with theoretical predictions. For example, the width of the monovalent ion condensation layer is roughly twice that for divalent ions. Of interest, evidence for a more ordered hydration shell around divalent cations appears in the X-ray scattering factors. It has also been demonstrated that separate scattering contributions from different types of counterions can be elucidated by tuning the X-ray energies above and below a particular ion's absorption threshold, in a technique termed 'anomalous SAXS' or ASAXS.[78,79] For example, the spatial distribution of Rb^+, Sr^{2+} and each in a mixture of these ions, could be compared through ASAXS methods.

5.5 Optical Methods: Lanthanide Luminescence

Optical absorption spectroscopies useful in elucidating metal ion coordination are generally limited to a subset of the transition ions, where d–d and ligand–metal charge-transfer bands can be diagnostic for coordination environments.[7] Of these, Co(II) [in the high-spin divalent form, in contrast to low-spin Co(III) of the well-used cobalt hexammine surrogate for hydrated Mg^{2+}] is perhaps the most applicable to nucleic acids. In the case of Co(II), however, the d–d transitions often used in monitoring metalloprotein active sites are mainly diagnostic for discriminating tetrahedral and octahedral metal coordination. Since there is little precedence for anything other than octahedral coordination of most metal ions to nucleic acids, this otherwise useful optical probe is less informative.

The optical properties of the f-block trivalent lanthanide ions [generally termed Ln(III)] are more environmentally responsive and have been developed as probes for coordination environments in proteins and nucleic acids.[91–100] The main ions used for these studies are Eu(III) and Tb(III). As with the d–d transitions of transition metals, absorption within the f-orbitals is formally forbidden and therefore weak. When observable, however, these features have very narrow lines that allow resolution between transitions that shift by a few tenths of a nanometer in response to changes in coordination environments. Most useful, the lanthanides have long-lived luminescence features that (a) allow the absorptive transitions to be detected *via* laser excitation profiles, (b) have lifetimes that are sensitive to quenching by ligands, most importantly water, and (c) can be sensitized by nearby organic ligands. In particular, luminescence lifetimes have been correlated with hydration, meaning that the number of displaced water molecules around the lanthanide in a biological binding site can be determined.[91–93]

Lanthanide ions that are bound to a nucleic acid and near a nucleobase can be excited via energy transfer from the nucleobase, providing an immediate method for detecting oligonucleotide-bound Ln(III) using a standard fluorimeter. This sensitized emission mainly results from interaction of Ln(III) with guanine bases, not from association with the phosphodiester backbone. Thus, detecting ion binding using only emission based on UV excitation will mainly sample ions that are bound near the nucleobases. A full excitation spectrum, which can be accomplished with laser excitation, may detect additional types of ions. Taken together, these properties make Ln(III) ions versatile optical probes for metal ion association and environment in nucleic acids.

Ln(III) ions, favoring larger hydration spheres with up to nine water ligands, have charge and coordination properties that differ significantly from those of $Mg(OH_2)_6^{2+}$.[94,95] However, the relationship between Ln(III) sites and Mg^{2+} association can be explored via competition of the optically detected Ln(III) by Mg^{2+}.

These favorable lanthanide spectroscopic properties have been used to measure the association of Eu(III) and Tb(III) with DNA and RNA oligonucleotides. In a recent example, the interaction of Tb(III) with human telomeric G-quartets provided evidence for encapsulation of two Tb(III) ions inside of the structure.[96] In a study of a Eu(III)-based cleavage agent, luminescence was used to quantitate binding to GppppG based on a red-shifted excitation band caused by phosphate coordination (Figure 5.8).[97] These data highlight the very narrow and sensitive $^5D_0-^7F_0$ transition. In yet another application, luminescence resonance energy transfer (LRET) can provide ion-partner distances.[100]

A set of studies has explored Eu(III) and Tb(III) binding to two RNA model systems: the anticodon loop of tRNA and a short RNA hairpin with a GAAA tetraloop (Figure 5.8). Apparent affinities of $K_{d,app,0.1\,M\,NaCl}$ of 1–10 μM were found for these interactions.[95,98] For both RNA interactions, luminescence lifetime measurements demonstrate dehydration of the ion in its RNA environment. The higher affinity site in the tRNA anticodon loop model $\{K_{d,app,0.1\,M\,NaCl}[Eu(III)] \approx 1.3\,\mu M\}$ is less dehydrated, with luminescence lifetimes corresponding to ~4 remaining aqua ligands. Moreover, based on excitation spectra, the Eu(III) 'site' in the GAAA tetraloop appears to be composed of two slightly different coordination environments, suggesting a heterogeneous population of two metal sites that have very similar properties, including common apparent affinities and hydration levels.

Competition studies indicated that apparent affinities of Mg^{2+} ions that replace Ln(III) in these RNAs (in 0.1 M NaCl) are weaker by 30–300-fold. Although Mg^{2+} ions localize at or near the Ln(III) sites, thermodynamic analysis indicates that in contrast to Eu(III), the Mg^{2+} remains hydrated. These

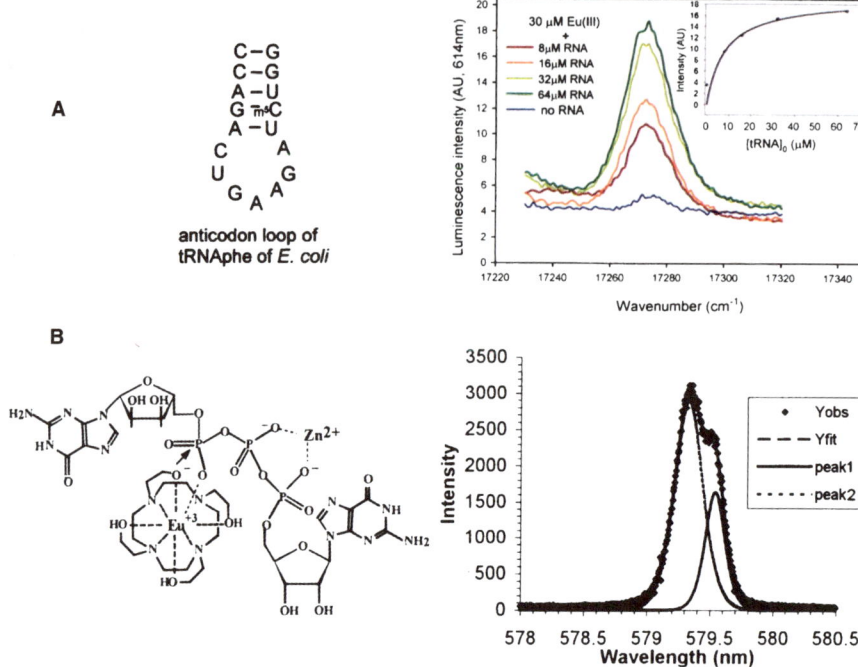

Figure 5.8 (A) Lanthanide luminescence spectroscopy demonstrating Eu(III) coordination to a tRNA anticodon loop model RNA. Luminescence from Eu(III) in solution is weak due to quenching from solvent. Eu(III) bound to RNA has a strong luminescence signal that increases in proportion to the number of bound Eu(III) ions. Fitting of these data provides an apparent affinity of ∼1.2 μM for the Eu(III)–RNA interaction in 0.1 M NaCl. Samples were 30 μM Eu(III) and 30–64 μM RNA in 0.1 M NaCl, 10 mM MES, pH 5.5. Reprinted from ref. 95, with permission of John Wiley & Sons, Inc. (B) Demonstration of the sensitivity of the Eu(III) $^7F_0$ to $^5D_0$ transition, obtained as an excitation spectrum, to the ion coordination environment. Excitation spectra are detected by monitoring λ_{em} at 614 nm for a sample with 20 μM Eu(III) complex (at left) and 20 μM GpppG. Eu(III) bound to the GpppG has a slightly red-shifted peak due to coordination to the phosphate group. The peak intensities, fitted over a range of GpppG concentrations, yield $K_d = 1.7 \times 10^{-5}$ for this interaction. Samples are at pH 6.5, 50 mM MES and 10 °C. Reprinted with permission from ref. 97. Copyright 2000 American Chemical Society.

are interesting details that demonstrate the balance of energetic contributions to apparent affinities in oligonucleotide–metal sites; the higher electrostatic contribution from the charge of the Ln(III) ions, along with possible contributions from RNA–Ln(III) bonding and entropy gain from removing counterions, apparently outweigh the penalty of dehydration in creating a higher affinity for these ions than for hydrated Mg^{2+}.

A thorough study of Tb(III) binding to the hammerhead ribozyme compared results from X-ray crystallography[99] and solution luminescence studies.[94] Although X-ray crystallography located electron density corresponding to up to three sites for bound Tb(III), the results from Eu(III) and Tb(III) luminescence were interpreted as a single binding event under solution conditions with $K_{d,app,10\,mM\,MgCl_2}[Tb(III)] \approx 4\,\mu M$. Structural and biochemical arguments support assignment to an interaction with a guanine residue, G5, which was one of the Tb(III) binding sites observed by crystallography. Based on lifetime measurements, both Eu(III) and Tb(III) ions in this site have lost only 3–4 water ligands (leaving hydration levels of 5–6). Arguments are made that the coordination environment of bound Ln(III) differs slightly from that modeled by X-ray crystallography. Of note, in this study the Ln(III) ions were not observed to be bound to what is otherwise considered a major cation site in the hammerhead ribozyme, the A9/G10.1 chelate that is observed by X-ray crystallography[99] and spectroscopic methods to be populated by Mg^{2+}, Mn^{2+} and other transition ions (Figure 5.2). This observation highlights the interesting selectivities that can be observed in the interactions of metal ions with structured oligonucleotides and the general necessity of critically considering the possible implications of metal ion substitution.

5.6 Vibrational Spectroscopy: IR, Raman and Resonance Raman

The vibrational bands of nucleobases and phosphodiester groups are sensitive to structural and electronic perturbations. Thus, IR and Raman spectroscopies have been utilized to investigate metal–nucleic acid interactions in DNA, RNA and nucleotide models.[101–112] Vibrational features are assigned to groups of atoms and can be diagnostic of sugar pucker, nucleobase protonation or metallation and phosphodiester geometry. The fast time-scale (picoseconds) of vibrational spectroscopy results in a superposition of spectra from all populations that are present in solution. Assignment of vibrational features to collections of atoms and coupled motions is a complex process that relies on both model systems and theoretical calculations, but large databases of IR and Raman features relevant to nucleic acids are available.[101–107] While sensitive to structural and electronic changes, a drawback of the application of these techniques to specific metal–nucleic acid interactions is the inability to assign features from individual sites in the absence of site-specific isotopic labeling or selective mutagenesis.

Because of low sensitivity and large background features from solvent, traditional IR spectroscopy generally requires samples to be in very high concentration or analyzed in the solid state. Resonance Raman, with UV excitation (UVRR), has significant advantages in sensitivity for nucleic acids in solution and selectivity for certain vibrational modes, simplifying the resulting spectra. In another variant with enhanced sensitivity, Raman microscopy or crystallography focuses on single crystals or condensed, oriented samples.[101,109,110] Using difference spectroscopy, changes due to ligand binding can be sensitively monitored with this technique.

A well-cited Raman spectroscopic study by Thomas and co-workers reports Raman signals from calf thymus DNA in the presence of several different divalent cations at high concentrations of both DNA and cation.[107] This study provides an overall binding avidity ranking between cation and DNA of $Pd^{2+} > Cu^{2+}, Co^{2+} > Ni^{2+}, Cd^{2+}, Mn^{2+} > Ca^{2+} > Mg^{2+}, Sr^{2+}, Ba^{2+}$. It was noted that this ranking roughly followed the tendency of the metal ion to cause denaturation of B-form DNA and also with the tendency of the metal ion to coordinate to the nucleobase rather than the phosphodiester backbone. A study by Navarro and co-workers, based on FTIR and FT-Raman spectra of solid and solution samples of metal ions with 5′-GMP and 5′-CMP, focuses on features assigned to different types of metal–phosphate interactions.[108] In the latter study, Mg^{2+} and Ca^{2+} are found to have predominant inner-sphere coordination to the 5′-phosphate, Co(II) and Cu(II) interact with the phosphate group through an intervening water and the 5′-phosphate is shielded by at least two water molecules from Na^+, Sr^{2+} and Ba^{2+}. It should be noted, however, that mixtures of species are evident in nearly all of the cases examined, indicating that the vibrational methods detect species existing in a complex equilibrium between these types of coordination modes.

Metal–RNA interactions have not been explored using vibrational spectroscopy to the same extent as DNA. At the time of writing, specific metal ion sites in structured RNAs remain to be investigated using IR-based methods. The unusual local geometries in RNA tertiary structure motifs do provide Raman markers, as recently demonstrated in studies of GNRA tetraloops, an IRES RNA and a hairpin derived from telomerase RNA.[111–113] Moreover, protonation of a single cytosine in the HDV ribozyme was recently detected via pH-dependent Raman crystallography.[110] These reports bode well for the future use of vibrational spectroscopy in discerning the details of metal ion coordination to complex RNA molecules.

5.7 Conclusion

Many spectroscopic tools are available to probe metal ions that are associated with nucleic acids. Using these techniques, it is possible to obtain very fine details of metal site occupation, specific ligation environment and structural responses of the nucleic acid to cation site occupation. Some overall trends can be noted based on the results of studies that are currently available. Results of solution NMR studies indicate that the lifetimes of cations on most sites in nucleic acids are very short. Vibrational spectra, giving a fast snapshot of all populations in solution, tend to show mixtures of ions associated with inner- and outer-sphere interactions. These observations and others are consistent with mobile, dynamic metal ion redistributions between hydrated, non-specific association and more specific interactions with nucleic acids. On the other hand, lanthanide luminescence data seem to show relatively homogeneous populations of Ln(III) ions in RNA binding sites, with partial ion dehydration.

The ability to detect NOEs between $Co(NH_3)_6^{3+}$ ligands and nucleic acids also points to a longer site occupancy for that trivalent cation. The apparent observation of a longer lived association between Ln(III) and $Co(NH_3)_6^{3+}$ ions with nucleic acids may be due to a generally higher affinity of trivalent cations, coupled to experimental conditions that might promote higher occupancy. Similarly, the very homogeneous site binding of Mn(II) ions observed in low-temperature EPR and EXAFS studies of hammerhead ribozymes indicates that these ions can be trapped in a low-energy, site-bound state with defined properties. In contrast, room temperature NMR studies of Mn(II)-induced paramagnetic broadening generally indicate high mobility of this ion, as is found for Mg^{2+}. This behavior could simply be an indication of temperature-dependent thermodynamic properties, but may also indicate that ion mobilities are coupled to other temperature-dependent motions in the biomolecule. One interesting horizon for the application of spectroscopic methods to detailed studies of metal–nucleic acid interactions is the further comparison of metal site population, nucleic acid dynamics and chemical reactivity in ribozymes.

Acknowledgements

The author is grateful to Janet Morrow, Sam Butcher, and Nancy Greenbaum for helpful comments. Support to V.J.D. from the National Institutes of Health (NIH GM58096) and the NSF Chemistry Division (CHE0111696) for work described in this chapter is gratefully acknowledged.

References

1. V. K. Misra and D. E. Draper, *Biopolymers*, 1998, **48**, 113.
2. V. K. Misra, R. Shiman and D. E. Draper, *Biopolymers*, 2003, **69**, 118.
3. A. M. Pyle, *J. Biol. Inorg. Chem.*, 2002, **7**, 679.
4. V. A. Bloomfield, D. M. Crothers and I. Tinoco Jr., *Nucleic Acids: Structures, Properties and Functions*, University Science Books, Sausalito, CA, 2000.
5. V. J. DeRose, S. Burns, N. K. Kim and M. Vogt, in *Comprehensive Coordination Chemistry II: from Biology to Nanotechnology*, J. McCleverty and T. J. Meyer (ed.), Elsevier, Oxford, 2004, Vol. **8**, 787–813.
6. E. Freisinger and R. K. O. Sigel, *Coord. Chem. Rev.*, 2007, **251**, 1834.
7. L. Que Jr., *Physical Methods in Bioinorganic Chemistry: Spectroscopy and Magnetism*, University Science Books, Sausalito, CA, 2000.
8. S. A. Woodson, *Curr. Opin. Chem. Bio.*, 2005, **9**, 104.
9. Y. Bai, M. Greenfield, K. J. Travers, V. B. Chu, J. Lipfert, S. Doniach and D. Herschlag, *J. Am. Chem. Soc.*, 2007, **129**, 14981.
10. R. Das, K. J. Travers, Y. Bai and D. Herschlag, *J. Am. Chem. Soc.*, 2005, **127**, 8272.
11. D. Grilley, A. M. Soto and D. E. Draper, *Proc. Natl. Acad. Sci. USA*, 2006, **103**, 14003.

12. J. Flinders and T. Dieckmann, *Prog. Nucl. Magn. Reson. Spectrosc.*, 2006, **48**, 137.
13. P. J. Lukavsky and J. D. Puglisi, *Methods Enzymol.*, 2005, **394**, 399.
14. R. L. Gonzalez Jr. and I. Tinoco Jr., *Methods Enzymol.*, 2001, **338**, 421.
15. J. Feigon, S. E. Butcher, L. D. Finger and N. V. Hud, *Methods Enzymol.*, 2001, **338**, 400.
16. Y. Tanaka and K. Taira, *Chem. Commun.*, 2005, 2069.
17. J. H. Davis, T. R. Foster, M. Tonelli and S. E. Butcher, *RNA*, 2007, **13**, 76.
18. H. Blad, N. J. Reiter, F. Abildgaard, J. L. Markley and S. E. Butcher, *J. Mol. Biol.*, 2005, **353**, 540.
19. D. O. Campbell, P. Bouchard, G. Desjardins and P. Legault, *Biochemistry*, 2006, **45**, 10591.
20. K. K. Griffiths and I. M. Russu, *J. Biomol. Struct. Dyn.*, 2006, **23**, 667.
21. F. Eckstein, *Annu. Rev. Biochem.*, 1985, **54**, 367.
22. M. Maderia, L. M. Hunsicker and V. J. DeRose, *Biochemistry*, 2000, **39**, 12113.
23. M. Maderia, T. E. Horton and V. J. DeRose, *Biochemistry*, 2000, **39**, 8193.
24. K. Suzumura, Y. Takagi, M. Orita and K. Taira, *J. Am. Chem. Soc.*, 2004, **126**, 15504.
25. A. Huppler, L. J. Nikstad, A. M. Allmann, D. A. Brow and S. E. Butcher, *Nat. Struct. Biol.*, 2002, **9**, 431.
26. N. J. Reiter, L. J. Nikstad, A. M. Allmann, R. J. Johnson and S. E. Butcher, *RNA*, 2003, **9**, 533–542.
27. J. S. Smith and E. P. Nikonowicz, *Biochemistry*, 2000, **39**, 5642.
28. M. Bachelin, G. Hessler, G. Kurz, J. G. Hacia, P. B. Dervan and H. Kessler, *Nat. Struct. Biol.*, 1998, **5**, 271.
29. J. Granot, J. Feigon and D. R. Kearns, *Biopolymers*, 1982, **21**, 181.
30. B. Halle and V. P. Denisov, *Methods Enzymol.*, 2001, **338**, 178.
31. W. H. Braunlin, *Adv. Biophys. Chem.*, 1995, **5**, 89.
32. D. M. Rose, M. L. Bleam, M. T. Record and R. G. Bryant, *Proc. Natl. Acad. Sci. USA*, 1980, **77**, 6289.
33. E. Berggren, L. Nordenskiold and W. H. Braunlin, *Biopolymers*, 1992, **32**, 1339.
34. S. S. Reid and J. A. Cowan, *Biochemistry*, 1990, **29**, 6025.
35. C. V. Grant, V. Frydman and L. Frydman, *J. Am. Chem. Soc.*, 2000, **122**, 11743.
36. J. A. Cowan, *J. Inorg. Biochem.*, 1993, **49**, 171.
37. J. S. Kieft and I. Tinoco Jr., *Structure*, 1997, **5**, 713.
38. S. E. Butcher, F. H. Allain and J. Feigon, *Biochemistry*, 2000, **39**, 2174.
39. S. Rudisser and I. Tinoco Jr., *J. Mol. Biol.*, 2000, **295**, 1211.
40. W. H. Braunlin, C. F. Anderson and M. T. Record Jr., *Biochemistry*, 1987, **26**, 7724.
41. Q. Xu, S. R. Jampani and W. H. Braunlin, *Biochemistry*, 1993, **32**, 11754.
42. C. V. Grant, V. Frydman, J. S. Harwood and L. Frydman, *J. Am. Chem. Soc.*, 2002, **124**, 4458.

43. F. C. Marincola, V. P. Denisov and B. Halle, *J. Am. Chem. Soc.*, 2004, **126**, 6739.
44. V. P. Denisov and B. Halle, *Proc. Natl. Acad. Sci. USA*, 2000, **97**, 629.
45. F. C. Marincola, M. Casu, G. Saba, C. Manetti and A. Lai, *Phys. Chem. Chem. Phys.*, 2000, **2**, 2425.
46. G. Wu, A. Wong, Z. Gan and J. T. Davis, *J. Am. Chem. Soc.*, 2003, **125**, 7182.
47. A. Wong and G. Wu, *J. Am. Chem. Soc.*, 2003, **125**, 13895.
48. R. Ida, I. C. Kwan and G. Wu, *Chem. Commun.*, 2007, 795–797.
49. M. L. Gill, S. A. Strobel and J. P. Loria, *J. Am. Chem. Soc.*, 2005, **127**, 16723.
50. M. L. Gill, S. A. Strobel and J. P. Loria, *Nucleic Acids Res.*, 2006, **34**, 4506.
51. S. Basu, A. Szewczak, M. Cocco and S. A. Strobel, *J. Am. Chem. Soc.*, 2000, **122**, 3240.
52. B. M. Hoffman, *Proc. Natl. Acad. Sci. USA*, 2003, **100**, 3575.
53. T. Prisner, M. Rohrer and F. MacMillan, *Annu. Rev. Phys. Chem.*, 2001, **52**, 279.
54. V. J. DeRose and B. M. Hoffman, *Methods Enzymol.*, 1995, **246**, 554.
55. P. Z. Qin and T. Dieckmann, *Curr. Opin. Struct. Biol.*, 2004, **14**, 350.
56. V. J. DeRose, *Curr. Opin. Struct. Biol.*, 2003, **13**, 317.
57. M. W. Makinen and D. Mustafi, *Met. Ions Biol. Syst.*, 1995, **31**, 89.
58. T. J. Smith, R. LoBrutto and V. L. Pecoraro, *Coord. Chem. Rev.*, 2002, **228**, 1.
59. O. Schiemann and T. F. Prisner, *Q. Rev. Biophys.*, 2007, **40**, 1.
60. T. E. Edwards and S. T. Sigurdsson, *Nature Protocols*, 2007, **2**, 1954.
61. N. K. Kim, A. Murali and V. J. DeRose, *Chem. Biol.*, 2004, **11**, 939.
62. T. A. Stitch, S. Lahiri, G. Yeagle, M. Dicus, M. Brynda, A. Gunn, C. Aznar, V. J. DeRose and R. D. Britt, *Appl. Magn. Reson.*, 2007, **31**, 321.
63. A. Danchin and M. Gueron, *Eur. J. Biochem.*, 1970, **16**, 532.
64. V. M. Marathias, K. Y. Wang, S. Kumar, T. Q. Pham, S. Swaminathan and P. H. Bolton, *J. Mol. Biol.*, 1996, **260**, 378.
65. T. E. Horton, D. R. Clardy and V. J. DeRose, *Biochemistry*, 1998, **37**, 18094.
66. L. M. Hunsicker and V. J. DeRose, *J. Inorg. Biochem.*, 2000, **80**, 271.
67. M. Vogt, S. Lahiri, C. G. Hoogstraten, R. D. Britt and V. J. DeRose, *J. Am. Chem. Soc.*, 2006, **128**, 16764.
68. S. R. Morrissey, T. E. Horton and V. J. DeRose, *J. Am. Chem. Soc.*, 2000, **122**, 3473.
69. C. G. Hoogstraten, C. V. Grant, T. E. Horton, V. J. DeRose and R. D. Britt, *J. Am. Chem. Soc.*, 2002, **124**, 834.
70. C. G. Hoogstraten and R. D. Britt, *RNA*, 2002, **8**, 252.
71. O. Schiemann, R. Carmieli and D. Goldfarb, *Appl. Magn. Reson.*, 2007, **31**, 543.
72. A. Potapov and D. Goldfarb, *Appl. Magn. Reson.*, 2006, **30**, 461.

73. N. Kisseleva, A. Khvorova, E. Westhof and O. Schiemann, *RNA*, 2005, **11**, 1.
74. S. A. Dikanov, B. D. Liboiron and C. Orvig, *J. Am. Chem. Soc.*, 2002, **124**, 2969.
75. J. E. Penner-Hahn, *Coord. Chem. Rev.*, 1999, **192**, 1101.
76. P. Glatzel and U. Bergmann, *Coord. Chem. Rev.*, 2005, **249**, 65.
77. J. Lipfert and S. Doniach, *Annu. Rev. Biophys. Biomol. Struct.*, 2007, **36**, 307.
78. R. Das, T. T. Mills, L. W. Kwok, G. S. Maskel, I. S. Millett, S. Doniach, K. D. Finkelstein, D. Herschlag and L. Pollack, *Phys. Rev. Lett.*, 2003, **90**, 188103.
79. K. Andresen, R. Das, H. Y. Park, H. Smith, L. W. Kwok, J. S. Lamb, E. J. Kirkland, D. Herschlag, K. D. Finkelstein and L. Pollack, *Phys. Rev. Lett.*, 2004, **93**, 248103.
80. B. K. Teo, J. R. Eisenberger, J. K. Barton and S. J. Lippard, *J. Am. Chem. Soc.*, 1978, **100**, 3325–3327.
81. I. V. Smirnov, F. W. Kotch, I. J. Pickering, J. T. Davis and R. H. Shafer, *Biochemistry*, 2002, **41**, 12133.
82. I. Y. Skuratovskii, S. S. Hasnain, D. G. Alexeev, G. P. Diakun and L. I. Volkova, *J. Inorg. Biochem.*, 1987, **29**, 249.
83. I. Sagi, Y. Hochman, G. Bunker, S. Carmeli and C. Carmeli, *Photosynth. Res.*, 1989, **57**, 275.
84. L. M. Hill, G. N. George, A. K. Duhme-Klair and C. G. Young, *J. Inorg. Biochem.*, 2002, **88**, 274.
85. L. Cunningham, J. Li and Y. Lu, *Nucleic Acids Symp. Ser. (Oxford)*, 1999, **41**, 70.
86. L. M. Hunsicker, Ph.D. Thesis, Texas A&M University, 2001.
87. M. Vogt, Ph.D. Thesis, Texas A&M University, 2004.
88. R. Das, L. W. Kwok, I. S. Millett, Y. Bai, T. T. Mills, J. Jacob, G. S. Maskel, S. Seifert, S. G. Mochrie, P. Thiyagarajan, S. Doniach, L. Pollack and D. Herschlag, *J. Mol. Biol.*, 2003, **332**, 311.
89. U. A. Perez-Salas, P. Rangan, S. Krueger, R. M. Briber, D. Thirumalai and S. A. Woodson, *Biochemistry*, 2004, **43**, 1746.
90. X. Fang, K. Littrell, X. J. Yang, S. J. Henderson, S. Siefert, P. Thiyagarajan, T. Pan and T. R. Sosnick, *Biochemistry*, 2000, **39**, 11107.
91. W. D. Horrocks Jr., *Methods Enzymol.*, 1993, **226**, 495.
92. G. R. Choppin and D. R. Peterman, *Coord. Chem. Rev.*, 1998, **174**, 283.
93. W. D. Horrocks Jr. and D. R. Sudnick, *Science*, 1979, **206**, 1194.
94. A. L. Feig, M. Panek, W. D. Horrocks Jr. and O. C. Uhlenbeck, *Chem. Bio.l*, 1999, **6**, 801.
95. C. Mundoma and N. L. Greenbaum, *Biopolymers*, 2003, **69**, 100.
96. E. Galezowska, A. Gluszynska and B. Juskowiak, *J. Inorg. Biochem.*, 2007, **101**, 678.
97. D. M. Epstein, L. L. Chappell, H. Khalili, R. M. Supkowski, W. D. Horrocks Jr. and J. R. Morrow, *Inorg. Chem.*, 2000, **39**, 2130.

98. N. L. Greenbaum, C. Mundoma and D. R. Peterman, *Biochemistry*, 2001, **40**, 1124.
99. A. L. Feig, W. G. Scott and O. C. Uhlenbeck, *Science*, 1998, **279**, 81.
100. F. Yuan, L. Griffin, L. Phelps, V. Buschmann, K. Weston and N. L. Greenbaum, *Nucleic Acids Res.*, 2007, **35**, 2833.
101. J. M. Benevides, S. A. Overman and G. J. Thomas, *J. Raman Spectrosc.*, 2005, **36**, 279.
102. W. L. Peticolas, *Methods Enzymo.l*, 1995, **246**, 389.
103. M. Banyay, M. Sarkar and A. Graslund, *Biophys. Chem.*, 2003, **104**, 477.
104. H. A. Tajmiriahi and T. Theophanides, *Can. J. Chem.*, 1983, **61**, 1813.
105. P. Carmona, R. Escobar, M. Molina and A. Rodriguez-Casado, *J. Raman Spectrosc.*, 1996, **27**, 817.
106. I. Mukerji and A. P. Williams, *Biochemistry*, 2002, **41**, 69.
107. J. Duguid, V. A. Bloomfield, J. Benevides and G. J. Thomas Jr., *Biophys. J.*, 1993, **65**, 1916.
108. M. de la Fuente, A. Hernanz and R. Navarro, *J. Biol. Inorg. Chem.*, 2004, **9**, 973.
109. P. R. Carey, *Annu. Rev. Phys. Chem.*, 2006, **57**, 527.
110. B. Gong, J. H. Chen, E. Chase, D. M. Chadalavada, R. Yajima, B. L. Golden, P. C. Bevilacqua and P. R. Carey, *J. Am. Chem. Soc.*, 2007, **129**, 13335.
111. V. Baumruk, C. Gouyette, T. Huynh-Dinh, J. S. Sun and M. Ghomi, *Nucleic Acids Res.*, 2001, **29**, 4089.
112. A. J. Hobro, M. Rouhi, E. W. Blanch and G. L. Conn, *Nucleic Acids Res.*, 2007, **35**, 1169.
113. V. Reipa, G. Niaura and D. H. Atha, *RNA*, 2007, **13**, 108.

CHAPTER 6
Metal Ions and the Thermodynamics of RNA Folding

DAVID P. GIEDROC AND NICHOLAS E. GROSSOEHME

Department of Chemistry, Indiana University, Bloomington, IN 47405–7102, USA

6.1 Introduction

Efforts to understand, in physical terms, how Mg^{2+} ions specifically stabilize the folded states of complex RNA molecules relative to component RNA duplexes go back over 35 years with early studies of the binding of and stabilization by, Mg^{2+} ions to transfer RNA (tRNA).[1–4] These studies appeared around the same time that the first crystallographic structures of tRNA[Phe] appeared[5,6] and were consistent with the presence of a mixture of strong and weak sites, with the strong 'sites' conjectured to be those that could be visualized in the crystal structure. Higher resolution structures of tRNA[Phe] have recently been solved and clearly reveal the presence of additional bound Mg^{2+} ions (11 in total);[7] this suggests that any attempt to correlate the number of Mg^{2+} binding sites observed in solution with the number observed in crystallographic structures[8] may be nothing more than coincidental.

That said, we are in the midst of a resurgence in our efforts to understand how Mg^{2+} ions stabilize RNA tertiary structure due, in part, to an explosion in the number of high-resolution structures of complex RNA molecules and ribonucleoprotein particles that have appeared over the last decade. This resurgence began with the first crystallographic structures of the hammerhead ribozyme in the mid-1990s,[9,10] to the structure of the P4–P6 domain of the self-splicing *Tetrahymena* group I intron in 1996[11] and a 23S ribosomal RNA fragment in 1999.[12] These structures were quickly followed by the ribosomal subunit structures,[13,14] a piece of transfer-messenger RNA (tmRNA),[15] a number of group I intron structures[16–19] (for a review, see ref. 20), riboswitches[21–23] (for a review, see ref. 24), A- and B-type ribonuclease P structures,[25,26] a number of additional ribozymes[27,28] (for a review, see ref. 29) and finally, coming nearly full circle, a new hammerhead ribozyme structure that that contains previously unrecognized tertiary structural interactions for this model RNA catalyst.[30,31] Indeed, with the near-atomic resolution structures of the 70S ribosome,[32–34] coupled with lower resolution structures of various conformational states, an unprecedented comprehensive molecular-level understanding of how this fascinating molecular machine catalyzes protein synthesis is within our grasp.[35–37] Many of these structures have been solved to sufficiently high resolution in the presence of Mg^{2+} to visualize the coordination spheres of individual bound Mg^{2+} ions (as discussed in Chapter 1). Presumably, these ions are bound to 'sites' with exceptionally long lifetimes so as to be characterized by significant occupancy in electron density maps.

Despite this wealth of structural information, only in several instances, including $tRNA^{Phe}$,[38,39] a 58-nucleotide 23S ribosomal RNA fragment,[40–42] bacterial[43,44] and ciliate[45–47] self-splicing group I introns, ribonuclease P,[48,49] and simple hairpin-type (H-type) RNA pseudoknots (of ~30 nucleotides or less),[50–52] have significant solution thermodynamic and/or computational studies been carried out that complement these high-resolution structural studies sufficiently to provide an understanding for how crystallographically observed Mg^{2+} ions stabilize the folded forms of these RNAs relative to their partially folded states. This chapter reviews those efforts. The reader is referred to several excellent reviews that have already appeared on this subject,[53–55] with what follows being an attempt to capture the essence of those earlier monographs, with additional emphasis placed here on new RNA systems and insights that have derived from new experimental approaches to this complex problem.

6.2 Physical Properties of RNA and Metal Ions Relevant to RNA Folding

6.2.1 Implications of RNA as a Folded Polyanion in Ion-RNA Interactions

Polymeric RNA is a polyanion with a net charge of −1 per phosphodiester linkage that gives rise to an extraordinary negative electrostatic potential around the macromolecule.[56] This potential is relatively smooth for 'random coil' single-stranded RNA and for RNA duplexes (helices and other secondary structural elements), with the potential larger for duplex *vs* single-stranded

RNA due to the fact that the phosphodiester linkages cannot move as far apart from one another as they can in 'unfolded' single-stranded RNA. When RNA molecules fold into complex three-dimensional shapes to create regions of tertiary structure, the negative electrostatic potentials in these regions can become very high, due to the proximity of phosphodiester groups on the RNA.[56] These 'pockets' of high negative electrostatic potential will interact even more strongly with counterions than do sites contained completely within secondary structural elements.[55]

In other words, the macromolecular charge density, dictated by the degree to which phosphate groups come into close proximity with one another, is lower for single-stranded nucleic acids than it is for duplexes, because the structure of the duplex imposes limits on the extent to which phosphate groups can move apart from one another.[53] In a folded RNA in which tertiary structure is superimposed on secondary structure, phosphate groups are in very close proximity and the associated increase in charge density translates into an increase in the number of positively charged species that might be associated with the phosphate groups or an increase in the apparent affinity of the particular 'site' for cations (see below). A corollary to this attraction of cations is that the anions derived from strongly dissociated salt solutions in water will be excluded from the surface of RNA. This preferential accumulation of cations and deficiency of anions at the nucleic acid–solvent interface not only strongly influence the electrostatic potential of the RNA, they contribute to it.[57] As such, these ions have a profound effect on the behavior of this polymer in aqueous salt solutions.

6.2.2 Physical Properties of Ions: Monovalent *versus* Di(multi)valent Ions

When salts are dissolved in aqueous solution, the positive and negative ions strongly dissociate and immediately acquire a shell of hydration.[57] This hydration shell plays an enormous role in modulating the physical properties and solution chemistries of these ions. The strength of these water–ion interactions is determined roughly by the size (volume, V) and valence (Z) of the ion as dictated by the charge density, $\zeta = Ze/V$; this, and also the polarizability of the ion, directly influence the magnitude of the unfavorable dissociation enthalpy of ion–water coordination bond.[58] Accordingly, the dissociation enthalpy becomes very large and unfavorable for small, highly charged cations such as Mg^{2+}, in which the unfavorable desolvation free energy is estimated to be $\sim 460\,kcal\,mol^{-1}$ for complete dehydration of Mg^{2+} on transfer to a vacuum. It is noted, however, that although this value can be used to accurately predict trends for a series of metals, the direct applicability of this value is convoluted by energies associated with a phase change and is much larger than the corresponding solution dehydration energy. In any case, Mg^{2+} forms stronger metal coordination bonds with water than do larger, more polarizable ions of the same valence. As a result, a series of metal ions at a constant ionic valence state, *e.g.* $Mg^{2+} \rightarrow Ca^{2+} \rightarrow Sr^{2+} \rightarrow Ba^{2+}$, ζ decreases by nearly fourfold due to the inverse relationship with V.[58] On the other hand, V increases significantly over the same range of ions and, in practice, it would seem difficult to discriminate experimentally between the relative importance of the role that charge density vs ion size plays in stabilizing RNA structure, particularly under conditions where fully hydrated ions are implicated in playing a major role in stabilizing the RNA (see Sections 6.3 and 6.4). Recent work on how divalent ions stabilize the natively folded form of the *Tetrahymena* group I intron ribozyme suggest that these two features may in fact represent two sides of the same coin.[58] Smaller ions with higher charge densities may approach the RNA more closely because the fully or partially desolvated volume of the ion is smaller; this close approach may more effectively counteract the like-charge repulsion of charged helices. The extent to which this picture characterizes the divalent metal-dependent stabilization of other complex RNAs is not yet known.

The above ideas are consistent with the formation of coordinate covalent bonds by metal ions as generalized by the hard/soft Lewis acid–base theory.[59,60] Generally, compact and highly charged Lewis acids (electron acceptors) interact more tightly with electron-donating ligands (Lewis bases) with large electronegativities, designated hard acids and bases, respectively. Conversely, soft acids (polarizable metal ions) tend to prefer coordination to softer (large and polarizable) bases. Although most metals and ligands discussed here fall within the 'hard' classification and have dominant ionic bonding character, it is important to keep in mind that changes of the identity of the metal ion or the oxidation state can significantly influence the covalency of the coordination bond and therefore impact the global thermodynamics. This is, of course, exactly the reasoning behind phosphorothioate rescue experiments.[61,62]

Monovalent ions, by virtue of their lower valence, $Z = 1$, are characterized by significantly smaller charge densities, ζ; this in turn, reduces the charge density of the ion relative to the corresponding divalent ions of the same volume (V) or ionic radius. Thus, monovalent cations shed their layer of hydration much more readily ($\sim 80 \, \text{kcal mol}^{-1}$ at 25 °C for complete dehydration upon transfer to a vacuum for K^+), although these energetic costs are still fairly high.[63] By the same token, trivalent ions, *e.g.* exchange-inert $Co(NH_3)_6^{3+}$ and analogous osmium and ruthenium hexamine ligand complexes, and spermidine, a trivalent polyamine, possess a higher ζ than do divalent ions of the same size. As a result of the differences in ζ among mono-, di- and multivalent ions, the relative effectiveness of these ions in stabilizing nucleic acid structures against thermal or chemical denaturation is always $M^{3+} > M^{2+} \gg M^+$ (M = metal).[46,64]

Complex folded RNA molecules often contain 'specific' ion binding sites that appear to discriminate between ions of different sizes and valences. These include a K^+ binding site near the RNA surface associated with an adenosine (AA) platform from the tetraloop–tetraloop receptor (TL–TLR) in the *Tetrahymena* group I intron P4–P6 domain[65,66] (see Figure 6.1B), and also a completely buried K^+ site in the 58-nucleotide ribosomal RNA (Figure 6.1A).[67] The molecular basis for this discrimination lies within specialized macromolecular coordination structures formed by chelated ions (see Section 6.4). The major favorable driving force that offsets the large energetic penalty (ΔG_{solv}) associated with desolvation is the electrostatic force governed by Coulomb's law (ΔG_{elec}), which becomes particularly large if the site is buried on folding due to the absence of any potential charge distribution through solvation layers. Consequently, specific Mg^{2+} ion binding sites will be dominated by coordination bonds to non-bridging phosphate oxygen atoms in solvent-inaccessible sites as to overcome the larger ΔG_{solv}, whereas K^+ sites can be formed with polar groups of only partial negative charge, e.g. ribose and keto oxygen atoms and nucleobase nitrogen atoms. If the volume of the monovalent ion becomes too large relative to K^+, e.g. Cs^+, ΔG_{elec} is insufficient to offset ΔG_{solv}, whereas smaller volume monovalent ions such as Li^+ and Na^+, ΔG_{elec} may well be

---

Figure 6.1 Recently solved crystallographic structures of complex RNA molecules that contain well-ordered Mg^{2+} (green spheres) and/or K^+ ions (blue–purple spheres). (A) 58-nucleotide rRNA fragment-L11 complex solved to 2.8 Å resolution (PDB code 1HC8).[67] Thirteen metals ions including 11 Mg^{2+}, 2 Os^{3+} (cyan spheres) and 1 K^+ are indicated. Right: zoom in on the Mg163–K–Mg167 region, with RNA-based coordination bonds shown (solvent water molecules not shown). Note that C22 OP1 bridges Mg163 and K^+ ion, while both non-bridging phosphate oxygen atoms of A23 make coordination bonds to two different metals (Mg167, K^+ ion). (B) *Azoarcus* pre-tRNA^{Ile} group I intron–two exon deoxy-ΩG complex solved to 3.1 Å resolution with 18 metal ions, refined as 13 Mg^{2+} (green spheres: M1, M3, M8–M18) and 5 K^+ ions (purple–blue spheres: M2, M4–M7) (1U68).[80] The five metal ion core is noted (M1–M3, M5, M12) and the TL–TLR region is highlighted; M4 is the K^+ ion that was first identified in the *Tetrahymena* group I intron in this position.[66] The active site metals, M1 and M2 (M2 is an Mg^{2+} ion in the riboΩG structure; PDB code 1ZZN) each make five RNA inner-sphere coordination bonds, with other Mg^{2+} ions containing 1–3 RNA-derived ligands; four of the five K^+ ions have four or more RNA-derived coordination bonds. (C) M-box Mg^{2+}-sensing riboswitch (*ykoK*) from *B. subtilis* solved to 2.6 Å resolution (1QB2).[74] The final model contains six Mg^{2+} ions (Mg1–Mg6) and 4 K^+ ions (K1–K4). Individual secondary structural regions are color-coded (left) with a five metal cluster (Mg1–Mg3, K1, K4) mediating long-range tertiary structural contacts between the base of P2 rich in non-canonical base pairs and L5 hairpin loop (residue numbers shaded as in the left panel). The metal coordination complexes for each of these metals is shown; RNA-derived oxygen (red); phosphorus (orange), nitrogen (blue) are shown as spheres, as are coordinated H_2O molecules (shaded yellow). Note that Mg1 contains four inner sphere ligands, while both non-bridging oxygen atoms of G100 and U24 make coordination bonds to two different metal ions (G100: Mg1, Mg2; U24: K4, Mg3); furthermore U23 O2P atom makes coordination bonds to both Mg2 and K4. K1 is octacoordinate with all but two ligands derived from O/N donor atoms from the RNA. (D) *T. tengcongensis glmS* ribozyme with the competitive inhibitor glucose-6-phosphate (G6P, shown in spacefill) bound in a post-cleavage state (2H0Z) solved to 2.7 Å resolution with seven metal ions (green spheres) resolved in the structure.[165] Note that M6 accepts two coordination bonds to the phosphate group of G6P. The main chain of the ribozyme is color-ramped from blue to red (5'–3').

Metal Ions and the Thermodynamics of RNA Folding 185

larger, depending on the degree to which the ion 'fits' into the site, although ΔG_{solv} will also be correspondingly larger and as a result the site will appear selective for K^+ over all other mono- and divalent ions. It is also noteworthy that the different coordination chemistries (*n*, coordination number; *d*, coordination bond length) of K^+ ($n=8$; $d=2.7$–3.3 Å) *vs* Mg^{2+} ($n=6$; $d=2.0$–2.3 Å) can also contribute to selectivity. Hence ion selectivity may be governed to a large extent by a delicate balance between two very large opposing free energy terms, ΔG_{solv} and ΔG_{elec}, the latter of which is strongly influenced by the extent to which the site is buried. In fact, recent studies of the ion selectivity of the isolated TL TLR interaction are less consistent with a strongly specific K^+ ion, at least with regard to the thermodynamic stabilization of this RNA–RNA interaction (see Section 6.4).[68]

6.2.3 RNA Folding is Generally Hierarchical

Quantitative efforts to understand how Mg^{2+} ions stabilize RNA structure benefits greatly from the fact that RNA folding is generally hierarchical, *i.e.*

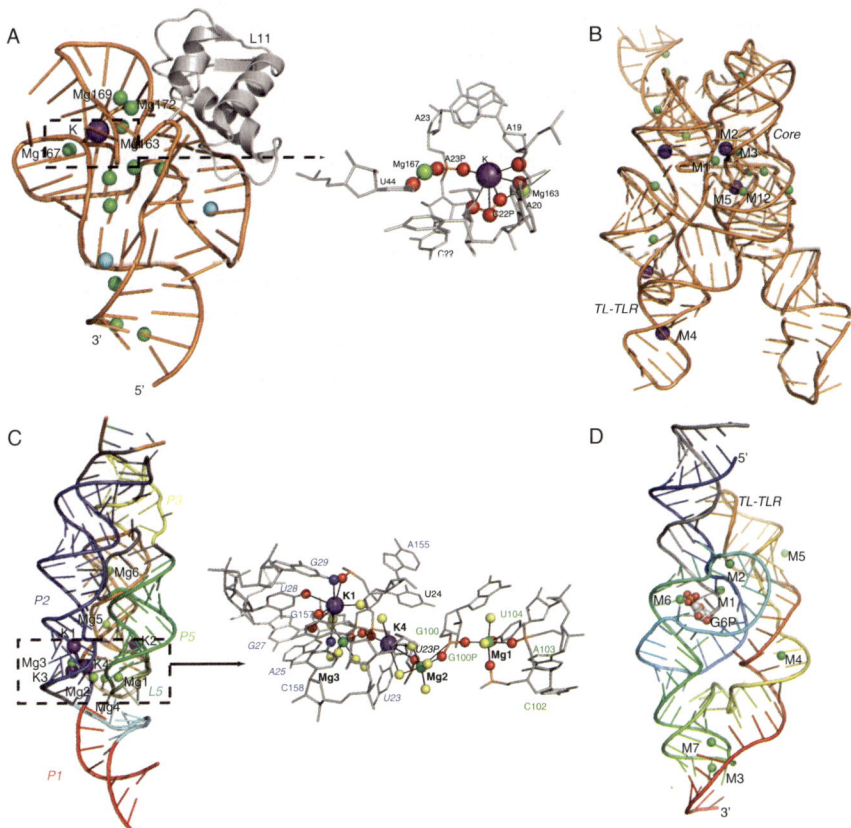

secondary structure generally folds before the establishment of tertiary structure in the molecule.[69,70] Microscopic reversibility posits that RNA unfolding would also be hierarchical, although recent chemical probing experiments suggest that this might be an oversimplifying assumption.[71] Secondary structure, however, is stable over a wide range of solution conditions and simply requires monovalent ion concentration in the 50–100 mM range to be sufficiently stable. RNA helices are strongly destabilized in the absence of any added cation (due to electrostatic repulsion of the phosphodiester groups), but do indeed form at very low ionic strength (*e.g.* <1 mM in monovalent cation). This feature of RNA (or more generally nucleic acid) helices directly enables the simplifying assumption that the 'unfolded states', a conformational ensemble of worm-like single-stranded chains with no base pairing, can be safely ignored over a wide range of solution conditions at ambient or near ambient temperature in the absence of chaotropic agents, and all one needs to consider is the effect of Mg^{2+} on altering the equilibrium between secondary and tertiary structure formation.

A comment is in order here about what is meant by the terms secondary structure and tertiary structure. Secondary structure refers to the formation of helical elements in the RNA and all associated loops (non-Watson–Crick base-paired and single-stranded regions) that are connected by these helical elements. In this definition, hairpin loops and bulged (internal) loops of defined structure, although often exhibiting non-canonical interactions that might be thought of as tertiary structure, are clearly elements of secondary structure. Tertiary structure is typically taken as the formation of non-covalent interactions that occur between elements of secondary structure, *i.e.* loop–loop or loop–stem and, in some cases, stem–stem interactions that are superimposed on an underlying helical secondary structure. An excellent example is the classic tetraloop–tetraloop receptor interaction in which a GAAA tetraloop interacts with an 11-nucleotide internal asymmetric loop.[11,72] A concise definition of secondary *vs* tertiary structure is important, because an RNA conformation or conformational ensembles are often representative of a defined thermodynamic 'state' in coupled equilibria that must be considered when trying to design experiments to understand the influence of Mg^{2+} on RNA structure and stability.

6.2.4 Intrinsic Competition of Monovalent and Divalent Ions for 'Sites' on RNA

As a polyanionic macromolecule, an RNA molecule has an abundance of potential non-specific cation binding sites as the result of simple electrostatics. Herein lies the single most complicating issue when it comes to trying to understand how Mg^{2+} ions, relative to monovalent ions, thermodynamically stabilize RNA. Although this will be more fully expanded upon below, the reason for this is simple: the interaction of a single Mg^{2+} ion with an RNA will occur with the *net* displacement of nominally two K^+ or Na^+ ions, irrespective of the molecular characteristics of the 'site'. Therefore, when considering how

Mg^{2+} ions thermodynamically stabilize an RNA molecule, any thermodynamic parameters that can be extracted from such an analysis will be conditional and strongly dependent on the concentration of monovalent ions.[73] In addition, as strongly dissociated ions in solution, Mg^{2+} and K^+/Na^+ 'feel' each other via long-range electrostatic repulsion; hence the competition between Mg^{2+} and K^+/Na^+ is not only at the 'site' level but also at a more fundamental physicochemical level.[53,54] Indeed, the concept of a metal ion binding 'site' in RNA is conceptually difficult to grasp and can take on a variety of forms, from specific coordination sites (discussed in this chapter and in Chapter 1) to completely delocalized binding (as discussed in Chapter 9), all of which make measurable contributions to the stabilizing effect of divalent cations on RNA structure.

6.3 Mg^{2+} Ions, Hydrodynamic Collapse and Dynamical Restriction

Shown in Figure 6.1 are ribbon representations of crystallographic structures of several large RNAs that have been solved during the past several years.

The remarkable feature of each of these structures is the extent to which individual helical elements pack on top of, or against, one another as well as the sharp transitions in polynucleotide chain direction that is required for RNA to fold into a compact conformation with a defined inside and outside. As a result, substantial structural information can often be elucidated using techniques that are capable of revealing the regions of RNA that become buried upon folding at nucleotide resolution and/or identify RNA-derived metal coordinating ligands. Examples of such techniques are in-line probing, which measures the spontaneous rates of strand scission of internucleotide linkages,[74,75] hydroxyl radical footprinting, which probes solvent accessibility,[76,77] and SHAPE (2′-hydroxyl acetylation analyzed by primer extension) chemistry,[78] while phosphorothioate rescue[61] and Tb(III) footprinting[79] can potentially identify Mg^{2+} ligands in the RNA. However, in order to define precisely the positions and coordination geometries of metal ion binding sites in RNA, crystallographic methods are required. Even so, these methods are somewhat subjective since the crystallographic resolution of many complex RNAs is often low, ≥ 3.0 Å resolution, which makes distinguishing non-nucleotide electron density as either water, monovalent (K^+) or divalent (Mg^{2+}) ions during the refinement of these structures very difficult (see Chapter 1 for further discussion).[80] A useful approach to distinguish monovalent from divalent ions in the native structure is through metal ion soaking experiments using a variety of heavier (more electron-dense) monovalent (Tl^+)[66,67], divalent (Mn^{2+}) and trivalent lanthanide (Tb^{3+}, Yb^{3+} or Eu^{3+}) metal ions coupled with known differences in coordination chemistry of K^+ vs Mg^{2+} (see above).[80] Additionally, knowledge of the themes in functional group preference for specific metal ions can also prove useful. Both mono- and divalent ions are capable of forming coordination bonds to base keto oxygen and nitrogen atoms, 2′ hydroxyl groups and both bridging and non-bridging phosphate oxygen atoms; by far the most common ligand for Mg^{2+} are the non-bridging oxygen atoms of the phosphate group, with each having a charge of ca. −1.

Well-ordered metal ions with high occupancies are often found at positions of the most negative electrostatic potential, where phosphate oxygen atoms from the same and/or neighboring strands come in very close physical proximity to one another. In nearly all cases, Mg^{2+} appears to be bound with at least one inner-sphere coordination bond, with 1–3 RNA-derived coordination bonds typical; however, Mg^{2+} ions with four (*e.g.* Mg1 in the Mg^{2+} sensing riboswitch)[74] or five (*e.g.* M1–M2 in the *Azoarcus* group I intron) are found, the latter of which is unprecedented in all RNA structures solved thus far.[80] The overwhelming majority of Mg^{2+} coordination bonds are made to non-bridging phosphate oxygen atoms with the uridine O4 keto oxygen sometimes present; this suggests that the sizable desolvation penalty associated with Mg^{2+} binding to RNA can be offset by strong Coulombic interactions with the RNA.[81] In contrast, K^+ binds to sites with a correspondingly greater number of RNA-derived ligands rich in electronegative base oxygen and nitrogen atoms in the major groove, in addition to bridging and non-bridging phosphate oxygens. It is important to point out, however, that crystallographic studies will miss most diffusely associated, highly mobile Mg^{2+} ions due to their relatively short lifetimes in fixed positions within the crystal lattice (see below).

6.3.1 Mg^{2+} Drives Hydrodynamic Collapse of Complex RNAs

The addition of cations to a secondary structure-only state is predicted to drive the formation of a compacted form of the RNA (Figure 6.2A). Application of a number of hydrodynamic techniques, including small-angle X-ray scattering[82–84] and sedimentation velocity centrifugation, coupled with time-resolved RNA footprinting,[85] is fully consistent with this picture, although the degree to which specific tertiary interactions are superimposed on this charge-neutralized and compacted state may well vary from RNA to RNA.[55,86] For example, in the P4–P6 domain of the group I intron, formation of the collapsed state is extremely rapid, occurring on the millisecond time-scale, with specific tertiary structural interactions occurring only much later; the collapsed state is 'non-specific' and simply provides a platform from which a search can be undertaken to form tertiary, structural specific interactions.[87–89] Analogous behavior might also characterize the 58-nt rRNA fragment, since a tertiary structural mutant unable to form tertiary structural interactions still undergoes measurable compaction.[90]

In contrast, for other RNAs, particularly those that fold without kinetic traps, including the bacterial group I intron from *Azoarcus*[44] (see Figure 6.1B) and the yeast bI5 intron,[91,92] the collapse is specific in the sense that many native-like interactions are indeed formed. Furthermore, despite the fact that the concentration of Mg^{2+} required to drive hydrodynamic compaction is often

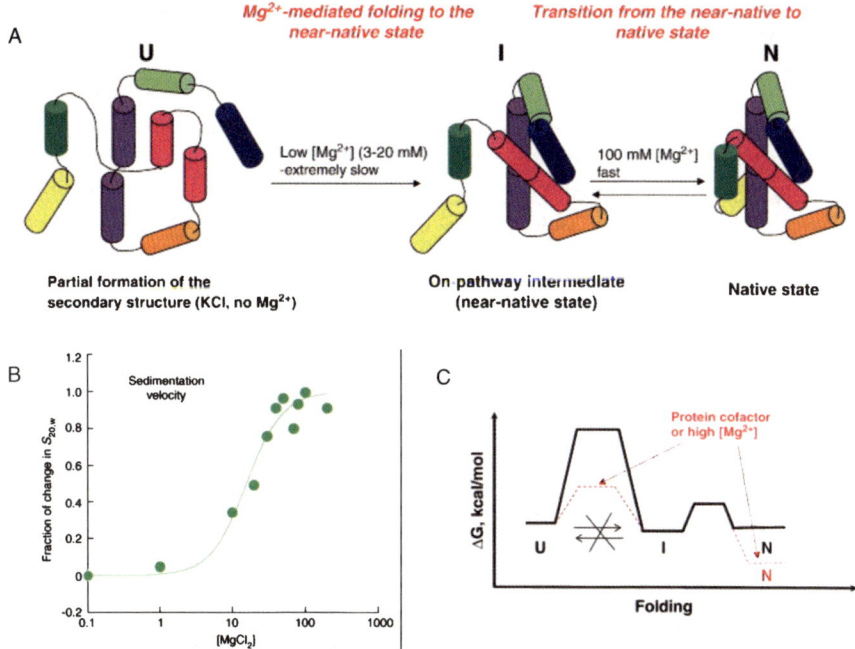

Figure 6.2 Example of the Mg^{2+}-induced collapse of a complex RNA.[172] (A) Illustration of a folding pathway of the yeast group II intron ai5γ-derived D135 ribozyme at 30 °C consistent with experiment. (B) Monitoring the change in global RNA compaction in the D135 RNA by sedimentation velocity as a change in $s_{20,w}$. The continuous line through the data represents a fit to the Hill equation with $c_{Mg}^{1/2} = 15 \pm 2$ mM and $n_{Hill} = 2.0 \pm 0.4$.[93,173] (C) Hypothetical free energy diagram for ribozyme folding to the near-native (collapsed) and native states at low (black lines) and high Mg^{2+} or protein cofactors (red lines).[173] Reprinted from refs. 172 and 173, with permission from Elsevier.

less than that required to obtain the final folded and catalytically active N-state, hydrodynamic collapse can be modeled as a single two-state folding transition. An example of the analysis of the Mg^{2+} dependence of the hydrodynamic collapse is shown for the ai5γ group II intron RNA in Figure 6.2 using a simple Hill model (see below), which yields $c_{Mg}^{1/2}$ of 15 mM and $n_{Hill} \approx 2$.

Note that this $c_{Mg}^{1/2}$ value is fairly high and full collapse is not obtained until ≥ 100 mM Mg^{2+} is added to these solutions (Figure 6.2B); hydrodynamic collapse is also extremely slow.[93] This indicates a likelihood that in the cell, some RNAs, *e.g.* the yeast mitochondrial group I (bI5) and group II (ai5γ) introns, require protein cofactors or chaperones to overcome the concentration of Mg^{2+} needed to obtain a fully folded and catalytically active N-state; this is exactly what has been found (Figure 6.2C).[94] The mechanism by which the specific protein cofactor CBP2 facilitates bI5 group I intron folding has recently been investigated by single-molecule fluorescence methods.[95]

6.3.2 Dynamical Restriction of RNA Conformational Fluctuations by Mg^{2+} Ions

Recent novel NMR experiments that exploit a large number of residual dipolar couplings (RDCs) have been developed that allow the site-specific investigation of RNA dynamics and structural flexibility in solution and also how Mg^{2+} ions might modulate these properties.[96–98] The HIV-1 transactivation response element (TAR) RNA,[99–102] a longstanding model RNA stem–loop that exploits 'induced fit'[103] to change its conformation adaptively in response to ligand binding (Figure 6.3), has recently been probed with these methods.

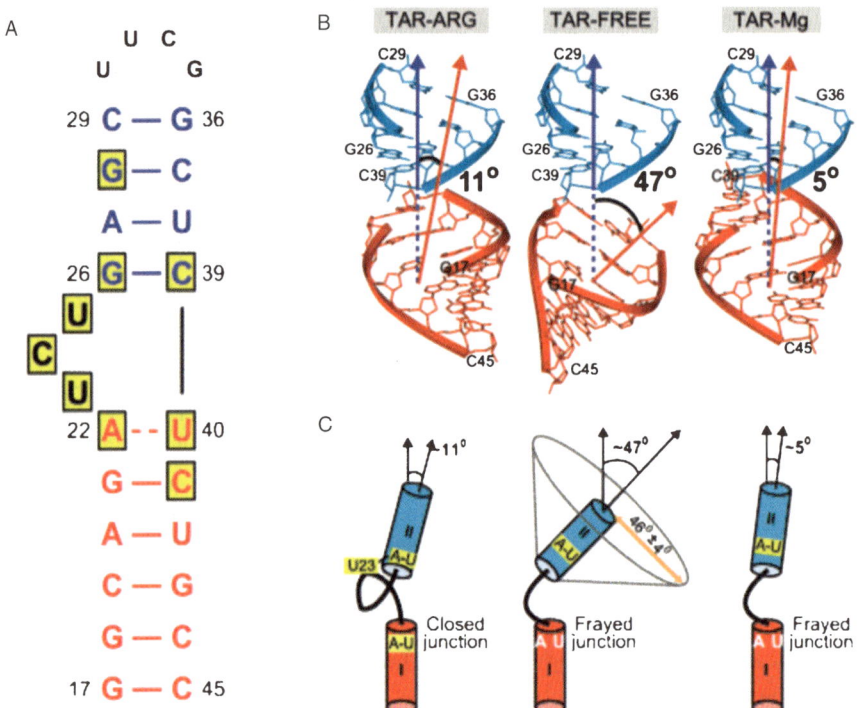

Figure 6.3 Mg^{2+} arrests global motions in the HIV-1 TAR hairpin in a manner similar to that of argininimide (ARG). (A) Secondary structure of the HIV-1 TAR hairpin used to probe global conformational dynamics using residual dipolar coupling methods, in which the naturally occurring six-nucleotide loop is replaced with a highly stabilizing UUCG tetraloop.[101] (B) The average RDC-derived inter-helical conformation is shown for TAR-ARG,[101] TAR-FREE,[99] and TAR-Mg.[100] The upper and lower stems are shaded red and blue, respectively, as in panel (A), with the helical axes of each shown; the inter-helical angles are also indicated on the structures. The upper (blue) stem is superimposed in all structures to highlight differences in the global conformation. (C) Cartoon representations of the differences between the global structures and conformational dynamics of TAR-ARG, free TAR and Mg^{2+} TAR.[101] Reprinted from ref. 101, with permission from Elsevier.

The HIV-1 TAR hairpin is composed of two helical domains separated by a 3-nucleotide UCU asymmetric loop (Figure 6.3A) that binds HIV-1 Tat, an arginine-rich protein. The binding of L-argininamide (ARG) mimics the structural changes induced into the TAR RNA by the Arg-rich peptide segment of Tat. In the absence of ARG and Mg^{2+}, the TAR RNA adopts a bent structure in which the two helical domains are characterized by a 47° bend angle with significant large amplitude (±46°) inter-helical domain motion. When ARG is bound, a nearly coaxially stacked helical junction results with (U38–A27)·U23 triple base pair in both HIV-1 and HIV-2 TARs (Figure 6.3B)[103,104] with greatly attenuated interdomain flexibility.[101] Strikingly, like the ARG structure, the structure of TAR determined in the presence of Mg^{2+} also reveals complete quenching of interdomain dynamics with a similar global structure, although the detailed structure around the helical junction is different in the two cases, *i.e.* there is no evidence for the (U38–A27)·U23 base triplet.[100,101] These perturbations in the structure and dynamics of TAR by ARG *vs* Mg^{2+} mirror those induced by the binding of ribosomal protein S15 *vs* Mg^{2+} to a three-helix junction rRNA fragment studied nearly 10 years earlier using single-molecule fluorescence resonance energy transfer (smFRET) experiments.[105]

Interestingly, the binding of the aminoglycoside neomycin B, a positively charged ligand containing five ammonium groups, to the TAR RNA also appears to induce a high degree of alignment of the two helical stems.[102] Although the crystal structure of TAR bound to divalent ions documents the presence of four bound metal (Ca^{2+} or Mg^{2+}) ions,[106] the structural and dynamical changes induced by the monovalent Na^+ ion are indistinguishable from that of Mg^{2+} (see below).[107] These findings are consistent with the idea that positively charged ligands, *i.e.* Mg^{2+} not only stabilize a particular conformation, but also significantly dampen global flexibility by stabilizing one or a small subset of conformations that are transiently accessible to the free TAR RNA. The picture that emerges from these studies is that the TAR RNA stem–loop readily adopts an electrostatically relaxed, bent and flexible conformation that is less strongly associated with counterions and also a globally rigid coaxial conformation that has stronger electrostatic potential and consequently a stronger association with counterions (see below).[107] This strong cation binding is due to the close approach of phosphate groups from both helical domains and the internal bulge that characterize the electrostatically restricted conformation (see Figure 6.3).

This view of how Mg^{2+} arrests global mobility in a simple two-helical domain RNA is reminiscent of the kink-turn (K-turn) motif in which a three-nucleotide bulge flanked on the 3′ side by two consecutive *trans*-Hoogsteen sugar edge A·G base pairs induces a sharp ∼60° bend in the helical axis of the RNA.[108] Originally discovered in the structures of the ribosomal subunits,[108] the K-turn is now known to be a ubiquitous RNA-folding element that typically functions as a binding site for proteins or mediates long-range RNA–RNA interactions.[109] In an isolated K-turn-containing RNA, Mg^{2+} shifts the equilibrium from a more extended RNA conformation to a dynamically restricted and tightly kinked conformation.[109,110] It seems likely that the electrostatic collapse of larger more complex RNAs may originate in part from the quenching of local dynamics in small RNA subdomains within secondary structural motifs by a localized association of Mg^{2+} ions.

6.4 Metal Ion Binding Models

6.4.1 Operational Classes of Ions that Associate with RNA

Draper *et al.* have outlined a conceptually straightforward formalism which groups metal ions that associate with RNA molecules into two structural classes that essentially define the extrema of a continuum of possible ion binding modes on RNA (Figure 6.4).[53,54]

These cation groups are represented by diffuse *vs* chelated ions, with an intermediate scenario consisting of partial dehydration of the RNA surface. As will become clear below, the binding of Mg^{2+} ions to distinct 'classes' of sites are not independent; on the contrary, one influences the other. The first class is composed of diffuse ions which are highly mobile ions that accumulate near the RNA surface due to favorable long-range Coulombic or electrostatic interactions. These ions remain fully hydrated and compose the 'ion(ic) atmosphere' around the RNA and are not bound in the usual sense of the word. Diffuse ions simply follow the electrostatic field of the RNA with the concentration maximum of those ions corresponding to the greatest electrostatic potential.[111] The free energy of the interaction of diffuse ions at a specific concentration, c_{Mg}, with an RNA macromolecule, ΔG_D, is the difference in the electrostatic free

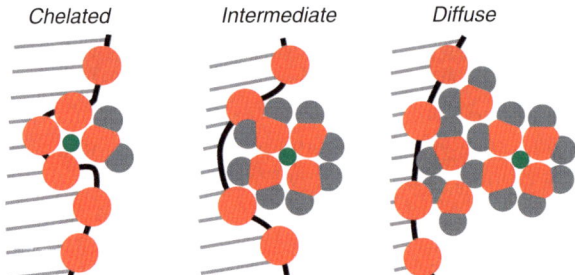

Figure 6.4 Conceptualized illustrations of potential ion environments in an RNA.[53] Represented are a Mg^{2+} ion (green filled circles), non-bridging phosphate oxygens (filled red circles) on the RNA surface (hatched) and water molecules. Left, Mg^{2+} is chelated and partially dehydrated by RNA phosphates; right, a diffusely associated Mg^{2+} ion remains fully hydrated but interacts with the RNA electrostatic field. A possible intermediate situation, in which Mg^{2+} retains one layer of hydrating water molecules that in turn interact with a partially dehydrated RNA surface, is shown in the middle. Reprinted from ref. 53, with permission from Cold Spring Harbor Laboratory Press.

energy of the RNA in the presence and absence of Mg^{2+}:[67]

$$\Delta G_D = \Delta G_{el}(c_{Mg}) - \Delta G_{el}(c_{Mg} = 0) \qquad (6.1)$$

Obviously, diffuse ions can contribute to the ion atmosphere around both the folded (N-state) and partially folded (I-states) and these free energies are denoted ΔG_D^N and ΔG_D^I, respectively (see below). For RNA, ΔG_D^N can be readily calculated to very high accuracy from a crystallographic structure using the non-linear Poisson–Boltzmann (NLPB) equation and thus provides a means to compare theory with experiment.[38] Both monovalent and divalent ions, e.g. Na^+ and Mg^{2+}, respectively, can function as diffuse ions; in fact, NLPB calculations suggest that consideration of only the differential interaction of diffusely associated monovalent and divalent ions with the folded N and partially folded I-states can effectively capture much of the Mg^{2+}-induced stabilization of the N- relative to the I-state(s) embodied in $\Delta\Delta G_{Mg}$ [see eqn (6.4)].[38,81] The reader is referred to Chapter 9 for further discussion of NLPB and for other theoretical treatments of cation localization.

Chelated ions define the other extreme of a continuum of ion binding sites in the Draper model. This class differs from diffuse ions in one important way; these ions must shed one or more water molecule from their first coordination sphere in order to make direct, or inner-sphere, coordination bonds with functional groups on the RNA (see Figure 6.1); the RNA has also lost its weakly bound (relative to metal ion) solvation shell.[81] Also, unlike diffuse ions, chelated ions do indeed 'bind' to defined structural sites on the RNA with an exquisite complementarity between ligand (a partially dehydrated metal ion of a particular size and charge distribution) and the macromolecular contacts contributed by the RNA (see Figure 6.1 and Chapter 1 for additional examples). These ions can be thought of as site-bound ions in the usual sense and are governed by the simple mass action equilibrium (Scheme 6.1).[67,81]

$$N + Mg^{2+} \underset{}{\overset{K_s}{\rightleftharpoons}} N-Mg^{2+}$$

Scheme 6.1

where K_s is the association equilibrium constant and the corresponding standard state free energy change of site-binding is $\Delta G_S^\circ = -RT \ln K_s$. Due to the specificity of these sites, it is assumed that they can only be formed on the N-state, thus the overall free energy of Mg^{2+} binding to the N-state, ΔG_{Mg}^N, is given by

$$\Delta G_{Mg}^N = \Delta G_D^N + \Delta G_S^N \qquad (6.2)$$

NLPB calculations suggest that the formation of these sites may well be energetically costly due to the large penalty for stripping even one of the water molecules from the first hydration shell.[112] The unfavorable contribution that ΔG_{Mg}^{solv} makes to the binding of chelated ions virtually ensures that these ions will only be found in the deepest wells of negative electrostatic potential (large favorable ΔG^{elec}) in order to offset this large unfavorable free energy change required by ion desolvation from[81]

$$\Delta G_S^\circ = \Delta G^{\text{elec}} + \Delta G_{\text{Mg}}^{\text{solv}} + \Delta G_{\text{RNA}}^{\text{solv}} + \Delta G_{\text{D,RNA}}^{\text{Na}} + \Delta G_{\text{D,RNA}}^{\text{Mg}} + \Delta G_{\text{Mg}}^{\text{trans}} \quad (6.3)$$

where ΔG^{elec} is the solvent-attenuated Coulombic attraction of Mg^{2+} to the site, $\Delta G_{\text{Mg}}^{\text{solv}}$ is the dehydration free energy of the ion on binding the RNA, $\Delta G_{\text{RNA}}^{\text{solv}}$ is the dehydration free energy of the RNA on Mg^{2+} binding, $\Delta G_{\text{D,RNA}}^{\text{Na}}$ and $\Delta G_{\text{D,RNA}}^{\text{Mg}}$ are the free energies associated with displacing the diffusely associated ions from the site and $\Delta G_{\text{Mg}}^{\text{trans}}$ is the free energy cost inherent to the macromolecular rigidification induced upon Mg^{2+} binding, which is predominantly entropic in nature due to reduction of translational entropy. The remarkable conclusion from NLPB calculations is that the physical location of most Mg^{2+} binding sites observed crystallographically, in tRNA^{phe} and also a 58-nucleotide 23S ribosomal RNA fragment (58-nt RNA; Figure 6.1A), can be rationalized on the basis of Mg^{2+}-induced stabilization by fully hydrated, diffusely associated ions; in fact, only in the case of one site on the 58-nt RNA (Mg167; see Figure 6.1A) is site-binding computed to be energetically favorable (negative sign on ΔG_S°) at 1.6 M monovalent salt concentration.[81] The extent to which this situation holds for other bound Mg^{2+} ions in other RNAs is unknown and would seem to depend critically on the incorporation of an accurate value for $\Delta G_{\text{Mg}}^{\text{solv}}$. This subject has been extensively discussed in recent years and the reader is referred to relevant reviews.[53,54,112]

6.4.2 General Considerations and Complications

Next, we discuss ligand (Mg^{2+}) binding models that can be used to investigate ion binding equilibria to RNA. The major complication is the linkage of Mg^{2+} binding to RNA folding; as a result, all models that ignore coupling of binding and folding do not yield thermodynamically rigorous binding parameters and serve only as phenomenological descriptions of the effects of Mg^{2+} on RNA stability.[113] Another major complication discussed above is the intrinsic (anti-cooperative) linkage between the mono- and divalent ions in stabilizing RNA.[38] A third complication involves the attempts to correlate the apparent Mg^{2+} binding affinities and stoichiometries to those Mg^{2+} ions observed in high-resolution crystallographic structures; unfortunately, this is not yet possible and is, in fact, fraught with peril.

It is first necessary to outline a coupled equilibrium scheme that can be used to quantify the effects of Mg^{2+} in stabilizing RNA structure (Figure 6.5).[53,54,114]

Given the fact that secondary structure is stabilized to a known degree by monovalent ions, we take stabilization of RNA structure to mean stabilization of RNA *tertiary* structure against thermal or chemical denaturation by the binding of some number of Mg^{2+} ions. At this point, it is not necessary to distinguish diffusely associated *vs* site-bound or chelated ions, only to

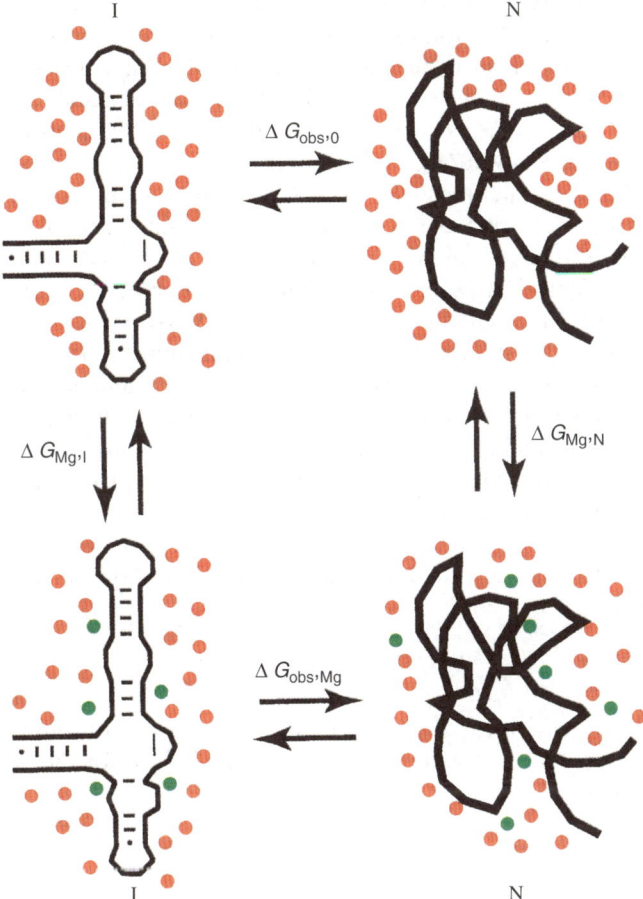

Figure 6.5 A thermodynamic cycle that can be used to relate the measurement of Mg^{2+}-induced stabilization of an RNA tertiary structure to measurements of Mg^{2+} interactions with RNA from experimental or computational methods.[54] The two left-hand diagrams represent an RNA fragment with only secondary structure present (the partially folded I-state); the right-hand structures are the same RNA in its native tertiary structure (N-state). K^+ (red filled circles) and Mg^{2+} ions (green filled circles) are indicated. The overall free energy change in moving from I-state RNA in the absence of Mg^{2+} (upper left) to the native RNA with Mg^{2+} (lower right) is independent of path; as a result, $\Delta\Delta G_{Mg} = \Delta G_{obs,Mg} - \Delta G_{obs,0} = \Delta G_{Mg,N} - \Delta G_{Mg,I}$. See text for details. Adapted from ref. 53, with permission from Cold Spring Harbor Laboratory Press.

recognize that there is a net Mg^{2+}-induced stabilization. Therefore, we need only consider two conformational states or ensembles. These are denoted I, an intermediate partially folded state(s) that is characterized by the full complement of secondary structure but fully lacking any tertiary structure and N, the native state or fully folded state. Unlike the N-state, the I-state very likely represents a diverse collection of rapidly interconverting microstates.

Employing this formalism makes two important points. First, this scheme allows us to define exactly what is meant by Mg^{2+}-induced stabilization of RNA tertiary structure, $\Delta\Delta G_{Mg}$, as the difference in the folding free energy between the I- and N-states at a given Mg^{2+} concentration as compared with $[Mg^{2+}] = 0$, $\Delta G_{obs,Mg} - \Delta G_{obs,0}$. Experimentally, $\Delta G_{obs,Mg}$ and $\Delta G_{obs,0}$ can, in favorable cases, be obtained by measuring the N-to-I equilibrium using thermal denaturation techniques as a function of Mg^{2+} concentration at a fixed monovalent salt concentration. Second, it permits two orthogonal approaches to $\Delta\Delta G_{Mg}$, since conservation of free energy requires that $\Delta\Delta G_{Mg} = \Delta G_{obs,Mg} - \Delta G_{obs,0} = \Delta G_{Mg,N} - \Delta G_{Mg,I}$ (see Figure 6.5). Hence experimental[114] or computational[81] methods that measure the *direct* binding of Mg^{2+} to the I- and N-states will allow a direct determination of $\Delta\Delta G_{Mg}$ without complications from Mg^{2+}-induced tertiary structural folding, in particular under conditions where $\Delta G_{obs,0}$ is experimentally inaccessible, *i.e.* in instances where tertiary structure fails to form in the absence of divalent ions.[114]

The free energies of Mg^{2+} binding to the N- and I-states, $\Delta G_{Mg,N}$ and $\Delta G_{Mg,I}$, respectively, are related to their respective partition functions, Z_N and Z_I, and hence:

$$\Delta\Delta G_{Mg} = -kT \ln(Z_N/Z_I) \qquad (6.4)$$

A suitable expression for Z_i as a function of Mg^{2+} concentration is provided by Γ_i, the preferential interaction coefficient of Mg^{2+} ions for the ith state, I or N, and is given by

$$\Gamma_i = \partial \ln Z_i / \partial \ln(c_{Mg}) \qquad (6.5)$$

where c_{Mg} is the free Mg^{2+} concentration. As required by eqns 6.3–6.5, $\Delta G_{Mg,N}$ and $\Delta G_{Mg,I}$ are obtained by integrating the area under a Γ vs $\ln c_{Mg}$ curve from $c_{Mg} = 0$ to some c_{Mg}:[114]

$$\Delta G_{Mg,i} = -RT \int \Gamma_i d \ln c_{Mg} \qquad (6.6)$$

It should be noted that the requirement for electrostatic neutrality in the bulk solution makes it impossible to determine the effects of changing Mg^{2+} concentration without changing the accompanying anion concentration (commonly Cl^- since Mg^{2+} is typically added as $MgCl_2$); however, the effects of Cl^- anion depletion from the RNA surface can be ignored provided the concentration of monovalent salt is large compared with the $MgCl_2$ such that the $[Cl^-]$ is effectively constant.[114]

The excess number of Mg^{2+} ions associated with the N-state relative to the I-state, $\Delta\Gamma(\Gamma_N - \Gamma_I)$, also somewhat carelessly termed the stoichiometry of Mg^{2+} binding, is simply a measure of the extent to which the N-state accumulates Mg^{2+} ions relative to the I-state and makes no inference as to the structural class or classes of bound ions to which these ions might conform (*e.g.* diffuse or chelated). Considering the thermodynamic cycle above (see Figure 6.5), $\Delta\Gamma$ can be determined directly by taking the difference of the direct solution binding measurements of Mg^{2+} to a suitable I-state model (Γ_I) and the N-state conformation (Γ_N) in solution through use of equilibrium dialysis, atomic emission spectroscopy[115] or fluorescent chelator competition experiments.[114,115] Alternatively, $\Delta\Gamma$ can be determined from the Wyman linkage relationship[116] as a measure of how the N-to-I unfolding equilibrium, governed by $K_{obs,Mg}$, is modulated by c_{Mg}:

$$\Delta\Gamma = \partial \ln K_{obs,Mg}/\partial \ln c_{Mg} = (-1/RT)(\partial \Delta G_{obs,Mg}/\partial \ln c_{Mg}) \quad (6.7)$$

where $\Delta G_{obs,Mg} = -RT \ln K_{obs,Mg}$. This analysis assumes a single two-state unfolding transition between the N- and I-states devoid of significantly populated intermediates. Note that $\Delta\Gamma$ is simply the slope of a plot of $\ln K_{obs,Mg}$ *vs* $\ln c_{Mg}$ [eqn (6.7)] and such a plot does not have to be linear; if it is curved, this would imply that $\Delta\Gamma$ is a function of the c_{Mg} over which the measurement was made, which can sometimes be the case (see below). Unfortunately, eqn (6.7) is also the standard implementation of the Hill equation, with the Hill coefficient, n_H defined as a *linear* slope of a $\partial \ln K_{obs,Mg}/\partial \ln c_{Mg}$ plot; n_H is then taken as equal to $\Delta\Gamma$. The important point is that $\Delta\Gamma$ can be a function of the c_{Mg} and is certainly influenced by the monovalent salt concentration. As a result, $\Delta\Gamma$ cannot be taken as a binding stoichiometry in the typical sense.[52,114]

Note that from eqns (6.1) and (6.2), $\Delta\Delta G_{Mg}$ can be recast as[67]

$$\Delta\Delta G_{Mg} = \Delta\Delta G_D + \Delta G_S^N \quad (6.8)$$

which simply attributes the net stabilization of RNA tertiary structure by Mg^{2+} as fractional contributions from differential accumulation of diffusely associated ions ($\Delta\Delta G_D$), which interact with both the N- and I-states, in addition to those due to site binding (ΔG_S^N) to the N-state.

6.4.3 Direct Binding Methods

Direct binding methods refer to those approaches that measure the binding of Mg^{2+} to RNA in solution. This free energy change, $\Delta G_{Mg,i}$, is given by the two vertical equilibria in Figure 6.5. One way to obtain $\Delta G_{Mg,i}$ is to titrate $MgCl_2$ into a solution of buffer (reference) *vs* buffer plus RNA in the presence of a known concentration of an indicator dye (D) of known affinity for Mg^{2+}, K_{Mg}, used to report on the free or bulk concentration of the metal ion. A common indicator dye for this purpose is 8-hydroxyquinoline-5-sulfonic acid (HQS). This approach is exactly analogous to chelator competition experiments that have been used to measure K_{Me} for metalloregulatory proteins, *i.e.* the macromolecule is spectroscopically silent and all that one is measuring is $D \cdot Mg^{2+}$ complex formation.[117] If the competitor macromolecule (RNA) binds the metal

(Mg^{2+}) as well as or better than the dye, there will be displacement or a 'lag' in the $D \cdot Mg^{2+}$ binding curve relative to the buffer-only titration (Figure 6.6A); when KCl or NaCl is in vast excess (hence Cl^- effects can be ignored), this displacement, denoted Δc_{Mg}, is $\sim \Gamma$. Measuring Γ for two thermodynamic

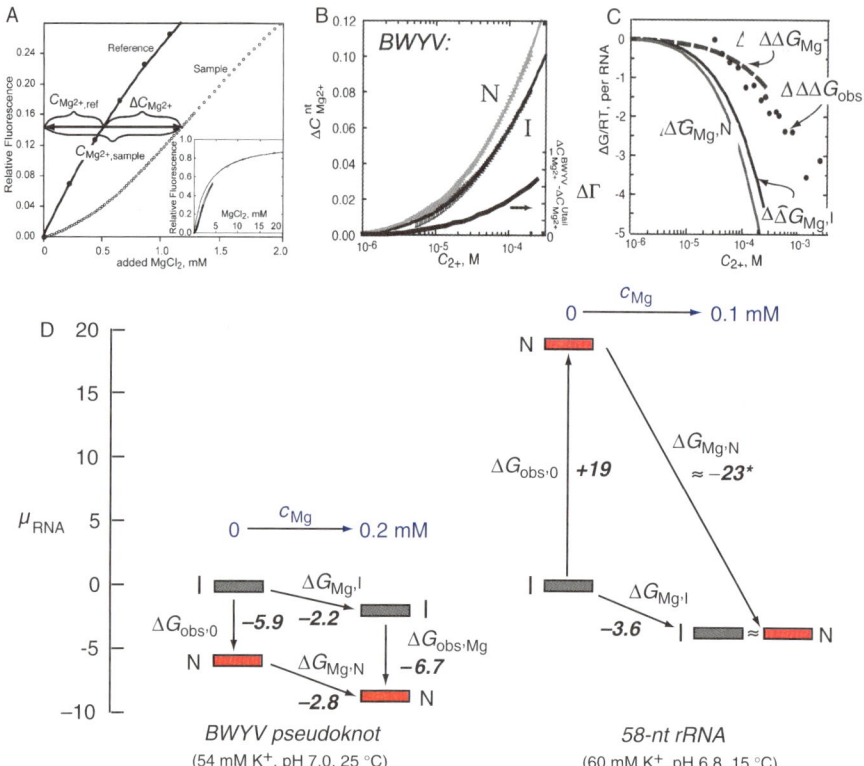

Figure 6.6 An approach to obtain Mg^{2+}–RNA interaction free energies using direct titration experiments.[114] (A) Example of a titration of the indicator dye, HQS, with $MgCl_2$ in presence (open symbols) and absence (closed symbols) of 4.18 mM A1088U 58-nt rRNA (in nucleotides) (60 mM K^+, pH 7.0). The solid line through the closed symbols represents a fit to a single-site binding model, $K = 337\,M^{-1}$. $\Delta c_{Mg^{2+}}$ is a direct measure of Mg^{2+} bound to the RNA at each c_{Mg} and thus is equal to Γ provided that the monovalent ion concentration is in large excess over the divalent ion concentration. (B) $\Delta c_{Mg^{2+}}$ (Γ) for N-state (gray binding curve) and I-state (black binding curve) models of the BWYV pseudoknot as a function of c_{Mg} with $\Delta\Gamma = \Delta c_{Mg^{2+}}(N) - \Delta c_{Mg^{2+}}(I)$ shown in the inset. (C) Plot of various ΔG_i (see Figure 6.5 for definitions) as a function of c_{Mg}. $\Delta\Delta G_{Mg} = \Delta G_{Mg,N} - \Delta G_{Mg,I}$ compares favorably from $\Delta\Delta G_{obs}$ as required by the free energy cycle shown in Figure 6.5. (D) Graphical representations of the free energy differences between the four states [I (gray bars) and N (red bars) in the presence and absence of Mg^{2+}) given at a single c_{Mg} (0.2 mM for the BWYV RNA and 0.1 mM for the 58-nt rRNA). Adapted from ref. 114, with permission from the National Academy of Sciences of the United States of America.

states, e.g. I- and N-states (see Figure 6.6B), allows one to calculate $\Delta\Gamma$ and thus $\Delta\Delta G_{Mg}$ at *any* c_{Mg} (see Figure 6.6C) using eqns (6.4)–(6.6).

$\Delta\Gamma$ can also be measured *indirectly* using the Wyman linkage relationship by analyzing the Mg^{2+} dependence of the thermal unfolding profiles from eqn (6.7). For the beet western yellows virus (BWYV) frameshifting pseudoknot (see Figure 6.8B for the structure),[118] $\Delta\Gamma \approx 0.7$ at 0.2 mM Mg^{2+} from direct titrations; note that from eqn (6.6), $\Delta\Delta G_{Mg}$, and thus $\Delta\Gamma$ is a logarithmic function of c_{Mg} and that this is near the maximum value of $\Delta\Gamma$, e.g. $\Delta\Gamma = 0.22$ at 0.03 mM Mg^{2+}. From linkage analysis, a linear analysis of a $\partial \ln K_{obs,Mg}/\partial \ln c_{Mg}$ plot over the same concentration range yields $\Delta\Gamma = 0.73$ ions and is in excellent agreement with $\Delta\Gamma$ measured directly at 0.2 mM Mg^{2+}. $\Delta\Gamma$ compares favorably with the results of calculations from the non-linear Poisson–Boltzmann (NLPB) equation; hence these data argue that the Mg^{2+}-induced stabilization of the pseudoknot relative to the partially folded S1 I-state is due to diffusely associated ions.[52] Also, for this particular RNA, the agreement between $\Delta\Delta G_{Mg}$ calculated from the vertical manifolds (from Mg^{2+} binding; $\Delta G_{Mg,N} - \Delta G_{Mg,I}$) and the horizontal manifolds (from measuring unfolding free energies using thermal melting profiles; $\Delta G_{obs,Mg} - \Delta G_{obs,0}$) (Figure 6.5) are in excellent agreement (-0.6 vs -0.8 kcal mol^{-1}), as expected for a closed thermodynamic cycle (see Figure 6.6C and D).

Parallel experiments were conducted with the 58-nt rRNA fragment (a U1061A substitution was used to stabilize the tertiary structure; this is the form of the RNA that was solved by X-ray crystallography bound to ribosomal protein L11;[12] see Figure 6.1A). Unlike the pseudoknot, this system is complicated by the fact that the tertiary structure of this RNA does not form at low (60 mM) monovalent salt concentrations in the absence of Mg^{2+} (thus $\Delta G_{obs,0}$ cannot be measured; see Figure 6.5) and uncertainties therefore arise in choosing an appropriate I-state model to measure the Mg^{2+} association with this state independent of Mg^{2+}-induced folding (to obtain $\Delta G_{I,Mg}$). In this case, the authors used A1088U RNA for the I-state model; this substitution prevents formation of a long-range triple base pair and blocks the formation of all detectable Mg^{2+}-induced tertiary structure. Interestingly, this mutation does not abrogate compaction of this RNA as measured by small-angle X-ray scattering, thus revealing that RNA compaction is uncoupled from tertiary structure formation in this RNA.[90] A similar degree of global compaction is observed for the WT (A1061U) RNA. This very fact highlights the robustness of the thermodynamic framework outlined in Figure 6.5 in uncoupling the effect of Mg^{2+} on specifically stabilizing tertiary structural folding from Mg^{2+} stabilization of the I-state.

$\Delta\Gamma$ for this thermodynamic system, obtained from the Wyman linkage relationship by measuring the Mg^{2+} dependence of the tertiary structure unfolding free energy obtained from optical melting profiles, is 3.9 ± 0.1 ions per RNA and compares favorably with direct binding methods at a c_{Mg} where the two binding curves are most different (~ 4.0 ions). More importantly, using $\Delta G_{obs,0} = +19$ kcal mol^{-1} obtained from an extrapolation from 1.6 M monovalent ion and 30% methanol, gives $\Delta G_{Mg,N} = -23$ kcal mol^{-1} when evaluated at 0.1 mM Mg^{2+}, which corresponds to the c_{Mg} at which half the RNAs have folded tertiary structure and half do not, thus $\Delta G_{obs,Mg} = 0 \, \text{kcal mol}^{-1}$ (Figure 6.6D).

The striking conclusion from this work is illustrated in the energy level diagrams for both RNAs (Figure 6.6D). The N-state of the BWYV pseudoknot is stable in the absence of Mg^{2+} ions, a result consistent with optically detected and calorimetric unfolding experiments,[51] whereas the 58-nt rRNA fragment is not. More importantly, $\Delta\Delta G_{Mg}$ is ~25-fold larger for the rRNA fragment, relative to the pseudoknot. This is well explained by NLPB calculations, which suggest much of this enhanced stability can be attributed to a single completely buried, chelated Mg^{2+} ion (Mg167; see Figure 6.1A),[81,90] On the other hand, the pseudoknot is stabilized by diffusely associated ions with no buried Mg^{2+} ions (but one fully hydrated Mg^{2+} ion present in the major groove of stem S1; see Figure 6.8B) evident in the crystal structure.[52]

The above thermodynamic framework permits a model-independent deconvolution of the extent to which the I- and N-states associate with Mg^{2+} and as such provides direct insight into the extent to which Mg^{2+} stabilizes the N-state relative to other partially folded I-states. A limitation of the above approach is the lack of access to suitable experimental (and computational) models of the I-state, since the structures of these are not known at high resolution.[90] We now discuss two simpler models of Mg^{2+}-induced folding of RNA tertiary structure that have been commonly employed and the limitations of each.

6.4.4 The Hill Model

The Wyman linkage model,[116] a popular manifestation of which is embodied in the Hill equation, considers Mg^{2+} as a ligand that binds to the n specific sites on the N-state not present on the I-state; as a result, Mg^{2+} binding to these n sites drives folding to the N-state (Scheme 6.2, where K_{eq} is taken as the average affinity of these sites for Mg^{2+} ions and n_H is the Hill coefficient).

$$I + n_H Mg^{2+} \xrightleftharpoons{K_{eq}} I\text{-}(Mg^{2+})_{n_H}$$

Scheme 6.2

The simplified partition function, Z, for this coupled equilibrium is[113]

$$Z = 1 + (K_{eq} c_{Mg})^{n_H} \qquad (6.9)$$

and the fraction of RNA molecules in the N-state is simply given by Θ_N:

$$\Theta_N = d \ln Z / d \ln c_{Mg} = \frac{(K_{eq} c_{Mg})^{n_H}}{1 + (K_{eq} c_{Mg})^{n_H}} \qquad (6.10)$$

with the free energy of Mg^{2+} binding at the folding transition midpoint equal to

$$\Delta G^H = -n_H RT \ln c_{Mg^{1/2}} \qquad (6.11)$$

where $c_{Mg}^{1/2}$ is the concentration of magnesium for which the RNA is half folded. n_H is often taken as a direct measure of $\Delta\Gamma$, the preferential interaction of Mg^{2+} to the N- relative to the I-state, but this may not be the case. The reason for this discrepancy is apparent upon careful inspection of Scheme 6.2: The model assumes infinitely positive cooperativity ($n_H = \Delta\Gamma$) and that Mg^{2+} binds very weakly, if at all, to the I-state, *i.e.* Mg^{2+} induces the folding transition. Hence use of eqn (6.10) to monitor the Mg^{2+} dependence of I- to N-state folding is purely a phenomenological description of the degree to which folding to the N-state is cooperative. Furthermore, mass action models of this kind fail since they ignore the contributions that diffuse ions make in governing the Mg^{2+}-induced stabilization of the N- *vs* the I-state, *i.e.* $\Delta\Gamma$ contains contributions from both diffuse and site-bound ions, in addition to the potentially very different electrostatic features of the two states. $\Delta\Gamma$ and ΔG^H are also strongly dependent on the initial solution conditions, *e.g.* monovalent ion concentration, since it is these conditions that dictate the chemical potential of the I-state. As a result, there is considerable uncertainty in equating n_H and $\Delta\Gamma$.

One way to minimize the degree to which diffuse ions contribute to $\Delta\Gamma$ is to perform experiments in very high concentrations of monovalent ions (>1.0–2.0 M), so that these ions make up the ion atmosphere and any additional sites can be attributed to 'chelated' or 'site-bound' divalent ions. This approach has been reported in several cases[74,115,119] and, in at least one case, the P4–P6 metal ion core,[115] suggest that the method has merit. The crystallographic structures of the intact bacterial intron (see Figure 6.1B)[16] and the isolated P4–P6 domain[11] reveals the presence of five Mg^{2+} ions in a core structure. Starting from the 2.0 M NaCl reference states, a Hill analysis of the folding of the RNA reveals $n_H = 1.8 \pm 0.1$ with a $c_{Mg}^{1/2}$ of 0.52 mM. Interestingly, a direct titration of the wild-type RNA compared with a mutant RNA that is unable to fold a structured metal ion core reveals a net of 1.9 ± 0.1 ions associated with the wild-type *vs* mutant RNA using either the Mg^{2+}-sensitive indicator dye, HQS or atomic emission spectroscopy (AES), which, like classical equilibrium dialysis, detects total bound Mg^{2+}.[115] In both cases, the binding data are well described by a reaction stoichiometry of $n = 1.9$ and a $c_{Mg}^{1/2}$ of 0.52 mM, in excellent agreement with the RNA folding data; therefore, in this case under these solution conditions, $n_H = \Delta\Gamma$. The reason for this agreement is that both folding and so-called 'ion counting' experiments are likely reporting on Mg^{2+} binding to the 2M Na^+ reference state, *i.e.* the *right-side* vertical equilibrium in Figure 6.5, $\Delta G_{Mg,N}$, where global folding from the I-state is *not* a dominant factor. This study suggests that the structural role of three of the five Mg^{2+} ions is easily satisfied by Na^+, with the other two sites bound far more preferentially by Mg^{2+} ions. Unfortunately, there is no way of rigorously establishing which two of the five ions in the crystal structure correspond to these two Mg^{2+} ions observed in solution without some nucleotide residue-specific footprinting experiments.

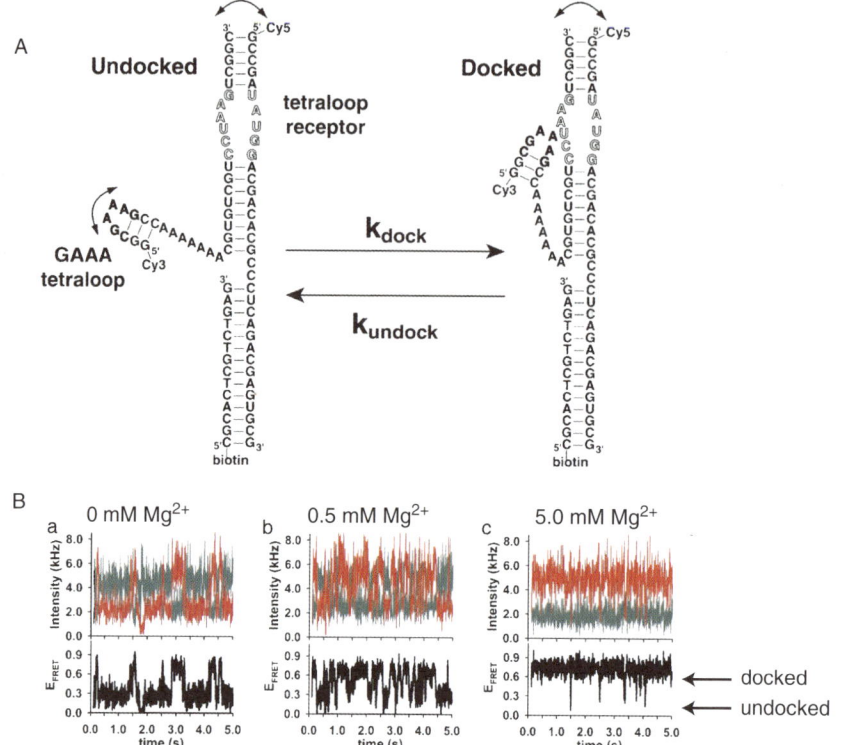

Figure 6.7 smFRET studies of the tetraloop-tetraloop receptor interaction. (A) Docking reaction of a model GAAA tetraloop (TL)–tetraloop receptor (TLR) system studied by Pardi and co-workers.[125] The 5′-biotin strand permits surface immobilization on streptavidin-doped glass slides for confocal microscopy. (B) Typical time-resolved fluorescence intensity and corresponding FRET efficiency, E_{FRET}, trajectories for the GAAA tetraloop–receptor system at 0 mM Mg^{2+} (left), 0.50 mM Mg^{2+} (middle) and 5.0 mM Mg^{2+} (right). The donor (Cy3) and acceptor (Cy5) signals are green and red, respectively. Two E_{FRET} states are resolved at all Mg^{2+} concentrations. The high E_{FRET} state ($E_{FRET} = 0.68$) is assigned as docked and the low E_{FRET} state ($E_{FRET} = 0.22$) is assigned as undocked. Adapted from ref. 125, with permission from the National Academy of Sciences of the United States of America.

In another example, the Hill model was recently used to investigate the effects of mono- vs di- and trivalent cations of the intramolecular docking–undocking equilibria of a unimolecular GAAA tetraloop–tetraloop receptor (TL–TLR) interaction using both stopped-flow ensemble kinetics and single-molecule FRET (smFRET) approaches (Figure 6.7A).

smFRET is now routinely applied to study folding and unfolding trajectories or conformational excursions of single RNA molecules, either in solution or immobilized on a surface, with the only requirement that FRET donor and acceptor probes be positioned in the RNA such that they are capable of reporting on distinct conformational states.[8,120–124] The crystal structure of the TL–TLR complex in the context of the P4–P6 domain of the group I intron exhibits a single bound K^+ ion[65,66] (Figure 6.1B) and therefore it was of interest to probe the extent to which monovalent vs di- and multivalent metal ions were capable of stabilizing the docked form of the RNA. Time-resolved smFRET trajectories reveal the presence of two primary FRET states, one with high FRET efficiency ($E_{FRET}=0.68$) and one with low FRET efficiency ($E_{FRET}=0.22$), assignable to the docked and undocked conformations, respectively (Figure 6.7B), and no evidence for significantly populated structural intermediates under these conditions.[125] The rate constants for docking, k_{dock}, and for undocking, k_{undock}, are obtained from

$$k_{dock} = 1/\langle \tau_{dock} \rangle \tag{6.12}$$

$$k_{undock} = 1/\langle \tau_{undock} \rangle \tag{6.13}$$

with $\langle \tau_{dock} \rangle$ and $\langle \tau_{undock} \rangle$, respectively, the latter of which is obtained from histogram plots of the number of events of time duration, t, of the docked or undocked states vs t, fitted to a single exponential function to obtain $\langle \tau_i \rangle$. In a two-state system, the equilibrium constant for docking, K_{dock}, is simply given by the ratio of the rate constants:

$$k_{dock} = k_{dock}/k_{undock} \tag{6.14}$$

As expected from simple electrostatics, $Co(NH_3)_6^{3+}$ was most effective at stabilizing the docked form ($c_{Mg}^{1/2} \approx 0.017$ mM at 0.1 M Na^+), followed by divalent ions ($c_{Mg}^{1/2} \approx 0.87$ mM for Mg^{2+} at 0.1 M Na^+), followed by monovalent ions ($c_{Mg}^{1/2} \approx 200$–220 mM for Na^+ and K^+). Interesting, Hill coefficients, n_H, were found to be ~ 1.0 for $Co(NH_3)_6^{3+}$ and Mg^{2+} and 1.9 ± 0.1 for Na^+ and K^+. These findings are exactly in accord with expectations for a single 'diffusely associated' fully hydrated Mg^{2+} ion being responsible for stabilizing the docked conformation relative to the undocked state, with the same degree of charge neutralization achieved by the cooperative binding of two monovalent ions to the same region of the docked conformation. Both k_{dock} and k_{undock} were found to be strong functions of $[Mg^{2+}]$, although k_{dock} was found to be far more sensitive; similar findings characterize the hairpin ribozyme.[126] Thus, in this case, the Hill analysis allows considerable insight into how Mg^{2+} stabilizes the docked TL–TLR complex relative to monovalent ions, despite the limitations of the model itself.

This intramolecular TL–TLR system has recently been used to explore the influence of hydrostatic pressure (P) and cosolutes (osmolytes) on the formation of RNA tertiary structure,[127] and complements recent comprehensive efforts to explore the Mg^{2+} dependence of the effect of osmolytes on tertiary and secondary structure using optical melting experiments.[128] For the TL–TLR, P up to 2.5 kbar was only modestly destabilizing ($\Delta\Delta G°_{dock} \leq 0.6$ kcal mol^{-1}). Increased P induces electrostriction of the solvent molecules around individual charged ions in solution; as a result, the P dependence of K_{dock} can be used to obtain estimates of ΔV, indicative of the stoichiometry of water release. Mg^{2+} interactions in the docked state make a favorable, but fairly small, contribution to ΔV, the origin of which is not fully understood. Cosolutes represent a complementary probe to P and can be used to investigate the influence of changing the water activity on RNA secondary and tertiary structure and will likely gain increased use as a means to modulate RNA–RNA interactions. A comprehensive analysis of the effect of a wide variety of osmolytes on a number of structurally diverse RNA molecules reveals that cosolutes are consistently destabilizing towards RNA secondary structure and can be stabilizing or destabilizing of tertiary structure. For example, urea is well known to destabilize RNA tertiary structure,[129] whereas trimethylamine oxide (TMAO)[130] and methanol[131] are often stabilizing. Mg^{2+} ions that neutralize a large fraction of negative charge on the RNA mitigate the effects of osmolytes on RNA tertiary structure; if Mg^{2+}-induced RNA folding significantly buries nucleobases in addition to the backbone, like that which occurs in complex RNAs, osmolytes will often be destabilizing of the N-state relative to the partially folded I-state.[128]

6.4.5 Diffuse Ion Association Models

Like the Hill treatment described above, this approach is also based on a phenomenological model in which the measurable parameter Δv is analogous to $\Delta \Gamma$.[132–134] The equilibrium thermal unfolding behavior of a folded RNA can be most simply modeled by a series of n sequential unfolding steps, each of which is governed by an equilibrium constant $K_1, K_2, \ldots, K_n$ according to Scheme 6.3

$$F \xrightleftharpoons{K_1} I_1 \xrightleftharpoons{K_2} I_2 \xrightleftharpoons{K_3} I_3 \cdots \xrightleftharpoons{K_n} U$$

Scheme 6.3

In this analysis, each unfolding step is assumed to be two-state and reversible, with each step governed by a measurable enthalpy of unfolding in the absence of added divalent metal ions, ΔH_0, and a melting temperature, T_m. A plot of T_m^{-1} vs [L] at a fixed concentration of monovalent ion that is far greater than the divalent ion concentration, [L], allows one to determine K_U and K_F, the K binding affinities of the two states that participate in the ith thermal unfolding equilibrium, according to the following equation:[132]

$$T_m^{-1} = T_{m0}^{-1} - (R/\Delta H_0) \ln(Z_F/Z_U) \qquad (6.15)$$

where

$$Z_i = \{1 + \mathrm{sqrt}(1 + 4[L]K_i)/2\}^m \qquad (6.16)$$

is the L (Mg^{2+}) binding polynomial for the ith (folded and unfolded) state. In eqn (6.16), m is the number of available binding sites for the ligand L. Since the model explicitly considers the influence of diffuse or highly mobile ions on the shift of a particular equilibrium, m is simply the number of phosphates weighted by the 'site-size' for L. For a divalent cation, e.g. Mg^{2+}, it is two phosphates, whereas that for a trivalent ion, e.g. $Co(NH_3)_6^{3+}$, is three phosphates. Analogous expressions have been derived for measuring the influence on an unfolding equilibrium for a trivalent ion.[64]

Knowledge of Z_i allows the derivation of expressions that can be obtained for ν_i, the binding density (or degree of binding) of M^{2+} or M^{3+} ions to the ith (folded or unfolded) state by simply taking the following derivative:

$$\nu_i = \partial \ln Z_i / \partial \ln [L] \qquad (6.17)$$

Here, the difference $\nu_F - \nu_U = \Delta\nu$ is equal to the net number of ions associated with folded state *versus* the unfolded state; $\Delta\nu$ is therefore analogous to $\Delta\Gamma$, albeit a model-dependent parameter in this case.

Although physically simplistic, this analysis qualitatively captures the influence of multivalent ions on this coupled unfolding equilibrium. For example, in the case of an autoregulatory pseudoknot that unfolds in two sequential unfolding transitions defined by K_1 and K_2 in which the longer stem S2 hairpin is an unfolding intermediate (Scheme 6.4),[64] K_F is significantly larger for the F vs S2 state with K_U comparable in both cases; further, the number of ions that participate in the unfolding reaction, $\Delta\nu$, for the F $\rightarrow$ S2 ($\Delta\nu = 1.1$) vs S2 $\rightarrow$ U ($\Delta\nu = 1.2$) transitions are more similar than different, even though there is a smaller number of phosphates modeled to participate in the first unfolding step *versus* the second (8 vs 14).[64]

$$F \underset{}{\overset{K_1}{\rightleftharpoons}} S2 \underset{}{\overset{K_2}{\rightleftharpoons}} U$$

Scheme 6.4

Similar findings characterize other hairpin-type (H-type) pseudoknots that stimulate ribosomal frameshifting.[135,136] This is consistent with the idea that the F form has a deeper well(s) of negative electrostatic potential, either near the helical junction region[137] or where loop L1 crosses the deep major groove of stem S2,[64] relative to the partially folded hairpin intermediate, and can therefore sequester more multivalent ions with stronger affinity at some fixed concentration of monovalent salt. The data further suggest that highly mobile, diffusely associated ions are primary players in modulating this coupled equilibrium; the recent analysis of Mg^{2+} binding by the BWYV pseudoknot (Figure 6.8B) using direct binding methods is fully consistent with this picture.[52]

6.4.6 Linkage of Protonation to Mg^{2+} Binding and RNA Folding

RNA differs strikingly from proteins by the near absence of functional groups with intrinsic pK_as near neutrality. However, it is now becoming increasingly clear that even in simple folded RNAs, intrinsic nucleobase pK_as can be strongly

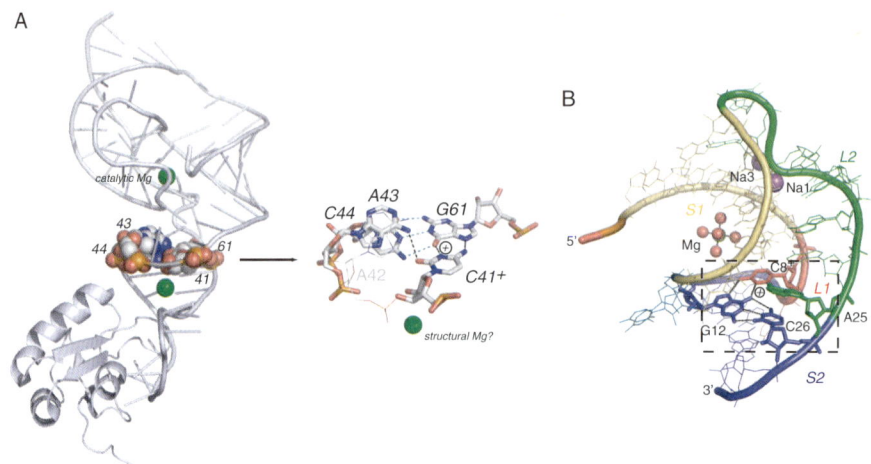

Figure 6.8 RNAs containing a major groove-derived protonated $A \cdot C^+ \cdot (G\text{-}C)$ base quadruple motif. (A) *Left*, Ribbon representation of the structure of the C75U pre-cleaved HDV genomic ribozyme (PDB accession code 1SJ3) solved in the presence of Mg^{2+}. Two Mg^{2+} ions were found in this structure: one near the scissile 5′-bridging oxygen atom (but quite far from the 2′-hydroxyl group of U–1) labeled a 'catalytic' ion, due to its proximity to the active site, with another near the base quadruple motif involving a protonated C41 (shown in *spacefill*), labeled the structural ion. *Right*, an expanded view of the base quadruple. Hydrogen bonds involving C41$^+$ are represented by the *black* dashed line, while Watson-Crick hydrogen bonds are represented by the *green* dashed line. (B) Crystal structure of the BWYV frameshifting pseudoknot solved to 1.25 Å (PDB accession code 1L2X).[137] The positions of three of metal ions visualized in the structure, including a fully hydrated Mg^{2+} ion in the major groove of S1, are shown as *spheres* and the base quadruple motif[51] is boxed. Pseudoknot secondary structural elements are shown: S1, stem 1; S2; stem 2; L1, loop 1; L2, loop 2).

shifted upwards.[138] In several cases, the thermodynamic effect of protonation at adenosine N1 (unperturbed pK_a of 3.8) and cytidine N3 (pK_a of 4.3) on the structure and stability of small- and medium-sized RNAs, *e.g.* pseudoknots[51,139] and hairpins,[140] in addition to complex folded RNAs, including the 58-nucleotide rRNA that binds ribosomal protein L11[42] and the hepatitis delta virus (HDV) genomic ribozyme (Figure 6.8A),[141] have been examined in some detail. The protonation of adenosine N1 will stabilize $A^+ \cdot C$ wobble pairing, and the protonation of cytidine N3 will stabilize Hoogsteen pairing between cytidine and guanosine, typically in context of a $C^+ \cdot (G\text{–}C)$ base triple [or a $A \cdot C^+ \cdot (G\text{–}C)$ base quadruple interaction] buried deep in the major groove (see Figure 6.8A, right).[51] The degree to which the latter interaction stabilizes a family of plant luteoviral frameshifting pseudoknots in the context of a base quadruple interaction has been evaluated for a number of these RNAs (Figure 6.8B).[51,139] These data reveal that protonation strongly stabilizes the N form, giving rise to a $\Delta\Delta G_{37}$ of ~ 3.5 kcal mol^{-1} for the BWYV RNA at 0.5 M monovalent ion concentration, with a folding pK_a estimated with a simple site-binding model of

~6.8.[51] Subsequent computational studies carried out in an effort to estimate the intrinsic pK_a for this group suggest a pK_a in excess of 10 for the related pea enation mosaic virus-1 (PEMV-1) pseudoknot[139,142] and also for the BWYV pseudoknot.[138] NMR experiments that quantify the intensity of the resonances associated with the downfield-shifted N4 amino protons of a related pseudoknot from sugar cane yellow leaf virus (ScYLV)[143] measured as a function of pH are consistent with this picture (D. P. Giedroc and P. V. Cornish, unpublished observations). Although no attempt has yet been made to examine the linkage of protonation of this site to Mg^{2+}-induced stability in these systems, which is well modeled by diffusely associated ions,[51,52] nonlinear Poisson–Boltzmann calculations of the protonated and unprotonated forms of the RNA calculations suggest that strong electrostatic stabilization with phosphate groups, base–base hydrogen bonding and a favorable desolvation free energy, may make significant contributions to raising the pK_a in this and other systems.[138]

The HDV genomic ribozyme is known to bind two Mg^{2+} ions, one at or near the self-cleavage site containing a protonated C75 (catalytic Mg^{2+}) and one found in the major groove below a positively charged $A \cdot C^+ \cdot (G\text{–}C)$ base quadruple that is isostructural[51] to that present in the ScYLV and BWYV pseudoknots (structural Mg^{2+}) some 15 Å distant (see Figure 6.8, left). Bevilacqua and coworkers have extensively examined the degree to which protonation of C75 and C41 is linked to catalysis and Mg^{2+}-dependent stability, respectively.[144–148] The WT RNA exhibits a striking pH dependence on the Mg^{2+} binding affinity to *both* the catalytic and structural metal sites; this linkage is abolished for the structural site *only* when the accepting G61 (G73 in the native sequence)–C44 base pair is replaced with a A61–U44 base pair that will not require protonation of C44 N3 to mediate pairing.[148] Interestingly, the thermodynamic coupling of Mg^{2+} binding to this structural site and protonation of the quadruple is favored by ~ -2.0 kcal mol^{-1}, with higher affinity at lower pH ($K_{d,app} = 0.14$ mM at 1.0 M NaCl). Although the origin of this effect is not known, the local charge neutralization required to accommodate inner-sphere coordination of this particular Mg^{2+} ion is one possibility, among others.[148] In contrast, proton and Mg^{2+} binding to the catalytic metal site is anticooperative;[144] a similar situation characterizes the $A^+ \cdot C$ wobble pair in the U6 spliceosomal RNA.[140] Studies such as these, while revealing the complexity of these linkage relationships, seem to establish that site-selective base protonation and Mg^{2+} binding to structural sites on RNA is often two sides of the same coin, in that both mitigate the effects of strong electrostatic repulsion when RNA folds. The catalysis of the HDV genomic ribozyme is discussed further in Chapter 8.

6.5 How Mg^{2+} Stabilizes RNA from Single-molecule Unfolding Studies Using Mechanical Force

The effect of mechanical force on unfolding RNA molecules using optical tweezers[149] has been pioneered by the Bustamante and Tinoco groups as a means to obtain insight into the kinetics and thermodynamics of RNA unfolding/folding.[150] In a typical implementation of this experiment, the free energy difference between relevant folded and partially folded states is obtained by measuring the kinetics of unfolding and refolding. Increasing the mechanical force acting on a single RNA molecule is the only method to do this and is now well grounded in the literature;[151–153] furthermore, mechanical force mimics the effect of enzymes (helicases, polymerases, *etc.*) and large ribonucleoprotein machines, *e.g.* the ribosome, on nucleic acid substrates.[124,150,154] In addition, the kinetics and thermodynamics are measured at physiological temperatures and pressures, unlike the thermal melting experiments discussed above, with a potentially short extrapolation back to zero force.

A common experimental set-up has been described by the Bustamante/Tinoco groups where the RNA is tethered to two beads via ~500-bp handles, with one bead immobilized by a dual-beam laser optical trap and the other mounted on a stage held by a micropipette (Figure 6.9A). When the stage is moved, tension is induced into the RNA; conformational changes in the RNA give rise to a change in extension, Δx, which in turn causes the trapped bead to move, thus changing the force.[155] Forces in the tens of piconewton range are accessible by this experiment, with the precision of holding a particular force constant to ± 0.1 pN. The ability of an RNA molecule to resist mechanically induced unfolding is directly related to the free energy difference in addition to the kinetics of unfolding/refolding that characterize a particular unfolding step. Since RNA unfolding (folding) is hierarchical,[69] then the first unfolding transition will be tertiary structural disruption followed by unfolding of the underlying secondary structural elements.[51,143] If the kinetic barrier to unfolding is low, then Δx for an unfolding transition will occur over a relatively narrow range of applied force; these are characteristics of 'compliant' structures that fold reversibly, with a transition state roughly midway between the F and U states; simple stem–loops exhibit this behavior,[156] at least for shorter helices.[157]

In other more complex RNAs, investigation of multiple force-ramp folding–unfolding trajectories often reveal that Δx for an unfolding transition occurs over a very wide range of forces on successive 'pulls'. In this case, unfolding of the tertiary structure is rate limiting and RNA unfolding is kinetically controlled. Here, in some trajectories, the unfolding force, F, needed to induce $\Delta x(F_{1/2})$ may exceed the force required to unfold the underlying tertiary structure. This would be evidence that the transition state for pseudoknot unfolding is structurally similar to the folded structure, *i.e.* the pseudoknot is a brittle[156] structure that resists mechanical deformation to varying degrees over a wide range of forces, but fractures completely once that force is reached. Thus, these tertiary structural interactions dramatically increase the kinetic stability of the pseudoknot by 'blocking the transmission of force'[155] to underlying secondary structural domains. As a result, the dependence of the rate constant for unfolding, k_{unfold}, will be relatively shallow. Since long-lived Mg^{2+} ions are often associated with tertiary structure, these ions function as

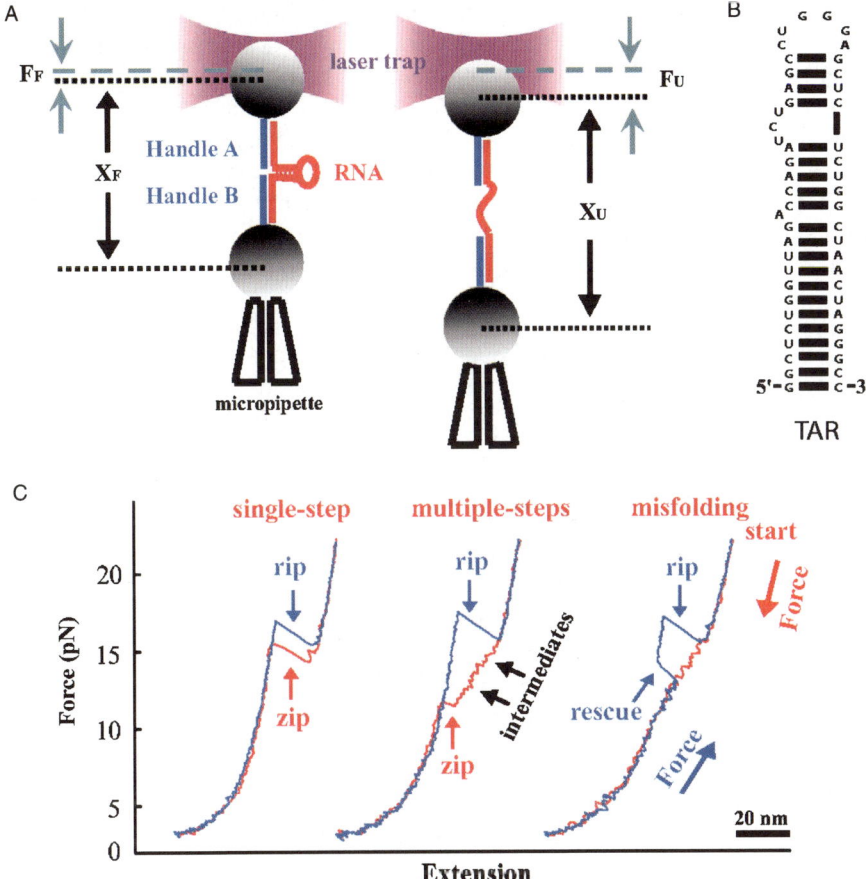

Figure 6.9 Typical optical-tweezers assay for studying folding, misfolding and rescue of HIV-1 TAR RNA.[164] (A) Experimental design of the assay. The TAR hairpin is flanked by dsDNA/RNA handles and tethered to two microspheres, one of which is held by a force-measuring trap and the other is held on a micropipette. By moving the micropipette, tension (*F*) is exerted on the molecule and the change in extension (*x*) is measured. (B) Native structure of the HIV-1 TAR hairpin. See Figure 6.3 for the solution structure of this RNA. (C) Representative classes of force-extension curves for TAR RNA that differ in the rate (slower → faster) at which the RNA is originally relaxed. In each experiment, force was first relaxed from 20 to 1 pN (*red*) and then was raised again to 20 pN (*blue*). Unfolding is indicated by rips that increase the extension, whereas folding and rescue are indicated by zips that shorten the extension. Arrows represent the fluctuations/intermediates observed on refolding. Reproduced from ref. 164 with permission from the National Academy of Sciences of the United States of America.

kinetic barriers to unfolding of RNA.[158] The key mechanistic feature of these experiments is in $x_{\text{unfold}}^{\ddagger}$ and $x_{\text{fold}}^{\ddagger}$. For a compliant structure, $x_{\text{unfold}}^{\ddagger} \approx x_{\text{fold}}^{\ddagger}$, *i.e.* the transition state is mid-way between the folded and unfolded states.[159] For a brittle structure, $x_{\text{unfold}}^{\ddagger}$ might be small relative to $x_{\text{fold}}^{\ddagger}$.

The ability to manipulate the folding landscape of single RNA molecules by force provides a new opportunity to investigate microscopic heterogeneity implicit in a folding landscape[156] and has now been applied to this study of simple RNA hairpins, frameshift-stimulating RNA pseudoknots,[160] a kissing loop interaction[155] and more complex RNAs, including the L-21 *Tetrahymena*[158] and *Azoarcus*[153] group I introns and the HCV IRES domain.[159] For complex RNAs, there is now considerable evidence consistent with the idea that the unfolding of tertiary structure,[158] or the unkissing of two kissing hairpin loops, (Figure 6.10) provides substantial kinetic barriers to force-induced unfolding

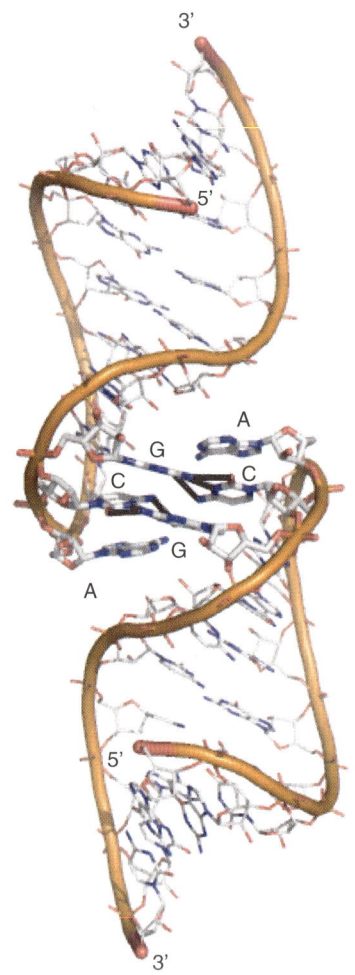

Figure 6.10 Representation of a kissing loop interaction formed between GACG loop sequences (PDB accession code 1F5U).[174] The C and G residues in the loop form canonical hydrogen bonding interactions as indicated, with the 5′ adenosines stacked on these base pairs. This structure has been shown to possess remarkable mechanical stability (see text for details).[155]

such that component secondary structures are stabilized to a degree far greater than they would be in isolation.[155] Another manifestation of this behavior is that unfolding–refolding trajectories often display substantial hysteresis, consistent with the fact that folding/unfolding trajectories are only at equilibrium at the slowest pulling/relaxation rates (see Figure 6.9).[156,159] In the case of the kissing hairpins, the forces over which the two hairpins unkiss from run to run is substantial, which gives a very shallow force dependence of the unkissing reaction, i.e. $x_{unkiss}^{\ddagger} \ll \Delta x$, consistent with designation of kissing hairpins as a brittle structure of unusual mechanical stability.[155] Here, the disruption of long-range tertiary structural interactions is rate limiting for unfolding in these cases. The relative influence of mono- vs divalent metal ions has not yet been investigated for kissing hairpin interaction; on the other hand, Mg^{2+} localization to a conserved asymmetric internal loop just below the kissing loops in the natural context arrests global mobility in a way that seems to regulate the nucleocapsid protein dependence of the process.[161]

It will be interesting to determine if ribosomal frameshift-stimulating mRNA pseudoknots behave as mechanically stable but brittle structures, in order to withstand the force of the elongating ribosome; this may increase the likelihood that the ribosome will shift translational reading frames, some, but not all, of the time, the frameshifting signal is encountered.[160,162,163] In this case, monovalent and divalent ions might be expected to function similarly, given the fact that pseudoknots appear to be stabilized exclusively by 'diffusely associated' ions from thermal unfolding experiments.[51,52,133,135] This is in contrast to the *Tetrahymena* group I intron, which was found to exhibit eight unfolding intermediates projected to correspond to identifiable kinetic barriers to mechanical unfolding.[158] In that work, the position of barriers along the force-unfolding–refolding trajectories were found to be strongly Mg^{2+} dependent and could be assigned to tertiary structural interactions; again, pronounced hysteresis could be observed in folding unfolding trajectories.

In general, simple RNA hairpins seem to be well described by a single reversible two-state, highly cooperative unfolding transition, with $-x_{fold}^{\ddagger} \approx x_{unfold}^{\ddagger}$,[159] this equality may be more strongly sequence dependent than had been otherwise appreciated.[157] Even in this simple case, recent unfolding–refolding experiments with a model RNA hairpin, the transactivation response region (TAR) hairpin (Figures 6.2A and 6.9B), suggest that this two-state folding behavior (if ramped from high force to low force, pH 8.0, 0.1 M K^+) is highly conditional and depends strongly on the force relaxation rate (Figure 6.9C).[164] Only at very slow relaxation rates was a single two-state unfolding transition observed in both the folding (zip) and unfolding (rip) directions (Figure 6.9C). At slightly faster pull rates, intermediates during folding were observed, with a single cooperative unfolding transition on the loading trajectory. At slightly faster unloading/reloading rates still, there was no evidence of a folding transition, with the unfolding trajectories characterized by a 'rescue' zip at $F < F_{1/2}$; characterization of mutant TAR RNAs coupled with measurement of Δx was consistent with the idea that the RNA had misfolded as the force was lowered and increased force was required to 'melt out' these non-native kinetically trapped structures to get to the N-state, which were then unfolded at $F<F_{1/2}$. Although it has not yet been reported if representative trajectories of all three classes are observed in the presence of Mg^{2+} vs K^+, it seems likely that this would be the case, given the findings that diffuse ion binding is necessary and sufficient to stabilize the bent structure of the TAR RNA hairpin.[107] These studies reveal the remarkable ability of mechanical force measurements to access various regions of the (un)folding landscape, from which a systematic analysis of the influence of monovalent vs divalent metal ions can now be undertaken for more complex RNA molecules.

6.6 Conclusion and Perspective

In this chapter, we have summarized salient features of RNA–Mg^{2+} interactions in complex RNAs and how these interactions function to either drive the folding of RNAs from an intermediate state that lacks some or all of the tertiary structure or stabilize the fully folded N-state relative to that can be achieved by monovalent ions. A recently described thermodynamic framework based on established paradigms in linkage theory, coupled with the utilization of direct methods to monitor Mg^{2+} binding readily, are significant steps forward in deconvoluting the complexity inherent in understanding any coupled equilibrium.[114] Computational approaches, e.g. the non-linear Poisson–Boltzmann model[38,56] and related Poisson–Boltzmann approaches,[138] coupled with the design of experiments that allow one to resolve the importance of divalent vs monovalent ions, will continue to play an ever increasing role in our understanding of this problem. The explosion of high-resolution structures of complex RNAs to a resolution sufficient to confidently identify specific features of the first coordination shell of both monovalent and divalent ions[80] will continue to inspire systematic attempts to understand the thermodynamic underpinnings of these interactions. This, in fact, is the greatest challenge. For example, the degree to which Mg^{2+} ions truly shed one or more tightly bound water molecules in solution when bound to RNA sites (see Figure 6.1C) and the degree to which metal-binding sites are selectivity bound to K^+ vs Mg^{2+} in solution[16] remain undetermined in most cases; however, what is clear is that a large fraction of new crystallographic structures are characterized by bound K^+ and Mg^{2+} ions with at least one and, in some cases, many more inner-sphere coordination bonds.[16,23,27,74,80]

In this regard, the structural and evolving biochemical characterization of two riboswitches that have appeared in the last year or so, the Mg^{2+}-sensing riboswitch[74] (Figure 6.1C) and glmS catalytic riboswitch (Figure 6.1D),[165–168] provide new opportunities to investigate old and new questions alike concerning the thermodynamics of Mg^{2+}–RNA interactions. For example, the glmS riboswitch appears fully folded in the absence of the activating small molecule metabolite, glucosamine-6-phosphate, GlcN6P,[167] and also competitive inhibitors, and may well be largely folded in the absence of divalent ions. This would allow the rigorous, model-independent deconvolution of the degree to which

Mg^{2+} binding is thermodynamically coupled to GlcN6P binding; the structure suggests that this coupling might be sizable, given the fact that at least of two of the inner-sphere ligands to a structural Mg^{2+} ion (M6; see Figure 6.1D) in the effector binding site is derived from the metabolite itself.[24] This is also true of the thiamine pyrophosphate (TPP)-sensing riboswitch[23,169] in which the entire ligand is buried and stabilized by hydrogen bonding, base stacking and inner-sphere coordination bonds formed by structural Mg^{2+} ions.[24]

The Mg^{2+}-sensing riboswitch, on the other hand, provides an exciting opportunity to compare and contrast Mg^{2+} binding sites that are required for Mg^{2+} sensing *versus* general RNA folding in the cell.[74] Mg^{2+} sensing sites may well have distinct structural and thermodynamic signatures which might effectively tune the Mg^{2+} binding affinity to the $10^4 M^{-1}$ range (in 2.1 M monovalent salt concentration) in driving the highly cooperative formation of the folded riboswitch. Although these sensing 'sites' are apparently not selective for Mg^{2+}, since Ca^{2+}, Sr^{2+}, Mn^{2+} and Co^{2+} induce the conformational change *in vitro* (at 10 mM), in the cell, these latter ions do not down-regulate *mgtE* gene expression *in vivo*.[74] This observation is consistent with the idea that the prevailing concentration of the metal ion in the cytoplasm of the cell will play a major role in setting the biological selectivity of this riboswitch rather than the intrinsic structure and thermodynamics of formation of the coordination chelate(s) itself.[170,171] In this way of thinking, it may not be the case that RNA Mg^{2+} sensing sites are easily distinguished from structural sites in other folded RNAs; in fact, a comparison of these metal binding sites with those in the 58-nucleotide RNA and *Azoarcus* group I intron seem to suggest many superficial similarities (Figure 6.1). We anxiously await the results of detailed thermodynamic studies along the lines outlined here that will provide support for or against this proposal.

Acknowledgements

We wish to thank Dr Wade Winkler, University of Texas Southwestern Medical School, for sharing the atomic coordinates for the Mg^{2+}-sensing riboswitch prior to release by the RCSB, and Dr David Draper, Johns Hopkins University, for comments on the manuscript. Financial support from the US National Institutes of Health (R01 AI040187) to D.P.G. is gratefully acknowledged.

References

1. G. Rialdi, J. Levy and R. Biltonen, *Biochemistry*, 1972, **11**, 2472.
2. R. Romer and R. Hach, *Eur. J. Biochem.*, 1975, **55**, 271.
3. A. Stein and D. M. Crothers, *Biochemistry*, 1976, **15**, 160.
4. A. Stein and D. M. Crothers, *Biochemistry*, 1976, **15**, 157.
5. J. D. Robertus, J. E. Ladner, J. T. Finch, D. Rhodes, R. S. Brown, B. F. Clark and A. Klug, *Nature*, 1974, **250**, 546.

6. S. H. Kim, F. L. Suddath, G. J. Quigley, A. McPherson, J. L. Sussman, A. H. Wang, N. C. Seeman and A. Rich, *Science*, 1974, **185**, 435.
7. H. Shi and P. B. Moore, *RNA*, 2000, **6**, 1091.
8. A. Y. Kobitski, A. Nierth, M. Helm, A. Jaschke and G. U. Nienhaus, *Nucleic Acids Res.*, 2007, **35**, 2047.
9. H. W. Pley, K. M. Flaherty and D. B. McKay, *Nature*, 1994, **372**, 68.
10. W. G. Scott, J. T. Finch and A. Klug, *Nucleic Acids Symp. Ser.*, 1995, 214.
11. J. H. Cate, A. R. Gooding, E. Podell, K. Zhou, B. L. Golden, C. E. Kundrot, T. R. Cech and J. A. Doudna, *Science*, 1996, **273**, 1678.
12. G. L. Conn, D. E. Draper, E. E. Lattman and A. G. Gittis, *Science*, 1999, **284**, 1171.
13. B. T. Wimberly, D. E. Brodersen, W. M. J. Clemons, R. J. Morgan-Warren, A. P. Carter, C. Vonrhein, T. Hartsch and V. Ramakrishnan, *Nature*, 2000, **407**, 327.
14. N. Ban, P. Nissen, J. Hansen, P. B. Moore and T. A. Steitz, *Science*, 2000, **289**, 905.
15. S. Gutmann, P. W. Haebel, L. Metzinger, M. Sutter, B. Felden and N. Ban, *Nature*, 2003, **424**, 699.
16. P. L. Adams, M. R. Stahley, M. L. Gill, A. B. Kosek, J. Wang and S. A. Strobel, *RNA*, 2004, **10**, 1867.
17. B. L. Golden, H. Kim and E. Chase, *Nat. Struct. Mol. Biol.*, 2005, **12**, 82.
18. F. Guo, A. R. Gooding and T. R. Cech, *Mol. Cell*, 2004, **16**, 351.
19. M. R. Stahley and S. A. Strobel, *Science*, 2005, **309**, 1587.
20. Q. Vicens and T. R. Cech, *Trends Biochem. Sci.*, 2006, **31**, 41.
21. R. T. Batey, S. D. Gilbert and R. K. Montange, *Nature*, 2004, **432**, 411.
22. A. Serganov, Y. R. Yuan, O. Pikovskaya, A. Polonskaia, L. Malinina, A. T. Phan, C. Hobartner, R. Micura, R. R. Breaker and D. J. Patel, *Chem. Biol.*, 2004, **11**, 1729.
23. A. Serganov, A. Polonskaia, A. T. Phan, R. R. Breaker and D. J. Patel, *Nature*, 2006, **441**, 1167.
24. C. A. Wakeman, W. C. Winkler and C. E. Dann III, *Trends Biochem. Sci.*, 2007, **32**, 415.
25. A. Torres-Larios, K. K. Swinger, A. S. Krasilnikov, T. Pan and A. Mondragon, *Nature*, 2005, **437**, 584.
26. A. V. Kazantsev, A. A. Krivenko, D. J. Harrington, S. R. Holbrook, P. D. Adams and N. R. Pace, *Proc. Natl. Acad. Sci. U.S.A.*, 2005, **102**, 13392.
27. A. Serganov, S. Keiper, L. Malinina, V. Tereshko, E. Skripkin, C. Hobartner, A. Polonskaia, A. T. Phan, R. Wombacher, R. Micura, Z. Dauter, A. Jaschke and D. J. Patel, *Nat. Struct. Mol. Biol.*, 2005, **12**, 218.
28. M. P. Robertson and W. G. Scott, *Science*, 2007, **315**, 1549.
29. W. G. Scott, *Curr. Opin. Struct. Biol.*, 2007, **17**, 280.
30. M. Martick and W. G. Scott, *Cell*, 2006, **126**, 309.
31. W. G. Scott, *Biol. Chem.*, 2007, **388**, 727.
32. B. S. Schuwirth, M. A. Borovinskaya, C. W. Hau, W. Zhang, A. Vila-Sanjurjo, J. M. Holton and J. H. Cate, *Science*, 2005, **310**, 827.

33. M. Selmer, C. M. Dunham, F. V. T. Murphy, A. Weixlbaumer, S. Petry, A. C. Kelley, J. R. Weir and V. Ramakrishnan, *Science*, 2006, **313**, 1935.
34. A. Korostelev, S. Trakhanov, M. Laurberg and H. F. Noller, *Cell*, 2006, **126**, 1065.
35. M. V. Rodnina, M. Beringer and W. Wintermeyer, *Trends Biochem. Sci.*, 2007, **32**, 20.
36. A. Korostelev and H. F. Noller, *Trends Biochem. Sci.*, 2007, **32**, 434.
37. A. Korostelev and H. F. Noller, *J. Mol. Biol.*, 2007, **37**,1058.
38. V. K. Misra and D. E. Draper, *J. Mol. Biol.*, 2000, **299**, 813.
39. V. M. Shelton, T. R. Sosnick and T. Pan, *Biochemistry*, 2001, **40**, 3629.
40. L. G. Laing and D. E. Draper, *J. Mol. Biol.*, 1994, **237**, 560.
41. D. E. Draper, Y. Xing and L. G. Laing, *J. Mol. Biol.*, 1995, **249**, 231.
42. G. L. Conn, R. R. Gutell and D. E. Draper, *Biochemistry*, 1998, **37**, 11980.
43. P. Rangan and S. A. Woodson, *J. Mol. Biol.*, 2003, **329**, 229.
44. S. Chauhan, G. Caliskan, R. M. Briber, U. Perez-Salas, P. Rangan, D. Thirumalai and S. A. Woodson, *J. Mol. Biol.*, 2005, **353**, 1199.
45. S. K. Silverman, M. L. Deras, S. A. Woodson, S. A. Scaringe and T. R. Cech, *Biochemistry*, 2000, **39**, 12465.
46. S. L. Heilman-Miller, D. Thirumalai and S. A. Woodson, *J. Mol. Biol.*, 2001, **306**, 1157.
47. E. Koculi, N. K. Lee, D. Thirumalai and S. A. Woodson, *J. Mol. Biol.*, 2004, **341**, 27.
48. X. Fang, T. Pan and T. R. Sosnick, *Biochemistry*, 1999, **38**, 16840.
49. X. W. Fang, T. Pan and T. R. Sosnick, *Nat. Struct. Biol.*, 1999, **6**, 1091.
50. D. P. Giedroc, C. A. Theimer and P. L. Nixon, *J. Mol. Biol.*, 2000, **298**, 167.
51. P. L. Nixon and D. P. Giedroc, *J. Mol. Biol.*, 2000, **296**, 659.
52. A. M. Soto, V. Misra and D. E. Draper, *Biochemistry*, 2007, **46**, 2973.
53. D. E. Draper, *RNA*, 2004, **10**, 335.
54. D. E. Draper, D. Grilley and A. M. Soto, *Annu. Rev. Biophys. Biomol. Struct.*, 2005, **34**, 221.
55. S. A. Woodson, *Curr. Opin. Chem. Biol.*, 2005, **9**, 104.
56. K. Chin, K. A. Sharp, B. Honig and A. M. Pyle, *Nat. Struct. Biol.*, 1999, **6**, 1055.
57. K. A. Dill and S. Bromberg, *Molecular Driving Forces*, Garland Science, New York, 2002.
58. E. Koculi, C. Hyeon, D. Thirumalai and S. A. Woodson, *J. Am. Chem. Soc.*, 2007, **129**, 2676.
59. R. G. Pearson, *J. Chem. Educ.*, 1968, **45**, 581.
60. R. G. Pearson, *J. Chem. Educ.*, 1968, **45**, 643.
61. J. L. Hougland, A. V. Kravchuk, D. Herschlag and J. A. Piccirilli, *PLoS Biol*, 2005, **3**, e277.
62. R. Knoll, R. Bald and J. P. Furste, *RNA*, 1997, **3**, 132.
63. D. E. Draper and V. K. Misra, *Nat. Struct. Biol.*, 1998, **5**, 927.

64. P. L. Nixon, C. A. Theimer and D. P. Giedroc, *Biopolymers*, 1999, **50**, 443.
65. J. H. Cate, A. R. Gooding, E. Podell, K. Zhou, B. L. Golden, A. A. Szewczak, C. E. Kundrot, T. R. Cech and J. A. Doudna, *Science*, 1996, **273**, 1696.
66. S. Basu, R. P. Rambo, J. Strauss-Soukup, J. H. Cate, A. R. Ferre-D'Amare, S. A. Strobel and J. A. Doudna, *Nat. Struct. Biol.*, 1998, **5**, 986.
67. G. L. Conn, A. G. Gittis, E. E. Lattman, V. K. Misra and D. E. Draper, *J. Mol. Biol.*, 2002, **318**, 963.
68. C. D. Downey, J. L. Fiore, C. D. Stoddard, J. H. Hodak, D. J. Nesbitt and A. Pardi, *Biochemistry*, 2006, **45**, 3664.
69. I. Tinoco Jr and C. Bustamante, *J. Mol. Biol.*, 1999, **293**, 271.
70. B. Onoa and I. Tinoco Jr, *Curr. Opin. Struct. Biol.*, 2004, **14**, 374.
71. K. A. Wilkinson, E. J. Merino and K. M. Weeks, *J. Am. Chem. Soc.*, 2005, **127**, 4659.
72. H. W. Pley, K. M. Flaherty and D. B. McKay, *Nature*, 1994, **372**, 111.
73. A. Laederach, I. Shcherbakova, M. A. Jonikas, R. B. Altman and M. Brenowitz, *Proc. Natl. Acad. Sci. U.S.A.*, 2007, **104**, 7045.
74. C. E. Dann 3rd, C. A. Wakeman, C. L. Sieling, S. C. Baker, I. Irnov and W. C. Winkler, *Cell*, 2007, **130**, 878.
75. G. A. Soukup and R. R. Breaker, *RNA*, 1999, **5**, 1308.
76. J. A. Latham and T. R. Cech, *Science*, 1989, **245**, 276.
77. T. Uchida, K. Takamoto, Q. He, M. R. Chance and M. Brenowitz, *J. Mol. Biol.*, 2003, **328**, 463.
78. E. J. Merino, K. A. Wilkinson, J. L. Coughlan and K. M. Weeks, *J. Am. Chem. Soc.*, 2005, **127**, 4223.
79. D. A. Harris, R. A. Tinsley and N. G. Walter, *J. Mol. Biol.*, 2004, **341**, 389.
80. M. R. Stahley, P. L. Adams, J. Wang and S. A. Strobel, *J. Mol. Biol.*, 2007, **372**, 89.
81. V. K. Misra and D. E. Draper, *Proc. Natl. Acad. Sci. U.S.A.*, 2001, **98**, 12456.
82. R. Russell, I. S. Millett, M. W. Tate, L. W. Kwok, B. Nakatani, S. M. Gruner, S. G. Mochrie, V. Pande, S. Doniach, D. Herschlag and L. Pollack, *Proc. Natl. Acad. Sci. U.S.A.*, 2002, **99**, 4266.
83. J. Lipfert, R. Das, V. B. Chu, M. Kudaravalli, N. Boyd, D. Herschlag and S. Doniach, *J. Mol. Biol.*, 2007, **365**, 1393.
84. J. Lipfert and S. Doniach, *Annu. Rev. Biophys. Biomol. Struct.*, 2007, **36**, 307.
85. C. Y. Ralston, B. Sclavi, M. Sullivan, M. L. Deras, S. A. Woodson, M. R. Chance and M. Brenowitz, *Methods Enzymol.*, 2000, **317**, 353.
86. Y. Bai, R. Das, I. S. Millett, D. Herschlag and S. Doniach, *Proc. Natl. Acad. Sci. U.S.A.*, 2005, **102**, 1035.
87. K. Takamoto, R. Das, Q. He, S. Doniach, M. Brenowitz, D. Herschlag and M. R. Chance, *J. Mol. Biol.*, 2004, **343**, 1195.

88. S. A. Woodson, *Nat. Struct. Biol.*, 2000, **7**, 349.
89. D. Thirumalai, N. Lee, S. A. Woodson and D. Klimov, *Annu. Rev. Phys. Chem.*, 2001, **52**, 751.
90. D. Grilley, V. Misra, G. Caliskan and D. E. Draper, *Biochemistry*, 2007, **46**, 10266.
91. A. E. Webb and K. M. Weeks, *Nat. Struct. Biol.*, 2001, **8**, 135.
92. K. L. Buchmueller and K. M. Weeks, *Biochemistry*, 2003, **42**, 13869.
93. L. J. Su, M. Brenowitz and A. M. Pyle, *J. Mol. Biol.*, 2003, **334**, 639.
94. H. R. Huang, C. E. Rowe, S. Mohr, Y. Jiang, A. M. Lambowitz and P. S. Perlman, *Proc. Natl. Acad. Sci. U.S.A.*, 2005, **102**, 163.
95. G. Bokinsky, L. G. Nivon, S. Liu, G. Chai, M. Hong, K. M. Weeks and X. Zhuang, *J. Mol. Biol.*, 2006, **361**, 771.
96. J. R. Tolman, H. M. Al-Hashimi, L. E. Kay and J. H. Prestegard, *J. Am. Chem. Soc.*, 2001, **123**, 1416.
97. C. Musselman, S. W. Pitt, K. Gulati, L. L. Foster, I. Andricioaei and H. M. Al-Hashimi, *J. Biomol. NMR*, 2006, **36**, 235.
98. M. H. Bailor, C. Musselman, A. L. Hansen, K. Gulati, D. J. Patel and H. M. Al-Hashimi, *Nat. Protocols*, 2007, **2**, 1536.
99. H. M. Al-Hashimi, Y. Gosser, A. Gorin, W. Hu, A. Majumdar and D. J. Patel, *J. Mol. Biol.*, 2002, **315**, 95.
100. H. M. Al-Hashimi, S. W. Pitt, A. Majumdar, W. Xu and D. J. Patel, *J. Mol. Biol.*, 2003, **329**, 867.
101. S. W. Pitt, A. Majumdar, A. Serganov, D. J. Patel and H. M. Al-Hashimi, *J. Mol. Biol.*, 2004, **338**, 7.
102. S. W. Pitt, Q. Zhang, D. J. Patel and H. M. Al-Hashimi, *Angew. Chem. Int. Ed.*, 2005, **44**, 3412.
103. J. D. Puglisi, R. Tan, B. J. Calnan, A. D. Frankel and J. R. Williamson, *Science*, 1992, **257**, 76.
104. M. Hennig and J. R. Williamson, *Nucleic Acids Res.*, 2000, **28**, 1585.
105. T. Ha, X. Zhuang, H. D. Kim, J. W. Orr, J. R. Williamson and S. Chu, *Proc. Natl. Acad. Sci. U.S.A.*, 1999, **96**, 9077.
106. J. A. Ippolito and T. A. Steitz, *Proc. Natl. Acad. Sci. U.S.A.*, 1998, **95**, 9819.
107. A. Casiano-Negroni, X. Sun and H. M. Al-Hashimi, *Biochemistry*, 2007, **46**, 6525.
108. D. J. Klein, T. M. Schmeing, P. B. Moore and T. A. Steitz, *EMBO J.*, 2001, **20**, 4214.
109. J. Liu and D. M. Lilley, *RNA*, 2007, **13**, 200.
110. T. A. Goody, S. E. Melcher, D. G. Norman and D. M. Lilley, *RNA*, 2004, **10**, 254.
111. V. K. Misra and D. E. Draper, *Biopolymers*, 1998, **48**, 113.
112. V. K. Misra, R. Shiman and D. E. Draper, *Biopolymers*, 2003, **69**, 118.
113. V. K. Misra and D. E. Draper, *J. Mol. Biol.*, 2002, **317**, 507.
114. D. Grilley, A. M. Soto and D. E. Draper, *Proc. Natl. Acad. Sci. U.S.A.*, 2006, **103**, 14003.

115. R. Das, K. J. Travers, Y. Bai and D. Herschlag, *J. Am. Chem. Soc.*, 2005, **127**, 8272.
116. J. Wyman Jr, *Adv. Protein Chem.*, 1964, **19**, 223.
117. M. L. VanZile, X. Chen and D. P. Giedroc, *Biochemistry*, 2002, **41**, 9765.
118. L. Su, L. Chen, M. Egli, J. M. Berger and A. Rich, *Nat. Struct. Biol.*, 1999, **6**, 285.
119. T. E. Horton, D. R. Clardy and V. J. DeRose, *Biochemistry*, 1998, **37**, 18094.
120. X. Zhuang, *Annu. Rev. Biophys. Biomol. Struct.*, 2005, **34**, 399.
121. G. Bokinsky and X. Zhuang, *Acc. Chem. Res.*, 2005, **38**, 566.
122. S. Myong, B. C. Stevens and T. Ha, *Structure*, 2006, **14**, 633.
123. P. V. Cornish and T. Ha, *ACS Chem. Biol.*, 2007, **2**, 53.
124. W. J. Greenleaf, M. T. Woodside and S. M. Block, *Annu. Rev. Biophys. Biomol. Struct.*, 2007, **36**, 171.
125. J. H. Hodak, C. D. Downey, J. L. Fiore, A. Pardi and D. J. Nesbitt, *Proc. Natl. Acad. Sci. U.S.A.*, 2005, **102**, 10505.
126. G. Bokinsky, D. Rueda, V. K. Misra, M. M. Rhodes, A. Gordus, H. P. Babcock, N. G. Walter and X. Zhuang, *Proc. Natl. Acad. Sci. U.S.A.*, 2003, **100**, 9302.
127. C. D. Downey, R. L. Crisman, T. W. Randolph and A. Pardi, *J. Am. Chem. Soc.*, 2007, **129**, 9290.
128. D. Lambert and D. E. Draper, *J. Mol. Biol.*, 2007, **370**, 993.
129. V. M. Shelton, T. R. Sosnick and T. Pan, *Biochemistry*, 1999, **38**, 16831.
130. T. C. Gluick and S. Yadav, *J. Am. Chem. Soc.*, 2003, **125**, 4418.
131. R. Shiman and D. E. Draper, *J. Mol. Biol.*, 2000, **302**, 79.
132. L. G. Laing, T. C. Gluick and D. E. Draper, *J. Mol. Biol.*, 1994, **237**, 577.
133. T. C. Gluick, N. M. Wills, R. F. Gesteland and D. E. Draper, *Biochemistry*, 1997, **36**, 16173.
134. P. L. Nixon and D. P. Giedroc, *Biochemistry*, 1998, **37**, 16116.
135. C. A. Theimer and D. P. Giedroc, *RNA*, 2000, **6**, 409.
136. C. A. Theimer and D. P. Giedroc, *J. Mol. Biol.*, 1999, **289**, 1283.
137. M. Egli, G. Minasov, L. Su and A. Rich, *Proc. Natl. Acad. Sci. U.S.A.*, 2002, **99**, 4302.
138. C. L. Tang, E. Alexov, A. M. Pyle and B. Honig, *J. Mol. Biol.*, 2007, **366**, 1475.
139. P. L. Nixon, P. V. Cornish, S. V. Suram and D. P. Giedroc, *Biochemistry*, 2002, **41**, 10665.
140. A. Huppler, L. J. Nikstad, A. M. Allmann, D. A. Brow and S. E. Butcher, *Nat. Struct. Biol.*, 2002, **9**, 431.
141. A. R. Ferre-D'Amare and J. A. Doudna, *J. Mol. Biol.*, 2000, **295**, 541.
142. P. L. Nixon, A. Rangan, Y. G. Kim, A. Rich, D. W. Hoffman, M. Hennig and D. P. Giedroc, *J. Mol. Biol.*, 2002, **322**, 621.
143. P. V. Cornish, M. Hennig and D. P. Giedroc, *Proc. Natl. Acad. Sci. U.S.A.*, 2005, **102**, 12694.

144. S. Nakano, D. M. Chadalavada and P. C. Bevilacqua, *Science*, 2000, **287**, 1493.
145. S. Nakano, D. J. Proctor and P. C. Bevilacqua, *Biochemistry*, 2001, **40**, 12022.
146. S. Nakano, A. L. Cerrone and P. C. Bevilacqua, *Biochemistry*, 2003, **42**, 2982.
147. P. C. Bevilacqua, T. S. Brown, D. Chadalavada, J. Lecomte, E. Moody and S. I. Nakano, *Biochem. Soc. Trans.*, 2005, **33**, 466.
148. S. Nakano and P. C. Bevilacqua, *Biochemistry*, 2007, **46**, 3001.
149. S. M. Block, *Nature*, 1992, **360**, 493.
150. I. Tinoco Jr., P. T. Li and C. Bustamante, *Q. Rev. Biophys.*, 2006, **39**, 325.
151. C. Cecconi, E. A. Shank, C. Bustamante and S. Marqusee, *Science*, 2005, **309**, 2057.
152. J. D. Wen, M. Manosas, P. T. Li, S. B. Smith, C. Bustamante, F. Ritort and I. Tinoco Jr., *Biophys. J.*, 2007, **92**, 2996.
153. C. Hyeon, R. I. Dima and D. Thirumalai, *Structure*, 2006, **14**, 1633.
154. C. Bustamante, Y. R. Chemla, N. R. Forde and D. Izhaky, *Annu. Rev. Biochem.*, 2004, **73**, 705.
155. P. T. Li, C. Bustamante and I. Tinoco Jr., *Proc. Natl. Acad. Sci. U.S.A.*, 2006, **103**, 15847.
156. J. Liphardt, B. Onoa, S. B. Smith, I. J. Tinoco and C. Bustamante, *Science*, 2001, **292**, 733.
157. M. T. Woodside, W. M. Behnke-Parks, K. Larizadeh, K. Travers, D. Herschlag and S. M. Block, *Proc. Natl. Acad. Sci. U.S.A.*, 2006, **103**, 6190.
158. B. Onoa, S. Dumont, J. Liphardt, S. B. Smith, I. Tinoco Jr. and C. Bustamante, *Science*, 2003, **299**, 1892.
159. C. Hyeon and D. Thirumalai, *Biophys. J.*, 2007, **92**, 731.
160. T. M. Hansen, S. N. Reihani, L. B. Oddershede and M. A. Sorensen, *Proc. Natl. Acad. Sci. U.S.A.*, 2007, **104**, 5830.
161. X. Sun, Q. Zhang and H. M. Al-Hashimi, *Nucleic Acids Res.*, 2007, **35**, 1698.
162. E. P. Plant, K. L. Jacobs, J. W. Harger, A. Meskauskas, J. L. Jacobs, J. L. Baxter, A. N. Petrov and J. D. Dinman, *RNA*, 2003, **9**, 168.
163. O. Namy, S. J. Moran, D. I. Stuart, R. J. Gilbert and I. Brierley, *Nature*, 2006, **441**, 244.
164. P. T. Li, C. Bustamante and I. Tinoco Jr., *Proc. Natl. Acad. Sci. U.S.A.*, 2007, **104**, 7039.
165. D. J. Klein and A. R. Ferre-D'Amare, *Science*, 2006, **313**, 1752.
166. D. J. Klein, S. R. Wilkinson, M. D. Been and A. R. Ferre-D'Amare, *J. Mol. Biol.*, 2007, **373**, 178.
167. J. C. Cochrane, S. V. Lipchock and S. A. Strobel, *Chem. Biol.*, 2007, **14**, 97.
168. R. A. Tinsley, J. R. Furchak and N. G. Walter, *RNA*, 2007, **13**, 468.
169. S. Thore, M. Leibundgut and N. Ban, *Science*, 2006, **312**, 1208.

170. D. P. Giedroc and A. I. Arunkumar, *Dalton Trans.*, 2007, **29**, 3107.
171. J. D. Helmann, *Mol. Cell*, 2007, **27**, 859.
172. O. Fedorova, C. Waldsich and A. M. Pyle, *J. Mol. Biol.*, 2007, **366**, 1099.
173. A. M. Pyle, O. Fedorova and C. Waldsich, *Trends Biochem. Sci.*, 2007, **32**, 138.
174. C. H. Kim and I. Tinoco Jr, *Proc. Natl. Acad. Sci. U.S.A.*, 2000, **97**, 9396.

CHAPTER 7
Metal Ions and RNA Folding Kinetics

SOMDEB MITRA AND MICHAEL BRENOWITZ

Department of Biochemistry, Albert Einstein College of Medicine, 1300 Morris Park Avenue, Bronx, NY 10461, USA

7.1 Introduction

7.1.1 A Brief Overview

In order to execute biological function, many RNA molecules must fold into unique three-dimensional structures.[1,2] Attaining these active conformations in a biologically relevant time span is crucial for normal cellular function. This task is by no means simple and has been aptly designated the 'RNA folding problem'.[3] Many decades of research have demonstrated the indispensability of cation-mediated charge neutralization and often the site-specific binding of cations in facilitating RNA folding. Although there is a dauntingly large set of pathways through which RNA molecules with even moderately complex structures can potentially reach their native conformations, only a small number of pathways dominate the folding of the RNA molecules that have been studied.[4-8]

Cations influence the choice of folding pathways by modulating the ensemble of unfolded molecules, affecting the stability of intermediate structures and energetically stabilizing the final native structure over alternate non-native forms.[9-11] In this chapter, we explore the contribution of cations to the time-dependent (kinetic) folding of RNA. The quantitative insight obtained from *in vitro* studies illuminates the physical principles governing the effects of cations on RNA structural dynamics. Although most of the studies in this field have been conducted *in vitro*, cations constitute an integral part of the cellular milieu and are necessary for intracellular stabilization of RNA structures. Hence the fundamental physical principles that emerge from *in vitro* studies can be extended to the physiological milieu. Inside the cell, other processes also influence RNA folding including coupling to transcription, protein binding and enzymatic activity.[12–15] Our view is that the kinetic folding pathways *in vivo* represent a subset of the alternative pathways that are studied *in vitro*.

7.1.2 RNA Folding is Hierarchical

7.1.2.1 Secondary Structure Formation

The folding of transcribed RNA is hierarchical. RNA secondary structure forms on the microsecond time-scale,[16–18] resulting in an array of Watson–Crick base-paired, double-stranded helices whose formation is highly favored by the base pairing and stacking.[19] Secondarily structured RNA is typically a loose collection of helices connected to each other by unpaired single-stranded regions, junctions or internal loops. The *in vitro* studies of RNA folding discussed in this chapter were conducted with molecules whose secondary structure is pre-annealed.[20,21] Such RNA molecules possessing secondary structure but lacking significant stable tertiary structure can be in extended, relaxed or compact conformations, depending on the ionic conditions.[21–28]

7.1.2.2 Tertiary Structure Formation

Tertiary contacts typically form following secondary structure formation or cation-mediated collapse.[22] Thus, tertiary contact formation occurs within a conformational restricted space. RNA compaction that is mediated by tertiary contact formation results in close packing of the duplex helices against each other. This packing can occur on time-scales as short as milliseconds and as long as hours, depending on the size and sequence of the molecule, and also the solution conditions.[29] Compact 'folded' RNA is stabilized by both long- and short-range interactions among different tertiary structural motifs and has the dimensions of the active native form.[19,30] Although favorable, the magnitude of the free energy change generated by tertiary structure formation from the secondary structure elements is much smaller than that accompanying the formation of the secondary structures from the extended chain.[31–37]

7.1.2.3 Parallel Folding Pathways and Kinetic Traps

The secondary structure that forms prior to the formation of tertiary contacts contains much, if not all, of the correct base pairing present in the native folded structure. However, in some cases rearrangement of base pairing accompanies the formation of the biologically active native structure.[38–40] The high stability of base pairing is thus a mixed blessing for RNA folding. On the one hand, it compensates for a major part of the entropic penalty arising from loss of flexibility of the linear polymeric form upon folding to a more rigid arrangement of small rod-like helices. On the other hand, alternative non-native base pairing patterns can have free energy values very close to that of the native form. Incorrectly base paired but energetically compatible structures can compete with the formation of the correctly base-paired native form and hence generate non-native low-energy basins, a type of 'kinetic trap' on the folding landscape of the RNA.[2,6,41–47] The kinetic traps that can limit the rate of folding of RNA molecules can possess native-like alternate topologies and/or non-native secondary or tertiary interactions.[48] Trapped molecules can take minutes if not hours to refold into the native structure. Hence the assembly of RNA molecules has been viewed as 'kinetic partitioning' in which a fraction of the molecules fold fast to the native state, either directly or through a series of structural intermediates, with the remaining fraction kinetically trapped in non-native structures from which they resolve out slowly, again, most likely through a set of intermediate structures.[5,6]

7.1.3 Why are Counterions Necessary for RNA Folding?

7.1.3.1 Stability of Secondary Structure Elements

As discussed extensively in earlier chapters, unfolded RNA can be visualized as a single-stranded polymeric chain with a backbone lined with negatively charged phosphate groups, formally bearing one negative charge per nucleotide. For the formation of secondary structure, the basic modules made up of A-form double-stranded helices, the negatively charged single strands have to overcome a tremendous amount of electrostatic repulsion (Coulombic force). Although this high energetic cost is partially defrayed by the large negative enthalpy change of base pairing, a small concentration of positively charged ions (counterions) is necessary to screen the strong repulsive electrostatic forces.[11,49,50] Completely unfolded RNA can adopt a wide array of conformations because of the flexibility of the backbone and intra-chain interactions, such as base stacking, that influences the shape and stability of the chain. It is challenging to quantitate the energetics of the transition from base-stacked structures to base-paired secondary structural modules.[51–57]

Although base-pair stability is dependent on counterion type and concentration,[58–61] the high stability of secondary structure even at the low ionic strengths[16,62] results in the 'unfolded' RNA being an ensemble of molecules with native or native-like secondary structure formed, at the initial conditions used for *in vitro* folding studies. Due to the limited availability of information on the influence of counterions on the kinetics of RNA secondary structure formation, we shall mainly focus on the effect of counterions on kinetics of tertiary structure formation. A complementary discussion of metal ions and the thermodynamics of RNA folding can be found in Chapter 6.[63] Chapter 1 explores the specific binding and coordination of cations with RNA.[64]

7.1.3.2 Role of Cations in Tertiary Structure Formation

The screening of backbone charge by counterions allows tertiary contacts to form. Clearly, the repulsive force between segments of the negatively charged backbone exceeds the attractive forces of tertiary interactions in the absence of counterions. While we briefly describe here the effects of surrounding cations on RNA structure, Chapter 9 provides a detailed account of the quantitative treatment of the counterion atmosphere around nucleic acids.[50] The high negative electric field potential near the surface of RNA causes cations in the solution to 'condense' around them, thereby generating a thermally fluctuating ion atmosphere surrounding each RNA molecule.[49,50,65] Except for some specific sites on a folded RNA molecule where the cations bind very strongly in their dehydrated or partially dehydrated form (see below and Chapter 1), the bulk of the ions in the condensed layer remain hydrated and non-specifically bound to the RNA. This 'diffusively bound' layer of ions exerts a dominant influence on the sensitivity of tertiary folding transitions to the counterion type and concentration.[66]

It has been suggested that strong correlation forces between ions in the condensed layer overcome the Coulombic repulsion and generate a net attractive force between the helices; the net attractive force is postulated to lead to the collapse of a population of rigid extended structures into a compact and ordered conformational ensemble.[67] However, studies of short DNA duplexes tethered by neutral linkers of variable length show that even at high ionic strength the duplexes only overcome electrostatic repulsion and form an ensemble of relaxed, randomly oriented molecules.[24] The extent to which counterion correlation forces can cause compaction of an RNA molecule to native like dimensions is an open question.

Upon relaxation from electrostatic repulsion following the addition of cations, the regions of RNA helix approach each other more frequently, thereby increasing the probability of tertiary contact formation among the different secondary structural motifs. Formation of tertiary contacts leads to an ensemble of structural intermediates, each of which can potentially interact with counterions in subtly different ways due to differences in their overall charge densities and also local geometry of the phosphodiester backbone. In other words, the counterion atmosphere affects the relative stabilities of the kinetic intermediates that appear during the course of folding. For instance, it was observed during Mg^{2+}-mediated refolding of the group I intron ribozyme from the protozoa *Tetrahymena thermophila* that when folding was initiated in a low salt background ($\sim 8\,mM$ Na^+ ions), tertiary contacts formed in a quasi-hierarchical manner, most of them adequately described by simple mono-exponential time–progress curves, with rate constants ranging from ~ 2 to $0.02\,s^{-1}$.[21,68] In contrast, when folding was initiated at a higher ionic strength ($200\,mM$ Na^+), not only did the rate constants increase by more than an order of magnitude (the fastest ones being $>30\,s^{-1}$), but almost all the time–progress curves displayed multiphasic behavior.[69] These differences show the important role played by counterion concentration in determining the initial conformational ensemble and also modulating the folding landscape; we will return to this point later.

In addition to neutralization of electrostatic charge, high-resolution crystal structures of large RNA molecules have revealed sites of binding of specifically coordinated partially dehydrated divalent ions (see refs. 70–73 and Chapter 1). Sites for direct coordination of multivalent counterions typically lie in deep pockets within the folded RNA structure. The electrostatic potential in those regions are estimated to be at least an order of magnitude higher, and the highly favorable electrostatic interactions of the cation with the chelating ligands on the RNA compensate for the highly unfavorable free energies of ion dehydration and repulsion of bound ions with the diffusive ions.[66] These calculations, based on the observation of bound ions in crystal structures, provide evidence for the importance of strongly bound counterions in stabilizing regions of high charge density in the folded conformation of an RNA molecule. This in turn indicates a net energetic stabilization of the native structure of RNA over the conformational ensemble of unfolded and partially folded intermediates.[74]

7.1.3.3 Mg^{2+} Ion-mediated Stability of RNA Structures

In this context, the unique attributes of Mg^{2+} ions in stabilizing compact RNA forms deserves a special note. Both theoretical and experimental studies have shown that although monovalent ions can overcome electrostatic repulsion to a significant extent and stabilize compact arrangements of RNA helices in the native structure, Mg^{2+} ions can have an over-riding effect on native tertiary structure stabilization.[34,62,75–77] Mg^{2+}-mediated tertiary structure stabilization has been demonstrated in RNA systems ranging from small and simple RNA molecules, such as hairpin ribozymes and riboswitches, to large and complex molecules, such as the ribosomal RNA and group I and II introns.[70–73,78–86] The importance of Mg^{2+} ion in RNA structure stabilization can be attributed to the different modes in which it can interact with an RNA molecule; simultaneously screening Coulombic repulsion between helices as 'diffuse ions', stabilizing pockets of high negative charge densities on the surface of RNA molecules as 'outer-sphere complexes' and directly contacting anionic ligands within folded RNA molecules as 'inner-sphere complexes'.[11,66,74,87–89]

The diffuse counterion atmosphere surrounding the RNA contains fully hydrated Mg^{2+} ions in which six water molecules are tightly coordinated around each Mg^{2+} ion in an octahedral arrangement $[Mg(H_2O)_6^{2+}]$, forming a highly ordered 'inner hydration shell' in addition to one or two outer layers of water molecules arranged loosely around the charged Mg^{2+} ion. 'Water-positioned' Mg^{2+} ions are crystallographically observed to accumulate in regions of high electrostatic potential such as the deep and narrow major groove of the A-form RNA helices and in non-canonical RNA structures. These Mg^{2+} ions lose their translational freedom and form 'outer-sphere complexes' where only their inner most hydration shell is retained and shared with the RNA ligands. Finally, strongly 'chelated' Mg^{2+} ions buried in deep solvent inaccessible regions have been observed in several high-resolution RNA crystal structures. Chelated

Mg^{2+} ions form 'inner-sphere complexes' by directly contacting the ligands in the RNA molecule, mainly the phosphates and sometimes the purine N7 or O6 and pyrimidine O4. Such Mg^{2+} chelates may not only stabilize regions containing closely packed phosphate groups buried within the interior of a compactly folded RNA molecule but also participate in the catalytic chemistry (see refs. 72 and 90–97 and Chapter 8).

Due to the unique properties and physiological relevance of Mg^{2+}, its interaction with RNA molecules has been extensively studied and it has been widely employed in cation-mediated RNA folding studies. Hence we will frequently cite time-dependent experimental studies that involve Mg^{2+}-mediated formation of RNA tertiary structures. Relative stabilities of the unfolded forms, kinetic intermediates and the native conformation of an RNA molecule, in both Mg^{2+} and monovalent ions, have important consequences on the kinetics of RNA folding and are discussed in the following section.

7.2 Cation-induced RNA Folding Kinetics

As introduced above, counterions interact with the unfolded, intermediate and folded structures in energetically distinct ways. The nature of these interactions depends not only on cation concentration but also on valence, hydrated radius and charge density.[11,24,77,93,98–104] The influence of ions on the kinetics of RNA folding is approached below from three perspectives: (1) the net energetic stability of the initial, intermediate and final states, (2) the position and plasticity of the transition state and (3) effects on the initial conformational ensemble and pathway partitioning.

7.2.1 Effects of Ionic Strength on Net Energetic Stability of Initial, Intermediate and Final States

7.2.1.1 Theoretical Background

It has been observed for some chemical reactions that a perturbation in the activation free energy ($\Delta G^{\circ\ddagger}$) induced by changes in solvent condition or structure correlates linearly with the difference in the equilibrium free energy of the reactants and the products ($\Delta G^\circ = -RT\ln K_{eq}$).[105] Leffer proposed that the energetic sensitivity of the transition state relative to that of the ground state can be quantitatively described from a rate–equilibrium free energy relationship (REFER) using a proportionality constant, $\alpha_x = (\partial \Delta G^{\circ\ddagger}/\partial x)/(\partial \Delta G^\circ/\partial x)$, where ∂x is a measure of the perturbation.[105] The perturbation is changes in properties such as ionic strength, denaturant concentration, pressure and temperature. The term α_x provides an estimate of the position of the transition state along the reaction coordinate and has a value between 1 and 0 depending on whether the transition state resembles the product or the reactant, respectively.[106]

Many reactions demonstrate a linear REFER, where α_x remains constant over a broad range of equilibrium free energy change.[107] Alternatively, some reactions are characterized by non-linear REFERs where perturbations leading to changes in $\Delta G°$ are manifested as changes in α_x values.[108] Non-linear REFERs can stem from a variety of effects of the perturbation on the system, including movement of the position of the most likely transition state ensemble (TSE) along the reaction coordinate, changes in the rate-limiting step, switch of reaction mechanism between parallel pathways and structural changes in the ground state. Before discussing cation-induced linear and non-linear REFERs in RNA folding reactions, we first review a few basic concepts connecting folding free energies to folding rates.

Consider that in the simplest case there exist only two species in the solution, the ensemble of unfolded (U) and folded native (N) molecules (Figure 7.1A).

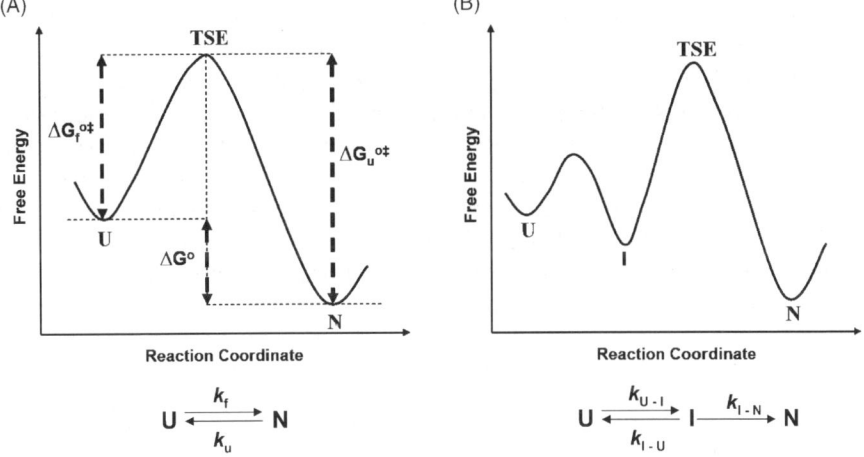

Figure 7.1 Schematic depiction of the relative free energies of unfolded (U), intermediate (I) and folded states (N), governing the kinetics of a folding reaction. (A) Relationship of the folding and unfolding activation energies ($\Delta G_f^{o\ddagger}$ and $\Delta G_u^{o\ddagger}$) to the equilibrium free energy change for folding ($\Delta G°$); $\Delta G° = \Delta G_f^{o\ddagger} - \Delta G_u^{o\ddagger}$. The greater this free energy difference (more negative values), the higher is the stability of the folded state over the unfolded state under a given set of thermodynamic conditions. From the van't Hoff relationship, $\Delta G°$ is related to the equilibrium constant K as $\Delta G° = -RT \ln K$. Since, $K = k_f/k_u$, where k_f and k_u are the microscopic rate constants for folding and unfolding, respectively, $\Delta G° = -RT \ln k_f/k_u$. (B) Free energy diagram for a sequential folding reaction where the unfolded state (U) is in rapid equilibrium with an on-pathway intermediate (I) that eventually gives rise to the folded state (N). k_{U-I} and k_{I-U} are the microscopic forward and reverse rates for the U–I equilibrium defined by the equilibrium constant $K_{I-U} = k_{U-I}/k_{I-U}$. k_{I-N} is the forward rate constant for the formation of the native state from the intermediate that dominates over the unfavorable unfolding reaction. The observed folding rate (k_{obs}) in this case is $k_{obs} = (k_{I-N}[U]K_{I-U})/(1 + K_{I-U})$.

The two species are separated by an activation energy barrier where the folding transition state ensemble (TSE) is located on the reaction coordinate at the highest point on the energy barrier. If the activation free energies going from U to N and N to U are $\Delta G_F^{o\ddagger}$ and $\Delta G_U^{o\ddagger}$, respectively, then the equilibrium free energy change for folding is given by $\Delta G° = \Delta G_F^{o\ddagger} - \Delta G_U^{o\ddagger}$. The greater the free energy difference, the higher is the stability of the folded over the unfolded state. Since $K = k_F/k_U$, where k_F and k_U are the microscopic rate constants for folding and unfolding, respectively, $\Delta G° = -RT \ln(k_F/k_U)$. Hence any solution condition that differentially affects the stability of the unfolded or folded states will perturb the equilibrium free energy difference between them and consequently the reaction kinetics.

Few RNA folding reactions are two-state. A more realistic case is shown in Figure 7.1B, where folding proceeds through at least one intermediate. As noted above, all large RNA molecules rapidly collapse to a compact intermediate structure consisting of loosely packed helices upon the addition of cations. Since the energetic force driving this collapse is neutralization of Coulombic repulsion, it is intuitive that cations preferentially stabilize compact intermediate structures. However, a subtle balance of initial, native and intermediate structure stabilities dictates folding speed. If an intermediate is strongly stabilized, its free energy is low, implying that the difference in the energy of the intermediates and the native state is small. Since the height of the energy barrier is directly proportional to the folding speed, greater stabilization of the intermediate results in slower folding. This is the case for multivalent ions that stabilize the folding intermediates to a much greater extent than monovalent ions alone. Consistent with this view, it was observed for the group I intron ribozyme from the large subunit (LSU) pre-rRNA of the protozoa *Tetrahymena thermophila* that incubating the ribozyme in a small amount (~ 100 mM) of monovalent ions prior to addition of high concentration of Mg^{2+} ions resulted in much faster folding of the RNA to the native state, as compared with the rate at which the RNA folds by addition of Mg^{2+} alone or when preincubated in a small amount of other multivalent ions.[77,104]

The scenario discussed above is also true when the counterion concentration (C) used to mediate kinetic refolding is greater than half the concentration of that same ion (C_m) required for mediating folding of the RNA to its native state. As shown in Figure 7.2B, in the absence of counterions, the free energy of the unfolded state is less than that of the native state (dashed line). On addition of cations, the free energies of the native and the intermediate states are lowered compared with that of the unfolded state (solid line). When C is less than C_m, the intermediates are stabilized to almost the same or even to a greater extent than the native state. As C becomes greater than C_m, the free energy of the native state is lowered compared with the unfolded and the intermediates, thereby making the refolding reaction favorable.

The kinetic intermediates that appear during folding can either be 'on pathway', leading to the native structure, or 'off-pathway', trapping the

(A)

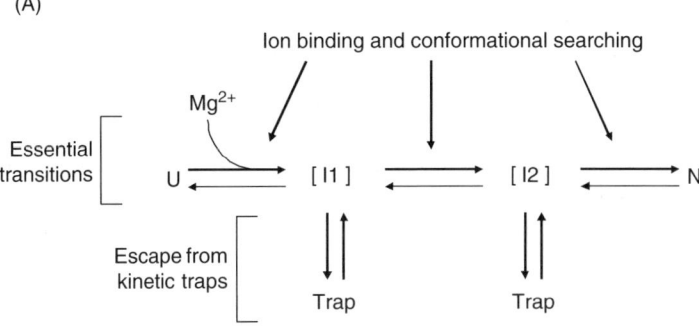

Current Opinion in Structural Biology

(B)

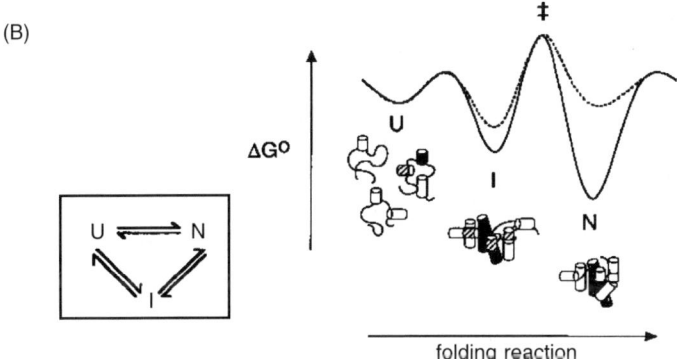

Figure 7.2 Effect of ions on the folding kinetics of the group I intron from *Tetrahymena thermophila*. (A) Sequential folding model for the L-21 *ScaI* ribozyme version of the intron, with on-pathway intermediates and kinetic traps. Reprinted from ref. 111, Copyright (1999), with permission from Elsevier. (B) Parallel folding model for the entire *Tetrahymena* group I intron where slowing of folding kinetics as a function of Mg^{2+} ions is explained on the basis of hyper-stabilization of a compact folding intermediate at high divalent ion concentrations. Adapted from ref. 109, with permission from the National Academy of Sciences, USA.

intermediates in non-native conformations that must at least partially unfold in order to refold to the native conformation. One of the several ways to test the presence of intermediates and their nature is to vary the concentration of the counterion used to refold the RNA molecule, the rationale being that ion concentrations can modulate the energy barriers to folding/unfolding by interacting with the different forms of the RNA molecule in energetically different ways. Studies exploring the effect of varying ion concentration (specifically Mg^{2+}) have revealed the following possible scenarios:

7.2.1.2 The Overall Folding Rate Decreases with Increasing Mg^{2+} Concentration

If there are competing low-energy basins on the folding energy landscape, then molecules may be trapped as non-native intermediates and thus diverge from the productive folding pathway (*e.g.* off-pathway intermediates). Both non-native and productive folding intermediates are predicted to be stabilized by increasing counterion concentrations. In this scenario, the energy barrier between the stabilized intermediates and the native state is raised, resulting in a net decrease in the overall folding rate with increasing concentration of counterions.

Examples of this folding behavior are well documented for the group I intron ribozyme from the LSU pre rRNA of the protozoa *Tetrahymena thermophila*. Both the intron and the L-21 *ScaI* ribozyme derived from it showed reduction in the overall folding rate with increasing Mg^{2+} concentration.[104,109–112] Several independent studies have shown that this large RNA molecule folds through a set of discrete intermediates which appear on time scales ranging from milliseconds to minutes.[4,5,20–22,113–115] The slowest folding step of the ribozyme is the consolidation of the P3–P7 domain that results from the formation of a long-range interaction, namely a pseudoknot structure, between two unpaired regions that are far apart in the linear sequence. This tertiary interaction forms in about 100 s, following addition of 10 mM Mg^{2+} ions at 37 °C in a buffer containing ∼10 mM monovalent cations.[40,116,117] However, the rate of appearance of catalytic activity for this ribozyme is much slower, corresponding to about 16 min under the same solution conditions.[112,117,118] Using a kinetic oligonucleotide hybridization assay and trans-oligonucleotide cleavage assay, it was shown that the rate of P3–P7 formation is essentially independent of the Mg^{2+} concentration, whereas the overall folding rate (monitored by appearance of the catalytically active ribozyme following addition of Mg^{2+}) decreased dramatically with increase in the Mg^{2+} concentration.[112] The Mg^{2+} dependence of the overall folding rate was sigmoidal, reaching saturation at ∼2 mM under the experimental conditions of this study. As summarized in Figure 7.2A, these results were interpreted as reflecting the preferential and cooperative population of an off-pathway intermediate by Mg^{2+}; the balance between the on- and off-pathway intermediates as a function of Mg^{2+} concentration determines the rate of productive RNA folding. Under these solution conditions, 2 mM Mg^{2+} was optimal for the formation of native ribozyme.

Similar behavior was demonstrated for the self-splicing version of the same intron (which includes parts of the flanking upstream and downstream exon sequences); the conversion rate of an intermediate to the native state decreased with increase in Mg^{2+} concentration.[109] However, the Mg^{2+} dependence was monotonic (as opposed to sigmoidal), which was interpreted as reflecting a relationship of the activation energy barrier separating the intermediate and the transition state to the free energy difference between the native and the intermediate states. This model predicts that slowing of the folding kinetics occurs as a result of hyper-stabilization of an on-pathway intermediate by Mg^{2+} that consequently lowers the free energy difference between the intermediate and the native states (Figure 7.2B).

7.2.1.3 The Overall Folding Rate Increases with Increasing Mg^{2+} Concentration

Since Mg^{2+} stabilizes folded RNA structures, it is intuitive that increasing Mg^{2+} concentration would increase the folding rate. However, this argument is valid only when folding is initiated by fast collapse, off-pathway intermediates are not present and productive intermediates interconvert through a slow Mg^{2+}-independent rate-limiting process. Examples of these scenarios are discussed below.

The folding rate of full-length Ribonuclease P RNA (RNase P RNA) derived from the bacterium *Bacillus subtilis* increases with increasing Mg^{2+} concentration up to 10 mM, as monitored by catalytic activity and circular dichorism.[119] The folding mechanism of the C domain of this RNA was subsequently shown to be free of kinetic traps and populated by two on-pathway intermediates, as concluded from the appearance of chevrons with roll-overs (Figure 7.3A), when the folding and unfolding rates are plotted as a function of Mg^{2+} concentration.[120] As shown schematically in Figure 7.3B, a chevron V-shaped plot represents unfolding rates on the left arm and folding rates on the right arm, separated by the vertex. The vertex is the slowest folding rate and typically corresponds to the midpoint (C_m) of the folding transition mediated by the ion concentration (in this case Mg^{2+}); unfolding is faster at Mg^{2+} concentrations below C_m and decreases with increasing Mg^{2+} concentration up to the C_m. The folding rate increases with increasing Mg^{2+} concentration above C_m. The unfolding and folding rates become constant below and above these two regimes of Mg^{2+} concentration on the left and right arms of the chevron, respectively. Such chevron plot rollovers indicate population of kinetic intermediates preceding Mg^{2+}-independent rate-limiting steps in folding.

These results were interpreted in terms of the model shown in Figure 7.3C, where the folding from the collapsed intermediate to the native state follows a sequential path involving two on-pathway kinetic intermediates that are separated by an Mg^{2+}-independent barrier. The population of the kinetic intermediates depend on the Mg^{2+} concentration and hence dictates the Mg^{2+}-dependent folding rates, whereas the slowest or rate-limiting step that involves interconversion of the two kinetic intermediates is independent of the Mg^{2+} concentration.[120]

Relatively simple RNA molecules also display increased folding rates with increasing Mg^{2+} concentration. Stopped-flow FRET was used to explore the kinetic folding behavior of a two-way (2WJ) and a four-way (4WJ) junction, revealing fast and slow phases during Mg^{2+}-dependent folding whose rates increase linearly with increasing Mg^{2+} concentration.[121] The results are consistent with a model, shown in Figure 7.4A and B, in which conversion between

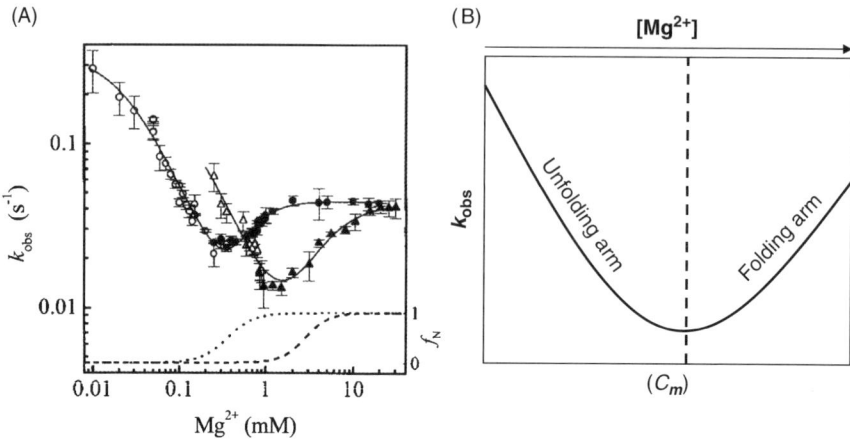

$$k_{obs} = \frac{[Mg^{2+}]^{n1}}{[Mg^{2+}]^{n1} + (K_{Mg1})^{n1}} \times kf + \frac{(K_{Mg2})^{n2}}{[Mg^{2+}]^{n2} + (K_{Mg2})^{n2}} \times ku$$

Figure 7.3 Effect of Mg^{2+} ions on the folding kinetics of the catalytic domain of RNaseP RNA from *Bacillus subtilis*. (A) Mg^{2+} dependence experiments conducted with 0.3 mM RNA in 20 mM Tris, pH 8.1 at 10 °C in 0 (● and ○) and 3 M urea (▲ and △). Folding (●, ▲) was initiated by adding Mg^{2+} to the RNA solution pre-equilibrated at the I_{eq} state, whereas unfolding (○, △) was initiated by diluting prefolded RNA 100-fold. The fraction of the N state determined from equilibrium Mg^{2+} titrations without urea (dotted line) and with 3 M urea (dashed line) is shown in the lower portion of the graph (right axis).[120] (B) A typical chevron-shaped plot showing the dependence of the observed folding rates (k_{obs}) on the Mg^{2+} ion concentration. (C) A sequential folding model for the C domain of *B. subtilis* RNaseP RNA. The Mg^{2+} dependence of the I_{eq}-to-N folding kinetics can be modeled as a three-step process involving two kinetic intermediates, I^{1k} and I^{2k}, separated by an Mg^{2+}-independent kinetic barrier to describe the Mg^{2+} chevron. K_{Mg1}, K_{Mg2} and $n1$, $n2$ are the Mg^{2+} midpoints and the Hill coefficients that define the relative populations of the I_{eq}-to-I^{1k} and the I^{2k}-to-N transitions, respectively; k_F and k_U are the folding and unfolding rates of the Mg^{2+}-independent, I^{1k}-to-I^{2k} transition. Both the I_{eq}-to-I^{1k} and I^{2k}-to-N transitions are assumed to be in fast equilibrium relative to the I^{1k}-to-I^{2k}. Adapted by permission from Macmillan Publishers Ltd, *Nature Structural Biology*, (ref. 120), Copyright (1999).

the extended (E) form and the docked (D) form occurs either directly (in the case of the 2WJ) or through an on-pathway intermediate (in the case of the 4WJ). The rate-limiting step of this reaction is the slow conformational search of an electrostatically screened intermediate for the native tertiary structure.

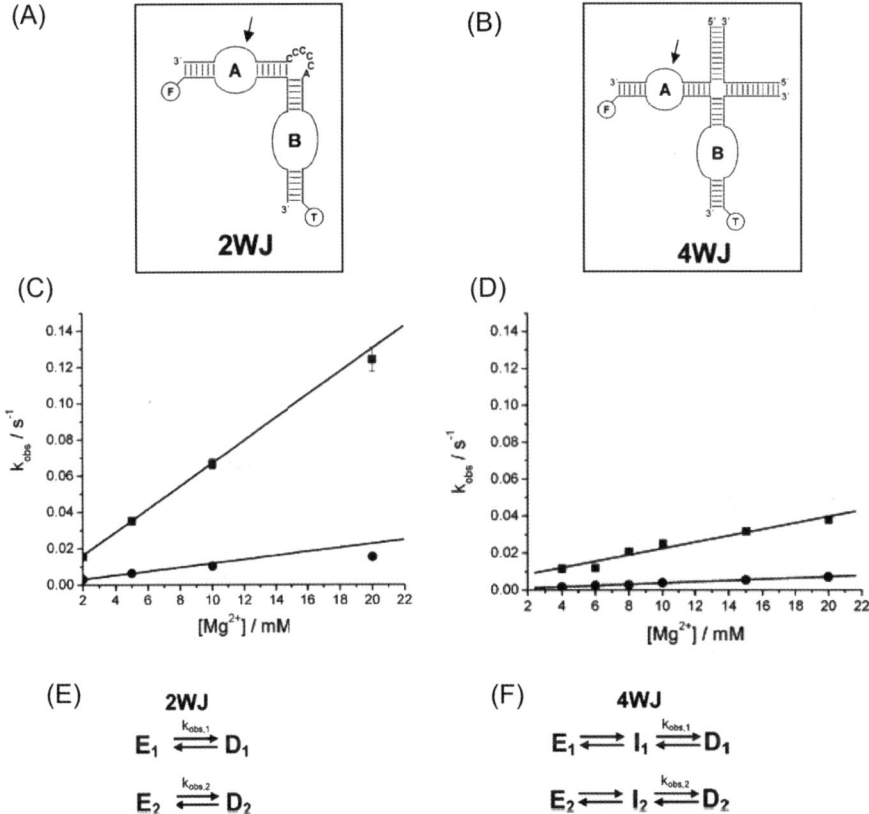

Figure 7.4 The two-way (A) and four-way (B) junction constructs used in the study. The arrows indicate potential cleavage site in each ribozyme. (C) and (D) show the dependence of the observed folding rates (k_{obs}) on Mg^{2+} for the two-way (C) and the four-way (D) junction ribozymes. (E) and (F) show models for the folding pathways of the 2WJ and 4WJ hairpin ribozymes. (E) Folding pathway of the minimal 2WJ ribozyme. E1, E2 and D1, D2 denote non-interconverting subpopulations of extended and docked ribozyme conformers, respectively, distinguished by different relaxation rates, $k_{obs,1}$ and $k_{obs,2}$ ($k_{obs,1} > k_{obs,2}$). (F) Folding of the natural 4WJ ribozyme. E1, E2 and D1, D2 are the same as in (E). I1 and I2 represent intermediates in each folding pathway. Reprinted with permission from ref. 121, Copyright 2005 American Chemical Society.

Single molecule fluorescence resonance energy transfer (FRET) and fluorescence correlation spectroscopy were used to show that cations alter the dynamics of conformational fluctuation by an RNA three-helix junction derived from the 30S ribosomal subunit.[122] As depicted in Figure 7.5A, these studies suggest a model in which the junction exists in a dynamic equilibrium between open and folded conformations, whose folding kinetics are modulated

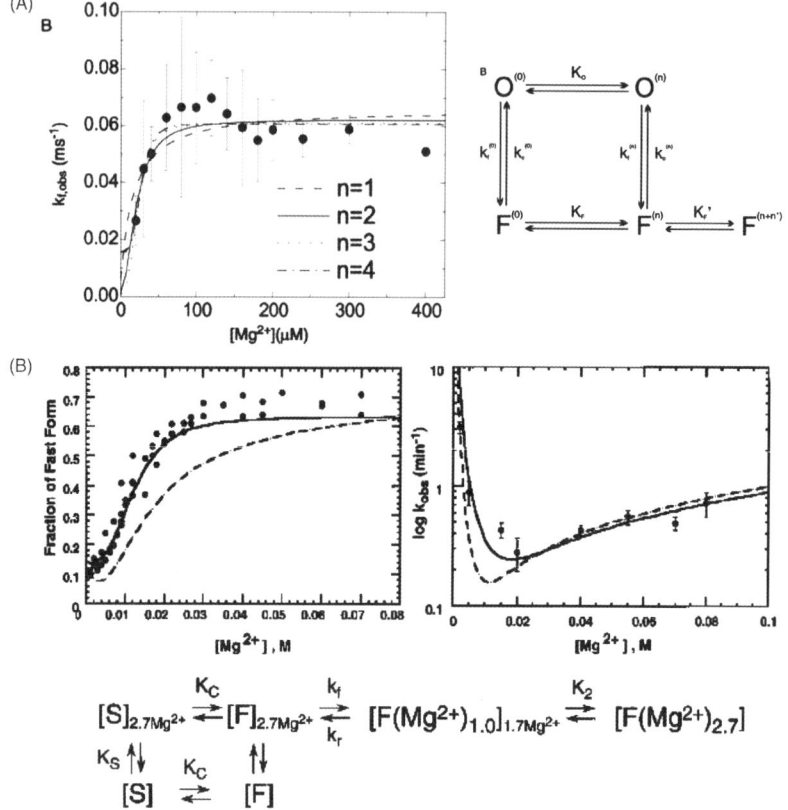

Figure 7.5 (A) The observed folding rate ($k_{f,obs}$) of the three-helix RNA junction as a function of [Mg^{2+}] fit to the equation $k_{f,obs} = \{k_f^{(0)} + k_f^{(n)} \cdot (K_o[Mg^{2+}])^n\}/1 + (K_o[Mg^{2+}])^n$, where, $k_f^{(0)}$, $k_o^{(0)}$, and $k_f^{(n)}$, $k_o^{(n)}$ denote rates of transitions between ion-free forms ($O^{(0)}$ and $F^{(0)}$) and n-ion-bound forms ($O^{(n)}$ and $F^{(n)}$). $k_{f,obs}$ increases rather steeply with [Mg^{2+}] up to the equilibrium constant and plateaus at higher [Mg^{2+}]. In the accompanying extended model scheme, it is assumed that n number of Mg^{2+} ions bind cooperatively to the junction. $k_f^{(0)}$ [$k_f^{(n)}$] and $k_o^{(0)}$ [$k_o^{(n)}$] are folding and opening rates for ion-free (ion-bound) junctions. K_O and K_f are binding constants for binding one Mg^{2+} ion to the open and folded junctions, respectively. Thus, $K_O = [O^{(1)}]/[O^{(0)}][Mg^{2+}]$ and $K_f = [F^{(1)}]/[F^{(0)}][Mg^{2+}]$. In the case where the folded form of the junction can be bound with more Mg^{2+} ions than the open form, a fifth state $F^{(n+n')}$ is introduced with additional binding constant $K_{f'}$. Similarly to K_O and K_f, $K_{f'} = [F^{(n+1)}]/[F^{(n)}][Mg^{2+}]$. Adapted from ref. 122, Copyright 2002 National Academy of Sciences, USA. (B) Effect of Mg^{2+} ions on the rates of exchange between the two conformers of the pseudoknot structure derived from the leader sequence of α mRNA of *E. coli*. Scheme of the mechanism that relates Mg^{2+} binding to the S–F equilibrium is shown at the top. Mg^{2+} shown outside brackets represents a weakly and non-specifically bound class of ions. Numbers of ions are taken from the fits assuming cooperative ion binding to the slow form. Global fits of this scheme to equilibrium and kinetic data are shown in the lower part. Solid lines represent fits where cooperative Mg^{2+} binding to the slow form is assumed, whereas dashed lines represent fits where independent Mg^{2+} binding to the slow form is assumed. Reprinted from ref. 100, Copyright (1997), with permission from Elsevier.

by cations; the folding rate increases and the opening rate decreases with increases in both Mg^{2+} and Na^+ concentration.

Our last example is a pseudoknot structure derived from the leader sequence of α mRNA of *Escherichia coli* that equilibrates between two stable alternative conformations, S and F (moving slow and fast in an electrophoretic field, respectively).[100] The interconversion kinetics of the S and F conformers resembles the similarly slow interconversion of the two productive kinetic intermediates in the folding pathway of the C domain of RNase P RNA.[120] Unlike the P RNA reaction, the rate of S to F conversion is linearly accelerated between 20 and 80 mM Mg^{2+}, suggesting that binding of a single Mg^{2+} ion is coupled to the rate-limiting step in this case (Figure 7.5B).

7.2.2 Position and Plasticity of the Transition State

7.2.2.1 Theoretical Background

The classical description of transition states reflects chemical reactions where only a few covalent bonds are broken or formed. In such reactions, the transition state refers to an intermediate structure corresponding to the energy maximum along a reaction coordinate between the reactants and products.[123] Multiple non-covalent (and sometimes covalent) interactions are broken or formed during macromolecular folding reactions; there can be multiple intermediate species of similar free energies sitting on a saddle point surrounding the energy maxima along the folding coordinate.[124] Such an ensemble of molecules is referred to as the transition state ensemble (TSE). It is usually difficult to define clearly the structural properties of the TSE because it is often heterogeneous in nature. The problem is aggravated in the case of large multi-domain RNA molecules, which may fold through multiple parallel folding pathways. Nevertheless, some of the classical methods used to probe transition states and activation energy barriers have been successfully applied to the analysis of RNA folding kinetics.[101,102,125,126]

The proportionality constant α_x proposed by Leffler gives an estimate of the position of the TSE.[105] A very similar and widely used parameter in the protein folding field is the Tanford β value (β_T).[127,128] β_T is obtained by measuring the activation energy and equilibrium free energy of unfolding as a function of denaturant concentration. β_T is expressed mathematically as $\beta_T = (\partial \Delta G_{\ddagger-F}/\partial[\text{denaturant}])/(\partial \Delta G_{U-F}/\partial[\text{denaturant}])$. The value of β_T is close to 0 if the TSE for the unfolding reaction is similar to the unfolded state and close to 1 if the TSE is similar to the native state. β_T changes upon solvent-induced perturbations that result in the movement of the TSE.[129,130]

Movement of the TSE along the folding coordinate follows the Hammond postulate:[131] "*If two states, as for example, a transition state and an unstable intermediate, occur consecutively during a reaction process and have nearly the same energy content, their interconversion will involve only a small reorganization of the molecular structures.*" The TSE can be imagined as a saddle on a reaction free energy landscape, situated at the highest point in the direction of the reaction and at the lowest point in any direction perpendicular to it. Thornton[132] proposed that if a linear perturbation is applied to the system, the position of the transition state will shift towards the more destabilized state in the direction of the reaction coordinate (Hammond behavior) and towards the more stabilized state along all other coordinates perpendicular to the reaction coordinate (anti-Hammond behavior). Hence the net effect on the structure of the TS will be a sum of the effects of the perturbation along all the coordinates. In the case of RNA folding, both denaturants and counterions have been used to perturb the folding energy landscape and observe its consequence on folding kinetics.[101,102,104,119,133–135] We shall limit our discussion to studies that explore the effects of counterions.

7.2.2.2 Hammond Behavior in Multivalent Cation-mediated RNA Folding

The dependence of the properties of the TSE on the charge density of the counterion has been explored for the folding kinetics of the *Tetrahymena* ribozyme.[101,102] It was observed that the rate at which a collapsed intermediate folded to the native state decreased with increase in polyamine charge. A comparable trend was observed for polyamine size. These observations were interpreted as reflecting increased stabilization of both the intermediate and the native state by counterions of higher charge and smaller size, resulting in a higher activation energy barrier for the transition. It was further proposed that the charge density (ζ, defined as $\zeta = Ze/V$, where Z is the valence of the counterion of volume V and e is the charge of a proton) on the counterion is a physiologically relevant parameter.

Measurement of the folding rates as a function of concentration for each counterion provided deeper insight into the behavior of the TSE. The dependence of the folding rate (k_{obs}) on counterion concentration (C) was grouped into three regimes: $C < C_m$, $C \geq C_m$ and $C \gg C_m$, where C_m is the counterion concentration at the midpoint of the equilibrium folding transition to the native state, mediated by the same counterion (Figure 7.6A). In the first regime, k_{obs} is dominated by the microscopic rate constant for unfolding (k_U) from the native state (N) to the intermediate (I). At low cation concentration, U is more stable than I and N. The stability of U decreases with increasing counterion concentration resulting in a decrease in k_U that is reflected as a decrease in k_{obs}. In the second regime, k_{obs} is dominated by the microscopic folding rate constant k_F; k_F increases as the stability of I and N over U increases until saturation of the counterion binding is reached. In the third regime, k_{obs} is dominated by the microscopic rate constant k_I. k_I reflects the slow transition from I to N. As opposed to the Tanford β values used in protein denaturation studies, in this case β was defined as $\beta = (\partial \ln k_{obs}/\partial \ln C)/(\partial \ln K/\partial \ln C)$ calculated separately for each regime (β_U, β_F and β_I) for each of the counterions studied (Figure 7.6B). On going from state A to B, a β value of 0 indicates a TSE

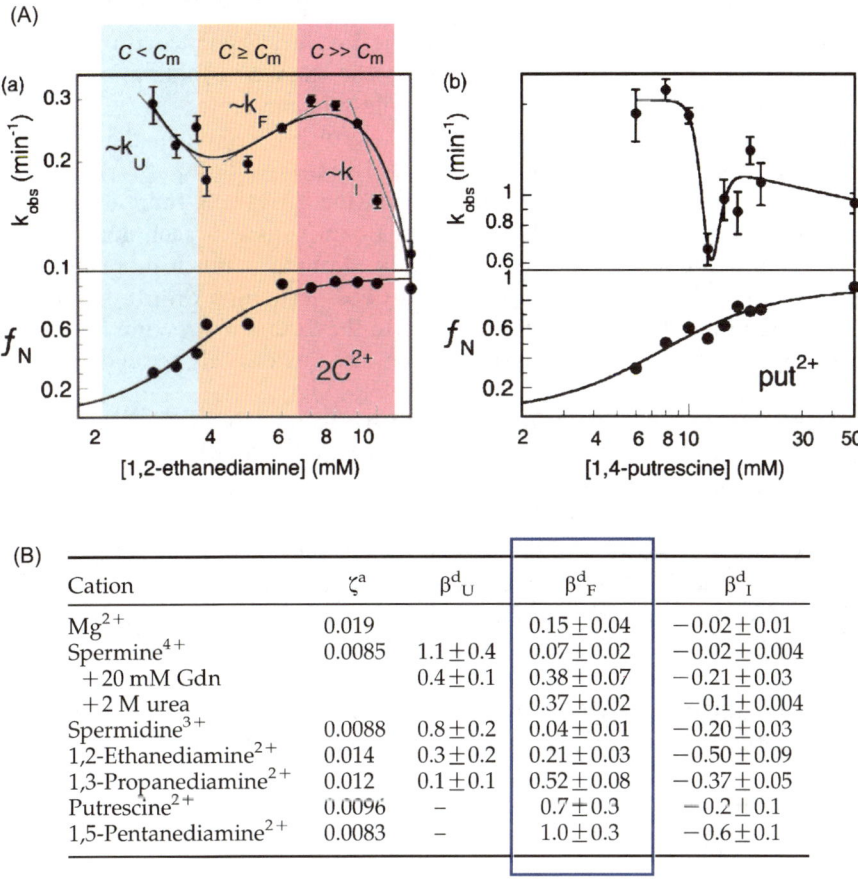

Figure 7.6 (A) Dependence of folding kinetics on polyamine concentration. Polyamine concentrations (C) versus observed folding rate (upper panels). Linear fits of $\ln k_{obs}$ versus $\ln C$ over regions where $k_{obs} \sim k_U$, k_F or k_I corresponding to the three regimes $C < C_m$, $C \geq C_m$ and $C \gg C_m$, respectively, where C_m is the counterion concentration at the midpoint of the equilibrium folding transition to the native state, mediated by the same counterion. The amplitudes of the folding transitions (lower panels) were fitted to the Hill equation. (Left) Ethane-1,2-diamine ($2C^{2+}$); (right) putrescine (put^{2+} or $4C^{2+}$). (B) Table showing the dependence of the β values for the three regimes on charge density (ζ). Adapted from ref. 102, Copyright (2006), with permission from Elsevier.

resembling state A, whereas a β value of 1 indicates that the TSE is similar to the B state. The β_F values, corresponding to the movement of the TSE in the regime where k_{obs} is dominated by k_F, indicate that with increasing ζ the TSE shifts more towards the unfolded state. This observation is consistent with the Hammond effect because both higher charge and charge density stabilize the native state, thereby shifting the TSE towards the destabilized unfolded state.

7.2.2.3 Early Transition State for the Folding of Large RNAs

Another approach, Φ-value analysis, uses the Φ-value, calculated as the ratio of the change in the activation free energy to the change in folding free energy ($\Delta\Delta G^{\ddagger}/\Delta\Delta G^{\circ\prime}$) of a mutant relative to the wild-type molecule.[124,136–138] Φ-value analysis indicates the extent to which the perturbed region in the molecule participates in the formation of the structured TSE; Φ-values range from 0 to 1, reflecting whether the perturbation results in the TSE being more denatured-like or more native-like, respectively. Fractional values in such analysis can stem from a number of effects and usually demand a much more complex interpretation; however, it functions as a robust technique with the simplistic assumption that Φ-value is linearly related to the extent of structure formation in the TSE.[124,139] Φ-value analysis has been used extensively to study protein folding and, to a more limited extent, RNA folding.

A Φ-value analysis was performed on the P4–P6 domain of the *Tetrahymena* ribozyme by following its Mg^{2+}-mediated folding with stopped-flow-fluorescence of an attached pyrene label.[140] These studies showed that mutation of the tertiary contacts that destabilize the folded RNA by $\sim 3\,\text{kcal}\,\text{mol}^{-1}$ ($\Delta\Delta G^{\circ\prime}$) yield less than a $\sim 0.4\,\text{kcal}\,\text{mol}^{-1}$ change in the activation free energy ($\Delta\Delta G^{\ddagger}$) (Figure 7.7A and B). The conclusion drawn from the calculated Φ-values, which are close to zero, is that this RNA folds through an early transition state in which the native tertiary contacts are not appreciably formed (Figure 7.7C). Investigation of the dependence of the folding kinetics on Mg^{2+} concentration revealed that the observed folding rate increased linearly between 1 and 30 mM with a slope of unity (Figure 7.7D). This result is consistent with the rate-limiting step of folding involving binding of only a single Mg^{2+} ion; this early TSE lacks tertiary interactions and is not stabilized by the multiple Mg^{2+} ions observed in the crystal structure of this RNA.[71]

7.2.2.4 Early Transition State for the Folding of Small Ribozymes

Early formation of the TSE has also been reported for the two-way junction form of the hairpin ribozyme; the rates of docking (k_{dock}) and undocking (k_{undock}) of the two helix–loop–helix domains (Figure 7.8A) oscillating between the enzymatically active folded state and an extended unfolded state were measured using single-molecule FRET.[141] Separate mutations in two of the three main tertiary contacts that stabilize the docked state of the ribozyme resulted in Φ-values of 0.3–0.4 over a two order of magnitude range of Mg^{2+} concentrations, whereas mutation of the third contact yielded a Φ-value of zero (Figure 7.8B). These data predict a TSE in which few of the tertiary contacts are at most partially formed. It was also observed that k_{dock} increased by over 40-fold with increase in [Mg^{2+}] from 1 to 500 mM, whereas k_{undock} remained essentially constant, showing only a slight but constant increase at very low [Mg^{2+}] (~ 1 mM) (Figure 7.8C). The significant uptake of Mg^{2+} ions in going from undocked to the transition state and the lack of Mg^{2+} uptake in going

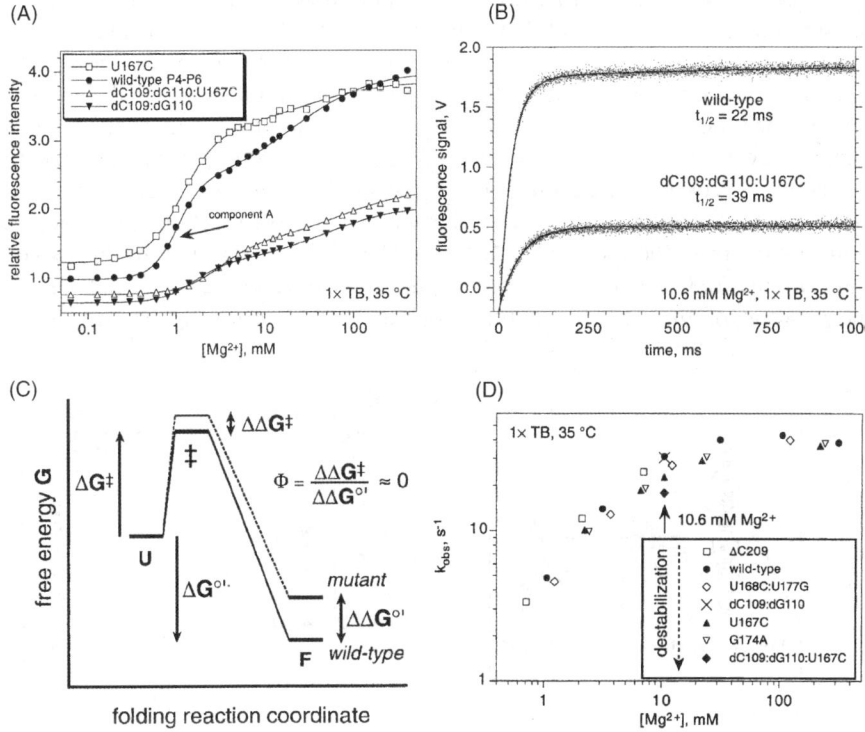

Figure 7.7 (A) Equilibrium fluorescence titration data for pyrene-labeled P4–P6 and several destabilized mutants in Tris–borate buffer. The major component of fluorescence change (component A) for wild-type pyrene-labeled P4–P6 is indicated by an arrow. (B) Fluorescence kinetics traces for Mg^{2+}-induced folding of pyrene-labeled P4–P6. Top trace: wild-type P4–P6. Bottom trace: thermodynamically destabilized dC109:dG110:U167C mutant. (C) A simple two-state energy diagram for P4–P6 tertiary folding. A single transition state (‡) is located between unfolded (U; secondary structure only) and folded (F; secondary plus tertiary structure) states. The dual observations of a large $\Delta G^{\ddagger}$ and an early transition state (low F) suggest that this model is incomplete. The folding reaction coordinate is not well defined, as the structure of P4–P6 at any point along the folding pathway other than the fully folded state is not known with precision. The free energy profile of a generic, destabilized mutant RNA is also shown; the magnitude of $\Delta \Delta G^{\ddagger}$ relative to $\Delta \Delta G^{o\prime}$ is exaggerated for clarity. The energy of the unfolded state is depicted as unperturbed by a mutation that affects $\Delta G^{o\prime}$, which is equivalent to the assumption that the tested mutations specifically affect interactions only in the folded state of P4–P6. (D) Dependence of the Mg^{2+}-induced folding rate of P4–P6 on $[Mg^{2+}]$ and on mutations that alter the tertiary folding energy $\Delta G^{o\prime}$ of P4–P6. Data are shown for wild-type P4–P6 (filled circles) and for six mutants (other symbols).[140] Reprinted with permission from the RNA Society.

from the transition state to the docked state suggests a TSE whose charge density is higher than the unfolded, but similar to, the folded state (Figure 7.8E).

Mg^{2+}-mediated folding in a background of 500 mM Na^+ revealed that k_{undock} starts to increase below 50 mM Mg^{2+} (as opposed to 0.1 mM Mg^{2+} in the absence of Na^+ ions) demonstrating that Mg^{2+} binding to the TSE is saturated at much higher concentrations in the presence of background monovalent ions due to competition between the cations for non-specific binding sites (Figure 7.8C and D). When folding is mediated exclusively by Na^+, equivalent cation uptake occurs transitioning from the extended state to the TSE and from the TSE to the docked state (Figure 7.8D). This observation suggests an ionic strength-induced shift in the position of the TSE along the reaction coordinate during Na^+-mediated

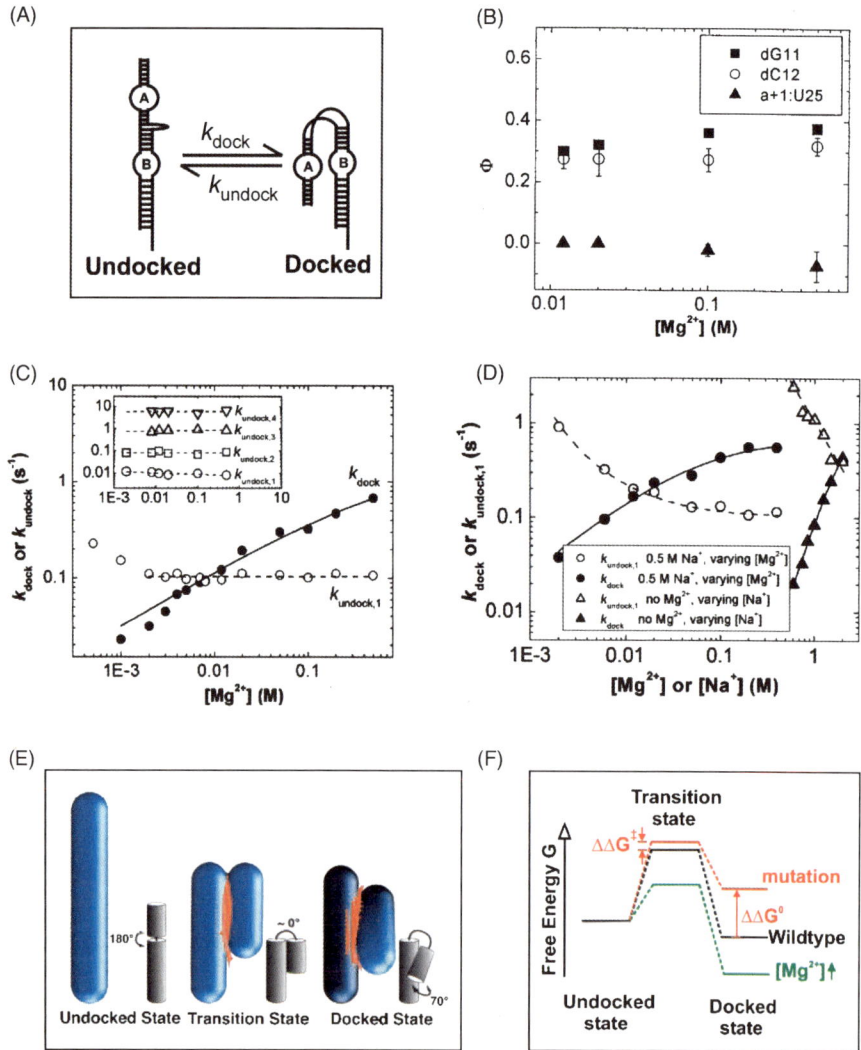

folding in contrast to the Mg^{2+}-mediated folding. Two models for the TSE were proposed based on non-linear Poisson–Boltzmann equation (Figure 7.8E). The first model has the two helical domains juxtaposed parallel to each other and in direct contact. This model is hypothesized to reflect the configuration for Mg^{2+}-mediated folding. The second model postulates parallel helices that are separated by 5 Å to accommodate hydrated metal ions. This model was deemed to rationalize better both Na^+- and Mg^{2+}-mediated folding. Overall, this study suggests that the folding TSE for this RNA is compact and, while electrostatically stabilized to the same extent as the native state, native tertiary contacts are only partially formed (Figure 7.8F).

7.2.2.5 Mg^{2+}-dependent Barriers on the Unfolding Reaction Coordinate

Individual RNA molecules can be mechanically unfolded by pulling at either of their two ends, one of which is tethered to a bead that is held in a force-measuring optical trap (laser tweezers) and the other to a second bead connected to a piezo-electric actuator through a micro-pipette.[142–144] Such studies provide spatial information on the location of the TSE since unfolding proceeds along a trajectory quantitatively defined by the end-to-end distance of the molecule and reveal the presence of kinetic intermediates and their molecular

---

Figure 7.8 (A) Schematic representation of the docking undocking transition of the hairpin ribozyme with respective rate constants k_{dock} and k_{undock}. (B) Dependence of the Φ-values on Mg^{2+} concentration for the dG11, dC12 and a + 1:U25 mutants. (C) Dependence of k_{dock} (●) and k_{undock} (○) of the wild-type ribozyme on [Mg^{2+}] at 37 °C. k_{dock} values are fitted to the Hill equation (solid line), $k_{dock} \sim k_{max}[Mg^{2+}]^n/([Mg^{2+}]^n + (Mg_{\frac{1}{2}})^n)$. Inset: [$Mg^{2+}$] dependence of all four undocking rate constants observed at 25 °C. (D) The [Na^+] dependence of k_{dock} and $k_{undock,1}$ (Δ, ▲) in the absence of Mg^{2+} and the [Mg^{2+}] dependence of k_{dock} and $k_{undock,1}$ in the presence of 500 mM Na^+ (○, ●). k_{dock} values are fitted to the Hill equation. (E) A theoretical model describing the electrostatic interactions of the hairpin ribozyme with metal ions. Domains A and B of the ribozyme modeled as two connected cylinders (blue) with their relative orientations depicted in gray. The 3D iso-concentration contour (red) at 3.0 M shows the accumulation of divalent metal ions at the domain interface in the docked state and the contact model of the transition state. On titration with Mg^{2+} alone, the most probable transition state is represented by the contact model, where domains A and B are in contact as shown. At 1 mM Mg^{2+} in the absence of Na^+, on titration with Na^+ alone or on Mg^{2+} titration in a background of 500 mM Na^+, the more appropriate transition state is a non-contact model where the two domains are only slightly separated. In all cases, docking likely occurs via an ensemble of transitions states that all satisfy these restrictions. (F) Free energy diagram of the docking and undocking transitions in dependence of either a mutation that disrupts a tertiary interaction or an increase in [Mg^{2+}]. The symbols are defined as $\Delta\Delta G^{\ddagger} = \Delta\Delta G_{dock}^{\ddagger}$ and $\Delta\Delta G^{\circ} = \Delta\Delta G_{dock}^{\ddagger} - \Delta\Delta G_{undock}^{\ddagger}$. Adapted from ref. 141, Copyright 2003 National Academy of Sciences, USA.

dimensions that are obscured in bulk measurements.[145,146] As the ends of the folded molecule are pulled apart at a constant velocity in a mechanical unfolding experiment, the pulling force gradually increases as a function of the increasing end-to-end distance. The resultant force–extension curves are characterized by features called 'rips' (plateaus interrupting a monotonically increasing force as a function of extension) that indicate release in the mechanical tension in the molecule corresponding to the complete unfolding of a structural domain. Each independently unfolding subdomain present in a complex RNA structure can produce a characteristic rip. When a series of rips are observed in a force–extension curve, each one is a distinct signature of the unfolding of the corresponding subdomain. Fitting the data to a suitable model estimates the structural properties of the partially unfolded intermediates and also helps in identifying the kinetic barriers associated with the unfolding and refolding of the molecule along the force trajectory.

A series of mechanical unfolding studies were conducted on an independently stabilized structural subdomain of the *Tetrahymena* ribozyme, the P5abc, in which the structural models of the subdomain were sequentially added.[142] As depicted in Figure 7.9A, first, the hairpin P5ab was studied, then another helix was added to generate the three-helix junction (P5abcΔA), and finally, the metal ion binding A-rich bulge was added to reconstitute the full P5abc structure. Mg^{2+} stabilized all three structures, as evidenced by the greater forces required to unfold all three forms in the presence of 10 mM Mg^{2+} compared with its absence. Moreover, when the hairpin and the P5abcΔA molecules (which unfolded at ∼14 and 12 pN, respectively) were held at a constant force over a period of time, they hopped back and forth between the fully folded and fully unfolded forms without detectable intermediates, in both the presence and absence of Mg^{2+} (Figure 7.9B and D). The folding and unfolding rates, calculated from the average dwell time in the compact and extended forms, respectively, at a given force, were used to derive the position of the TSE on the reaction coordinate. The TSE located at ∼12 nm is equidistant from the folded and the unfolded states, suggesting that in the absence of stable tertiary contacts the helices are malleable to reversible mechanical deformation.

In contrast, P5abc with its Mg^{2+}-stabilized tertiary structure, did not display this 'hopping behavior' at any unfolding force in the presence of Mg^{2+}. In the presence of Mg^{2+} ions, P5abc unfolded rapidly and irreversibly at high mechanical forces of ∼19±3 pN, consistent with rigidity under mechanical stress but facile fracturing with slight mechanical deformation (Figure 7.9C and E). A small fraction (4%) of the force extension curves demonstrated rips at low forces (∼13 pN) revealing the presence of an additional kinetic barrier. As summarized in Figure 7.9F, the two kinetic barriers were assigned to the Mg^{2+}-dependent tertiary interactions that stabilize the subdomain. The high kinetic barriers to unfolding are not present in the absence of Mg^{2+}; P5abc does oscillate between the folded and a partially unfolded state but passes through intermediates. The position of the TSE for the unfolding of P5abc, as estimated from the probability

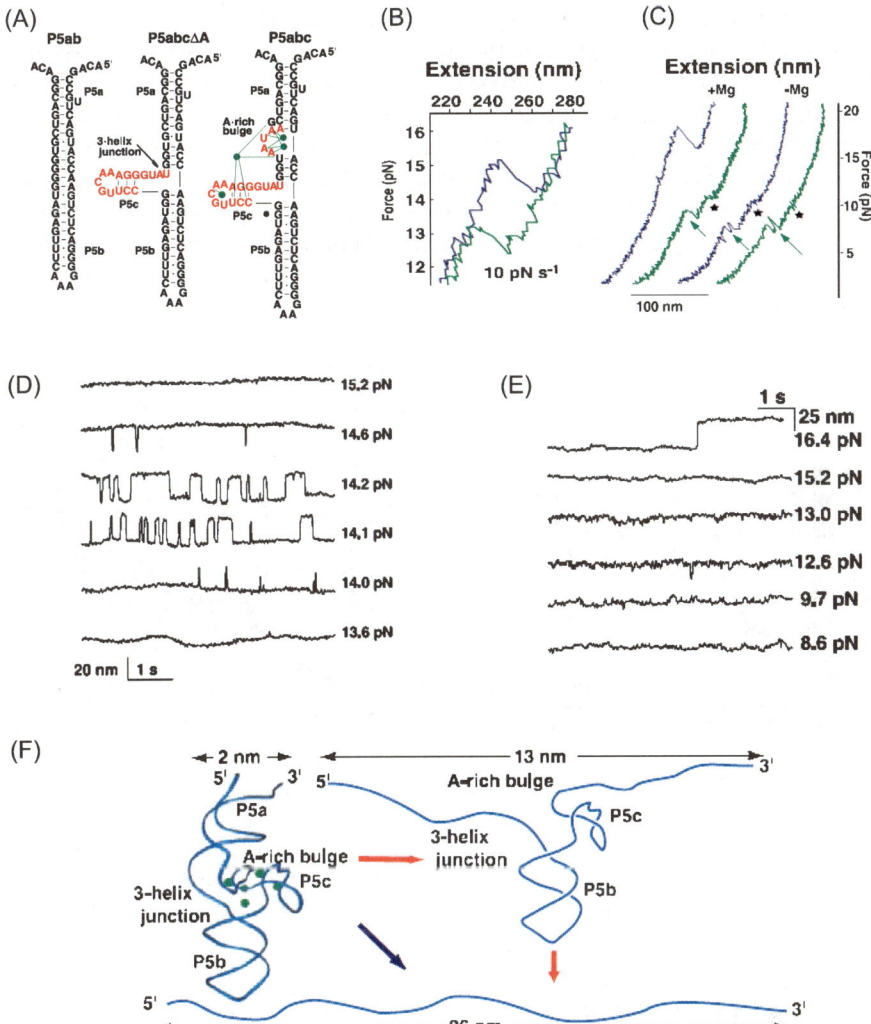

Figure 7.9 (A) Sequence and secondary structure of the P5ab, P5abcΔA and P5abc RNAs. The green dots represent magnesium ions that form bonds (green lines) with groups in the P5c helix and the A-rich bulge. (B) Detail of the stretching (blue) and relaxing (green) force–extension curves for the P5abcΔA molecule taken at low and high loading rates in 10 mM Mg^{2+}. (C) Comparison of P5abc force–extension curves in the presence and absence of Mg^{2+}. (D) Length versus time traces of the RNA hairpin at various constant forces in 10 mM Mg^{2+}, showing the 'hopping' behavior. (E) Length versus time traces of P5abc in 10 mM Mg^{2+}, showing rapid and irreversible unfolding at high mechanical forces of ∼19±3 pN. (F) Model for P5abc's unfolding by force in Mg^{2+}. From ref. 142, reprinted with permission from the AAAS.

of the molecule unfolding at a given force, is approximately 1.6 nM from the folded state (and hence resembling more the folded state).

A key finding was that even though the activation free energy ($\Delta G^{\ddagger}$) for breaking the Mg^{2+}-stabilized tertiary contact was much smaller than that for opening the helices, the former process is rate limiting. This paradoxical observation was rationalized by the fact that the force required to unfold a given domain is inversely proportional to the distance on the reaction coordinate of the TSE from the folded state. Hence the force required to unfold P5abc in Mg^{2+} is much greater than that required to unfold P5ab and P5abcΔA, both lacking the Mg^{2+}-stabilized tertiary contact, by the virtue of the former's more native-like TSE (Figure 7.9F).

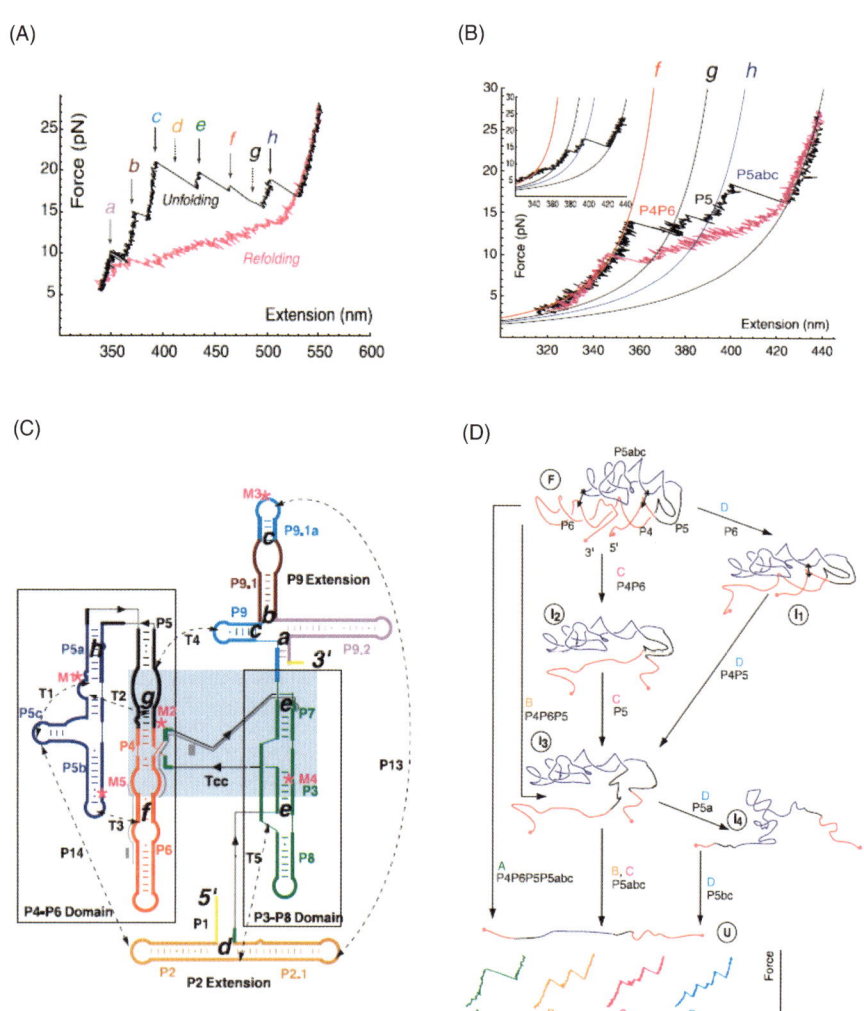

Mechanical unfolding studies of the entire *Tetrahymena* ribozyme conducted in the presence of Mg^{2+} using laser optical tweezers identified six major 'rips' in the force extension curve in the presence of Mg^{2+} ions.[125] Since the rips were not observed during unfolding or folding in the absence of Mg^{2+}, they are ascribed to Mg^{2+}-dependent kinetic barriers to mechanical unfolding. Refolding in the presence of Mg^{2+} ions lacks the distinct transitions; only the extended plateau (between 10 and 13 pN) also observed during unfolding and refolding in the absence of Mg^{2+} was observed (Figure 7.10A). This transition is thought to represent progressive unwinding/winding of secondary structure. Thus, secondary structure formation during force quenched refolding was stated to be independent of Mg^{2+}-generated barriers.

As shown in Figure 7.10B, the independently folding P4–P6 domain alone displayed four different types of unfolding force extension curves, two of which (g and h) were well correlated with the disruption of Mg^{2+}-mediated or -stabilized tertiary interactions. As summed up in the unfolding map for the P4–6 domain (Figure 7.10D), there are at least four independent unfolding trajectories, involving at least three intermediates and two Mg^{2+}-dependent barriers. Additional kinetic barriers were identified by pulling on progressively larger molecules that contained the peripheral secondary structure elements and the catalytic core helices of the ribozyme in addition to the P4–P6 domain (Figure 7.10C). The TSE

Figure 7.10 (A) Representative unfolding (black) and refolding (pink) force–extension curves of the L-21 RNA displaying six unfolding events (rips). Letters and arrows correspond to the positions assigned to the kinetic barriers. The unfolding curve shown here does not display barriers *d* and *g*, indicated by the dashed arrows. (B) Force–extension curves for the P4–P6 domain: black, unfolding curve; pink, refolding curve. The solid lines are WLC curves for double-stranded RNA–DNA handles and an increasing number of RNA nucleotides as the unfolding progresses. The unfolding of the subdomains releases 50 nucleotides (*f*, red, P4–P6 helices), then 30 nucleotides (*g*, black, P5 helix), then 70 nucleotides (*h*, blue, P5abc three-helix junction). The color code of the WLC curves corresponds to the motifs broken; it is identical with the motif color code shown in (C). Inset: unfolding curves of P4–P6 in the presence of 10 μM antisense DNA oligonucleotide I that prevents T3 and hence disrupts barrier *f* (shown in C). (C) Secondary structure of the L-21 ribozyme. The two main domains, P4–P6 and P3–P8, are boxed; a light blue box indicates the catalytic core. Gray lines label sequences targeted by complementary DNA oligonucleotides; dashed lines are tertiary contacts (T) and base-paired regions (P); M labels are site-directed mutations (M1, A to U at site 186; M2, C to G at 260; M3, UGC to ACG at 348 to 350; M4, U to A at 273; M5, cyclic permutation at 148); cc is the catalytic core. The letters *a–h* indicate the proposed positions of the kinetic barriers. (D) Mechanical unfolding map of the P4–P6 domain. Four different unfolding trajectories were detected under the given experimental resolution. F, folded; U, unfolded; I, intermediates: I1 (P6 unfolded), I2 (P6–P4 unfolded), I3 (P6–P4–P5 unfolded); I4, P6–P4–P5–P5a unfolded. Sample force-extension curves of the different trajectories are shown in the bottom panel. Trajectory C was observed most frequently. P4–P6 colors are same as in (C). From ref. 125, reprinted with permission from the AAAS.

for the catalytic core was estimated to be located at ~1.5 nm from the folded state (hence more native like) as is the Mg^{2+}-stabilized P5abc subdomain.[142] These two observations support the proposal that conserved base pairing and tertiary interactions within the core also produce a 'brittle' structure that yield smoothly, but only at very high mechanical forces. The unfolding barriers arising predominantly from secondary structure disruption were reversible characterized by 'hopping' when held at a constant force, unlike the Mg^{2+}-stabilized tertiary interactions that unfold in a concerted fashion. The overall conclusion drawn from these studies is that kinetic barriers corresponding to discrete Mg^{2+}-dependent tertiary interactions are rate limiting in mechanical unfolding. In contrast, the thermal stability of the entire ribozyme is predominantly dictated by the base-paired secondary structure elements.

7.2.3 Effects on the Initial Conformational Ensemble and Pathway Partitioning

7.2.3.1 Background Information

RNA folding is generally considered to occur by following multiple parallel pathways on a rugged landscape.[4,5,69,113–115,147] The pathways are populated by intermediates with distinct structural and electrostatic characteristics. Partitioning of the folding flux among pathways occurs at both early and late stages of the reaction.[9,104,114] Since the intermediates populated during early and late stages of folding have different structural and electrostatic properties, the nature of the forces needed to stabilize them is most likely different. Hence, the forces that partition the reaction flux most likely have different origins at different stages.

For example, neutralization of the negatively charged polyanionic backbone is crucial for the secondary structural modules to approach each other during the earliest steps in RNA folding.[22,25–27,107] Hence electrostatic forces are expected to play a dominant and decisive role in determining the starting point for the molecule on the upper broad region of the folding funnel. In this region, multiple conformations are sampled and preferential stabilization of some conformers over others can channel the respective molecules into distinct folding pathways. These pathways are well separated by energy barriers whose heights are proportional to the ionic strength of the solution. Similarly, the electrostatic environment that an RNA molecule experiences during the course of folding will determine the relative stabilities of both the productive (on pathway) and non-productive (off-pathway) intermediates along a given pathway.[115] Almost all large RNA structures contain specific divalent ion binding sites whereas only a few examples of specific monovalent ion binding sites in RNA structures have been documented.[72,98,148–152] Divalent ions usually stabilize both native and non-native intermediates and hence slow folding by attenuating the ability of RNA molecules to resolve out of kinetic traps.[47,112] In contrast, an electrostatic environment generated predominantly by monovalent ions is usually less efficient in stabilizing the kinetic traps and hence speeds up folding along productive pathways.[69,103,153]

7.2.3.2 Effects of Metal Ions on the Folding of an Independently Stable Subdomain

In a study exploring the effects of temperature and background ionic strength on the kinetic folding pathways of the independently folding P4–P6 domain of the *Tetrahymena* ribozyme, it was shown that folding was biphasic when initiated by adding Mg^{2+} in a low ionic strength buffer.[147] The slow phase disappeared with increasing concentration of background monovalent ions (Figure 7.11A). Contrary to expectation, folding of the 'fast' phase was slower at higher temperatures (Figure 7.11B). As summarized in Figure 7.11C, these investigators proposed that some tertiary interactions are present at higher ionic strengths in the absence of Mg^{2+} ions; favorable interactions in the partially unfolded (Mg^{2+} free) state were proposed to facilitate fast folding whereas the weakening of such interactions by decreasing the background ionic strength and increasing the temperature slowed folding. The two observed phases were viewed to reflect partitioning of molecules along parallel pathways with the choice of pathway dictated by the conformational heterogeneity in the unfolded state.

7.2.3.3 Cation-induced Conformational Heterogeneity in the Initial State

The ionic strength-dependent conformational bias in the unfolded state of the *Tetrahymena* ribozyme was explored using single-molecule FRET.[113] Conformational heterogeneity could populate different starting points at the top of a folding funnel and by doing so bias RNA molecules to particular folding pathways. RNA was incubated with different amounts of Na^+ ions in the unfolded state and folding was initiated by adding Mg^{2+} ions with simultaneous rapid adjustment of the salt concentration to a common final solution condition. Since the initial conditions are different but folding occurs in an identical electrostatic environment, the differences in folding rates, if any, would be due to the differences in the distribution of structures in the unfolded state ensemble.

The results of these studies indicate the presence of at least three different salt-dependent conformational ensembles (Figure 7.12A). Preincubation at low Na^+ concentrations (<150 mM) yielded slow folding ($k \approx 0.02\,\mathrm{s}^{-1}$) to mostly misfolded (catalytically inactive) RNA. Preincubation at intermediate Na^+ concentrations (150–250 mM) yielded folding along both productive and non-productive pathways; catalytically active molecules fold rapidly ($k \approx 1\,\mathrm{s}^{-1}$). Preincubation at high Na^+ concentrations (>250 mM) resulted in predominantly fast folding to the catalytically active native

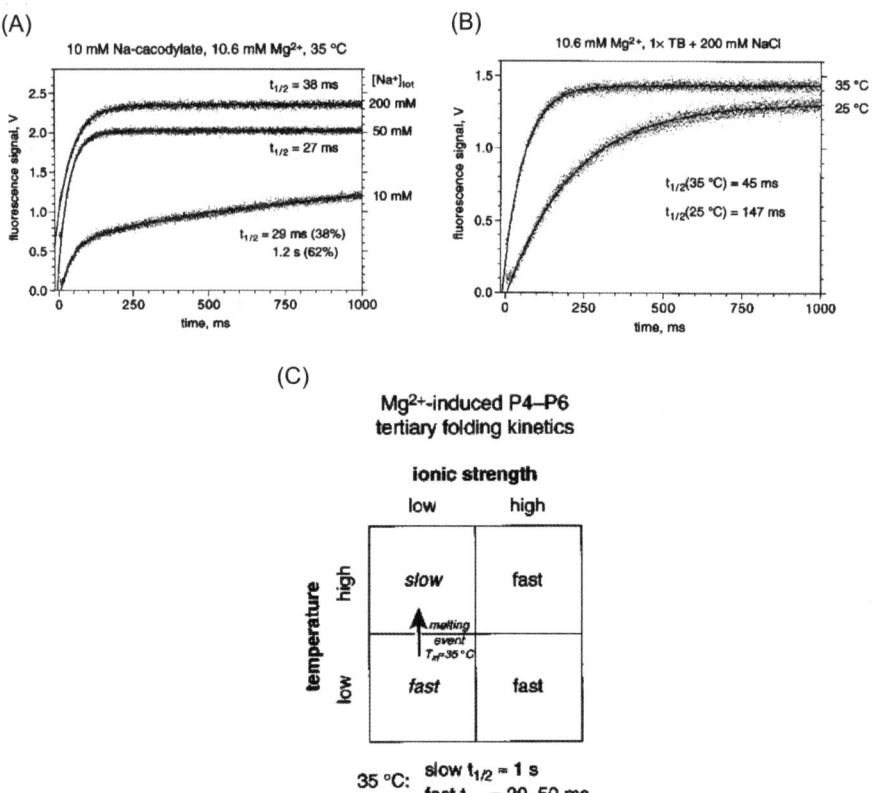

Figure 7.11 (A) Fluorescence kinetic traces for Mg^{2+}-induced folding of pyrene-labeled P4–P6 in sodium cacodylate buffer with varying [Na^+]. Three experiments are shown, in 10 mM sodium cacodylate buffer supplemented with NaCl to total [Na^+] of 10, 50 or 200 mM (*i.e.* 10 mM sodium cacodylate with additional 0, 40 or 190 mM NaCl). (B) Temperature dependence of Mg^{2+}-induced folding of pyrene-labeled P4–P6 in Tris–borate buffer with additional 200 mM NaCl. (C) Summary of buffer- and temperature-dependent kinetics for Mg^{2+}-induced P4–P6 tertiary folding. The slow folding component is observed only at particularly low ionic strength and higher temperature. In low ionic strength 10 mM sodium cacodylate, the narrow (5–10 °C) temperature range over which the kinetics change from fast to slow indicates that a melting event is responsible for the kinetics switch. Addition of ~30–50 mM total Na^+ to the cacodylate buffer provides sufficient ionic strength such that the slow component vanishes and folding is entirely fast. Reprinted with permission from ref. 147, Copyright 2000 American Chemical Society.

structure (Figure 7.12B and C). As shown in the schematic folding landscape (Figure 7.12D), the initial Na^+ concentration determines the partitioning of molecules between three distinct folding channels right from the earliest steps in folding.

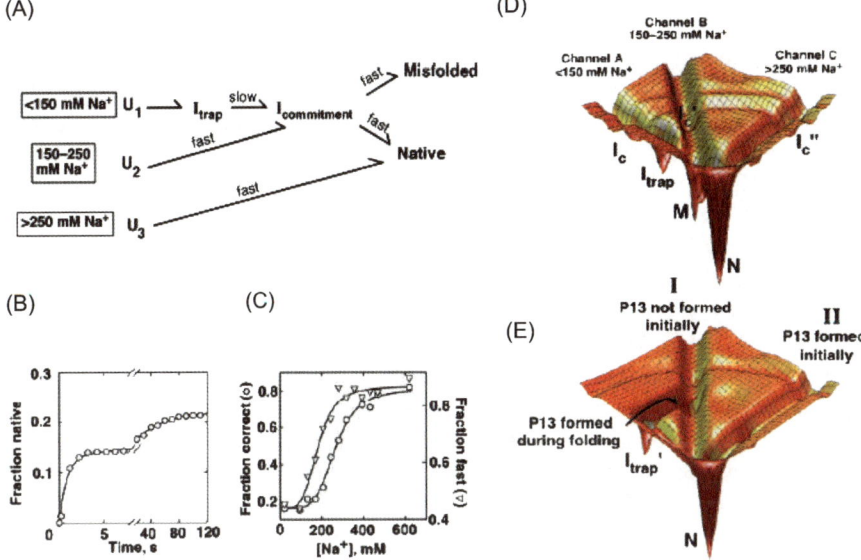

Figure 7.12 (A) Schematic model describing the effects of monovalent ions in generating the initial conformational heterogeneity. Different Na$^+$ concentrations cause three different starting states to be differentially populated that then fold along discrete pathways through the landscape. With low concentrations of Na$^+$ in the preincubation, a significant fraction (0.55) starts to fold from U1, proceeds through the late-folding intermediate I$_{trap}$ and slowly escapes this intermediate to partition between native and long-lived misfolded forms. With intermediate Na$^+$ concentrations in the preincubation, folding starts from U2; I$_{trap}$ is avoided but the partitioning between the native and long-lived misfolded forms is the same as at low Na$^+$, suggesting the presence of the common folding intermediate I$_{commitment}$. With high Na$^+$ in the preincubation, U3 is populated and essentially all of the ribozyme folds fast and correctly, thereby avoiding both I$_{trap}$ and I$_{commitment}$. (B) Upon initiation of folding by flowing a buffer containing 5 mM Mg^{2+} over the surface-attached ribozyme, replacing the 170 mM Na$^+$ solution present initially, the fraction of molecules that attained FRET values of ~0.9, indicating native-state formation, is plotted against folding time. The curve represents a fit by two exponentials with rate constants of 1.21 ± 0.07 and 0.025 ± 0.001 s^{-1}. The combined amplitude of these two folding phases, 0.219 ± 0.003, represents the fraction of ribozyme that folded to the native state without forming the long-lived misfolded state. (C) The fraction of ribozyme that folded correctly (○) and the fraction of the correct-folding molecules that folded fast (~1 s^{-1}, ▽), are plotted against initial Na$^+$ concentration. (D) A schematic of channels for the wild-type ribozyme. The dependence of folding properties on initial conditions indicates the presence of free energy barriers between the channels. The simplest model is shown, in which there are three pathways, A–C, which lie in channels A, B and C, respectively. Pathways A and B predominate at low initial Na$^+$ concentration, pathway B at intermediate Na$^+$ concentration and pathway C at high Na$^+$ concentration. I$_c$ is a collapsed intermediate that forms along pathway A and analogous intermediates are postulated along pathways B and C. (E) Effects of P13 formation during folding starting with a preformed P3. The landscape for folding of the U273A ribozyme is depicted for simplicity, in which the long-lived misfolded form is not formed significantly. Adapted from ref. 115, Copyright 2002 National Academy of Sciences, USA.

Two regions of the RNA molecules are conspicuously prestructured at the higher concentrations of Na^+. A long-range pseudoknot interaction that structures the catalytic core of the ribozyme (P3 domain) is preformed in Na^+ and is therefore a likely candidate to guide the molecules into fast and correct folding pathways. The authors used a mutant ribozyme in which the correct pseudoknot structure is stabilized relative to the competing incorrect form, the U273A mutant,[154] to show that whereas stabilization of this interaction indeed results in an increased fraction of correctly folded molecules, it is neither sufficient nor necessary for fast folding. A long-range peripheral tertiary contact (P13) is present at Na^+ concentrations that support fast folding. However, disruption of this contact in the context of the stabilized pseudoknot surprisingly resulted in fast folding at both high and low salt concentrations. This dual effect was proposed to arise from a P13-generated, salt-dependent kinetic barrier to folding, as depicted in the schematic folding landscape for the U273A mutant ribozyme shown in Figure 7.12E. Early formation of P13 at high salt

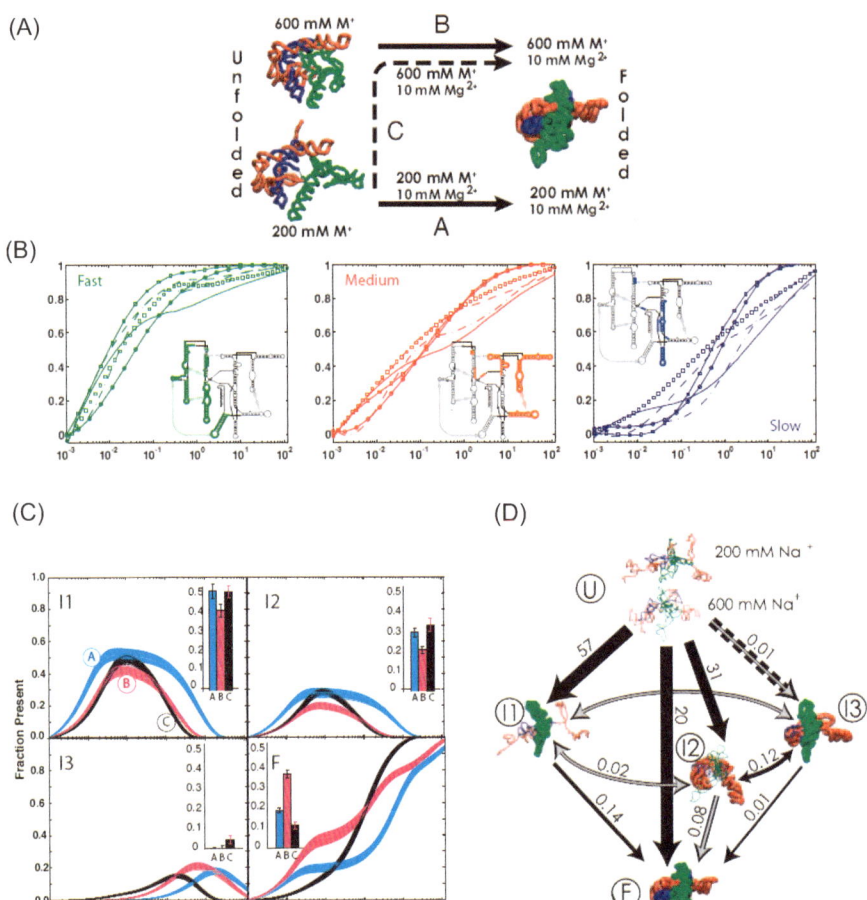

concentration most likely preorganizes the core for fast Mg^{2+}-mediated folding whereas in a low-salt background P13 forms during folding and either stabilizes fast-forming kinetic traps in other regions of the molecule or creates topological barriers to core folding. Abolition of P13 in a low-salt solution would ameliorate the hypothetical barriers and thus expedite folding. It is clear that the salt present in the initial solution produces a heterogeneous ensemble of structured 'unfolded' molecules that guide the molecules into alternative folding pathways.

7.2.3.4 Modulation of the Electrostatic Environment During Mg^{2+}-mediated RNA Folding

Time-resolved hydroxyl radical footprinting follows the ensemble average of changes in the solvent-accessible surface of the phosphodiester backbone of RNA with single nucleotide spatial and millisecond time resolution.[155] The time–progress curves provided by time-resolved footprinting can be used to

Figure 7.13 (A) A summary of the experimental scheme of the analyzed folding reactions. Reactions A and B (solid arrows) are folding experiments in which Mg^{2+} is added over a constant concentration of Na^+ or K^+. In reaction C (dashed arrow), Mg^{2+} and either Na^+ or K^+ are added simultaneously to initiate folding. (B) Average time–progress curves summarizing the clusters distinguished by k-means clustering of the 25 ribozyme protections analyzed for each folding reaction: Reactions A and B (as in Figure 7.13A) are in 200 mM Na^+ (solid line), in 600 mM Na^+ (open squares), in 200 mM K^+ (dashed line) and in 600 mM K^+ (filled circles); reaction C (as in Figure 7.13A) is 200 mM Na^+ (filled circles) and K^+ (filled squares) adjusted to 600 mM concurrently with Mg^{2+}. The clusters are characterized by fast (green), medium (red) and slow (blue) initial rates. The secondary structure insets show the structural localization of the clusters. (C) Time evolution of the intermediates (I1, I2 and I3) and the final state (F) observed for folding reactions A (cyan), B (magenta) and C (black) (as described in Figure 7.13A). The width of the curves illustrates propagated errors on the time evolution of the intermediates. Insets: the relative initial fluxes and propagated error for each reaction calculated from the ratios of the measured rate constants. (D) The kinetic models for the Mg^{2+}-induced folding of wild-type ribozyme. The structured regions of each representative species are modeled from the •OH reactivity patterns and rendered as bold ribbons. The thickness of the arrows is proportional to the resolved rate. Black arrows indicate that there is no significant difference in rate measured for the different salt concentrations. Gray–black arrows indicate differences in the resolved rates between the two salt concentrations indicated (black and gray represent 200 and 600 mM Na^+, respectively). The rates that are dramatically different for folding in the presence of K^+ rather than Na^+ are shown by black dashed arrows. The numbers accompanying arrows indicate the resolved rates (in s^{-1}) for the folding transition conducted in the presence of 200 mM Na^+. Adapted from ref. 9, Copyright 2007 National Academy of Sciences, USA.

resolve 'structural kinetic' folding models that define the predominant reaction intermediates and their structures and interconversion rates among the species present in solution.[4] Whereas the *Tetrahymena* ribozyme was shown to fold by parallel pathways from the earliest steps, the reaction flux passes through relatively few of the possible pathways.

This approach was used to explore the effects of monovalent ion electrostatic environments on the Mg^{2+}-mediated kinetic folding pathways of the *Tetrahymena* ribozyme.[9] In order to generate different initial conformational ensembles, the monovalent ion or 'salt' concentration in the initial solution was varied. As shown in the experimental scheme shown in Figure 7.13A, folding was initiated by adding Mg^{2+} and allowed to proceed either by maintaining the same salt concentration as in the initial solution or by simultaneously increasing both the salt and Mg^{2+} concentrations. This experimental approach allows reactions initiated from different conditions to fold under a common regime (reactions B and C) and the reactions initiated from the common initial condition to fold under different regimes (reactions A and C).

A common kinetic model configuration was obtained for the Mg^{2+}-mediated folding of the wild type *Tetrahymena* ribozyme for reactions conducted when either Na^+ or K^+ as the background monovalent cation, at 200 or 600 mM salt concentration, and when folding was initiated by the concomitant addition of monovalent cations along with Mg^{2+} (Figure 7.13B). The kinetic intermediates that define the model have structured P4–P6, the periphery and both P4–P6 and the periphery. What does change among the reactions is the partitioning of the folding flux among the possible pathways (Figure 7.13C). The affiliation of groups of tertiary contacts (that often represent independent structural modules or subdomains) within a given temporal cluster is invariant with the differences in the salt concentration in the initial solution or during folding (Figure 7.13B and D). Since the structural features of the folding intermediates are derived from the tertiary interactions present in a given cluster of time–progress curves, it was proposed that the topology and the stability of the native state define the dominant pathways by which the RNA folds.

This hypothesis was tested by studying the Mg^{2+}-mediated folding of a ribozyme in which the tetraloop–tetraloop receptor interaction stabilizing the P4–P6 domain was perturbed by mutation. This mutant ribozyme folds to the native state but with reduced stability.[156] As with the wild-type ribozyme, a common kinetic model configuration was resolved for folding under different experimental conditions. However, the structural characteristics of the folding intermediates are dramatically different from those of the wild type. Indeed, the stability of the native state is a determinant of the folding mechanism.[9]

This study also looked closely at the initial partitioning of the folding flux among the parallel pathways. This partitioning varied with the initial solution conditions and was insensitive to whether folding was initiated by Mg^{2+} alone or by the concomitant addition of monovalent cations (Figure 7.13C and D). This result supports the hypothesis that the initial partitioning of the reaction flux into discreet folding pathways is guided by the distribution of conformations present in the initial state ensemble. On the other hand, the interconversion of the kinetic intermediates and their passage to the final native state are dependent on the electrostatic environment present during folding. The conclusions of this study are that (i) the overall stability of a given native conformation determines the dominant folding pathways through sets of discrete structural intermediates, (ii) the conformations present in the initial ensemble at the top of the folding funnel fall into discrete folding pathways and (iii) once a molecule has been channeled into a given folding pathway, electrostatic forces determine the magnitude of the subsequent free energy barriers, both within the pathway and between two adjacent pathways. Under some conditions, folding pathways behave as channels with little or no interconversion of molecules among the dominant pathways.

7.3 Conclusion

Study of the role of electrostatic forces in stabilizing RNA structure and determining the height of folding activation energy barriers is an active field of research, with new information steadily flowing into the literature. A variety of novel techniques have been developed and applied to the RNA folding problem that monitor time-dependent changes in the conformation of RNA in an electrostatic environment ranging from detection of their global dimension to the solvent accessibility of single nucleotides in ensemble measurements and single molecule measures. Significant progress is also seen in the quantitative evaluation of the observable properties, facilitating the extraction of information that is revealing the behavior of discrete populations within an ensemble. These advances have revealed unprecedented details of the mechanisms of effect of metal ions on the kinetic folding pathways at the molecular level.

Theoretical challenges in estimating the contributions of ions to structural stability, conformational dynamics and folding energy barriers are rapidly being met with steady developments in *in silico* analysis. Although most of the studies to date have focused on RNA molecules ranging from small hairpins to ribozymes of intermediate size, the stage is set for extending the studies to monitor ion-dependent kinetics of assembly and conformational changes of large RNA–protein complexes such as the ribosome. The ultimate technical challenge in the field is likely to be detection of conformational dynamics in the physiological ionic milieu, at the molecular level, within a live cell during transcription of an RNA molecule.

Acknowledgement

The writing of this article was supported by grant PO1-GM66275 from the National Institutes of Health.

References

1. R. F. Gesteland, T. R. Cech and J. F. Atkins, *The RNA World*, 2006, Vol. **43**, Cold Spring Harbor Laboratory Press, Cold Spring Harbor, NY.
2. J. R. Fresco, *RNA Structure and Fun*ction, 1998, Vol. **35**, Cold Spring Harbor Laboratory Press, Cold Spring Harbor, NY.
3. D. Herschlag, *J. Biol. Chem.*, 1995, **270**, 20871.
4. A. Laederach, I. Shcherbakova, M. Liang, M. Brenowitz and R. Altman, *J. Mol. Biol.*, 2006, **358**, 1179.
5. J. Pan, D. Thirumalai and S. Woodson, *J. Mol. Biol.*, 1997, **273**, 7.
6. D. Thirumalai and C. Hyeon, *Biochemistry*, 2005, **44**, 4957.
7. T. Sosnick and T. Pan, *Curr. Opin. Struct. Biol.*, 2003, **13**, 309.
8. Z. Xie, N. Srividya, T. Sosnick, T. Pan and N. Scherer, *Proc. Natl. Acad. Sci. USA*, 2004, **101**, 534.
9. A. Laederach, I. Shcherbakova, M. Jonikas, R. Altman and M. Brenowitz, *Proc. Natl. Acad. Sci. USA*, 2007.
10. S. Woodson, *Curr. Opin. Chem. Biol.*, 2005, **9**, 104.
11. D. Draper, D. Grilley and A. Soto, *Annu. Rev. Biophys. Biomol. Struct.*, 2005, **34**, 221.
12. T. Pan and T. Sosnick, *Annu. Rev. Biophys. Biomol. Struct.*, 2006, **35**, 161.
13. J. Grohman, M. Del Campo, H. Bhaskaran, P. Tijerina, A. Lambowitz and R. Russell, *Biochemistry*, 2007, **46**, 3013.
14. R. Schroeder, A. Barta and K. Semrad, *Nat. Rev. Mol. Cell. Biol.*, 2004, **5**, 908.
15. S. Koduvayur and S. Woodson, *RNA*, 2004, **10**, 1526.
16. P. Cole and C. Crothers, *Biochemistry*, 1972, **11**, 4368.
17. D. Crothers, P. Cole, C. Hilbers and R. Shulman, *J. Mol. Biol.*, 1974, **87**, 63.
18. E. Brauns and R. Dyer, *Biophys. J.*, 2005, **89**, 3523.
19. P. Brion and E. Westhof, *Annu. Rev. Biophys. Biomol. Struct.*, 1997, **26**, 113.
20. P. Zarrinkar and J. Williamson, *Science*, 1994, **265**, 918.
21. B. Sclavi, M. Sullivan, M. Chance, M. Brenowitz and S. Woodson, *Science*, 1998, **279**, 1940.
22. L. Kwok, I. Shcherbakova, J. Lamb, H. Park, K. Andresen, H. Smith, M. Brenowitz and L. Pollack, *J. Mol. Biol.*, 2006, **355**, 282.
23. S. Chauhan, G. Caliskan, R. Briber, U. Perez-Salas, P. Rangan, D. Thirumalai and S. Woodson, *J. Mol. Biol.*, 2005, **353**, 1199.
24. Y. Bai, R. Das, I. Millett, D. Herschlag and S. Doniach, *Proc. Natl. Acad. Sci. USA*, 2005, **102**, 1035.
25. K. Takamoto, R. Das, Q. He, S. Doniach, M. Brenowitz, D. Herschlag and M. Chance, *J. Mol. Biol.*, 2004, **343**, 1195.
26. R. Das, L. Kwok, I. Millett, Y. Bai, T. Mills, J. Jacob, G. Maskel, S. Seifert, S. Mochrie, P. Thiyagarajan, S. Doniach, L. Pollack and D. Herschlag, *J. Mol. Biol.*, 2003, **332**, 311.
27. R. Russell, I. Millett, M. Tate, L. Kwok, B. Nakatani, S. Gruner, S. Mochrie, V. Pande, S. Doniach, D. Herschlag and L. Pollack, *Proc. Natl. Acad. Sci. USA*, 2002, **99**, 4266.

28. R. Russell, I. Millett and S. Doniach, *Nature Struct. Biol.*, 2000, **7**, 367.
29. D. Thirumalai, N. Lee, S. Woodson and D. Klimov, *Annu. Rev. Phys. Chem.*, 2001, **52**, 751.
30. J. Pan and S. Woodson, *J. Mol. Biol.*, 1999, **294**, 955.
31. D. Turner and N. Sugimoto, *Annu. Rev. Biophys. Biophys. Chem.*, 1988, **17**, 167.
32. Z. Lu, D. Turner and D. Mathews, *Nucleic Acids Res.*, 2006, **34**, 4912.
33. D. Mathews and D. Turner, *Curr. Opin. Struct. Biol.*, 2006, **16**, 270.
34. A. Stein and D. Crothers, *Biochemistry*, 1976, **15**, 160.
35. I. J. Tinoco and C. Bustamante, *J. Mol. Biol.*, 1999, **293**, 271.
36. A. Banerjee, J. Jaeger and D. Turner, *Biochemistry*, 1993, **32**, 153.
37. L. Jaeger, E. Westhof and F. Michel, *J. Mol. Biol.*, 1993, **234**, 331.
38. F. Kramer and D. Mills, *Nucleic Acids Res.*, 1981, **9**, 5109.
39. M. Wu and I. J. Tinoco, *Proc. Natl. Acad. Sci. USA*, 1998, **95**, 11555.
40. J. Pan and S. Woodson, *J. Mol. Biol.*, 1998, **280**, 597.
41. S. Walstrum and O. Uhlenbeck, *Biochemistry*, 1990, **29**, 10573.
42. V. Emerick and S. Woodson, *Biochemistry*, 1993, **32**, 14062.
43. S. Woodson and V. Emerick, *Mol. Cell. Biol.*, 1993, **13**, 1137.
44. S. Woodson and T. Cech, *Biochemistry*, 1991, **30**, 2042.
45. J. R. Fresco, A. Adains, R. Ascione, D. Henley and T. Lindahl, *Tertiary Structure in Transfer Ribonucleic Acids*, Cold Spring Harbor Laboratory Press, Cold Spring Harbor, NY, 1966, Vol. 31, p. 527.
46. E. Hawkins, S. Chang and W. Mattice, *Biopolymers*, 1977, **16**, 1557.
47. D. Thirumalai and S. Woodson, *RNA*, 2000, **6**, 790.
48. R. Russell, R. Das, H. Suh, K. J. Travers, A. Laederach, M. A. Engelhardt and D. Herschlag, *J. Mol. Biol.*, 2006, **363**, 531.
49. G. Manning, *Q. Rev. Biophys.*, 1978, **11**, 179.
50. J. M. Schurr, in *Metal Ion–Nucleic Acid Interactions*, N. V. Hud, ed., Royal Society of Chemistry, Cambridge, 2008, Chapter 9.
51. T. Dewey and D. Turner, *Biochemistry*, 1979, **18**, 5757.
52. T. Dewey and D. Turner, *Biochemistry*, 1980, **19**, 1681.
53. I. J. Tinoco, P. Borer, B. Dengler, M. Levin, O. Uhlenbeck, D. Crothers and J. Bralla, *Nat. New Biol.*, 1973, **246**, 40.
54. J. Barrio, G. Tolman, N. Leonard, R. Spencer and G. Weber, *Proc. Natl. Acad. Sci. USA*, 1973, **70**, 941.
55. P. Doty, H. Boedtker, J. Fresco, B. Hall and R. Haselkorn, *Ann. N. Y. Acad. Sci.*, 1959, **81**, 693.
56. J. Fresco, B. Alberts and P. Doty, *Nature*, 1960, **188**, 98.
57. S. Freier, R. Kierzek, J. Jaeger, N. Sugimoto, M. Caruthers, T. Neilson and D. Turner, *Proc. Natl. Acad. Sci. USA*, 1986, **83**, 9373.
58. Z. Tan and S. Chen, *Biophys. J.*, 2006, **90**, 1175.
59. A. Williams, C. Longfellow, S. Freier, R. Kierzek and D. Turner, *Biochemistry*, 1989, **28**, 4283.
60. S. Nakano, M. Fujimoto, H. Hara and N. Sugimoto, *Nucleic Acids Res.*, 1999, **27**, 2957.

61. M. Serra, J. Baird, T. Dale, B. Fey, K. Retatagos and E. Westhof, *RNA*, 2002, **8**, 307.
62. P. Cole, S. Yang and D. Crothers, *Biochemistry*, 1972, **11**, 4358.
63. D. P. Giedroc and N. E. Grossoehme, in *Metal Ion–Nucleic Acid Interactions*, ed. N. V. Hud, Royal Society of Chemistry, Cambridge, 2008, Chapter 6.
64. C. Hsiao, E. Tannenbaum, H. VanDeusen, E. Hershkovitz, G. Perng, A. R. Tannenbaum and L. D. Williams, in *Metal Ion–Nucleic Acid Interactions*, ed. N. V. Hud, Royal Society of Chemistry, Cambridge, 2008, Chapter 1.
65. G. Manning, *Biophys. Chem.*, 1977, **7**, 189.
66. D. Draper, *RNA*, 2004, **10**, 335.
67. V. Murthy and G. Rose, *Biochemistry*, 2000, **39**, 14365.
68. B. Sclavi, S. Woodson, M. Sullivan, M. Chance and M. Brenowitz, *Methods Enzymol.*, 1998, **295**, 379.
69. T. Uchida, K. Takamoto, Q. He, M. Chance and M. Brenowitz, *J. Mol. Biol.*, 2003, **328**, 463.
70. J. Cate, A. Gooding, E. Podell, K. Zhou, B. Golden, C. Kundrot, T. Cech and J. Doudna, *Science*, 1996, **273**, 1678.
71. J. Cate, R. Hanna and J. Doudna, *Nat. Struct. Biol.*, 1997, **4**, 553.
72. P. Adams, M. Stahley, A. Kosek, J. Wang and S. Strobel, *Nature*, 2004, **430**, 45.
73. D. Klein, P. Moore and T. Steitz, *RNA*, 2004, **10**, 1366.
74. V. Misra, R. Shiman and D. Draper, *Biopolymers*, 2003, **69**, 118.
75. R. Römer and R. Hach, *Eur. J. Biochem.*, 1975, **55**, 271.
76. B. Laggerbauer, F. Murphy and T. Cech, *EMBO J.*, 1994, **13**, 2669.
77. S. Heilman-Miller, D. Thirumalai and S. Woodson, *J. Mol. Biol.*, 2001, **306**, 1157.
78. A. Pyle, O. Fedorova and C. Waldsich, *Trends Biochem. Sci.*, 2007, **32**, 138.
79. A. Serganov, A. Polonskaia, A. Phan, R. Breaker and D. Patel, *Nature*, 2006, **441**, 1167.
80. M. Erat and R. Sigel, *Inorg. Chem.*, 2007, **46**, 11224.
81. M. Erat, O. Zerbe, T. Fox and R. Sigel, *ChemBioChem*, 2007, **8**, 306.
82. J. Cate and J. Doudna, *Structure*, 1996, **4**, 1221.
83. R. Hanna and J. Doudna, *Curr. Opin. Chem. Biol.*, 2000, **4**, 166.
84. J. Lemay, J. Penedo, R. Tremblay, D. Lilley and D. Lafontaine, *Chem. Biol.*, 2006, **13**, 857.
85. T. Wilson, M. Nahas, L. Araki, S. Harusawa, T. Ha and D. Lilley, *Blood Cells Mol. Dis.*, 2007, **38**, 8.
86. A. Serganov, Y. Yuan, O. Pikovskaya, A. Polonskaia, L. Malinina, A. Phan, C. Hobartner, R. Micura, R. Breaker and D. Patel, *Chem. Biol.*, 2004, **11**, 1729.
87. D. Grilley, V. Misra, G. Caliskan and D. Draper, *Biochemistry*, 2007, **46**, 10266.
88. A. Soto, V. Misra and D. Draper, *Biochemistry*, 2007, **46**, 2973.

89. D. Grilley, A. Soto and D. Draper, *Proc. Natl. Acad. Sci. USA*, 2006, **103**, 14003.
90. E. Doherty and J. Doudna, *Annu. Rev. Biochem.*, 2000, **69**, 597.
91. J. Hougland, A. Kravchuk, D. Herschlag and J. Piccirilli, *PLoS Biol.*, 2005, **3**, e277.
92. S. Shan and D. Herschlag, *RNA*, 2002, **8**, 861.
93. J. O'Rear, S. Wang, A. Feig, L. Beigelman, O. Uhlenbeck and D. Herschlag, *RNA*, 2001, **7**, 537.
94. S. Shan, A. Kravchuk, J. Piccirilli and D. Herschlag, *Biochemistry*, 2001, **40**, 5161.
95. S. Shan and D. Herschlag, *Biochemistry*, 1999, **38**, 10958.
96. S. Shan, A. Yoshida, S. Sun, J. Piccirilli and D. Herschlag, *Proc. Natl. Acad. Sci. USA*, 1999, **96**, 12299.
97. T. McConnell, D. Herschlag and T. Cech, *Biochemistry*, 1997, **36**, 8293.
98. A. Ke, F. Ding, J. Batchelor and J. Doudna, *Structure*, 2007, **15**, 281.
99. Y. Bai, M. Greenfeld, K. Travers, V. Chu, J. Lipfert, S. Doniach and D. Herschlag, *J. Am. Chem. Soc.*, 2007, **129**, 14981.
100. T. Gluick, R. Gerstner and D. Draper, *J. Mol. Biol.*, 1997, **270**, 451.
101. E. Koculi, C. Hyeon, D. Thirumalai and S. Woodson, *J. Am. Chem. Soc.*, 2007, **129**, 2676.
102. E. Koculi, D. Thirumalai and S. Woodson, *J. Mol. Biol.*, 2006, **359**, 446.
103. E. Koculi, N. Lee, D. Thirumalai and S. Woodson, *J. Mol. Biol.*, 2004, **341**, 27.
104. S. Heilman-Miller, J. Pan, D. Thirumalai and S. Woodson, *J. Mol. Biol.*, 2001, **309**, 57.
105. J. E. Leffler, *Science*, 1953, **117**.
106. I. Sánchez and T. Kiefhaber, *Biophys. Chem.*, 2003, **100**, 397.
107. C. D. Johnson, *Chem. Rev.*, 1975, **75**, 755.
108. W. P. Jencks, *Chem. Rev.*, 1985, **85**, 511.
109. J. Pan, D. Thirumalai and S. Woodson, *Proc. Natl. Acad. Sci. USA*, 1999, **96**, 6149.
110. P. Zarrinkar and J. Williamson, *Nucleic Acids Res.*, 1996, **24**, 854.
111. D. Treiber and J. Williamson, *Curr. Opin. Struct. Biol.*, 1999, **9**, 339.
112. M. Rook, D. Treiber and J. Williamson, *Proc. Natl. Acad. Sci. USA*, 1999, **96**, 12471.
113. R. Russell and D. Herschlag, *J. Mol. Biol.*, 1999, **291**, 1155.
114. R. Russell and D. Herschlag, *J. Mol. Biol.*, 2001, **308**, 839.
115. R. Russell, X. Zhuang, H. Babcock, I. Millett, S. Doniach, S. Chu and D. Herschlag, *Proc. Natl. Acad. Sci. USA*, 2002, **99**, 155.
116. P. Zarrinkar and J. Williamson, *Nat. Struct. Biol.*, 1996, **3**, 432.
117. D. Treiber, M. Rook, P. Zarrinkar and J. Williamson, *Science*, 1998, **279**, 1943.
118. M. Rook, D. Treiber and J. Williamson, *J. Mol. Biol.*, 1998, **281**, 609.
119. T. Pan and T. Sosnick, *Nat. Struct. Biol.*, 1997, **4**, 931.
120. X. Fang, T. Pan and T. Sosnick, *Nat. Struct. Biol.*, 1999, **6**, 1091.

121. G. Pljevaljčić, D. Klostermeier and D. Millar, *Biochemistry*, 2005, **44**, 4870.
122. H. Kim, G. Nienhaus, T. Ha, J. Orr, J. Williamson and S. Chu, *Proc. Natl. Acad. Sci. USA*, 2002, **99**, 4284.
123. H. Eyring, *J. Chem. Phys.*, 1935, **3**.
124. A. Fersht, *Curr. Opin. Struct. Biol.*, 1995, **5**, 79.
125. B. Onoa, S. Dumont, J. Liphardt, S. Smith, I. J. Tinoco and C. Bustamante, *Science*, 2003, **299**, 1892.
126. L. Bartley, X. Zhuang, R. Das, S. Chu and D. Herschlag, *J. Mol. Biol.*, 2003, **328**, 1011.
127. C. Tanford, *Adv. Protein Chem.*, 1970, **24**, 1.
128. C. Tanford, *Adv. Protein Chem.*, 1968, **23**, 121.
129. A. Matouschek and A. Fersht, *Proc. Natl. Acad. Sci. USA*, 1993, **90**, 7814.
130. A. Matouschek, D. Otzen, L. Itzhaki, S. Jackson and A. Fersht, *Biochemistry*, 1995, **34**, 13656.
131. G. S. Hammond, *J. Am. Chem. Soc.*, 1955, **77**.
132. E. R. Thornton, *J. Am. Chem. Soc.*, 1967, **89**.
133. V. Shelton, T. Sosnick and T. Pan, *Biochemistry*, 1999, **38**, 16831.
134. T. Goody, S. Melcher, D. Norman and D. Lilley, *RNA*, 2004, **10**, 254.
135. T. Wilson and D. Lilley, *RNA*, 2002, **8**, 587.
136. A. Matouschek, J. J. Kellis, L. Serrano and A. Fersht, *Nature*, 1989, **340**, 122.
137. A. Fersht, A. Matouschek, J. Sancho, L. Serrano and S. Vuilleumier, *Faraday Discuss.*, 1992, **183**.
138. A. Fersht, A. Matouschek and L. Serrano, *J. Mol. Biol.*, 1992, **224**, 771.
139. A. Fersht, *Proc. Natl. Acad. Sci. USA*, 2004, **101**, 17327.
140. S. Silverman and T. Cech, *RNA*, 2001, **7**, 161.
141. G. Bokinsky, D. Rueda, V. Misra, M. Rhodes, A. Gordus, H. Babcock, N. Walter and X. Zhuang, *Proc. Natl. Acad. Sci. USA*, 2003, **100**, 9302.
142. J. Liphardt, B. Onoa, S. Smith, I. J. Tinoco and C. Bustamante, *Science*, 2001, **292**, 733.
143. M. Manosas, J. Wen, P. Li, S. Smith, C. Bustamante, I. J. Tinoco and F. Ritort, *Biophys. J.*, 2007, **92**, 3010.
144. S. Smith, Y. Cui and C. Bustamante, *Science*, 1996, **271**, 795.
145. P. Li, C. Bustamante and I. J. Tinoco, *Proc. Natl. Acad. Sci. USA*, 2007, **104**, 7039.
146. J. Wen, M. Manosas, P. Li, S. Smith, C. Bustamante, F. Ritort and I. J. Tinoco, *Biophys. J.*, 2007, **92**, 2996.
147. S. Silverman, M. Deras, S. Woodson, S. Scaringe and T. Cech, *Biochemistry*, 2000, **39**, 12465.
148. P. Adams, M. Stahley, M. Gill, A. Kosek, J. Wang and S. Strobel, *RNA*, 2004, **10**, 1867.
149. S. Basu, R. Rambo, J. Strauss-Soukup, J. Cate, A. Ferré-D'Amaré, S. Strobel and J. Doudna, *Nat. Struct. Biol.*, 1998, **5**, 986.
150. Y. Wang, M. Lu and D. Draper, *Biochemistry*, 1993, **32**, 12279.
151. D. Draper and V. Misra, *Nat. Struct. Biol.*, 1998, **5**, 927.

152. G. Conn, A. Gittis, E. Lattman, V. Misra and D. Draper, *J. Mol. Biol.*, 2002, **318**, 963.
153. I. Shcherbakova, S. Gupta, M. Chance and M. Brenowitz, *J. Mol. Biol.*, 2004, **342**, 1431.
154. J. Pan, M. Deras and S. Woodson, *J. Mol. Biol.*, 2000, **296**, 133.
155. I. Shcherbakova, S. Mitra, R. Beer and M. Brenowitz, *Nucleic Acids Res.*, 2006, **34**, e48.
156. I. Shcherbakova and M. Brenowitz, *J. Mol. Biol.*, 2005, **354**, 483.

CHAPTER 8
Metal Ions in RNA Catalysis

JOHN K. FREDERIKSEN,[a,†] ROBERT FONG[a,†] AND JOSEPH A. PICCIRILLI[a, b, c]

[a] Department of Biochemistry and Molecular Biology; [b] Department of Chemistry; and [c] Howard Hughes Medical Institute, The University of Chicago, 929 E. 57th Street, Chicago, IL 60637, USA

8.1 Introduction

The ability of catalytic RNA to adopt defined three-dimensional architectures and perform catalysis is critically dependent on the presence of metal ions. Therefore, any attempts to achieve a detailed mechanistic understanding of ribozyme function require a consideration of how metal ions interact with RNA to promote its proper folding, stabilize its catalytically competent structure, and stimulate catalysis. As the role of metal ions in RNA folding and structure is the emphasis of other chapters in this volume, the present discussion focuses on the ways in which metal ions exert their effects along the reaction coordinate of RNA-catalyzed reactions.

After the discovery that RNA could catalyze chemical reactions,[1] it was thought that all ribozymes would be obligate metalloenzymes, employing metal ions both to stabilize structure and participate in chemistry.[2] Given the polyanionic nature of RNA polymers, it was clear that cations would be required for the folding and stabilization of the three-dimensional structures of ribozymes, in which negatively charged phosphate groups are often brought into close proximity. Consistent with this hypothesis, all ribozymes studied to date require the presence of cations in some form for function. The relatively restricted functional group repertoire

---

[†] These authors contributed equally to this work.

RSC Biomolecular Sciences
Nucleic Acid–Metal Ion Interactions
Edited by Nicholas V. Hud
© Royal Society of Chemistry 2009
Published by the Royal Society of Chemistry, www.rsc.org

available to RNA, coupled with pK_a values of free RNA nucleosides that lie outside the range considered feasible for general acid–base catalysis, engendered the additional expectation that metal ions would be required to promote catalysis by interacting with the chemical transition state. The folded structures of ribozymes were then thought to provide a scaffold for positioning catalytically essential divalent metal ions within the active site. Initial observations that ribozymes seemed to require the presence of divalent metal ions for catalytic activity supported this notion. As the first of the known transition state metal ion interactions were identified in the group I intron, Steitz and Steitz proposed a 'two-metal ion' model for large ribozyme catalysis by mechanistic analogy with *Escherichia coli* DNA polymerase and alkaline phosphatase.[3] This model has gained increasing experimental support[4,5] and remains the theoretical framework within which experimental observations relating to ribozyme catalysis are interpreted.

As the condition space for ribozyme catalysis began to be explored more widely, it was discovered that three members of the class of small ribozymes could function in the absence of divalent metal ions if molar concentrations of monovalent cations were present.[6] For example, the hammerhead sequence $HH_{16.1}$ was found to react in 2 M Li_2SO_4 alone at a rate within threefold of that measured in the presence of 2 M Li_2SO_4 and 10 mM $MgCl_2$. Similar results were obtained when assaying cleavage by the hairpin and *Neurospora* VS ribozymes, which in some cases actually reacted faster in buffers with high monovalent ion concentrations than in standard Mg^{2+}-containing buffers.[6] Subsequent experiments with the hairpin ribozyme revealed activity in the presence of the 'exchange-inert' coordination complex cobalt hexammine $[Co^{3+}(NH_3)_6]$,[7] an observation that has also been made for catalysis by the hammerhead[8] and *glmS*[9] ribozymes. Taken together, these results prompted a re-examination of the concept that all ribozymes are strict metalloenzymes as they demonstrated that the small ribozymes could promote catalysis often rather robustly without the participation of divalent metal ions. The absolute requirement for cations of any kind implied that metal ions probably serve a structural role in stabilizing the active ribozyme architecture or contribute in a very indirect way to catalysis by electrostatic shielding, but the actual catalytic machinery that operates directly on atoms of the transition state resides wholly within the RNA itself. Experimental observation of pK_a-shifted nucleobases within folded RNAs made plausible the suggestion that ribozymes could employ a general acid–base strategy using their ionizable functional groups.[10–15] Studies using chemical modification of the scissile phosphate in conjunction with mutations of a putative catalytically important nucleobase have provided convincing validation for this proposal.[16] Catalysis by the large ribozymes, however, remains absolutely dependent upon the presence of divalent metal ions, and this mechanistic distinction between the two ribozyme classes suggests a mechanistic paradigm with respect to the role of metal ions that will be discussed more fully later in this chapter.

One formidable hurdle impeding the elucidation of the specific role of metal ions in promoting RNA catalysis involves differentiating those ions from the large number of diffusely bound ions associated with the anionic phosphodiester backbone. To that end, several structural, spectroscopic, computational and biochemical techniques have been employed successfully to identify specific metal ion binding sites within RNA. However, even after these metal ion binding sites have been identified, the question remains as to whether the bound metal ion participates directly in catalysis or only stabilizes the RNA's tertiary structure. This question can be addressed only by experiments that interrogate the transition state of the chemical reaction. Metal ion rescue experiments that probe reaction kinetics using chemically modified RNA have emerged as the gold standard for identifying and characterizing these functionally important metal ion–RNA interactions. This methodology has been used to great effect in developing an atomic resolution model of the active site of the *Tetrahymena* group I intron, the ribozyme for which the catalytic mechanism and the role of metal ions have been characterized most extensively. Subsequent metal ion rescue experiments have revealed metal ions that participate directly in catalysis within the active sites of RNase P, the group II intron and the spliceosome.

Identification of these active site metal ions has simultaneously provided important insights into the mechanisms of catalysis by the large ribozymes and focused our attention on a more fundamental physical organic chemical level of analysis. In attempting to understand the precise nature of the catalytic contribution of active site metal ions on a quantitative level, much work has been done to study phosphoryl transfer reactions using simple model compounds with properties that can be varied systematically and with which chemistry is always the rate-determining step. Insights from these studies have been used to construct theoretical models for understanding ribozyme catalysis in general and the mechanistic role of metal ions in particular. The measurement of kinetic isotope effects within these model systems has further refined our understanding of the transition state structure of these reactions and has the advantage of being able to probe the reaction coordinate with a minimal chemical perturbation. Our current understanding at this level of the role of metal ions in catalysis and its theoretical implications will be discussed in a subsequent section of this chapter.

8.2 Metal Ions and Their Ability to Facilitate RNA Catalysis in Biological Systems

To understand why particular metal ions are employed by Nature or have been found experimentally to be competent to perform catalysis within RNA active sites, we must examine the properties of metal ions that are relevant to this role and the extent to which different metal ions possess these properties. In general, the relative abundance of a particular metal ion, coupled with its physicochemical properties, determines its biological role. The combination of these factors explains why ribozyme active sites favor Mg^{2+} so heavily as a catalytic cofactor.

Magnesium is extremely abundant in the Earth's crust in mineral form and in seawater in the form of dissolved salts. Cellular concentrations of ions reflect those of the natural environment and indeed magnesium is the fourth most abundant ion in cellular metabolism.[17] Although other ions such as sodium or potassium are more abundant within cells, their chemical properties make them less suitable for interacting with RNA functional groups in a manner useful for facilitating catalysis directly. Correspondingly, other metal ions with more suitable properties (Mn^{2+}, for example, has physicochemical properties very similar to those of Mg^{2+}) that are able to substitute for magnesium in the function of obligate metalloribozymes are not as abundant. In addition, because metal ions are typically required in millimolar concentrations for their structural or catalytic roles, the solubility of the ion at physiologic pH is an important factor. Thus, whereas sodium, magnesium and zinc ions are soluble in water at millimolar concentrations, other ions such as iron are not (Table 8.1).

The ability of a metal ion to promote catalysis directly requires that it be able to coordinate in a dynamic manner to ligands that are involved in the transition state of the chemical reaction. Therefore, the properties of a metal ion that govern its ability to coordinate and polarize specific ligands, and also the kinetic and thermodynamic profiles of these interactions, are important factors in determining the suitability of a particular metal ion for a catalytic role. When present within active sites, metals typically form thermodynamically stable, yet kinetically labile, interactions that allow them to be held firmly in place while simultaneously undergoing the changes in ligand coordination required by the chemical mechanism of the reaction.

Metals can be classified into three groups according to their kinetics of ligand exchange. Alkali and alkaline earth metals, with the exceptions of magnesium and beryllium, tend to have extremely rapid ligand exchange rates, on the order of 10^8 s^{-1} or faster (Table 8.1). This property accounts for the utility of Ca^{2+} ($k_{ex} \approx 10^9 s^{-1}$), for example, as a signaling molecule in biological systems. Magnesium falls into the group of ions with intermediate ligand exchange rates ($k_{ex} \approx 10^5 s^{-1}$), in which the rate of exchange is also largely independent of ligand identity. The third group includes ions such as Al^{3+}, Be^{2+}, Cr^{3+} and Co^{3+} that exhibit extremely low rates of ligand exchange, rendering them less suitable for the dynamic requirements of catalytic function.

Metal ions also vary in their binding preference for different types of ligands (Table 8.1). Much of this preference can be accounted for by the 'hardness' and 'softness' concept used to describe the interaction between a Lewis acid (the metal ion) and a Lewis base (the ligands). Atoms are described as hard if they are small and difficult to polarize and soft if they are larger and more easily polarized. Pearson has developed a quantitative description of the concept by defining the absolute hardness (η) of an element as being proportional to the difference between its ionization energy and its electron affinity.[18–22] In general, 'hard' metal ions such as Mg^{2+}, Mn^{2+}, Ca^{2+}, Na^+, and K^+ have a higher affinity for hard ligands and tend to form interactions dominated by electrostatics, whereas 'soft' metal ions including Cd^{2+}, Hg^{2+} and Pb^{2+} tend to form more covalent interactions with soft ligands. Most biologically functional metal ion–ligand interactions occur between harder ligands and borderline to hard metal ions, whereas soft ions tend to be toxic as they form bonds to softer

Table 8.1 Physicochemical properties of some monovalent and divalent cations.

Ion	Abundance in modern ocean/mM[a]	Ionic radius/Å[b]	Charge density $(q^2/r)^c$	$\Delta H_{hydration}$/kcal mol^{-1}[d]	$k_{ex}(H_2O)$/ s^{-1}[e]	Coordination number[b]	First pK_a of $[M(H_2O)_x]^g$	Pearson hardness $(\eta)^i$	Transport number[j]
Li$^+$	2.6×10^{-4}	0.59–0.92	1.1–1.7	−514.1	5×10^8	4, 6, 8	13.8	35.12	13–22
Na$^+$	470	0.99–1.18	0.8–1.0	−405.4	8×10^8	4, 6, 8	14.48	21.08	7–13
K$^+$	10	1.37–1.51	0.7	−320.9	10^9	4, 6, 8	—	13.64	4–6
Be^{2+}	$<10^{-10}$	0.27–0.45	8.9–14.8	−2487	10^2	4, 6	6.5^h	67.84	—
Mg^{2+}	50	0.57–0.89	4.5–7.0	−1922.1	6×10^5	4, 6, 8	11.42	32.55	12–14
Ca^{2+}	10	1.00–1.23	3.3–4.0	−1592.4	3×10^8	6, 8, 10	12.7	19.52	8–12
Mn^{2+}	3.6×10^{-6}	0.66–0.96	4.2–6.1	−1845.6	2×10^7	4, 6, 8	10.6	9.02	—
Zn^{2+}	10^{-4}	0.60–0.90	4.4–6.7	−2044.3	5×10^7	4, 6, 8	8.96–9.60	10.88	10–13
Cd^{2+}	$<10^{-10}$	0.78–1.10	3.6–5.1	−2384.9	4×10^8	4, 6, 8	10.08–11.7	10.29	—
Co^{2+}	3.1×10^{-6}	0.58–0.75	5.3–6.9	−2054.3	2×10^6	4, 6	9.65	8.22	—
Ni^{2+}	10^{-6}	0.55–0.69	5.8–7.3	−2105.8	3×10^4	4, 6	9.86	8.50	—
Fe^{2+}	10^{-4}	0.63	6.3	−1920.0	4.4×10^{6f}	4	9.5	7.24	—
Pb^{2+}	$<10^{-10}$	0.98–1.49	2.7–4.1	−1479.9	—	4, 6, 8, 10, 12	7.8	8.46	—

[a]Data from ref. 201.
[b]Data from ref. 202. Ionic radius varies with coordination number, with higher coordination numbers giving rise to larger ionic radii.
[c]Calculated from ionic radii.
[d]Data from ref. 203.
[e]Data from refs. 204 and 205
[f]Data from ref. 206.
[g]Data from ref. 207.
[h]Data from ref. 208.
[i]Data from ref. 22.
[j]Data from ref. 205.
Transport number describes the total number of solvent molecules that travel in close association with an ion.

ligands such as thiol groups within cells. The phosphate oxygens of nucleic acids are among the hardest known biological ligands and the transition states of ribozyme-catalyzed phosphoryl transfer reactions involve the development of negative charge on these oxygen atoms that appropriately positioned metal cations can stabilize. Hence, harder metal ions such as magnesium are ideally suited to facilitate catalysis within ribozyme active sites.

Although metal ions can bind to ligands by direct inner sphere coordination, they often interact using water molecules in their hydration sphere. With respect to a role in catalysis, metal ions can contact atoms in the transition state through hydrogen bonds mediated by these organized water molecules. Alternatively, metal ions can alter the pK_a of a bound water molecule through bond polarization, thus enhancing its nucleophilicity as required by a hydrolytic mechanism. A metal ion's affinity for water, and also its ability to stabilize a bound anionic ligand, is related to its charge density. Magnesium has a relatively small ionic radius (Table 8.1) and thus a high charge density and therefore would be predicted to interact strongly with water. Its relatively high hydration enthalpy and transport number (Table 8.1) confirm this. The ability of a metal cation to facilitate the autoionization of bound water, reflected in the pK_a of its hydrated complex (Table 8.1), is a manifestation of its Lewis acidity. The Irving–Williams series[23] provides a rank order of the Lewis acidity of the first d-series divalent metal ions, many of which are found at enzyme active sites: $Ca^{2+} < Mg^{2+} < Mn^{2+} < Fe^{2+} < Co^{2+} < Ni^{2+} < Cu^{2+} > Zn^{2+}$. Most of the transition metals in the series are redox active and therefore are not generally employed for the coordination of potentially sensitive RNA functional groups. Zn^{2+} is frequently involved in reactions of carbonyl functionalities, whereas Mg^{2+}, despite its weaker Lewis acidity, is preferred for catalysis of phosphoryl transfer reactions due to its hardness.

Finally, metal ions that participate in catalysis need to be specifically positioned to do so by functional groups within the catalytic macromolecule. In the context of metal ion catalysis by ribozymes, the RNA must create binding pockets of sufficiently negative electrostatic potential and precise ligand topology to orient the metal ions appropriately for their catalytic roles. As metal ions exhibit differing ionic radii, preferred coordination geometries (Table 8.1), and ligand preferences, the RNA must evolve in a manner tailored to the requirements of the metal ion it chooses to use. The choice of metal ion is conditioned not only upon its physicochemical properties, but also upon its abundance and availability.

The ability of RNA to form architectures capable of specifically ligating functional metal ions other than magnesium has been demonstrated by several *in vitro* evolution studies. In 1993, Joyce's group isolated variants of the *Tetrahymena* group I ribozyme capable of cleaving an RNA substrate in the presence of 10 mM Ca^{2+} as the sole divalent metal ion, with catalytic efficiencies reaching $10^4 M^{-1} min^{-1}$.[24] The selected variants maintained the ability to splice efficiently in Mg^{2+} and Mn^{2+}, suggesting that they had become modified in such a way as to be able to replace Mg^{2+} or Mn^{2+} with Ca^{2+} within their active sites.

In 2005, Breaker's group undertook the evolution of allosteric ribozymes that could respond to a wide range of divalent metal ions, in part to reveal the range of cationic binding pockets that RNA is capable of forming, and attempted to correlate systematically the metal ion selectivity with the physicochemical properties of the metal ions.[25] As it had been shown previously that stem II of the hammerhead ribozyme could be engineered to serve as an allosteric effector of its cleavage activity, a pool of hammerhead constructs with a 40-nucleotide randomized stretch within stem II was subjected to positive selection for cleavage activity in the presence of various monovalent and divalent cations. In addition, a negative selection step, in which variants that were active in $MgCl_2$ were discarded, enriched for ribozymes capable not only of functioning in the presence of other metal ions, but also of discriminating between Mg^{2+} and other cationic effectors. Several classes of ribozymes were isolated that exhibited cleavage activity in the presence of Cd^{2+}, Co^{2+}, Mn^{2+}, Ni^{2+}, and Zn^{2+}, with some representatives able to induce self-cleavage with rates approaching that of the original ribozyme. No Ca^{2+}, Sr^{2+} or monovalent-responsive ribozymes were isolated, suggesting that it may be difficult for this particular RNA to construct binding pockets capable of distinguishing these ions from Mg^{2+}. Further characterization of one class of the selected ribozymes showed that whereas they could be activated by the divalent ions listed above, a variety of other divalent ions tested could not provide allosteric activation. This implied that the RNA binding pockets created by the engineered ribozymes were sensitive to specific physicochemical features of the metal ions, rather than their divalent character only.

Analysis of the ion selectivity of the allosteric ribozymes in terms of the physicochemical properties of the ions revealed some broad correlations. In general, the ribozymes tended to favor ions with smaller ionic radii, a similar range of hydration enthalpies, a lower pK_a of bound water molecules, and borderline hardness. Although these correlations were imperfect, they confirm that the structure of an RNA biopolymer and the metal ions that it employs for function share an intimate relationship that is evolutionarily tuned to enable efficient function within the cellular environment.

8.3 Mechanistic Roles for Metal Ions in Ribozyme Reactions

As phosphoryl transfer reactions are ubiquitous in biology, extensive efforts have been made using physical organic chemistry of model compounds to provide a detailed mechanistic description of this fundamental chemical process. Such model studies, typically employing simple phosphate esters or dinucleotides, offer the advantage of being able to vary systematically reaction center atoms or groups relatively easily and to monitor reaction progress under conditions in which chemistry is rate limiting.[26–34] Insights gained from these studies have been particularly valuable for understanding ribozyme mechanisms in general and the role of metal ions in ribozyme reactions in particular.

Ribozymes carry out their phosphoryl transfer reactions by one of two general mechanisms that differ in the identity of the nucleophile attacking the scissile phosphate. The small ribozymes (hammerhead, hairpin, HDV, glmS, VS) catalyze endonucleolytic RNA cleavage using a 2′-hydroxyl group to mount an intramolecular attack at the cleavage site phosphate. In contrast, the large ribozymes (group I and group II introns, RNase P, spliceosome) catalyze intermolecular attack by water or a 2′- or 3′-hydroxyl group. As metal ions have the ability to coordinate active site ligands and thus alter their electronic properties, they can exert effects in the reaction transition states, providing significant rate accelerations. Here, we discuss the specific roles that metal ion coordination can play in facilitating phosphoryl transfer reactions. We also provide quantitative estimates of their contributions to reactivity using insights gained from model studies.

8.3.1 Lewis Acid Stabilization of the Leaving Group

Nucleophilic attack at a phosphate center, such as that catalyzed by ribozymes, results in the expulsion of an oxyanion leaving group. Stabilization of this leaving group by the reduction of its negative charge is thus a possible strategy to enhance reactivity and is thought to be the most important source of rate acceleration provided by ribozymes.[35,36] Leaving group activation can be accomplished by protonation using a general acid, by hydrogen bonding (with proton transfer ultimately occurring via solvent), or by metal cation coordination. This last strategy, shown to be more important for catalysis by the large ribozymes, depends on the ability of the metal ion to act as a Lewis acid to withdraw electron density from its coordinated ligand (Figure 8.1A).

As discussed in the previous section, the shift in the pK_a of water upon its coordination to a particular metal ion is a measure of the Lewis acidity of the metal, and in the case of Mg^{2+} this shift is on the order of 4 log units [the pK_a of $Mg(H_2O)_6^{2+}$, including a statistical correction for the 12 ionizable protons, is 12.6]. Various lines of evidence suggest that Mg^{2+} coordination induces a similar pK_a shift in alcohols. Model compound studies have quantified the dependence of the reaction rate of phosphate ester cleavage on the leaving group pK_a (β_{lg}), providing a value of −0.97 for diester hydrolysis[30] and −1.28 for base-catalyzed RNA cleavage.[35] Taken together, the catalytic advantage of metal ion coordination of the leaving group can be estimated to be $10^{[(0.97 \text{ or } 1.28) \times 4]} \approx 10^4$–$10^5$-fold.

8.3.2 Nucleophile Activation

The deprotonated, oxyanionic form of a hydroxyl group is significantly more nucleophilic than its protonated, neutral counterpart. Hence, increasing the fractional concentration of the oxyanionic form of the nucleophile by facilitating its deprotonation should increase the phosphoryl transfer reaction rate in reactions such as base-catalyzed RNA cleavage, in which the oxyanion is the active nucleophilic form. In addition to providing an increased oxyanion concentration by lowering the pK_a, metal ion coordination simultaneously decreases nucleophilicity. Thus, a metal-coordinated alkoxide $Mg(OR)^+$ is less

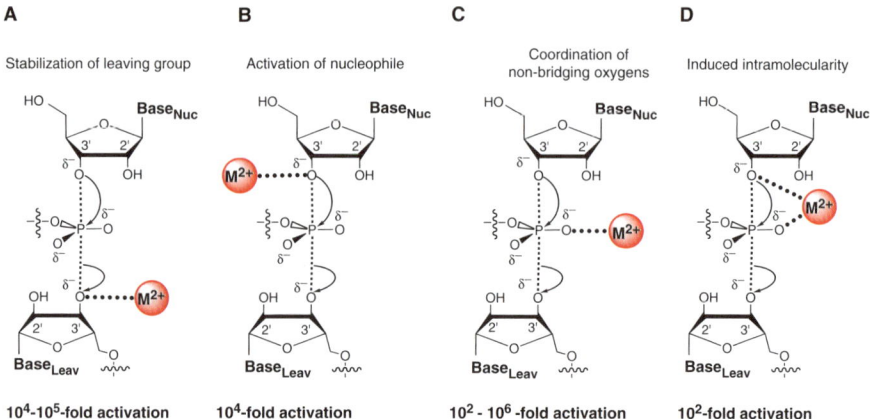

Figure 8.1 Strategies used by divalent metal ions to accelerate nucleophilic phosphotransesterification reactions. In the reaction shown, $Base_{Nuc}$ attacks a phosphodiester linkage via nucleophilic attack of a 3′-hydroxyl group, resulting in the expulsion of the leaving group $Base_{Leav}$. The reaction proceeds through a trigonal bipyramidal transition state. (A) Lewis acid stabilization of the leaving group 3′ oxygen. (B) Activation of the incoming 3′ oxygen nucleophile. (C) Coordination of the non-bridging oxygens. (D) Induced intramolecularity, in which simultaneous coordination of the nucleophile and a non-bridging oxygen maintain the geometry required for in-line nucleophilic attack.

nucleophilic than the free alkoxide RO⁻; nevertheless, $Mg(OR)^+$ has greater nucleophilicity than the corresponding neutral alcohol ROH. In the case of phosphoester hydrolysis, where H_2O is the attacking nucleophile, catalysis can be provided by the enhanced nucleophilicity of the metal-coordinated hydroxyl as compared with H_2O alone. The increased pK_a of $Mg(OH)^+$ compared with water of $12.6 - (-1.3) = 13.9$ in conjunction with the β_{nuc} of 0.3 measured for phosphodiester hydrolysis[31] allows an estimate of $10^{(0.3 \times 13.9)} = 10^4$-fold for the rate acceleration that can be provided in this case (Figure 8.1B).

8.3.3 Coordination of Non-bridging Phosphoryl Oxygens

As nucleophilic attack on a phosphate center progresses, significant negative charge develops on the non-bridging phosphoryl oxygens due to the disruption of the p_π–d_π orbital overlap that delocalizes the formal negative charge in the ground state.[37] It follows that hydrogen bond donation, proton donation, or electrostatic interactions with a positively charged group could facilitate catalysis by stabilizing this developing negative charge. Metal ion coordination of one non-bridging oxygen has been estimated to provide a 10^2-fold rate acceleration, a value that can increase to 10^6-fold upon coordination of both non-bridging oxygens[38] (Figure 8.1C). The 10^5-fold faster reaction of ribonucleoside

3′-alkyl phosphotriesters provides an independent estimate of the extent to which neutralization of the non-bridging phosphoryl oxygen can enhance reaction rate.[39,40]

8.3.4 Induced Intramolecularity

Generating the trigonal bipyramidal transition state in a phosphoryl transfer reaction requires an in-line geometry of nucleophilic attack. To the extent that simultaneous coordination of the attacking nucleophile and the non-bridging phosphoryl oxygen (such as that observed in the classic two-metal ion mechanism of phosphoester transfer) could assist in promoting the proper orientation for reaction, this could provide an additional source of catalysis. A study of the hydrolysis of a series of phosphorylated pyridines provides an estimate of 10^2-fold for the catalytic contribution of such induced intramolecularity[41] (Figure 8.1D).

8.3.5 Indirect Catalytic Contributions

Beyond these direct roles for metal ions in facilitating ribozyme catalysis through inner sphere coordination of reaction center atoms, metal ions can also enhance reactivity through a number of indirect routes. For example, a hydrated metal ion could act as a general acid or general base to facilitate proton transfer along the reaction coordinate. General acid–base catalysis can stabilize the developing negative charge in the transition state, activate weak nucleophiles, and stabilize poor leaving groups. In general, functional groups possessing pK_as near physiologic pH are considered optimal general acids or bases, as they can maintain a significant fraction of the reacting species in the catalytically active protonation state. As such, RNA has traditionally been viewed as less able than proteins to take advantage of general acid–base catalysis since RNA does not possess any functional groups with pK_as near neutrality. However, in an elegant theoretical analysis that has gained experimental support in studies of the HDV ribozyme, Bevilacqua demonstrated the feasibility of employing RNA nucleobases to perform general acid–base catalysis.[42] Specifically, he showed that the penalty for employing functional groups with non-optimal pK_as is less severe than would be predicted by considering the fractional population of the catalytically active species alone. This occurs as a result of the non-zero α and β values for proton transfer during phosphoester cleavage.

This insight suggests the possibility that a hydrated metal ion could function as a general acid or base, donating or accepting a proton via its bound water molecules. Although such a role has been described for hydrated zinc ions in model studies of dinucleotide hydrolysis,[43] hydrated magnesium possesses a significantly higher pK_a, and thus would require significant intra-enzyme pK_a shifting to provide comparable general acid–base-mediated rate acceleration.

Indeed, RNA nucleobase functional groups theoretically are better suited to donate and accept protons, possibly explaining the observation that the small ribozymes currently thought to employ general acid–base mechanisms do not require divalent ions for activity. Nevertheless, a potential role for a hydrated magnesium ion as a general acid or base in HDV ribozyme catalysis has been inferred based on both crystallographic and biochemical evidence.[15,44,45]

A recent review discussing alternative roles for metal ions in RNA catalysis suggests expanding our traditional understanding of metal ion-mediated catalysis to include indirect mechanistic roles that do not require direct ligation of transition state functional groups by metal ions.[46] Drawing upon ideas arising from the study of protein metalloenzymes, the authors proposed that divalent metal ions could provide structural or electrostatic transition state stabilization for ribozymes through outer sphere coordination. In addition, divalent metal ions might also provide long-range electrostatic stabilization or promote nucleobase pK_a shifting to facilitate a general acid–base mechanism. Other cationic species could, in principle, substitute for divalent metal ions that function in these indirect capacities. Taken together, these alternative mechanisms could explain why the small ribozymes are cation-dependent, without being dependent specifically on divalent cations.

8.4 Experimental Techniques for Detecting Catalytic Metal Ion–Ligand Interactions in Ribozymes

Identifying metal ion–ligand interactions important for ribozyme catalysis entails locating functional groups within ribozymes that can bind metal ions and determining whether those groups participate in catalysis. The search for metal ion–ligand partners can focus on the ribozyme itself or on an oligonucleotide substrate, if a *trans*-cleavage assay is available. Within the ribozyme, non-bridging phosphate oxygens are the most likely candidate ligands to bind and position metal ions in the active site. In turn, these metal ions activate attacking oxygen nucleophiles and stabilize oxyanion leaving groups in the pentavalent transition state. Several methods are available for determining whether and how metal ions bind to these functional groups during ribozyme catalysis.

8.4.1 X-ray Crystallography

Crystal structures of ribozymes and ribozyme–substrate complexes have provided a wealth of information about where and how metal ions bind within ribozyme active sites. If the resolution is sufficiently high and the electron density sufficiently localized, individual metal ions (usually Mg^{2+} and K^+) can be discerned. The relatively low electron density of these ions makes unambiguous identification difficult, however, as water molecules can exhibit a similar pattern of electron density. In such cases, soaking the crystals with heavier transition metals such as Mn^{2+}, Pb^{2+}, Co^{3+}, or Os^{3+} can highlight metal binding sites and assist in phasing.

To date, crystal structures of several ribozymes in various states have been solved, including the group I[4, 47–50] and group II[51] introns, RNase P,[52–54] the ribosome,[55,56] and small ribozymes like the hammerhead,[57–59] HDV,[45,60,61] hairpin,[62] and glmS[63] ribozymes. When evaluating the roles of metal ions in these structures, it should be remembered that each structure represents a static model of a *ground state* of the ribozyme. These models are useful in that positions and coordination distances of metal ions can help to assign possible roles in catalysis. Moreover, even if no metal ions are detected in a ribozyme active site, the overall architecture of the site can often suggest where catalytic metal ions may reside. However, crystal structures do not always reveal the numbers and locations of metal ions within an active site, nor do they establish the roles of these metal ions in catalysis. For example, crystal structures of some protein restriction endonucleases have been solved that contain zero, one, two, or even three metals residing in the active site.[64–66] The significant variability in the numbers and positions of Mg^{2+} ions within the various group I ribozyme structures[48–50] further highlights the need for biochemical assays that test hypotheses derived from crystal structures.

8.4.2 Spectroscopic Methods – Electron Paramagnetic Resonance (EPR) and Nuclear Magnetic Resonance (NMR)

EPR and NMR methods have been employed in several instances to identify bound metal ions in folded RNAs and to characterize metal ion binding sites. As both of these methods are described in detail elsewhere in this volume, this section will highlight some specific literature examples of their use. EPR experiments investigating divalent metal ion binding within RNA rely on the ability of the paramagnetic ^{55}Mn atom (spin-5/2) to substitute functionally for ^{24}Mg, which is spectroscopically silent (^{25}Mg is paramagnetic, but its low natural abundance precludes it from routine use as an EPR probe). The Mn^{2+} EPR signal changes slightly but reproducibly depending upon the specific coordination environment surrounding the Mn^{2+} atom. Thus, either alone or in conjunction with other spectroscopic techniques, Mn^{2+} EPR can yield information about the geometry of a specific divalent metal ion binding site. This information can include details about the number and identity of the surrounding ligands, and also whether they coordinate *via* inner or outer sphere interactions. The hammerhead ribozyme has been the ribozyme most extensively studied by EPR spectroscopic methods. DeRose and colleagues employed EPR spectroscopy to identify and characterize a high affinity Mn^{2+} binding site within constructs of the minimal hammerhead ribozyme.[67,68] Through the use of electron spin-echo envelope modulation (ESEEM) spectroscopy, an advanced EPR technique, this binding site was localized to a region between bases A9 and G10.1,[69] where a Mn^{2+} ion was also identified crystallographically.[70,71] The binding site has also been identified in the tertiary-stabilized hammerhead ribozyme,[72] in which Mn^{2+} binds to the site nearly two orders of magnitude more tightly than in the minimal hammerhead ribozyme.[73] EPR spectroscopy has also been applied to the study of metal ion–ligand interactions within RNA nucleotides,[74] tRNA,[74,75] and the hairpin[76] and group I[77] ribozymes.

NMR spectroscopy has been used to determine the three-dimensional solution structures of several small RNAs,[78–83] and also to monitor dynamic interdomain interactions within those RNAs. The binding of divalent metal ions to RNA induces conformational changes and alters the chemical shifts of NMR-active nuclei. Provided that specific chemical shifts can be assigned reliably NMR is thus a powerful tool with which to monitor metal ion-induced conformational changes and the geometry of metal ion binding sites. The ability of Mg^{2+} to induce conformational rearrangements within folded RNAs has been demonstrated for a number of RNAs, including the P5abc domain from the group I ribozyme[84] and the transactivation response element (TAR) RNA from HIV-1.[85] Site-specific substitutions such as phosphorothioate and spin-labeled nuclei (^2H, ^{13}C, ^{15}N) can simplify the analysis of NMR spectra of RNA, and also provide additional distance constraints between nuclei. For example, the high-affinity metal ion binding site within the minimal hammerhead ribozyme was investigated using ^{31}P NMR in conjunction with phosphorothioate substitutions placed within the site.[86] The coordination environment of this binding site was elucidated by combining EPR and NMR techniques that monitored binding of Mn^{2+} in ^2H_2O to a minimal hammerhead containing [^{15}N]guanine at G10.1.[69] Mn^{2+} may also be used in a purely NMR experiment to monitor the paramagnetic line broadening of NMR signals that occurs as Mn^{2+} comes within close proximity of a metal ion binding site. This strategy, along with intermolecular nuclear Overhauser effects (NOEs) and Mg^{2+}-induced chemical shift alterations, was used to identify metal ion binding sites within the hairpin ribozyme[76] and an adenine-sensing riboswitch.[87]

8.4.3 Computational and Database Methods

The advent of supercomputers has facilitated the development of several computational approaches aimed at identifying metal ion binding sites in folded RNAs. Although they do not provide direct evidence of metal binding, these methods have yielded insights into how RNAs construct metal binding sites. The results derived from computational methods can direct experimental attention towards potential metal binding sites that may not be apparent from a crystal structure alone.

The electrostatic potential surface of folded RNAs exerts a large influence on the distribution of metal ions over that surface. Consequently, much effort has focused on developing an accurate description of the interactions between the potential surface of RNA and ions and water molecules in the surrounding solvent. Numerical solutions to the non-linear Poisson–Boltzmann equation (NLPB) have provided perhaps the most realistic picture of solvent–RNA interactions. The NLPB describes the charge distribution over an arbitrary molecular surface based on the molecular structure and the dielectric properties of the molecule and surrounding solvent. Solutions to the NLPB for specific RNA structures can be used to map regions of positive and negative electrostatic potential onto RNA surfaces. Regions of especially high negative electrostatic potential correlate well with crystallographic metal binding sites.[88,89] Thus, assuming an RNA structure is available, NLPB calculations can facilitate predictions of the locations of metal binding sites.

Another, less computationally intensive approach analyzes the particular microenvironments associated with known metal binding sites and uses that knowledge to predict new metal binding sites.[90] The program, called FEATURE, is based on an algorithm developed for protein microenvironments.[91] It applies numerous binding site parameters (functional groups, nucleobase identities, and so on) to determine the locations of diffuse and site-bound metal ions in a structured RNA. Thus, the method takes into account the biochemical and structural properties of RNA metal binding sites, in addition to charge. The program correctly identified crystallographic site- and diffuse-bound metal ions in the *Tetrahymena* group I ribozyme P4–P6 domain and in a 58-nt RNA derived from *E. coli* 23S rRNA. The authors view the program as a tool for the rapid screening of RNA structural motifs.

Web-based databases of metal ion binding sites provide a quick way to identify and visualize hundreds of RNA and RNA–protein structures that contain metal ions. The Metalloprotein Database and Browser (MDB; http://metallo.scripps.edu/)[92] allows the user to search the Protein Data Bank (PDB) for metal binding sites based on several different parameters including metal or ligand identity, number of ligands, and metal–ligand distance. A Java-based applet allows visualization and manipulation of metal sites and includes tools for measuring distances and torsion angles. Quantitative tools are also available for assistance in solving X-ray crystal structures and evaluating designed metal sites. Recently, an RNA-specific metal ion database has been created that offers similar features but also incorporates RNA structural motifs known to bind metal ions. As of February 2007, the Metals in RNA (MeRNA) database (http://merna.lbl.gov/)[93] includes 389 PDB entries and 16, 122 instances of 23 distinct metal ions. MeRNA may also be used in conjunction with the Structural Classification of RNA (SCOR) database[94] to correlate metal binding site geometry with surrounding RNA tertiary structure motifs.

8.4.4 Nucleotide Analogue Interference Mapping (NAIM)

A NAIM experiment can rapidly screen whether substitution of specific functional groups within an RNA interferes with metal ion binding. The RNA of interest is transcribed in the presence of the four wild-type NTPs and a specific nucleotide α-thiotriphosphate. The phosphorothioate nucleotide is included in the transcription reaction at a concentration such that on average, each full-length RNA transcript contains a single modified nucleotide. The resulting population of transcripts is then subjected to some biochemical challenge, such as the ability to fold, perform catalysis, or bind a given macromolecule. The 'winners' – those modified RNA molecules able to perform the challenge – are isolated and sequenced by iodine-mediated cleavage of the phosphorothioate linkage. At positions where the modification has interfered with an RNA's ability to perform the challenge, gaps appear in the sequencing ladders. Thus, a single NAIM experiment can screen the functionality of a particular atom or group of atoms at every position in an RNA molecule. This makes the approach ideal as a tool for the rapid screening of functional groups within RNA.

The availability of many different phosphorothioate-tagged nucleotide triphosphate analogues[95] has allowed researchers to investigate almost every type of functional group within an RNA molecule. Since non-bridging phosphate oxygens are the most common RNA metal ligands, phosphorothioate modifications within ribozymes are readily studied. Unfortunately, a NAIM experiment yields information only on the functionality of pro-R_P non-bridging oxygens, because RNA polymerases incorporate only S_P α-thiotriphosphate nucleotides into a growing transcript. Nonetheless, even interferences at pro-R_P non-bridging oxygens can suggest candidate ligands for more detailed analysis, and basic metal rescue experiments can also be performed[96] (see below). So far, almost every ribozyme has been subjected to some form of NAIM analysis.

8.4.5 Lanthanide Ion-induced Cleavage and Luminescence

Another biochemical method for locating metal ion binding sites within folded RNAs exploits the ability of lanthanide metal ions to localize preferentially to regions of Mg^{2+} binding. Lanthanide ions are thought to reflect more accurately Mg^{2+} coordination geometry compared with other heavy metal ion probes of RNA structure, such as Pb^{2+}.[97] Like Mg^{2+}, lanthanide ions strongly prefer oxygen ligands, but coordinate oxygen many times more tightly than Mg^{2+}.[98-100] In addition, lanthanide aqua ions have pK_a values close to neutrality and hence can easily abstract protons from 2′-hydroxyl groups to promote RNA strand scission. Thus, lanthanide ions compete with Mg^{2+} for binding sites on RNA molecules and promote relatively site-specific cleavage of the RNA backbone.

In a typical lanthanide cleavage experiment, an RNA molecule folded in Mg^{2+} is exposed to increasing concentrations of lanthanide ion, usually Tb^{3+} or Eu^{3+}, in the 10–1000 μM range. The lanthanide-induced cleavages are mapped to specific residues on the RNA by denaturing gel electrophoresis. In this way, researchers have predicted the locations of metal binding sites in several ribozymes, including the group II intron,[101] RNase P,[102] HDV,[103] and hairpin ribozymes.[104] The technique is not of sufficient resolution to determine the identity of particular RNA ligands that bind metal ions or participate in catalysis. Instead, lanthanide cleavage maps can suggest regions of RNA that may be of interest for closer investigation. They are also used to monitor the formation of tertiary structure and how conformational changes alter that structure.[97,103]

Lanthanide ions also possess luminescent properties that can provide further information about site-specific metal ion binding to RNA. When lanthanide ions are excited directly by a laser, they emit fluorescent light whose rate of decay strongly depends on the coordination environment of the ion.[105] Bound hydroxide provides a vibrational pathway for fluorescence decay,[105,106] and hence lanthanide fluorescence is largely quenched in aqueous solution. However, the fluorescent lifetime increases significantly upon replacement of water molecules in the lanthanide coordination sphere with D_2O or RNA ligands.[105–107] Thus, lanthanide luminescence is a sensitive probe of the coordination environment of the metal ion. This method has been employed to detect and characterize metal ion binding in tRNA,[98,99] RNA hairpin loops,[108] the U2–U6 spliceosomal RNAs,[106] and the hammerhead, hairpin, and HDV ribozymes.[104,107,109]

8.4.6 Heavy Atom Isotope Effects

Heavy atom isotope effects on chemical reactions provide information about the structure of a chemical transition state with respect to the degree of bonding to the nucleophile, the leaving group, and the surrounding atoms. Isotopic substitution represents the smallest possible perturbation of a catalytic system, yet can exert significant effects on chemical reactivity.[110] Hence, isotopically-labeled substrates are a powerful tool for probing the structures of enzymatic transition states. Isotope effects are expressed as a ratio of the chemical reaction rate involving the lighter isotope to the corresponding rate involving the heavier isotope ($^{light}k/^{heavy}k$). The isotope effect is said to be 'normal' if this ratio is greater than 1 (favoring the lighter isotope) and 'inverse' if it is less than 1 (favoring the heavier isotope). The isotopic substitution can affect both the equilibria and rates of chemical reactions, giving rise to equilibrium and kinetic isotope effects, respectively. Kinetic isotope effects are termed primary isotope effects if the isotopically substituted atom is directly involved in bond formation or cleavage. Such effects are associated with isotopic substitution of nucleophilic and leaving group atoms. Secondary isotope effects arise from isotopic substitution of atoms not directly involved in bond formation or cleavage. These atoms may participate in the transition state structure or may reside far from the reaction site. The magnitude of an isotope effect is proportional to the relative mass difference between the two isotopes. Thus, larger effects are observed when comparing hydrogen and deuterium (a 100% mass difference) than when comparing ^{12}C and ^{13}C (an 8.3% difference) or ^{16}O and ^{18}O (a 12.5% difference). As a result, isotope effects involving atoms other than hydrogen are relatively subtle and require extremely precise experimental methods to measure.

Although several studies have used isotope effects to investigate different ribozyme mechanisms, relatively few have examined the specific role of metal ions in ribozyme transition state structures. In general, metal ion coordination of an attacking nucleophile such as oxygen creates a 'stiffer', more strongly bonded transition state. Such an environment usually favors heavier isotopes in a given reaction.[110–112] As a result, metal ion coordination tends to make inverse contributions to the overall isotope effect of an oxygen nucleophile ($^{16}k/^{18}k$). This observation may be used as a signature for metal ion involvement in a given transition state. For example, the nucleophilic isotope effect for hydroxide-catalyzed hydrolysis of thymidine 5′-p-nitrophenylphosphate (T5PNP) has been measured as 1.068 ± 0.007 in $H_2{}^{18}O$.[113] When Mg^{2+} is used to catalyze the same reaction, the isotope effect drops to 1.027 ± 0.013, presumably the result of the Mg^{2+} ion coordinating the water nucleophile and thereby creating a stronger bonding environment. The value of the Mg^{2+}-catalyzed isotope effect on T5PNP hydrolysis was found to be similar to that for the $H_2{}^{18}O$ nucleophile in the hydrolysis reaction catalyzed by RNase P (1.030 ± 0.012). This observation was taken as evidence for direct Mg^{2+} coordination of the water nucleophile during RNase P catalysis.

Heavy atom isotope effects therefore have significant potential for identifying catalytic metal ions and elucidating their roles in stabilizing ribozyme transition states. Recent developments in the organic synthesis of isotopically labeled oligonucleotide substrates will allow the extension of isotope effect analysis to other transition state atoms, most notably the 2′- and 3′-hydroxyl groups.[114] Isotopic substitution of these groups may make possible a deeper understanding of the role metal ions play in activating RNA hydroxyl groups for intramolecular cyclization reactions.

8.4.7 Metal Ion Rescue Experiments

Metal ion rescue experiments can circumvent the experimental problems presented by the 'sea of metal ions' that are diffusely associated with the RNA by focusing attention on the interaction of one atom with one or more metal ions. A quantitative analysis of such experiments has elucidated the metal ion–ligand interactions that stabilize the transition states of several ribozyme-catalyzed reactions. Metal ion rescue experiments take advantage of the differential affinity of a metal ion for a particular ligand versus an atomic substitution of the ligand. Replacement of the original ligand with an atom that interacts poorly with the metal ion diminishes the biological activity of interest. Restoration of activity upon addition of a different metal ion that interacts more strongly with the substitution suggests that the original metal ion–ligand interaction has been re-established or 'rescued'. Metal ion rescue experiments thus provide functional evidence of specific metal ion–ligand interactions that contribute to the biological functions of RNA. The technique has been used to identify catalytic metal ion–ligand interactions in several ribozymes, including the group I intron,[115–124] the group II intron,[125–127] bacterial RNase P,[128–133] and the spliceosome.[134–136] The validity of the approach has been demonstrated by the accurate predication of catalytic metal ion–ligand interactions evident in three different group I ribozyme crystal structures.[4,49,50,123]

In RNA biochemistry, metal ion rescue experiments most often address whether a given functional group coordinates a metal ion to effect catalysis. The functional group is almost always oxygen, whether as a 2′-OH group, a 3′-bridging oxygen, or a non-bridging phosphate oxygen, while the initial metal ion is usually Mg^{2+}. In place of oxygen, the atomic perturbation substitutes sulfur or nitrogen, two ligands that do not interact particularly well with Mg^{2+}. The rescuing metal ions are typically transition metals such as Mn^{2+}, Zn^{2+}, Cd^{2+}, or Co^{2+}. The higher polarizability and d-orbitals of these metal ions allow them to interact much more strongly with sulfur or nitrogen than Mg^{2+}, which prefers to bind oxygen ligands almost exclusively.

To report accurately the role of a metal ion in a given ribozyme reaction, a metal ion rescue experiment must satisfy certain thermodynamic conditions. First, the assay must monitor the same transition from ground state to transition state, regardless of the presence of an atomic substitution or rescuing metal ion. Occasionally, the altered conditions may change the initial ground state compared with the original, unmodified reaction, or they may accelerate a nonchemical step unexpectedly.[123] Care must be taken, then, to identify experimental conditions under which the chemical step of the reaction is rate-limiting for all modifications and metal ions employed. Second, the non-specific effects of the rescuing metal ion must be taken into account. Even under conditions in which chemistry is rate limiting, a rescuing metal ion may confer an inherent catalytic advantage to an unmodified ribozyme reaction. To control for this possibility, the experiment must monitor the reaction of both modified and unmodified ribozymes in the presence of both metal ions. The rates of these reactions can then be used to calculate a relative rate, k_{rel}, that controls for any non-specific effects of the rescuing metal ion. For a ribozyme system in which sulfur substitutes for an oxygen ligand and Mn^{2+} is the rescuing metal ion, the relative rate is given by

$$k_{rel} = \frac{k_S^{Mn^{2+}}/k_S^{Mg^{2+}}}{k_O^{Mn^{2+}}/k_O^{Mg^{2+}}}$$

The magnitude of the rescue is therefore proportional to the Mn^{2+} rate acceleration of the sulfur reaction divided by the Mn^{2+} acceleration inherent to the oxygen reaction. If this value is significantly larger than unity, a metal ion rescue of the sulfur reaction is said to have occurred.

8.4.8 Metal Rescue in Practice: The *Tetrahymena* Group I Ribozyme Active Site

The group I ribozyme derived from the protozoan *Tetrahymena thermophila* catalyzes a phosphotransesterification reaction between an exogenous guanosine nucleophile and a substrate that forms base pairs with the ribozyme:

$$CCCUCUA + G \rightarrow CCCUCU + GA$$

278   Chapter 8

The ribozyme is active at millimolar concentrations of Mg^2 or Mn^{2+}, whereas neither Ca^{2+} nor monovalent ions alone support catalysis. Metal ion rescue experiments coupled with quantitative kinetic analyses have facilitated the development of a detailed picture of the metal ion–ligand interactions that stabilize the transition state of this reaction. To illustrate how metal ion rescue experiments are used in practice, the following sections will describe the results that support each of the metal ion–ligand interactions depicted in Figure 8.2.

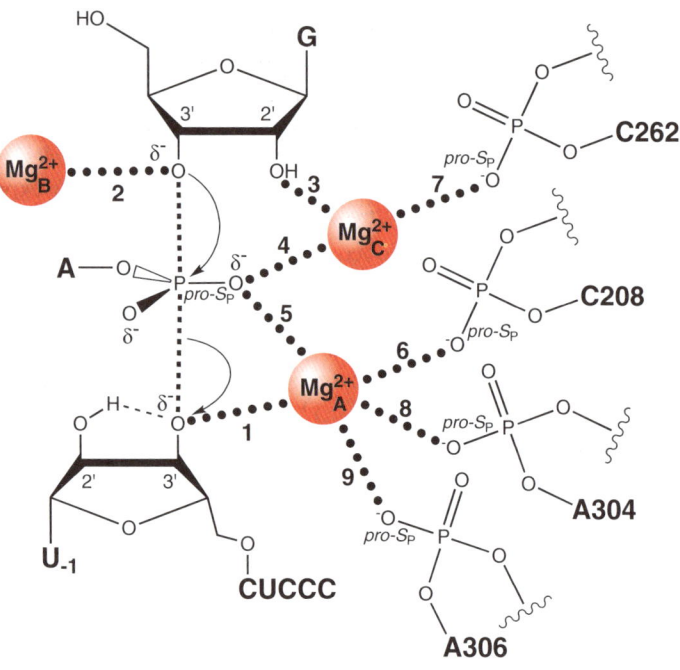

Interaction	Reference
1	Ref. 115
2	Ref. 117
3	Ref. 116, 118
1, 5	Ref. 120
4, 5	Ref. 121
6	Ref. 122, 124
7	Ref. 123
8	Ref. 124
9	Ref. 124

Figure 8.2 Magnesium ion–ligand interactions (numbered 1–9) that stabilize the transition state of the *Tetrahymena thermophila* group I ribozyme.

8.4.8.1 Metal A Coordinates the 3'-Oxygen of the Leaving Group

The ability to synthesize oligonucleotides in which sulfur replaces a 3'-bridging oxygen atom[137] led to the first definitive link between an RNA functional group and a catalytic metal ion in the *Tetrahymena* active site. The study employed the substrate d(CCCUCU$_{3'X}$A), where X is either oxygen or sulfur, in the *Tetrahymena* reaction.[115] In the presence of 12 mM Mg^{2+}, the reactivity of the sulfur substrate was about 1000-fold lower than the reactivity of the all-oxygen substrate. Although higher concentrations of Mg^{2+} stimulated the sulfur reaction modestly, the relative catalytic defect between the two substrates was unchanged. However, a mixture of 10 mM Mg^{2+} and 2 mM Mn^{2+} improved the reaction rate of the sulfur substrate such that it reacted only eightfold slower than the oxygen substrate. The more thiophilic Zn^{2+} ion also supported comparable catalysis with the sulfur substrate. Although Cd^{2+} supported a much lower catalytic rate in this study, subsequent work showed that lower concentrations of Cd^{2+} could rescue the reaction.[120,121] Higher Mg^{2+} concentrations inhibited the ability of Mn^{2+} to stimulate the sulfur substrate reaction, consistent with Mn^{2+} binding to a Mg^{2+} site to rescue reactivity. Taken together, the results strongly supported the existence of a divalent metal ion that binds to the 3'-oxygen of the U_{-1} leaving group in the *Tetrahymena* ribozyme reaction (interaction **1** in Figure 8.2). In later studies, this metal ion became known as metal A (M_A) shown in Figure 8.2.

It is useful to highlight some of the controls of this study, since they illustrate the practical concerns about metal ion rescue experiments that must be considered. First, the all-DNA substrate was used because previous results had shown that the chemical step of its reaction was rate limiting over all concentrations of guanosine and substrate. In addition, the binding constants of the unmodified and sulfur-substituted substrates for the ribozyme were measured and found to be similar. These controls ensured that any differences in the reaction rates between the two substrates could be ascribed to differences in chemistry only. Second, the reaction was performed with the all-oxygen substrate both in the presence of 12 mM Mg^{2+} alone and in the presence of 10 mM Mg^{2+} plus 2 mM Mn^{2+}. This experiment controlled for any effect that the presence of Mn^{2+} might have had on reaction of the unmodified substrate. Mn^{2+} did, in fact, increase the reaction rate of the all-oxygen substrate by about threefold. In contrast, the metal rescue arose from the approximately 400-fold Mn^{2+} stimulation of the reaction of the sulfur-substituted substrate. Taking into account the inherent threefold stimulation of the all-oxygen reaction, Mn^{2+} accelerated the reaction of the sulfur substrate by >100-fold. Overall, then, the study included controls that (1) ensured that the assay monitored the same chemical transition in both metal ions and (2) accounted for any unforeseen effects of the rescuing metal ion.

8.4.8.2 Metal B Interacts with the 3'-OH Nucleophile of Guanosine

Sulfur substitution of the 3'-bridging oxygen also led to the discovery of a second metal ion involved in catalysis by the *Tetrahymena* ribozyme. This time, however, the study focused on metal ion binding to the 3'-OH nucleophile of the attacking guanosine (interaction **2** in Figure 8.2).[117] Since sulfur is not an especially good nucleophile at phosphorus centers, the assay monitored the reverse reaction according to

$$\text{CCCUCU} + \text{IpU/IpsU} \rightarrow \text{CCCUCUU} + \text{I}_{3'\text{OH/SH}}$$

where IpU and IpsU are inosine–uridine dinucleotides with either oxygen or sulfur, respectively, at the 3'-bridging positions. In this reaction, the terminal 3'-OH of CCCUCU acts as the nucleophile and attacks the scissile phosphate of the dinucleotide. In the presence of 110 mM Mg^{2+}, IpU dinucleotide was a substrate in the reverse reaction, whereas no detectable activity was observed with the IpsU substrate. The addition of various soft metal ions promoted the reaction of the IpsU substrate. However, only Cd^{2+} produced a true metal rescue, accelerating the reaction rate with IpsU significantly above the rate for the IpU reaction in Cd^{2+}. This acceleration was diminished by using Mg^{2+} as a competitive inhibitor, suggesting that Cd^{2+} bound in a site that was also accessible to Mg^{2+}. These results provided evidence for a metal ion that interacted with the 3'-OH leaving group in the reverse reaction. By the principle of microscopic reversibility, the metal ion (metal B, M_B in Figure 8.2) would coordinate the 3'-OH nucleophile in the forward reaction.

8.4.8.3 Metal C Coordinates the 2'-OH of Guanosine

Initial evidence for the participation of metal C in catalysis (M_C in Figure 8.2) came from an assay in which 2'-deoxy-2'-aminoguanosine ($G_{2'NH_2}$) replaces guanosine as the cofactor in the reaction. The study used a *cis*-acting group I ribozyme derived from the *nrdB* gene of bacteriophage T4.[116] Splicing of this construct with $G_{2'NH_2}$ was about 30-fold slower than with unmodified guanosine when only Mg^{2+} was present. In contrast, Mn^{2+} reduced this catalytic defect to about 3-fold, accelerating the reaction with $G_{2'NH_2}$ almost 10-fold over the Mg^{2+}-catalyzed rate. Later experiments using the *Tetrahymena* ribozyme confirmed the presence of the interaction between M_C and the 2'-OH of the guanosine nucleophile (interaction **3** in Figure 8.2) and characterized further the binding energetics of this interaction.[118]

At this point, three metal ion interactions within the *Tetrahymena* active site had been identified, but whether each interaction resulted from binding of a distinct ion remained unclear. The observed metal rescues could have arisen from one or two metal ions coordinating two functional groups simultaneously. M_C, for example, might also coordinate the nucleophilic 3'-OH of guanosine, without the need to postulate the existence of a distinct M_B. To determine the number of metal ions that bound within the *Tetrahymena* active site, Shan *et al.*

developed methods for 'fingerprinting' rescuing Mn^{2+} ions for metal sites A, B, and C.[119] The logic of the approach hinged on the determination of different apparent Mn^{2+} affinities associated with rescue of specific modifications (sulfur or nitrogen) whose presence inhibited reaction in Mg^{2+} alone. Thus, a Mn^{2+} ion that rescued the reaction of a 3′-bridging sulfur substrate (M_A site) would bind with an affinity distinct from a Mn^{2+} ion that rescued the reaction of $G_{2'NH_2}$ (M_C site). These affinities were treated as thermodynamic fingerprints specific for the interaction of a metal ion with a particular site. If Mn^{2+} rescued the deleterious effects of two different modifications with the same affinity in each case, the simplest interpretation would be that the same metal ion coordinated both functional groups simultaneously. In fact, three distinct Mn^{2+} affinities were found for rescue of each of the three sites that had been so far discovered. This result was interpreted as supporting a model of the *Tetrahymena* active site in which three distinct metal ions participated in catalysis.

This approach in which different metal ions were associated with different rescuing metal ion affinities proved invaluable in the elucidation of the role of the scissile *pro-S*$_P$ non-bridging oxygen. Previous work had shown that sulfur substitution at this position resulted in a 1000-fold decrease in reaction rate, but softer metal ions could not rescue this thio effect. In contrast, when both the 3′-bridging and scissile *pro-S*$_P$ non-bridging oxygens were replaced with sulfur, both Zn^{2+} and Cd^{2+} stimulated the reaction about 200-fold over Mn^{2+}.[120] Whether this result was due to one or two rescuing metal ions was unclear. However, further investigation suggested that two distinct thiophilic metal ions at sites A and C were required to rescue the reaction of the double sulfur substrate.[121] Knowing the Mg^{2+} and Mn^{2+} affinities for each metal site, Shan *et al.* performed competition experiments to determine the identity of the two rescuing metal ions. The results led to a model in which both M_A and M_C bound the scissile *pro-S*$_P$ oxygen in the transition state (interactions **4** and **5** in Figure 8.2).[121]

8.4.8.4 *Identification of Ribozyme Ligands that Position the Catalytic Metal Ions*

All of the studies discussed so far focused on metal ion interactions with atoms of the *Tetrahymena* substrate. However, little was known about the functional groups within the ribozyme itself that coordinated and positioned the three metal ions within the active site. A previous NAIM study had found that sulfur substitution of a number of individual *pro-R*$_P$ non-bridging oxygens inhibited the *Tetrahymena* splicing reaction.[96] The addition of Mn^{2+} had rescued splicing of some of the sulfur-modified ribozymes, but the precise nature of this rescue was not explored in detail. However, the NAIM data, combined with phylogenetic analysis of the group I intron catalytic core, provided a starting point from which investigators could begin a search for potential ligands. As the most likely ligand candidates, the non-bridging phosphate oxygens of several conserved nucleotides within the catalytic core received special attention. Through a combination of chemical synthesis and DNA splint-mediated ligation,[138] mutant ribozymes were constructed that contained sulfur substitutions of individual *pro-R*$_P$ or *pro-S*$_P$

non-bridging oxygens. With the modified ribozymes in hand, the already established metal ion rescue assays could be used to determine whether the substitution disrupted binding of one of the known catalytic metal ions.

To date, three studies have identified specific nonbridging phosphate oxygens within the *Tetrahymena* ribozyme that coordinate one of the three catalytic metal ions. Phosphorothioate substitution of the C208 *pro-S*$_P$ oxygen resulted in a thio effect that was partially rescued upon addition of Mn^{2+} ions.[122] Analysis of the reaction of the mutant ribozyme with a 3′-bridging sulfur substrate led to a link between the C208 *pro-S*$_P$ oxygen and M_A (interaction **6** in Figure 8.2). Subsequent work identified the C262 *pro-S*$_P$ nonbridging oxygen as a ligand for M_C (interaction **7** in Figure 8.2).[123] Analysis of the C262 *pro-S*$_P$ phosphorothioate ribozyme was complicated by the unexpected Cd^{2+} rescue of a rate-limiting conformational change. As a result, the authors had to determine reaction conditions under which the chemical steps of the wild-type and mutant ribozyme reactions would be rate-limiting. Once this had been accomplished, metal ion affinity assays were used to link the C262 *pro-S*$_P$ oxygen to M_C and to rule out interactions with M_A or M_B. Finally, the *pro-S*$_P$ oxygens of A304 and A306 were identified recently as two additional ligands for M_A (interactions **8** and **9**, respectively, in Figure 8.2).[124]

8.4.8.5 Comparison with Structural Data

The metal ion rescue experiments described above were all begun prior to the publication of the group I ribozyme crystal structures.[4,49,50] Thus, it is informative to compare the results of the biochemical analyses with the locations of metal ions determined crystallographically. The highest resolution structure, the 3.4 Å structure of the group I intron from *Azoarcus sp. BH72*, was crystallized with both exons in a stage that putatively mimics the splicing reaction just after

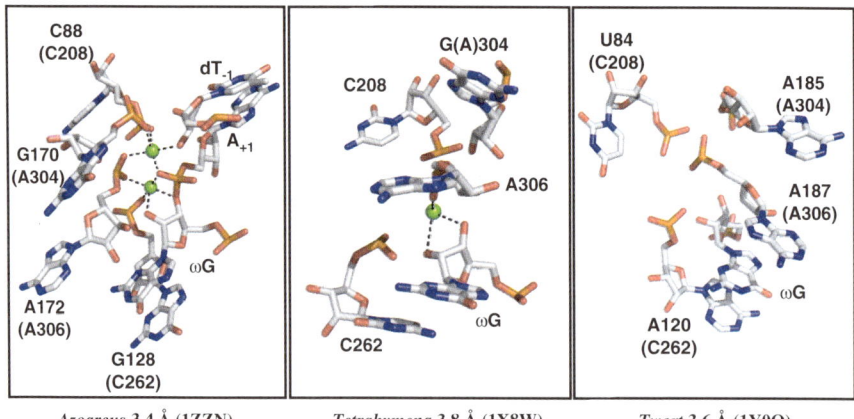

Figure 8.3 Active site structures of three different group I ribozymes. Magnesium ions are shown as green spheres. The dashed black lines denote putative metal ion-ligand interactions. Nucleotide numbers in parentheses refer to homologous *Tetrahymena* nucleotides.

the first chemical step.[4] The structural data suggest that the active site contains two metal ions (Figure 8.3, left panel). One of these lies within coordination distance of the pro-S_P nonbridging oxygens of C88 (*Tetrahymena* C208), G170 (*Tetrahymena* A304), and A172 (*Tetrahymena* A306). Two other close contacts to this metal ion include the 3'-OH group of U_{-1} (dT-1 in the structure) and the pro-R_P nonbridging oxygen of the scissile phosphate (corresponding to the pro-S_P nonbridging oxygen in the forward reaction). This metal ion therefore appears to correspond to M_A. The other metal ion lies within coordination distance of the pro-S_P nonbridging oxygen of G128 (*Tetrahymena* C262), the 2'-OH group of ωG (analogous to the exogenous guanosine nucleophile), and the pro-R_P nonbridging oxygen of the scissile phosphate. These contacts support the assignment of this metal ion as M_C. Other ligands that appear to coordinate the two metal ions include the pro-R_P and pro-S_P nonbridging oxygens of residues A172 and U173 (*Tetrahymena* U307), respectively.

The other group I structures also have features consistent with the active site model deduced from the metal ion rescue experiments. A single metal ion is present in the active site of the 3.8 Å structure of the *Tetrahymena* ribozyme.[50] This ion corresponds to M_C, as it appears to coordinate the C262 pro-S_P oxygen and the 2'-OH of ωG, as well as the pro-R_P nonbridging oxygen of A306 (Figure 8.3, center panel). The 3.6 Å structure of the group I intron from bacteriophage *Twort* shows no metals at the active site.[49] However, the pro-S_P nonbridging oxygen of A120 (*Tetrahymena* C262) and the 2'-OH group of ωG are positioned so that they could conceivably coordinate an ion corresponding to M_C (Figure 8.3, right panel).

Overall, then, the group I crystal structures support the existence of M_A and M_C and many of their proposed ligands. However, M_B does not appear in any of the crystal structures yet solved. The absence of M_B could result from a number of possibilities: (i) M_B binds an order of magnitude less tightly than M_A or M_C,[119] implying that a metal ion cannot localize sufficiently to the M_B site in the crystal to yield unambiguous electron density; (ii) a metal ion corresponding to M_B may be present but disordered and therefore not observable at the resolutions of the crystal structures; (iii) the ribozyme conformations within the crystal structures are incompatible with the binding of M_B; (iv) M_B is recruited only in ribozymes containing sulfur or nitrogen substitutions; or (v) M_B may be required for the catalytic mechanisms of some group I introns (*Tetrahymena*) but not others (*Azoarcus, Twort*). Further work will be needed to distinguish between these various possibilities and to confirm the functionality of the other surrounding metal ion ligands. However, metal ion rescue experiments have clearly demonstrated their usefulness in detecting many of the individual metal ion–ligand interactions that facilitate RNA catalysis.

8.5 The Large Ribozymes: A 'Two-Metal Ion' Mechanistic Paradigm

While the small ribozymes have been shown to retain significant catalytic activity in the presence of monovalent cations alone, the large ribozymes (RNase P, the group I and group II introns, the spliceosome, and the ribosome) have an absolute requirement for divalent metal ions. Accumulated biochemical and structural evidence has demonstrated that these ribozymes are true metalloenzymes, utilizing metal ion interactions with reactive functional groups in the transition state to facilitate catalysis. Based on early experimental observations concerning the catalytic roles of metal ions in RNase P and the *Tetrahymena* group I ribozymes, Steitz and Steitz predicted in a seminal 1993 paper that large ribozymes would employ the same 'two-metal ion' mechanism of catalysis that had been characterized biochemically and structurally for protein enzymes involved in phosphoryl transfer processes, such as DNA polymerase and alkaline phosphatase.[3] In this mechanistic model, one metal ion stabilizes the leaving group oxygen, a second metal ion activates the attacking nucleophile, and both metals coordinate one of the non-bridging phosphate oxygens to stabilize the developing negative charge of the transition state. The group I intron (Figure 8.4A) was the first and, until very recently, the only large ribozyme for which interactions proposed in the two-metal ion model have been confirmed experimentally (although it remains unclear whether two or three metal ions mediate catalysis in group I introns; see above). The details of the active site metal ion interactions for the group I intron have been described above. We now consider what is known concerning the role of metal ions in the catalytic mechanisms of the other large ribozymes.

8.5.1 RNase P

RNase P (Figure 8.4B) is a ubiquitous ribonucleoprotein complex involved in the maturation of transfer RNA (tRNA) in all kingdoms of life and is required for cell viability. It catalyzes the endonucleolytic cleavage, using water as the attacking nucleophile, of a precursor sequence in pre-tRNAs to generate tRNAs with mature 5' ends. Whereas the RNase P complex of higher species often contains a number of proteins associated with the conserved RNA molecule, bacterial RNase P contains only a single small protein subunit. In addition, the RNA component of the bacterial complex has been shown to be able to catalyze its cleavage reaction *in vitro* in the absence of any protein, making it a true ribozyme. RNase P has the distinction of being the only ribozyme to act as a multiple turnover catalyst *in vivo* and as a result much attention has been focused towards elucidating its mechanism.

The catalytic RNA component of RNase P is composed of two independently folding domains. The specificity or S-domain[54,139] includes a number of helices and a large non-Watson–Crick region that interacts with the T-stem and loop of a bound pre-tRNA substrate. The catalytic or C-domain[139,140] includes the active site and can function independently to cleave model substrates.[141,142] Early on it was shown that divalent metal ions were absolutely required for RNase P catalysis.[143,144] Whereas Mg^{2+} and Mn^{2+} could support fairly robust catalysis, various combinations of other divalent ions have since been found to allow reaction, albeit at more modest levels.[143,145–147] Using a photo-crosslinking methodology that could experimentally distinguish the substrate binding and

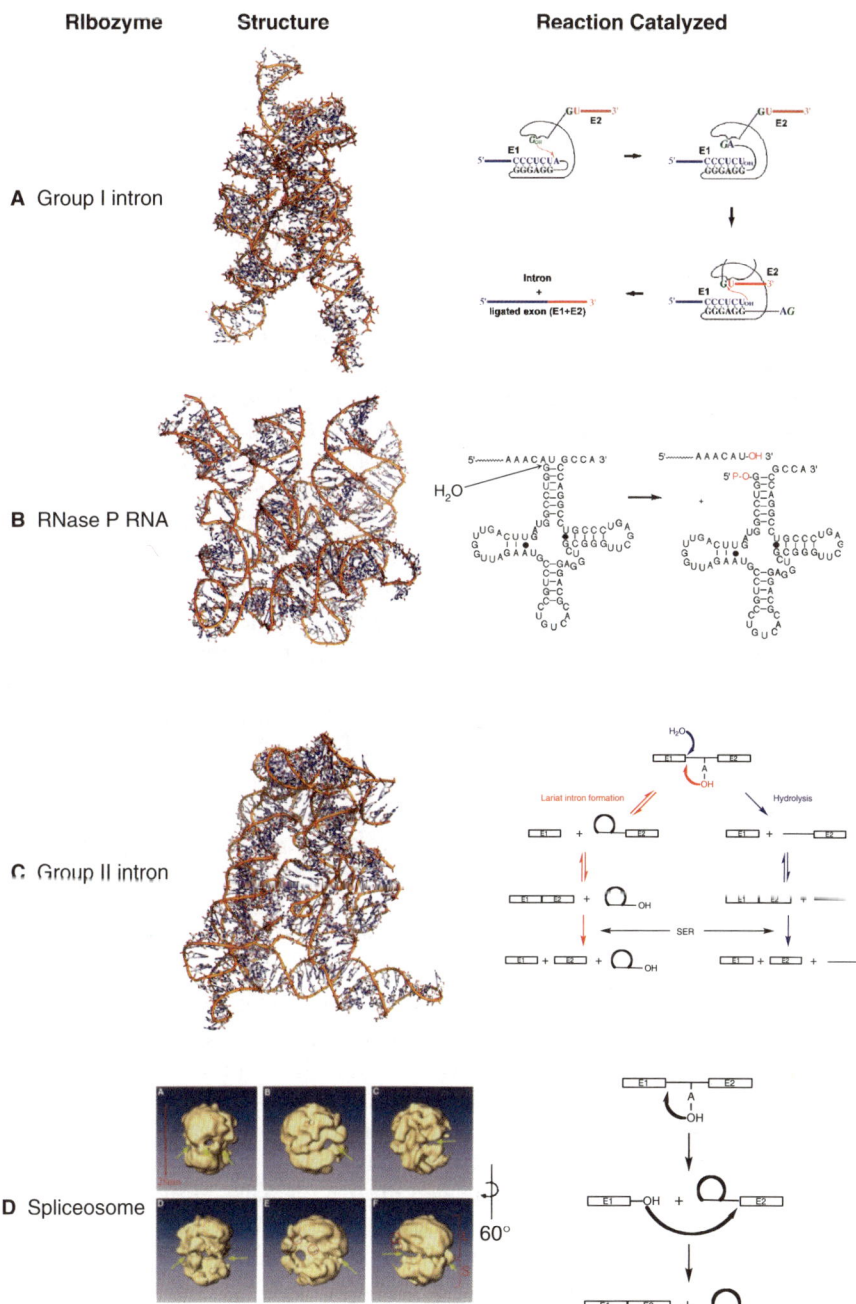

Figure 8.4 The large ribozymes and their corresponding reactions. **(A)** Group I intron (structure derived from PDB file 1ZZN).[4] **(B)** RNase P (structure derived from PDB file 2A64).[53] **(C)** Group II intron (structure derived from PDB file 3BWP).[51] **(D)** Spliceosome (cryo-electron microscopic reconstruction taken from ref. 200, with permission).

catalytic steps, Pace's group demonstrated that the requirement for divalent metal ions was primarily catalytic monovalent ions alone could mediate the formation of the enzyme–substrate complex, but divalent ions were required to perform cleavage.[143] The dependence of the observed reaction rate on magnesium concentration was determined in a later study, and Hill analysis indicated involvement of at least three bound Mg^{2+} ions in the reaction, suggesting a multiple metal ion mechanism for catalysis.[148] In addition, a 2′-H modification of the substrate at the cleavage site caused a 3400-fold drop in rate, reduced the Hill slope of the magnesium dependence to 2, and decreased the apparent magnesium affinity from the micromolar to the millimolar range. These observations suggested that the 2′-OH group at that position could serve as a ligand for one of the catalytically important metal ions.[148]

In efforts to identify metal ion interactions with the scissile phosphate in the transition state of the *E. coli* RNase P reaction, a sulfur substitution of the *pro*-R_P oxygen at the cleavage site was introduced and shown to decrease the observed reaction rate by four orders of magnitude.[128,129] The addition of Mn^{2+} or Cd^{2+} was able to restore most activity, with the Cd^{2+} stimulation exhibiting a dependence on two rescuing metal ions, consistent with coordination of the *pro*-R_P oxygen at the cleavage site by two catalytically important metal ions in the transition state of the reaction. Functionally important metal ion coordination to this non-bridging oxygen has been similarly demonstrated for *Bacillus subtilis*[130] and eukaryal[132,149] RNases P, suggesting an evolutionarily conserved RNA-based catalytic mechanism. The *pro*-S_P oxygen and 3′-oxygen leaving group at the cleavage site phosphate center were also probed by sulfur substitution and were shown to be positions important for catalysis. However, the addition of softer metal ions was unable to rescue the deleterious effect of either of these modifications and promote cleavage at the proper site.[150] Although these results are not inconsistent with transition state metal ion coordination at these positions, they do not provide any definitive experimental validation for such a model. Measurement of kinetic isotope effects for catalysis by RNase P in the presence of ^{18}O-enriched H_2O has, however, shown that a Mg^{2+}–hydroxide complex is the catalytic species that acts as the nucleophile in the rate-limiting transition state.[113]

The transition state model for RNase P catalysis that emerges from these studies is shown in Figure 8.5B and is consistent with the direct involvement of at least two, but more likely three, metal ions in the reaction mechanism. The presence of metal ion contacts to the water nucleophile, cleavage site 2′-OH, and the *pro*-R_P oxygen at the scissile phosphate have been confirmed experimentally. The coordinations to the *pro*-S_P oxygen and the leaving group non-bridging oxygen have been inferred from the deleterious effect of sulfur substitution at these positions, but remain unconfirmed. In addition, the unambiguous assignment of the individual metal ions to their particular combination of roles depicted in Figure 8.5B has not yet been achieved as it has for the group I intron.

Additional work has identified functional groups within the conserved P4 helix of the C-domain as important sites for metal ion coordination within RNase P. Phylogenetic comparative analysis suggested that the P4 helix harbors residues necessary for ribozyme function,[151] and NAIM studies have

Metal Ions in RNA Catalysis

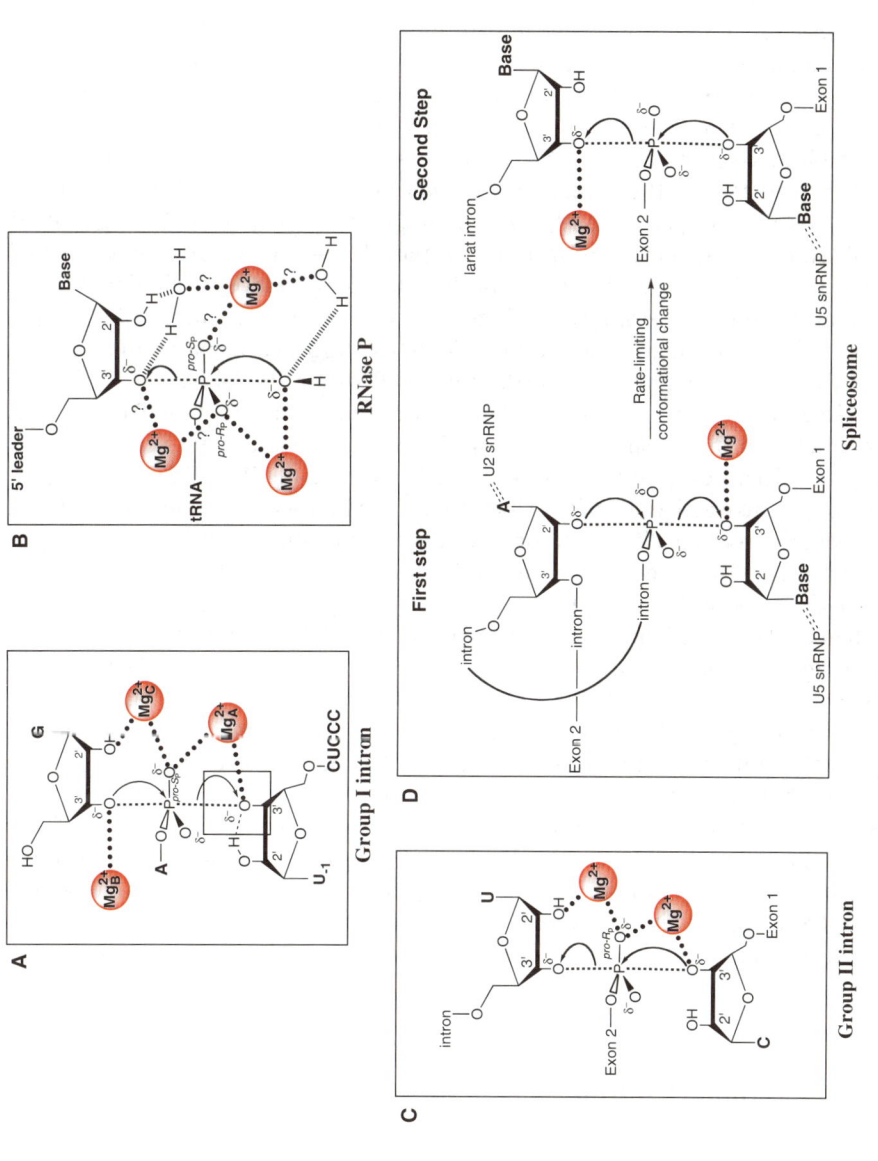

Figure 8.5 Transition state models for the large ribozymes.

implicated both non-bridging phosphate oxygens and purine N7 atoms within P4 as metal ion ligands.[152–154] These studies laid the groundwork for more detailed analyses of RNase P ribozymes bearing site-specific atomic substitutions at the conserved residues in question. Using self-cleaving ribozyme constructs derived from *E. coli* RNase P, Christian *et al.* introduced a number of site-specific phosphorothioate substitutions within the P4 helix.[131] Sulfur substitution at four sites resulted in more than a 1000-fold decrease in the reaction rate in the presence of Mg^{2+} alone. However, the addition of Mn^{2+} or Cd^{2+} rescued reactivity significantly at only two sites – the R_P and S_P non-bridging phosphate oxygens of A67. Analogous experiments using phosphorothioate-substituted RNase P derived from *B. subtilis* gave a similar result for the S_P non-bridging oxygen of A49, the homologue to A67 in *E. coli* RNase P.[155] Subsequent experiments using *E. coli* RNase P also investigated the effects of site-specific 7-deaza modifications of the adenine residues surrounding A67.[133] Metal ion rescue experiments performed on ribozymes containing both 7-deaza and A67 S_P phosphorothioate modifications suggested a model in which at least two, and possibly three, metal ions bind within the P4 helix.[133]

However, whether the proposed metal ions that bind within the P4 helix also interact with the scissile phosphate in the transition state is not yet clear. Two crystal structures of different RNase P RNAs place the P4 helix in the vicinity of the putative active site of the ribozyme.[52,53] Although the resolutions of these structures were insufficient to observe Mg^{2+} ions directly, the residues implicated previously as metal ion ligands within P4 appear to be distant from the active site. Consistent with this model, a crosslinking study suggests that whereas the P4 helix lies in proximity to the pre-tRNA substrate, the P4 metal ion ligands are several nucleotides distant from the scissile phosphate.[156] Structural mutations within P4 involving the bulged uridine U69 did alter the affinity of the Cd^{2+} ion that rescues reactivity of a substrate bearing an R_P phosphorothioate at the scissile phosphate. In addition, changing the position of the bulged U69 also appeared to change the position of the pre-tRNA substrate within the ground state active site. These observations led the authors to propose that the P4 metal ion binding site helps position the substrate within the active site for optimal catalytic metal ion coordination.[156] As yet, however, the RNA ligands within RNase P that position the catalytic metal ions have not been identified functionally, although possible candidates within one crystal structure were observed in heavy atom derivatives.[53]

8.5.2 Group II Intron

Defined as an independent structural intron class in 1982,[157] hundreds of group II introns (Figure 8.4C) have since been identified in the organelles of plants, fungi, and algae, and more recently in bacteria. The ability of these introns to catalyze their own excision was first demonstrated in 1986 for the ai5γ and bi1 introns from the mitochondrial genome of *Saccharomyces cerevisiae*.[158,159] Although most group II introns require protein cofactors for efficient splicing *in vivo*, the ability of these particular introns to splice with reasonable efficiency *in vitro* in the absence of protein, albeit under non-physiologic ionic conditions, indicates that the catalytic machinery resides within the RNA polymer.

Extensive biochemical characterization primarily of the ai5γ group II intron has yielded important insights into the mechanism of the natural splicing pathway and revealed a number of catalytic activities in a variety of different *cis*- and *trans*-splicing contexts. The catalytic versatility of the group II intron renders it useful for the investigation of fundamental principles of RNA catalysis. In addition, extensive structural and mechanistic similarities to the spliceosome make the group II intron an excellent model system to gain insights about the chemistry and orchestration of pre-mRNA splicing. Finally, group II introns have been shown to be mobile genetic elements capable of inserting into ectopic sites by retrohoming and retrotransposition.[160–162] As studies in numerous organisms have revealed the extent to which such mechanisms contribute to genetic diversity and have suggested practical applications to gene therapy, our ability to understand on a fundamental level how group II introns catalyze their own splicing takes on added importance.

The natural pathway of group II intron splicing involves two successive transesterification reactions. In the first step, the 2′-OH group of a bulged adenosine mounts a nucleophilic attack on the 5′ splice site, liberating the 5′ exon and generating a branched lariat-3′ exon intermediate. The 3′-OH group at the terminus of the 5′ exon then attacks the 3′ splice site in the second splicing step to generate spliced exons and lariat intron. The first splicing step has been shown to be reversible, with implications for intron mobility and a potential proofreading mechanism for proper splice site selection.[163] A hydrolytic reaction at the 5′ splice site has been shown *in vitro* to compete with the first step branching reaction.[164] Whereas this hydrolysis generally occurs with very low efficiency *in vivo*, it is significant for introns lacking the highly conserved branch-point adenosine residue.[165] In addition, the free intron is able to catalyze the hydrolytic cleavage of spliced exons at their ligation junction in a reaction called spliced exons reopening, or SER.[166]

Early studies of *cis*-splicing of the group II intron demonstrated an absolute requirement for divalent metal ions. In the first study of catalysis by the intron, a reaction buffer containing 10 mM Mg^{2+}, 2 mM spermidine, and 40 mM Tris, pH 7.6–7.8, was employed.[159,167,168] Later, it was demonstrated that increasing the Mg^{2+} concentration to 100 mM could substitute for the spermidine requirement.[169,170] As spermidine was well known to stabilize RNA tertiary structure, it could be inferred that the requirement for divalent ions reflected in part their role in maintaining the three-dimensional intron architecture required for catalysis. Whether these metal ions were specifically required for transition state stabilization at the scissile phosphate, however, was not yet established. However, the observations that no amount of added monovalent cation could overcome the divalent metal ion requirement, and that every identified group II-catalyzed reaction required divalents regardless of what domains or substructures were involved, were highly suggestive. The addition of molar concentrations of monovalent ions in the presence of the required divalents was shown to provide further enhancement of splicing efficiency.[171,172] Interestingly, the distribution of products from competing pathways was shown to depend on the identity of the monovalent cation and on its associated counteranion. Specifically, $(NH_4)_2SO_4$ favored branched lariat formation, whereas KCl favored 5' splice site hydrolysis. NH_4Cl promoted both pathways with about equal efficiency.[164,169,170]

Initial hints that site-bound metal ions might participate in group II intron catalysis came from an observation made by Jacquier's group concerning the metal ion dependence of the first splicing step. In this study, it was shown that under conditions in which reaction chemistry was likely to be rate limiting, manganese was able to accelerate the first transesterification step by two orders of magnitude when it was introduced in the reaction buffer in place of magnesium.[173] In the absence of manganese, magnesium titration of the reaction did not exhibit saturation up to 150 mM Mg^{2+}. However, in the presence of 10 mM $MnCl_2$, the magnesium dependence of the reaction was saturable with a K_{Mg} dependent on the monovalent cation concentration in the reaction. Taken together, these results suggested the presence of a specific divalent metal ion binding site important for catalysis that bound Mn^{2+} more tightly than Mg^{2+}. In the presence of manganese (which could occupy this site), the additional magnesium dependence presumably reflected a structural requirement for this ion. Consistent with this hypothesis, manganese titration in the presence of high concentrations of monovalent cations [0.5 M $(NH_4)_2SO_4$] and magnesium (100 mM), present to fulfill the structural requirement, yielded an apparent K_{Mn} of 15 mM. Several hypotheses were advanced by the authors concerning the potential function of this site-bound metal ion, including a role in the activation of the branch-point 2'-OH nucleophile. Such a role in nucleophile activation during the first splicing step remains to be demonstrated conclusively, however.

Metal ion specificity switch experiments have identified metal ions activating the leaving group in both splicing steps. Incorporation of a 3'-phosphorothiolate linkage at the 5' splice site (in which sulfur replaces the 3'-oxygen leaving group) abolished *cis*-splicing in the presence of magnesium alone.[126] Reactivity of this mutant ribozyme was restored robustly upon addition of 10–20 mM $MnCl_2$, $ZnCl_2$ or $CdCl_2$. Analogous thio substitution of the leaving group oxygen at the 3' splice site in the full-length intron, however, failed to produce a deleterious effect on second step splicing in the presence of magnesium. Although this result may have indicated that metal ion coordination to the leaving group was not required for second step catalysis, it was also possible that a non-chemical step was rate-limiting. The latter possibility was particularly likely in light of a well-characterized conformational change that had been shown to occur between the two splicing steps. A *trans*-assay for the second step of splicing, in which the catalytic apparatus could be folded separately into its active conformation before the addition of substrate, provided a potential means to bypass this rate-limiting conformational change and interrogate the chemical step. In this 'tripartite' assay, a 3' splice site oligonucleotide is combined with the 5' exon and a ribozyme containing all but the 3'-terminal six nucleotides of the intron.[174] Bridging sulfur substitution within the 3′ splice site oligonucleotide resulted in a 100-fold loss of splicing activity that was restored to within 3–4-fold by the addition of 10 mM $MnCl_2$.[127] $CdCl_2$ and $CoCl_2$ also rescued activity strongly, whereas $ZnCl_2$ had a more modest effect. Taken together, these studies demonstrated that metal ion coordination to the leaving group facilitated catalysis in both splicing steps of the group II intron.

A more detailed kinetic characterization of the tripartite assay for second step splicing confirmed that the metal rescue of the 3′ splice site sulfur substitution observed previously reflected metal ion coordination to the 3′-oxygen leaving group in the transition state of the chemical step.[127] Additionally, metal rescue of a 2′-aminocytidine modification revealed transition state metal ion coordination to the 2′-OH group at the 3′ splice site. This metal ion exhibited the same thermodynamic signature as the one interacting with the phosphoryl oxygen leaving group, rescuing reactivity with a similar ∼3 mM apparent affinity. This observation raised the likely possibility that the same metal ion interacts with both positions in the transition state, a model that still awaits definitive confirmation by double substitution experiments. Finally, while sulfur substitution of the non-bridging phosphoryl oxygens had previously revealed a role for the *pro*-R_P oxygen in catalysis,[175] the inability to rescue activity of this mutant with softer divalent metal ions has precluded its definitive assignment as a transition state ligand.

The use of metal ion rescue experiments to identify nucleophile activation by metal ions during group II intron splicing poses a unique methodological challenge. Due to the poor nucleophilicity of sulfur at phosphate centers,[20,176] thiol substitution of the nucleophilic hydroxyl group in either splicing step would engender an absolute block to reaction. However, the *in vitro* reversibility of each splicing step provides a means to circumvent this problem. By the principle of microscopic reversibility, the forward and reverse reactions must progress through the same transition state, so that a metal ion–nucleophile interaction in the forward reaction should be identical to a metal ion–leaving group interaction in the reverse reaction.

Our laboratory employed this strategy in a recent study examining reverse second step splicing of the group II intron.[5] This reverse reaction, in which spliced exons are incubated with full-length intron, occurs *via* two competing pathways: a hydrolytic pathway (the SER reaction), in which water or hydroxide acts as the nucleophile, and a transesterification pathway, in which the terminal 3′-OH group of the intron acts as the nucleophile. The latter pathway represents the reverse of the natural splicing pathway and has been shown to require the branched lariat form of the intron rather than the linear intron.[173] Since we sought to identify a metal ion–nucleophile interaction in the second step of the natural splicing pathway, we incubated in our assay the lariat form of the full-length intron with an exon 1–exon 2 construct bearing a sulfur linkage at the leaving group position of the splice junction. This modification abolished reverse splicing activity as compared with reaction of the all-oxygen-containing exon 1–exon 2 construct. Reactivity could be restored by addition of

MnCl$_2$ and ZnCl$_2$, demonstrating metal ion coordination to the leaving group oxygen in the reverse second step reaction and thus, by the principle of microscopic reversibility, to the nucleophile in the forward exon ligation reaction.

Prior to identification of this second metal ion within the group II intron active site, the group I intron was the only ribozyme for which the 'two-metal ion mechanism' was demonstrated conclusively either biochemically or structurally. As structural information at atomic resolution for the full-length group II intron has remained elusive until recently, biochemical probing by atomic mutagenesis has provided our only glimpse into the catalytic mechanism of the splicing pathway. The accumulated insight from the experiments outlined above has led to the transition state model for group II intron splicing shown in Figure 8.5C. Extension of the two-metal ion strategy to the description of the group II intron splicing pathway now suggests a mechanistic paradigm for catalysis by the large ribozymes. The large ribozymes catalyze nucleotidyl transfer from 3'-oxygen leaving groups to water or to a 2'- or 3'-oxygen nucleophile. In contrast, the small ribozymes catalyze endonucleolytic 2'-O-transphosphorylations. As uncatalyzed 2'-O-transphosphorylations have been shown to occur 10^6-fold more rapidly than nucleotidyl transfer,[177,178] the obligate divalent metal ion requirement of the large ribozymes likely reflects the need for greater catalytic potency to facilitate reaction on a physiologic timescale. Concomitantly, the larger size of the obligate metalloribozymes may provide sufficient structural scaffolding to create an active site capable of precisely positioning multiple metal ions to interact with atoms of the scissile phosphate in the transition state.

Very recently, Pyle's group has solved the crystal structure of an intact group II intron isolated from the extremophilic bacterium *Oceanobacillus iheyensis*.[51] The construct employed contains all six intron domains and was crystallized in a postcatalytic state after having undergone both splicing steps. The 3.1 Å structure provides an intriguing glimpse into the complex architecture of the folded intron. The intricate network of tertiary interactions that organizes the intron around the catalytic domain V includes many interactions that have been predicted biochemically and phylogenetically, as well as several that have not yet been identified experimentally. Unfortunately, domain VI is disordered in the structure, precluding visualization of the branch point adenosine. Nevertheless, the demonstration of two metal ions bound to key regions of domain V within the structure is potentially suggestive. The first metal ion forms apparent inner sphere contacts to backbone ligands within a bulged region of domain V spanning nucleotides 375–379. The second metal ion forms three inner sphere contacts to phosphoryl oxygens within the essential domain V catalytic triad. One of these three ligands corresponds to the *pro-S*$_P$ oxygen of A816 in the ai5γ group II intron, which coordinates a metal ion important for the first step of splicing.[125] These two metals are separated by a distance of 3.9 Å in the crystal structure, which corresponds well with the optimum separation invoked for a classic two-metal ion mechanism.[3] Furthermore, the two metal ions are positioned on an exposed surface of domain V that would be accessible to the 5′ and 3′ splice sites. While these observations are not sufficient to provide a structural snapshot of a two-metal ion active site such as that achieved for the group I intron, they are nevertheless highly intriguing and consistent with a two-metal ion mechanism for group II intron splicing.

8.5.3 Spliceosome

Since the discovery 30 years ago that eukaryotic RNA messages are interrupted by non-coding regions,[179,180] extensive study of the pre-mRNA splicing pathway has revealed that the process is catalyzed by an elaborate and dynamic ribonucleoprotein complex.[181] Consisting of five snRNAs (U1, U2, U4, U5, U6) and over 200 identified proteins, the spliceosome (Figure 8.4D) is arguably the most complex macromolecular assembly in the cell.[182] Because the spliceosome catalyzes a central step in gene expression, and since alternative splicing has been shown to be an epigenetic regulatory mechanism of fundamental importance in biology, understanding the mechanistic basis of spliceosomal function is critical. Detailed mechanistic studies remain challenging, however, due to the complexity and fluidity of the splicing machinery. The ability to reconstitute pre-mRNA splicing *in vitro* from purified components has not yet been achieved, and structural information at atomic resolution is not imminently likely. Despite this, many fundamental aspects of the splicing pathway have been elucidated, including in part the role of metal ions.

Central to the splicing pathway, the highly conserved spliceosomal snRNAs facilitate splice site recognition, conformational rearrangements, and possibly reaction chemistry. Specifically, U1 and U2 snRNAs recognize the 5′ splice site and branch site adenosine, respectively. U6 then displaces U1, forming basepairing interactions with U2 that bring the 5′ splice site and branch site into apposition to allow for the first transesterification step. The resulting 5′ exon and lariat intron-3′ exon are oriented in part by U5 to complete the second transesterification, yielding spliced exons and liberating the lariat intron. Extensive structural and mechanistic similarities to the group II intron suggested that the catalytic apparatus of the spliceosome would reside within its RNA components, and that metal ions would participate in its chemical reaction pathway.

Definitive demonstration of metal ion catalysis by the spliceosome was achieved in 1997 by metal ion rescue experiments performed in mammalian splicing extract.[134] In this study, a splicing substrate bearing a 3′-thio leaving group at the 5′ splice site was unreactive in 1.5–3.5 mM $MgCl_2$, but did undergo the first step of splicing in the presence of 1.5 mM $MnCl_2$. Native gel analysis of splicing complexes confirmed that the sulfur substitution did not affect spliceosome assembly, but rather some subsequent step, most likely the first transesterification reaction. Analogous sulfur substitution of the 3′-oxygen leaving group at the 3′ splice site was performed in an attempt to reveal metal ion coordination during the second splicing step. However, splicing efficiency was unaffected by this modification, suggesting one of two alternative explanations. Either metal ion coordination to the leaving group was not required for second step splicing, or a non-chemical step insensitive to sulfur modification limited the reaction. This asymmetry of metal ion rescue signatures between the two splicing steps paralleled that observed initially for group II intron splicing,[126] strengthening the evolutionary relationship between the two splicing machineries and prompting a similar experimental reinvestigation of the second splicing step.

In a later study, a *trans*-assay for the second step of pre-mRNA splicing analogous to the tripartite reaction employed in the study of group II intron splicing was developed to bypass a potential rate-limiting conformational change between the two splicing steps.[135] Specifically, a pre-mRNA substrate lacking the 3′ exon and several adjacent intron nucleotides was allowed to undergo first step splicing in HeLa nuclear extract to generate free 5′ exon and lariat intron. A 3′ splice site substrate containing the last five nucleotides of the intron followed by the 3′ exon was then added to facilitate the second step of intron splicing. When the 3′-oxygen leaving group in this substrate was replaced by sulfur, second step reactivity was inhibited in the presence of magnesium. Reactivity was restored in the presence of manganese, providing evidence for functionally important metal ion coordination in the second splicing step.

The current model for the transition states of pre-mRNA splicing that emerges from this work is shown in Figure 8.5D and involves metal ion coordination of the leaving group in both splicing steps. As yet, metal ion coordination to the nucleophile in either step has not been demonstrated. In contrast to group II intron splicing, the steps of spliceosomal pre-mRNA splicing have not been shown to be reversible, precluding biochemical probing of the reverse reactions to reveal these potential metal ion interactions. However, given the extensive mechanistic similarities to the group II intron that have been revealed thus far, it is likely that the spliceosome reaction proceeds by a two-metal ion mechanism such as that recently demonstrated for group II intron splicing.[5]

The search for ligands that position these active site metal ions takes on a particular significance with respect to our understanding of pre-mRNA splicing. As *in vitro* splicing has only been achievable in nuclear extracts, the question of whether the catalytic apparatus of the spliceosome resides wholly in its RNA component, or whether some of the many spliceosomal proteins contribute to catalysis, remains open. In an attempt to address this question, Manley's group demonstrated that when a protein-free human U2/U6 complex was incubated with an RNA oligonucleotide bearing the branch site consensus sequence, a reaction occurred in which the branch site adenosine became linked covalently *via* its 2′-OH group to the highly conserved AGC region of U6.[183] Although this compelling result demonstrated that the spliceosomal snRNAs possessed an intrinsic catalytic activity chemically related to splicing, the reaction observed did not yield the natural splicing product and occurred with extremely low efficiency. This might represent the limit of activity that can be reconstituted in the absence of protein and may suggest a role for the spliceosomal proteins in orienting the various snRNAs for reaction or in facilitating necessary conformational changes.

An alternative strategy for addressing the question of whether the spliceosome is a ribozyme would be to demonstrate that the ligands to the catalytic metal ions reside within the snRNAs. Several positions within U6 have been shown to exhibit splicing defects upon phosphorothioate substitution and thus are candidates for more detailed analysis by metal ion rescue experiments. To date, only a single position within U6 has been shown to exhibit a metal ion rescue signature upon phosphorothioate substitution. Specifically, sulfur substitution of the *pro-S*$_P$ oxygen of U80 was shown by Lin's group to inhibit first step splicing in the presence of magnesium but not cadmium.[184] This result demonstrated that U80 serves as a ligand for a functionally important metal ion in the first splicing step, but whether this metal ion is an active site catalytic metal ion is not yet known.

8.6 The Small Ribozymes: Endonucleolytic Cleavage and General Acid–Base Catalysis

As mentioned earlier in this chapter, all ribozymes were thought to be obligate metalloenzymes whose catalytic activity depended on divalent metal ions, most notably Mg^{2+}. This view has changed with the discovery that some ribozymes catalyze phosphodiester cleavage in high concentrations of monovalent cations alone.[9,15,185,186] Importantly, each of these ribozymes catalyzes the same intramolecular transesterification reaction – attack of a 2'-OH group on the adjacent phosphate to yield a 2',3'-cyclic phosphate leaving group (Figure 8.6). This observation has given rise to models in which these ribozymes (called 'endonucleolytic' ribozymes) effect catalysis *via* a general acid–base mechanism involving a catalytic nucleobase. Support for these models has come from studies of the hepatitis delta virus (HDV) ribozyme,[13,14,16] the hairpin ribozyme,[187–190] the VS ribozyme,[191,192] and most recently the hammerhead ribozyme,[57,193,194] which suggest that critical nucleobases participate in general acid–base catalysis. In light of this evidence, metal ions must assume a catalytic role in the mechanisms of endonucleolytic ribozymes distinct from that observed for the large ribozymes. As noted for the transition states of the large ribozymes, metal ions help to position external nucleophiles and directly coordinate functional groups that undergo charge buildup. For the endonucleolytic ribozymes, metal ions have been implicated in catalysis by the hammerhead and HDV ribozymes. Due to space constraints, we offer only minimal coverage of the hammerhead ribozyme, but we refer the reader to several excellent papers.[57,195,196]

For the HDV ribozyme, experiments have identified a role for a divalent metal ion–water complex in general acid–base catalysis. In the presence of 10 mM Mg^{2+}, the HDV reaction rate increases log-linearly between pH 4 and 6 and becomes independent of pH above pH 7.[15] In contrast, in 1 M NaCl in the

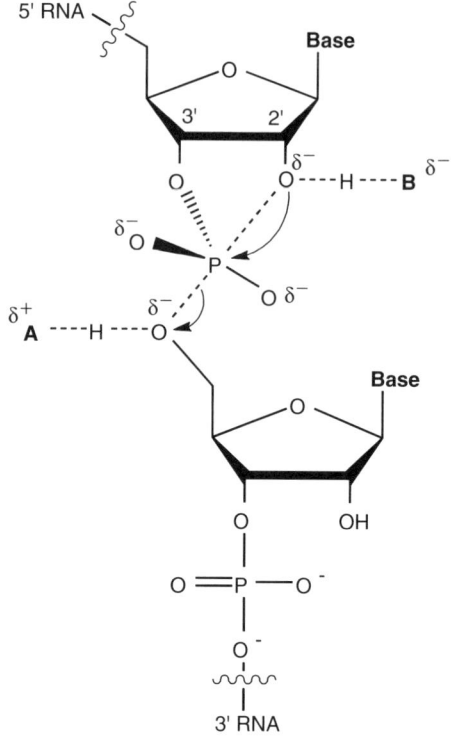

Figure 8.6 Transition state model for general acid–base-catalyzed intramolecular cleavage of RNA *via* nucleophilic attack of a 2′-hydroxyl group on the adjacent phosphodiester linkage. In this model, the general acid **A** protonates the 5′ leaving group oxygen, while the general base **B** deprotonates the 2′-hydroxyl group nucleophile.

absence of divalent metal ions, the reaction rate decreases log-linearly from pH 6 to 8.[15] These observations have been interpreted as suggesting that a divalent metal ion–hydroxide complex acts as a general base to abstract a proton from the nucleophilic 2′-OH group. In this model, the overall pH dependence of the HDV reaction in the presence of divalent metal ions increases log-linearly between pH 4 and 6. This region of the pH dependence profile reflects titration of the general base $Mg(OH_2)_6^{2+}$, while the cytosine nucleobase that functions as the general acid remains protonated. Above pH 6, the increase in the concentration of metal ion–hydroxide is offset by a decrease in the concentration of the protonated nucleobase. The resulting reaction rate is therefore independent of pH above pH 6. Additional evidence for this model comes from a Hill analysis showing that maximum reactivity is associated with the uptake of at least one functional magnesium ion, whose binding affinity increases with pH.[15]

NMR,[197] X-ray crystallographic,[45] and footprinting[103,198] analyses of the HDV ribozyme also strongly suggest the presence of a metal ion near the active site (Figure 8.7). This metal ion probably binds to the active site *via* outer sphere

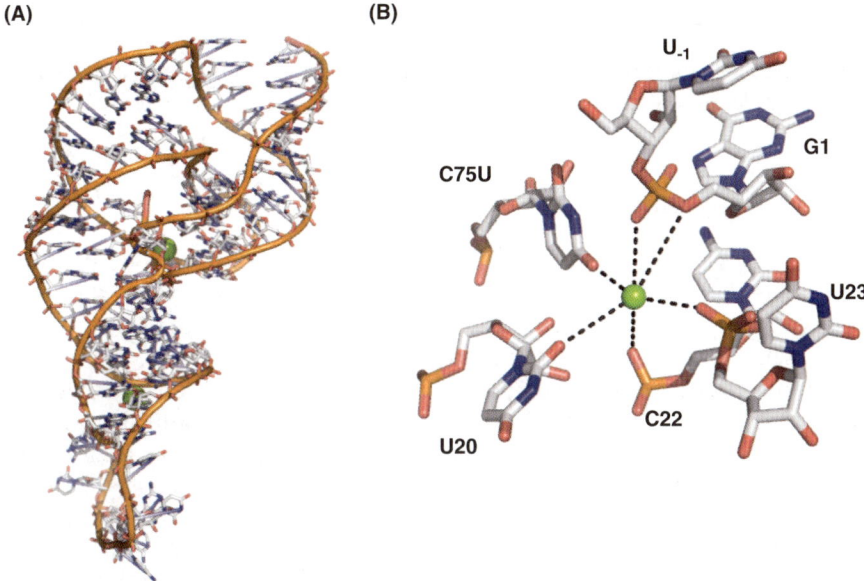

Figure 8.7 Three-dimensional structure of the HDV ribozyme in the pre-cleaved state (derived from PDB file 1SJ3).[45] Magnesium ions are shown as green spheres. (A) View of the overall structure. (B) Close-up view of the active site. The dashed black lines denote outer sphere metal ion–ligand interactions.

coordination, since the exchange-inert complex $Co(NH_3)_6^{3+}$ apparently competes with it for binding to the active site.[15,44] However, alternative catalytic models have been proposed in which the metal ion acts as a general acid, with cytosine 75 functioning as the general base.[45] This model is based on the position of the Mg^{2+} ion in the crystal structure of an HDV precursor in which the catalytic C75 has been mutated to uridine to prevent cleavage.[45] In this structure, the Mg^{2+} ion lies within outer sphere coordination distance of several functional groups, most notably the 5′-oxygen leaving group (Figure 8.7B). Crystal structures of the self-cleaved form of the ribozyme have not helped to identify the catalytic role of this metal ion, since these structures showed no metal ions within the active site.[60,61] Hence, determination of the precise role of this metal ion in the HDV reaction mechanism (as a general acid or a general base) still awaits experimental verification.

The ability of monovalent cations to support catalysis by endonucleolytic ribozymes also suggests a second possible mechanistic function for metal ions. Rather than directly binding transition states or participating in general acid–base mechanisms, metal ions in some cases may promote catalysis solely through long-range electrostatic effects. These effects would manifest as metal ion-induced pK_a shifts of nucleobase functional groups, and as the ability of outer sphere coordinated metal ions to promote catalysis by organizing favorable solvent interactions.[46] Indeed, metal ions themselves are not always required for ribozyme activity, as evidenced by the modest reaction rates achieved for some ribozymes in the presence of organic cations.[199] Thus, whereas some ribozymes absolutely require divalent metal ions for catalytic activity, others may rely on metal ions only to provide an appropriate structural and electrostatic environment for catalysis.

8.7 Conclusions

A rich structural and biochemical literature has helped to elucidate the transition state coordination environment for catalytic metal ions within the active sites of the large ribozymes. For these ribozymes, divalent metal ions play some of the same roles (Lewis acid stabilization of the leaving group, nucleophile activation, transition state stabilization) that have been observed in protein metalloenzymes. Elegant biochemical experiments have identified transition state ligands and allowed the estimation of the catalytic power afforded by divalent metal ions to large ribozyme transition states. The small ribozymes at first glance do not appear to fit the description of the typical metalloenzyme. Their catalytic mechanisms do not require the presence of divalent metal ions and appear to rely on nucleobase-mediated general acid–base catalysis. Although the metal ions make no obligate interactions with the chemical transition state, they still contribute to catalysis in a hidden but essential manner. The cations form an ion atmosphere around the folded RNA, screening the negative charges of the phosphodiester backbone to allow the molecule to adopt three-dimensional structures that facilitate catalysis. This atmosphere of ions retained by the folded RNA may contribute indirectly to catalysis by enhancing substrate binding, by allowing acquisition of conformations amenable to catalysis, and by stabilizing the conformation in the transition state relative to the ground state. Additionally, the ion atmosphere could contribute to transition state stabilization by providing electrostatic stabilization of the negative charges that usually develop in the transition states of ribozyme-catalyzed reactions.

From these analyses, a mechanistic paradigm has emerged: large ribozymes use multiple metal ions directly in catalysis, whereas the endonucleolytic ribozymes have no obligate need for direct involvement of metal ions, possibly reflecting a need for less catalytic power. The uncatalyzed endonucleolytic cleavage of RNA (internal 2′-O-transphosphorylation) occurs 10^6-fold faster than uncatalyzed nucleotidyl transfer, reflecting the proximity of the nucleophilic 2′-hydroxyl group to the phosphorus reaction center. Perhaps the use of multiple metal ions by large ribozymes, then, reflects the need for greater catalytic power to promote nucleotidyl transfer reactions on a time-scale amenable to life.

Our understanding of ribozyme catalysis in general and the role of metal ions in particular will continue to advance as we gain new insights from studies of phosphoryl transfer using model compounds, particularly with the refinement of powerful approaches to probe fine details of transition state structure such as measurement of heavy atom isotope effects. Concurrently, the application of creative biochemical and spectroscopic techniques both to identify functional metal ions and characterize their precise catalytic roles at the active sites of ribozymes will continue to enhance our understanding of these versatile metalloenzymes. For example, metal ion rescue experiments involving doubly substituted ribozymes have promise for identifying additional ligands for catalytic metal ions, allowing us to expand functional transition state models and gain new insight into the fundamental question of how RNA positions and harnesses the catalytic power of metal ions.

References

1. K. Kruger, P. J. Grabowski, A. J. Zaug, J. Sands, D. E. Gottschling and T. R. Cech, *Cell*, 1982, **31**, 147.
2. A. M. Pyle, *Science*, 1993, **261**, 709.
3. T. A. Steitz and J. A. Steitz, *Proc. Natl. Acad. Sci. USA*, 1993, **90**, 6498.
4. M. R. Stahley and S. A. Strobel, *Science*, 2005, **309**, 1587.
5. P. M. Gordon, R. Fong and J. A. Piccirilli, *Chem. Biol.*, 2007, **14**, 607.
6. J. B. Murray, A. A. Seyhan, N. G. Walter, J. M. Burke and W. G. Scott, *Chem. Biol.*, 1998, **5**, 587.
7. A. Hampel and J. A. Cowan, *Chem. Biol.*, 1997, **4**, 513.
8. M. Roychowdhury-Saha and D. H. Burke, *RNA*, 2007, **13**, 841.
9. A. Roth, A. Nahvi, M. Lee, I. Jona and R. R. Breaker, *RNA*, 2006, **12**, 607.
10. P. C. Bevilacqua, T. S. Brown, S. Nakano and R. Yajima, *Biopolymers*, 2004, **73**, 90.
11. E. M. Moody, J. T. Lecomte and P. C. Bevilacqua, *RNA*, 2005, **11**, 157.
12. P. Legault and A. Pardi, *J. Am. Chem. Soc.*, 1997, **119**, 6621.
13. I. H. Shih and M. D. Been, *Proc. Natl. Acad. Sci. USA*, 2001, **98**, 1489.
14. A. T. Perrotta, I. Shih and M. D. Been, *Science*, 1999, **286**, 123.
15. S. Nakano, D. M. Chadalavada and P. C. Bevilacqua, *Science*, 2000, **287**, 1493.
16. S. R. Das and J. A. Piccirilli, *Nat. Chem. Biol.*, 2005, **1**, 45.
17. J. A. Cowan, in *The Biological Chemistry of Magnesium*, J. A. Cowan ed., VCH, New York, 1995, pp. vii,viii.
18. R. G. Parr and R. G. Pearson, *J. Am. Chem. Soc.*, 1983, **105**, 7512.
19. R. G. Pearson, *J. Am. Chem. Soc.*, 1963, **85**, 3533.
20. R. G. Pearson, *Science*, 1966, **151**, 172.
21. R. G. Pearson, *Proc. Natl. Acad. Sci. USA*, 1986, **83**, 8440.
22. R. G. Pearson, *Inorg. Chem.*, 1988, **27**, 734.
23. H. Irving and R. J. P. Williams, *J. Chem. Soc.*, 1953, 3192.
24. N. Lehman and G. F. Joyce, *Nature*, 1993, **361**, 182.
25. M. Zivarts, Y. Liu and R. R. Breaker, *Nucleic Acids Res.*, 2005, **33**, 622.
26. J. B. Thomson, B. K. Patel, V. Jimenez, K. Eckart and F. Eckstein, *J. Org. Chem.*, 1996, **61**, 6273.

27. A. C. Hengge, A. E. Tobin and W. W. Cleland, *J. Am. Chem. Soc.*, 1995, **117**, 5919.
28. A. C. Hengge and W. W. Cleland, *J. Am. Chem. Soc.*, 1991, **113**, 5835.
29. A. J. Kirby and M. Younas, *J. Chem. Soc. B*, 1970, 1187.
30. A. J. Kirby and M. Younas, *J. Chem. Soc. B*, 1970, 510.
31. A. J. Kirby and M. Younas, *J. Chem. Soc. B*, 1970, 1165.
32. S. A. Khan and A. J. Kirby, *J. Chem. Soc. B*, 1970, 1172.
33. A. J. Kirby and A. G. Varvoglis, *J. Chem. Soc. B*, 1968, 135.
34. A. J. Kirby and A. G. Varvoglis, *J. Am. Chem. Soc.*, 1967, **89**, 415.
35. M. Kosonen, E. Youseti-Salakdeh, R. Stromberg and H. Lonnberg, *J. Chem. Soc., Perkin Trans.*, 1997, **2**, 2661.
36. G. J. Narlikar, V. Gopalakrishnan, T. S. McConnell, N. Usman and D. Herschlag, *Proc. Natl. Acad. Sci. USA*, 1995, **92**, 3668.
37. R. S. Alexander, Z. F. Kanyo, L. E. Chirlian and D. W. Christianson, *J. Am. Chem. Soc.*, 1990, **112**, 933.
38. T. Lonnberg and H. Lonnberg, *Curr. Opin. Chem. Biol.*, 2005, **9**, 665.
39. M. Kosonen, K. Hakala and H. Lonnberg, *J. Chem. Soc., Perkin Trans.*, 1998, **2**, 663.
40. M. Kosonen and H. Lonnberg, *J. Chem. Soc., Perkin Trans.*, 1995, **2**, 1203.
41. D. Herschlag and W. P. Jencks, *Biochemistry*, 1990, **29**, 5172.
42. P. C. Bevilacqua, *Biochemistry*, 2003, **42**, 2259.
43. T. C. Bruice, A. Tsubouchi, R. O. Dempcy and L. P. Olson, *J. Am. Chem. Soc.*, 1996, **118**, 9867.
44. S. Nakano, A. L. Cerrone and P. C. Bevilacqua, *Biochemistry*, 2003, **42**, 2982.
45. A. Ke, K. Zhou, F. Ding, J. H. Cate and J. A. Doudna, *Nature*, 2004, **429**, 201.
46. R. K. Sigel and A. M. Pyle, *Chem. Rev.*, 2007, **107**, 97.
47. P. L. Adams, M. R. Stahley, M. L. Gill, A. B. Kosek, J. Wang and S. A. Strobel, *RNA*, 2004, **10**, 1867.
48. P. L. Adams, M. R. Stahley, A. B. Kosek, J. Wang and S. A. Strobel, *Nature*, 2004, **430**, 45.
49. B. L. Golden, H. Kim and E. Chase, *Nat. Struct. Mol. Biol.*, 2005, **12**, 82.
50. F. Guo, A. R. Gooding and T. R. Cech, *Mol. Cell.*, 2004, **16**, 351.
51. N. Toor, K. S. Keating, S. D. Taylor and A. M. Pyle, *Science*, 2008, **320**, 77.
52. A. Torres-Larios, K. K. Swinger, A. S. Krasilnikov, T. Pan and A. Mondragon, *Nature*, 2005, **437**, 584.
53. A. V. Kazantsev, A. A. Krivenko, D. J. Harrington, S. R. Holbrook, P. D. Adams and N. R. Pace, *Proc. Natl. Acad. Sci. USA*, 2005, **102**, 13392.
54. A. S. Krasilnikov, X. Yang, T. Pan and A. Mondragon, *Nature*, 2003, **421**, 760.
55. N. Ban, P. Nissen, J. Hansen, P. B. Moore and T. A. Steitz, *Science*, 2000, **289**, 905.

56. B. T. Wimberly, D. E. Brodersen, W. M. Clemons Jr., R. J. Morgan-Warren, A. P. Carter, C. Vonrhein, T. Hartsch and V. Ramakrishnan, *Nature*, 2000, **407**, 327.
57. M. Martick and W. G. Scott, *Cell*, 2006, **126**, 309.
58. H. W. Pley, D. S. Lindes, C. DeLuca-Flaherty and D. B. McKay, *J. Biol. Chem.*, 1993, **268**, 19656.
59. W. G. Scott, J. T. Finch and A. Klug, *Nucleic Acids Symp. Ser.*, 1995, 214.
60. A. R. Ferre-D' Amare and J. A. Doudna, *J. Mol. Biol.*, 2000, **295**, 541.
61. A. R. Ferre-D' Amare, K. Zhou and J. A. Doudna, *Nature*, 1998, **395**, 567.
62. P. B. Rupert, A. P. Massey, S. T. Sigurdsson and A. R. Ferre-D' Amare, *Science*, 2002, **298**, 1421.
63. D. J. Klein and A. R. Ferre-D' Amare, *Science*, 2006, **313**, 1752.
64. E. A. Galburt and B. L. Stoddard, *Biochemistry*, 2002, **41**, 13851.
65. A. Pingoud and A. Jeltsch, *Nucleic Acids Res.*, 2001, **29**, 3705.
66. R. A. Kovall and B. W. Matthews, *Curr. Opin. Chem. Biol.*, 1999, **3**, 578.
67. T. E. Horton, D. R. Clardy and V. J. DeRose, *Biochemistry*, 1998, **37**, 18094.
68. S. R. Morrissey, T. E. Horton, C. V. Grant, C. G. Hoogstraten, R. D. Britt and V. J. DeRose, *J. Am. Chem. Soc.*, 1999, **121**, 9215.
69. M. Vogt, S. Lahiri, C. G. Hoogstraten, R. D. Britt and V. J. DeRose, *J. Am. Chem. Soc.*, 2006, **128**, 16764.
70. H. W. Pley, K. M. Flaherty and D. B. McKay, *Nature*, 1994, **372**, 68.
71. W. G. Scott, J. B. Murray, J. R. Arnold, B. L. Stoddard and A. Klug, *Science*, 1996, **274**, 2065.
72. A. Khvorova, A. Lescoute, E. Westhof and S. D. Jayasena, *Nat. Struct. Biol.*, 2003, **10**, 708.
73. N. Kisseleva, A. Khvorova, E. Westhof and O. Schiemann, *RNA*, 2005, **11**, 1.
74. C. G. Hoogstraten, C. V. Grant, T. E. Horton, V. J. DeRose and R. D. Britt, *J. Am. Chem. Soc.*, 2002, **124**, 834.
75. C. G. Hoogstraten and R. D. Britt, *RNA*, 2002, **8**, 252.
76. S. E. Butcher, F. H. Allain and J. Feigon, *Biochemistry*, 2000, **39**, 2174.
77. F. H. Allain and G. Varani, *Nucleic Acids Res.*, 1995, **23**, 341.
78. C. Cheong, G. Varani and I. Tinoco Jr., *Nature*, 1990, **346**, 680.
79. T. Dieckmann, E. Suzuki, G. K. Nakamura and J. Feigon, *RNA*, 1996, **2**, 628.
80. S. E. Butcher, T. Dieckmann and J. Feigon, *EMBO J.*, 1997, **16**, 7490.
81. D. G. Sashital, V. Venditti, C. G. Angers, G. Cornilescu and S. E. Butcher, *RNA*, 2007, **13**, 328.
82. C. W. Hilbers, A. Heerschap, C. A. Haasnoot and J. A. Walters, *J. Biomol. Struct. Dyn.*, 1983, **1**, 183.
83. P. J. Salemink, T. Swarthof and C. W. Hilbers, *Biochemistry*, 1979, **18**, 3477.
84. M. Wu and I. Tinoco Jr., *Proc. Natl. Acad. Sci. USA*, 1998, **95**, 11555.

85. H. M. Al-Hashimi, S. W. Pitt, A. Majumdar, W. Xu and D. J. Patel, *J. Mol. Biol.*, 2003, **329**, 867.
86. M. Maderia, L. M. Hunsicker and V. J. DeRose, *Biochemistry*, 2000, **39**, 12113.
87. J. Noeske, H. Schwalbe and J. Wohnert, *Nucleic Acids Res.*, 2007, **35**, 5262.
88. K. Chin, K. A. Sharp, B. Honig and A. M. Pyle, *Nat. Struct. Biol.*, 1999, **6**, 1055.
89. V. K. Misra and D. E. Draper, *Proc. Natl. Acad. Sci. USA*, 2001, **98**, 12456.
90. D. R. Banatao, R. B. Altman and T. E. Klein, *Nucleic Acids Res.*, 2003, **31**, 4450.
91. S. C. Bagley and R. B. Altman, *Protein Sci.*, 1995, **4**, 622.
92. J. M. Castagnetto, S. W. Hennessy, V. A. Roberts, E. D. Getzoff, J. A. Tainer and M. E. Pique, *Nucleic Acids Res.*, 2002, **30**, 379.
93. L. R. Stefan, R. Zhang, A. G. Levitan, D. K. Hendrix, S. E. Brenner and S. R. Holbrook, *Nucleic Acids Res.*, 2006, **34**, D131.
94. P. S. Klosterman, M. Tamura, S. R. Holbrook and S. E. Brenner, *Nucleic Acids Res.*, 2002, **30**, 392.
95. S. R. Das, R. Fong and J. A. Piccirilli, *Curr. Opin. Chem. Biol.*, 2005, **9**, 585.
96. E. L. Christian and M. Yarus, *Biochemistry*, 1993, **32**, 4475.
97. M. R. Hargittai and K. Musier-Forsyth, *RNA*, 2000, **6**, 1672.
98. M. S. Kayne and M. Cohn, *Biochemistry*, 1974, **13**, 4159.
99. D. E. Draper, *Biophys. Chem.*, 1985, **21**, 91.
100. E. Nieboer, *Struct. Bonding*, 1975, **22**, 1.
101. R. K. Sigel, A. Vaidya and A. M. Pyle, *Nat. Struct. Biol.*, 2000, **7**, 1111.
102. N. M. Kaye, N. H. Zahler, E. L. Christian and M. E. Harris, *J. Mol. Biol.*, 2002, **324**, 429.
103. D. A. Harris, R. A. Tinsley and N. G. Walter, *J. Mol. Biol.*, 2004, **341**, 389.
104. N. G. Walter, N. Yang and J. M. Burke, *J. Mol. Biol.*, 2000, **298**, 539.
105. W. D. Horrocks, G. F. Schmidt, D. R. Sudnick, C. Kittrell and R. A. Bernheim, *J. Am. Chem. Soc.*, 1977, **99**, 2378.
106. F. Yuan, L. Griffin, L. Phelps, V. Buschmann, K. Weston and N. L. Greenbaum, *Nucleic Acids Res.*, 2007, **35**, 2833.
107. A. L. Feig, M. Panek, W. D. Horrocks Jr. and O. C. Uhlenbeck, *Chem. Biol.*, 1999, **6**, 801.
108. N. L. Greenbaum, C. Mundoma and D. R. Peterman, *Biochemistry*, 2001, **40**, 1124.
109. R. A. Tinsley and N. G. Walter, *Biol. Chem.*, 2007, **388**, 705.
110. A. G. Cassano, V. E. Anderson and M. E. Harris, *Biopolymers*, 2004, **73**, 110.
111. H. Taube, *Annu. Rev. Nucl. Sci.*, 1956, **6**, 277.
112. H. Taube, *J. Phys. Chem.*, 1954, **58**, 523.
113. A. G. Cassano, V. E. Anderson and M. E. Harris, *Biochemistry*, 2004, **43**, 10547.

114. Q. Dai, J. K. Frederiksen, V. E. Anderson, M. E. Harris and J. A. Piccirilli, *J. Org. Chem.*, 2008, **73**, 309.
115. J. A. Piccirilli, J. S. Vyle, M. H. Caruthers and T. R. Cech, *Nature*, 1993, **361**, 85.
116. A. S. Sjogren, E. Pettersson, B. M. Sjoberg and R. Stromberg, *Nucleic Acids Res.*, 1997, **25**, 648.
117. L. B. Weinstein, B. C. Jones, R. Cosstick and T. R. Cech, *Nature*, 1997, **388**, 805.
118. S. O. Shan and D. Herschlag, *Biochemistry*, 1999, **38**, 10958.
119. S. Shan, A. Yoshida, S. Sun, J. A. Piccirilli and D. Herschlag, *Proc. Natl. Acad. Sci. USA*, 1999, **96**, 12299.
120. A. Yoshida, S. Sun and J. A. Piccirilli, *Nat. Struct. Biol.*, 1999, **6**, 318.
121. S. Shan, A. V. Kravchuk, J. A. Piccirilli and D. Herschlag, *Biochemistry*, 2001, **40**, 5161.
122. A. A. Szewczak, A. B. Kosek, J. A. Piccirilli and S. A. Strobel, *Biochemistry*, 2002, **41**, 2516.
123. J. L. Hougland, A. V. Kravchuk, D. Herschlag and J. A. Piccirilli, *PLOS Biol.*, 2005, **3**, 1536.
124. M. Forconi, J. Lee, J. K. Lee, J. A. Piccirilli and D. Herschlag, *Biochemistry*, 2008, **47**, 6883.
125. P. M. Gordon and J. A. Piccirilli, *Nat. Struct. Biol.*, 2001, **8**, 893.
126. E. J. Sontheimer, P. M. Gordon and J. A. Piccirilli, *Genes Dev.*, 1999, **13**, 1729.
127. P. M. Gordon, E. J. Sontheimer and J. A. Piccirilli, *Biochemistry*, 2000, **39**, 12939.
128. J. M. Warnecke, J. P. Furste, W. D. Hardt, V. A. Erdmann and R. K. Hartmann, *Proc. Natl. Acad. Sci. USA*, 1996, **93**, 8924.
129. Y. Chen, X. Li and P. Gegenheimer, *Biochemistry*, 1997, **36**, 2425.
130. J. M. Warnecke, R. Held, S. Busch and R. K. Hartmann, *J. Mol. Biol.*, 1999, **290**, 433.
131. E. L. Christian, N. M. Kaye and M. E. Harris, *RNA*, 2000, **6**, 511.
132. T. Pfeiffer, A. Tekos, J. M. Warnecke, D. Drainas, D. R. Engelke, B. Seraphin and R. K. Hartmann, *J. Mol. Biol.*, 2000, **298**, 559.
133. E. L. Christian, N. M. Kaye and M. E. Harris, *EMBO J.*, 2002, **21**, 2253.
134. E. J. Sontheimer, S. Sun and J. A. Piccirilli, *Nature*, 1997, **388**, 801.
135. P. M. Gordon, E. J. Sontheimer and J. A. Piccirilli, *RNA*, 2000, **6**, 199.
136. S. L. Yean, G. Wuenschell, J. Termini and R. J. Lin, *Nature*, 2000, **408**, 881.
137. R. Cosstick and J. S. Vyle, *Nucleic Acids Res.*, 1990, **18**, 829.
138. M. J. Moore and P. A. Sharp, *Science*, 1992, **256**, 992.
139. A. Loria and T. Pan, *RNA*, 1996, **2**, 551.
140. T. Pan, *Biochemistry*, 1995, **34**, 902.
141. T. Pan and M. Jakacka, *EMBO J*, 1996, **15**, 2249.
142. X. W. Fang, T. Pan and T. R. Sosnick, *Nat. Struct. Biol.*, 1999, **6**, 1091.
143. D. Smith, A. B. Burgin, E. S. Haas and N. R. Pace, *J. Biol. Chem.*, 1992, **267**, 2429.

144. K. J. Gardiner, T. L. Marsh and N. R. Pace, *J. Biol. Chem.*, 1985, **260**, 5415.
145. E. Kikovska, N. E. Mikkelsen and L. A. Kirsebom, *Nucleic Acids Res.*, 2005, **33**, 6920.
146. S. Cuzic and R. K. Hartmann, *Nucleic Acids Res.*, 2005, **33**, 2464.
147. M. Brannvall and L. A. Kirsebom, *Proc. Natl. Acad. Sci. USA*, 2001, **98**, 12943.
148. D. Smith and N. R. Pace, *Biochemistry*, 1993, **32**, 5273.
149. B. C. Thomas, J. Chamberlain, D. R. Engelke and P. Gegenheimer, *RNA*, 2000, **6**, 554.
150. J. M. Warnecke, E. J. Sontheimer, J. A. Piccirilli and R. K. Hartmann, *Nucleic Acids Res.*, 2000, **28**, 720.
151. J. L. Chen and N. R. Pace, *RNA*, 1997, **3**, 557.
152. N. M. Kaye, E. L. Christian and M. E. Harris, *Biochemistry*, 2002, **41**, 4533.
153. A. V. Kazantsev and N. R. Pace, *RNA*, 1998, **4**, 937.
154. M. E. Harris and N. R. Pace, *RNA*, 1995, **1**, 210.
155. S. M. Crary, J. C. Kurz and C. A. Fierke, *RNA*, 2002, **8**, 933.
156. E. L. Christian, K. M. Smith, N. Perera and M. E. Harris, *RNA*, 2006, **12**, 1463.
157. F. Michel, A. Jacquier and B. Dujon, *Biochimie.*, 1982, **64**, 867.
158. A. C. Arnberg, G. Van der Horst and H. F. Tabak, *Cell*, 1986, **44**, 235.
159. R. Van der Veen, A. C. Arnberg, G. van der Horst, L. Bonen, H. F. Tabak and L. A. Grivell, *Cell*, 1986, **44**, 225.
160. M. J. Curcio and M. Belfort, *Cell*, 1996, **84**, 9.
161. T. H. Eickbush, *Curr. Biol.*, 1999, **9**, R11.
162. Y. Aizawa, Q. Xiang, A. M. Lambowitz and A. M. Pyle, *Mol. Cell*, 2003, **11**, 795.
163. K. Chin and A. M. Pyle, *RNA*, 1995, **1**, 391.
164. D. L. Daniels, W. J. Michels Jr. and A. M. Pyle, *J. Mol. Biol.*, 1996, **256**, 31.
165. V. T. Chu, Q. Liu, M. Podar, P. S. Perlman and A. M. Pyle, *RNA*, 1998, **4**, 1186.
166. M. Podar, P. S. Perlman and R. A. Padgett, *Mol. Cell. Biol.*, 1995, **15**, 4466.
167. C. L. Peebles, P. S. Perlman, K. L. Mecklenburg, M. L. Petrillo, J. H. Tabor, K. A. Jarrell and H. L. Cheng, *Cell*, 1986, **44**, 213.
168. C. Schmelzer and R. J. Schweyen, *Cell*, 1986, **46**, 557.
169. C. L. Peebles, E. J. Benatan, K. A. Jarrell and P. S. Perlman, *Cold Spring Harbor Symp. Quant. Biol.*, 1987, **52**, 223.
170. K. A. Jarrell, C. L. Peebles, R. C. Dietrich, S. L. Romiti and P. S. Perlman, *J. Biol. Chem.*, 1988, **263**, 3432.
171. U. Kuck, I. Godehardt and U. Schmidt, *Nucleic Acids Res.*, 1990, **18**, 2691.
172. U. Schmidt, B. Riederer, M. Morl, C. Schmelzer and U. Stahl, *EMBO J*, 1990, **9**, 2289.

173. E. Deme, A. Nolte and A. Jajcquier, *Biochemistry*, 1999, **38**, 3157.
174. A. Bar-Shalom and M. J. Moore, *Biochemistry*, 2000, **39**, 10207.
175. R. A. Padgett, M. Podar, S. C. Boulanger and P. S. Perlman, *Science*, 1994, **266**, 1685.
176. C. L. Dantzman and L. L. Kiessling, *J. Am. Chem. Soc.*, 1996, **118**, 11715.
177. R. R. Breaker, G. M. Emilsson, D. Lazarev, S. Nakamura, I. J. Puskarz, A. Roth and N. Sudarsan, *RNA*, 2003, **9**, 949.
178. G. K. Schroeder, C. Lad, P. Wyman, N. H. Williams and R. Wolfenden, *Proc. Natl. Acad. Sci. USA*, 2006, **103**, 4052.
179. L. T. Chow, R. E. Gelinas, T. R. Broker and R. J. Roberts, *Cell*, 1977, **12**, 1.
180. S. M. Berget, C. Moore and P. A. Sharp, *Proc. Natl. Acad. Sci. USA*, 1977, **74**, 3171.
181. J. P. Staley and C. Guthrie, *Cell*, 1998, **92**, 315.
182. T. W. Nilsen, *Bioessays*, 2003, **25**, 1147.
183. S. Valadkhan and J. L. Manley, *Nature*, 2001, **413**, 701.
184. S. L. Yean, G. Wuenschell, J. Termini and R. J. Lin, *Nature*, 2000, **408**, 881.
185. J. B. Murray, A. A. Seyhan, N. G. Walter, J. M. Burke and W. G. Scott, *Chem. & Biol.*, 1998, **5**, 587.
186. A. T. Perrotta and M. D. Been, *Biochemistry*, 2006, **45**, 11357.
187. Y. I. Kuzmin, C. P. Da Costa, J. W. Cottrell and M. J. Fedor, *J. Mol. Biol.*, 2005, **349**, 989.
188. Y. I. Kuzmin, C. P. Da Costa and M. J. Fedor, *J. Mol. Biol.*, 2004, **340**, 233.
189. L. L. Lebruska, I. I. Kuzmine and M. J. Fedor, *Chem. Biol.*, 2002, **9**, 465.
190. A. R. Ferre-D' Amare, *Biopolymers*, 2004, **73**, 71.
191. A. C. McLeod and D. M. Lilley, *Biochemistry*, 2004, **43**, 1118.
192. T. J. Wilson, A. C. McLeod and D. M. Lilley, *EMBO J*, 2007, **26**, 2489.
193. J. Han and J. M. Burke, *Biochemistry*, 2005, **44**, 7864.
194. M. Roychowdhury-Saha and D. H. Burke, *RNA*, 2006, **12**, 1846.
195. J. A. Nelson and O. C. Uhlenbeck, *Mol. Cell.*, 2006, **23**, 447.
196. S. Wang, K. Karbstein, A. Peracchi, L. Beigelman and D. Herschlag, *Biochemistry*, 1999, **38**, 14363.
197. Y. Tanaka, M. Tagaya, T. Hori, T. Sakamoto, Y. Kurihara, M. Katahira and S. Uesugi, *Genes Cells*, 2002, **7**, 567.
198. M. Matysiak, J. Wrzesinski and J. Ciesiolka, *J. Mol. Biol.*, 1999, **291**, 283.
199. D. J. Earnshaw and M. J. Gait, *Nucleic Acids Res.*, 1998, **26**, 5551.
200. M. Azubel, S. G. Wolf, J. Sperling and R. Sperling, *Mol. Cell*, 2004, **15**, 833.
201. T. Pan, D. T. Long, O. C. Uhlenbeck, in *The RNA World*, 1st edn., ed. R. F. Gesteland and J. F. Atkins, Cold Spring Harbor Laboratory Press, Cold Spring Harbor, NY, 1993, pp. 271–302.
202. R. Shannon, *Acta. Crystallogr., Sect. A*, 1976, **32**, 751.
203. C. S. G. Phillips and R. J. P. Williams, *Inorganic Chemistry*, Oxford University Press, New York, 1965.

204. J. A. Cowan, in *The Biological Chemistry of Magnesium*, J. A. Cowan ed., VCH, New York, 1995, pp. 1–23.
205. D. Smith, in *The Biological Chemistry of Magnesium*, J. A. Cowan ed., VCH, New York, 1995, pp. 111–135.
206. D. T. Richens, *The Chemistry of Aqua Ions: Synthesis, Structure and Reactivity: a Tour Through the Periodic Table of the Elements*, Wiley, Chichester, 1997.
207. A. L. Feig, O. C. Uhlenbeck, in *The RNA World*, 2nd edn., ed. R. F. Gesteland, T. R. Cech, and J. F. Atkins, Cold Spring Harbor Laboratory Press, Cold Spring Harbor, NY, 1999, pp. 287–319.
208. J. E. Huheey, *Inorganic Chemistry: Principles of Structure and Reactivity*, HarperCollins College Publishers, New York, 1993.

CHAPTER 9
Polyanion Models of Nucleic Acid–Metal Ion Interactions

J. MICHAEL SCHURR

Department of Chemistry, University of Washington, Seattle, WA 98195-1700, USA

9.1 Introduction

9.1.1 Background and Objectives

DNAs and RNAs exhibit certain striking properties that arise primarily from their large electric charges. This chapter concerns theoretical concepts, models and results that guide our thinking about such properties and in favorable cases provide quantitatively useful predictions.

The theory of fully ionized electrolytes commenced with the derivation of the non-linear Poisson–Boltzmann (NLPB) equation for the mean electrostatic potential surrounding an ion in solution.[1] The theory of rod-like polyions began somewhat later with mathematical solutions and analyses of the NLPB equation for uniformly charged cylinders in the absence of added salt.[2–5] Subsequent progress was limited by a lack of analytical solutions of the NLPB equation under more general conditions, specifically in the presence of added salt, and for more realistic shapes and charge distributions. Extensive numerical solutions of the NLPB were often required in order to advance. A much simpler quasi-chemical theory of counterion condensation (CC) for linear polyions,[6,7] and its extensions to predict various polyion properties of DNA yielded valuable new insights, imparted considerable impetus to the field at the time, and continues to impact the field up to the present. Significant parallel development of thermodynamic expressions to connect the results of NLPB and CC theories with the measured properties of solutions also took place at the same time. In addition, numerous comparative studies of CC and NLPB theories with regard to counterion condensation phenomena were undertaken, partly in an effort to establish a more fundamental basis for CC theory. The early work on polyion theory was reviewed in books by Rice and Nagasawa[8] and Oosawa[9] and in a review article by Katchalsky.[10] The work on relevant thermodynamic relations was reviewed in a book by Eisenberg[11] and in several review articles.[12–15] Critical comparisons of NLPB and CC theories with each other and with the results of Monte Carlo (MC) simulations were also reviewed.[12–16] A critical review of selected experimental and theoretical studies of polyions up to about 1992 is provided in a book by Schmitz.[17]

Progress since 1985 has come in part from the availability of sufficient computer power to test the validity of the NLPB equation itself via MC simulations and to numerically solve the NLPB equation for polyions with realistic shapes and charge distributions in a timely manner. Roughly, developments in this area since 1985 can be divided into six main categories: (i) numerical evaluations of mean electrostatic potentials, small ion distributions and electrostatic free energies of various DNAs, RNAs and their complexes with other ionic species by NLPB, MC and other methods; (ii) extensions of CC theory in various ways; (iii) derivations and applications of thermodynamic expressions to connect the results of NLPB and CC theories to measured properties; (iv) development and application of MC and other methods to account for small ion–small ion correlations due to Coulombic and hard-core repulsions in the ion atmosphere near the polyion surface; (v) studies of interactions between DNA duplexes, not only under conditions where the ion atmosphere is diffuse, NLPB theory applies and repulsions prevail, but also under conditions where the ion atmosphere is more highly localized and structured, NLPB theory fails and attractive interactions between DNAs prevail, leading to compact arrays; and (vi) studies of dynamic and electrokinetic properties of DNA. With regard to category (v), theoretical illumination of the factors responsible for the observed attractions between DNA molecules has occurred only in the last decade.

The primary goal here is to provide an introductory review of the foundations and selected theoretical advances pertaining to equilibrium properties of polyions. Space limitations preclude discussion of important recent work on dynamic and electrokinetic properties of DNA.[18–24] The emphasis will be on the nature of the approximations in the models themselves and on certain results and their meaning, with little discussion of the analytical or numerical methods used to obtain them. A lengthy introduction is presented with the aim of defining the basic terminology and describing both the relevant phenomena and our current understanding in a completely descriptive manner, before addressing the theoretical foundations and physical models more formally. The intent of this review is to enable a non-expert to engage the voluminous and rapidly expanding literature concerning the effects of electrostatics on DNA and RNA behavior.

9.1.2 Polyanions and Their Compensating Charges

Duplex DNA is the exemplar of a highly charged semi-flexible polyanion. Over the pH range from about 4.0 to 10.5, its intrinsic (electronically bound) charge consists entirely of single electronic charges on each of the two phosphate groups of every base pair. In the absence of compensating positive charges, the free energy cost of forcing so much negative charge into such a small space would be prohibitive. Hence, in any of its solution or condensed phases, a duplex DNA is generally accompanied by sufficient excess positively charged ions (counterions) in its vicinity to neutralize completely its phosphate charges. This principle is general and applies to any double- or single-stranded DNA or RNA.

Neutralizing metal cations may be bound to specific sites on a DNA or RNA molecule, as discussed in other chapters, or instead may be associated non-specifically with the polyanion as part of its diffuse *ion atmosphere*. The ion atmosphere of a DNA or RNA refers both to the region wherein its neutralizing counterions are localized and to the small ions of either charge that reside therein, regardless of whether they are confined to specific sites or distributed throughout. Specific binding differs from diffuse binding only by its smaller confinement volume, its larger occupation probability per unit volume and in some cases a significant electronic contribution to its stability. Binding to specific sites is usually somewhat selective for counterions of a particular size and charge, whereas non-specific association via the ion atmosphere is to some extent independent of counterion size and exhibits a strong preference for higher valence. Despite its largely non-specific nature, the ion atmosphere modulates many important physical properties.

9.1.3 The Ion Atmosphere

Consider a dilute DNA or RNA immersed in a salt solution. Within the ion atmosphere, the mean concentrations of small ions vary gradually from their values in the bathing salt solution toward their respective values at the polyion surface. Determining the mean small ion concentrations and their mutual correlations everywhere within the ion atmosphere of a highly charged polyion is a thorny and venerable problem at the core of polyelectrolyte theory. So far, there has emerged no simple and accurate solution to this problem that is applicable to all circumstances. However, numerous experiments and both MC and molecular dynamics (MD) simulations and new analytical approaches have provided considerable insight, as well as quantitatively useful results in many cases.

With respect to a bathing salt solution, the ion atmosphere of DNA has everywhere both an excess of positive counterions and a deficit of negative ions (co-ions). The mean charge density at any point in the ion atmosphere is just the difference between the mean concentrations of the counterion and co-ion charges. In the case of a monovalent bathing salt, such as KCl, the concentration of neutral salt at any point in the ion atmosphere can be regarded as that of the co-ions (Cl$^-$), so the net density of positive charges is simply the protonic charge times the *excess* concentration of counterions (K$^+$) above that of the neutral salt *at that same point*. The integral of this excess counterion charge density over the ion atmosphere yields a total positive charge that compensates the intrinsic charge on the DNA. Although the total concentration of small ions at any point in the ion atmosphere generally exceeds that of the bathing salt solution, the neutral salt is significantly *excluded* from the ion atmosphere of DNA or any other polyion. This phenomenon is akin to the Donnan effect in a macroscopic system.

When the counterion valence is +1 or +2, the ion atmosphere is entirely or in part diffuse, extending out some distance much greater than a small ion diameter from the DNA surface. Correlations between the positions of +1 counterions in the diffuse part of the ion atmosphere normally have negligible net consequences.[25] However, when the counterion valence is +3 or greater, then the ion atmosphere is pulled in much closer to the polyanion surface and strong correlations between the counterion positions due to both Coulombic and hard-core repulsions cause damped oscillations of the counterion density along a normal to the surface, as well as substantial corrugations in the counterion–counterion pair correlation function parallel to the surface, even when the charge of the polyanion is uniformly distributed over that same surface. Because the intrinsic charges on DNAs and RNAs have a valence of -1, some *local* overcharging (beyond neutralization) is inevitably encountered with counterions of +2 or higher valence. In this case, the local *charge per unit length* of a dilute rod-like polyanion plus any given configuration of its ion atmosphere undulates between positive and negative values as a function of distance along the polyanion. This charge undulation is not truly periodic, but instead dephases over some moderately short range. It does not appear in the mean charge density, which is an average over different configurations with different phases of the undulations, but instead is manifested by corrugations in the counterion–counterion pair correlation function parallel to the polyion surface. The rather different structure and internal correlations of the ion atmosphere in the case of +3- and higher valent counterions in comparison with those in the case of +1-valent counterions apparently determine whether two neighboring DNA molecules experience a mutual repulsion or instead a mutual attraction.

9.1.4 Equilibrium Polyion Properties and Phenomena

Any polyion plus its ion atmosphere has a substantial *electrostatic free energy* relative to that of the corresponding neutral polymer of the same structure without an ion atmosphere. For a given bathing salt solution, this electrostatic free energy naturally depends on the number and positions of the intrinsic polyion charges, the internal dielectric constant and shape of the polyion. Generally, it will vary with changes in structure or environment of the polyion, such as occur during (i) a B to Z transition of duplex DNA, (ii) a melting transition or (iii) the relative approach of two duplex DNAs toward one another. It will also vary with titration events that introduce additional negative or positive intrinsic charges. This electrostatic free energy also generally depends on the salt concentration(s) and dielectric constant of the bathing salt solution, and also the counterion valence(s). Thus, by varying ionic conditions in the bathing salt solution, one may alter the electrostatic free energies and also the difference in electrostatic free energy between two different DNA structures and thereby shift their mutual equilibrium. For example, the electrostatic free energies of both duplexes and single strands rise with decreasing salt concentration, but that of the duplexes rises much faster, so eventually the duplexes become unstable even at room temperature. In fact, on lowering the bathing salt concentration from 0.2 to 10^{-5} M, the melting temperature of an unstrained native DNA falls from about 86 to 9 °C.[26] The electrostatic free energy also contributes to the equilibrium constant for site-specific counterion binding, such as the binding of an Mg^{2+} ion to multiple phosphate groups of a DNA or RNA. An essential point is that binding constants for charged ligands of any kind, including regulatory proteins, are strongly modulated by varying ionic conditions in the bathing salt solution.

DNA and other highly charged linear polyanions in solution exhibit the phenomenon of *counterion condensation*.[6,7] In many (but not all) experiments, the behavior of a DNA is quantitatively practically indistinguishable from that of a molecule with the same shape and flexibility, but with a much smaller intrinsic charge. Such experiments include measurements of colligative properties, such as the osmotic pressure and Donnan effect and also assessments of DNA–DNA self-repulsions, as manifested in the bending rigidity and thermodynamic second virial coefficient. Experimentally, it appears that *weakly charged* polyions with sufficient distance between their intrinsic charges release *all* of their neutralizing counterions to participate in the Donnan equilibrium and other colligative properties. It also appears that their self-repulsions arise from simple screened Coulomb interactions between *all* of their intrinsic charges. In contrast, dilute duplex DNAs of sufficient length in 10^{-5}–10^{-3} M monovalent salt appear to release only about 25% of their neutralizing counterions to participate in those same properties. By inference, an unreleased fraction of the neutralizing counterions must remain associated in some way with the DNA, unable to contribute significantly to its colligative properties, and is referred to as the *condensed* fraction. Moreover, this condensed fraction remains surprisingly constant over a wide range of monovalent salt concentration from $\sim 10^{-5}$ up to ~ 0.01 M, before *declining* appreciably at higher salt concentration.[27]

For any long linear polyion in sufficiently dilute salt, the release of its counterions appears to be governed primarily by its reduced linear charge density, $\xi = L_B/(|z_C|b)$, where b is the mean axial spacing between intrinsic charges, z_C is the counterion valence and the Bjerrum length is $L_B = e^2/(\varepsilon kT)$, where e is the protonic charge, ε is the dielectric constant, k is Boltzmann's constant and T is absolute temperature. According to CC theory, which applies only under low salt conditions, when $\xi < 1.0$ all counterions that are not specifically bound to the polyion are released, but when $\xi > 1.0$, a fraction (f) of the counterions remains condensed, so that the *effective* reduced linear charge density of the polyion *plus its condensed ions* is 1.0.[6,7] Hence the condensed fraction of the counterions is just $f = 1 - \xi^{-1}$. For DNA, $\xi = 4.2$, so a substantial fraction, $f = 0.76$, of its univalent counterions remains condensed at all salt concentrations up until ~ 0.001 M, before f begins to decline.[27]

A compelling demonstration of the onset of counterion condensation was provided by a surprisingly abrupt decrease in electrophoretic mobility of the linear polycation, 6,6-ionene, when ξ was increased from 0.82 to 1.85 through the region where $\xi = 1.0$.[28] When $\xi > 1.0$, the condensed fraction of the counterions increased even more rapidly with increasing ξ than would be expected for a rigid polyion. This discrepancy could be attributed quantitatively to a coupling between counterion condensation, which decreases the bending rigidity and bending fluctuations, which enhance counterion condensation, of that rather flexible species.[29]

Unlike their neutral counterparts, all polyions exert a *Donnan effect*, whenever they are confined (possibly by a semi-permeable membrane) to a region that is in equilibrium with a much larger volume of a bathing salt solution.[30] The released counterions accompanying the polyions act to suppress the entry of neutral salt from the bathing solution into the polyion compartment. This effect is due entirely to the dual requirements for (i) equality of the chemical potentials of both counterions and co-ions on either side of the semi-permeable membrane and (ii) electro-neutrality of the polyion-containing region. It is important to distinguish between the number of counterions that are released by the polyion, but remain in the polyion compartment to preserve electro-neutrality, and the somewhat smaller number that is transferred as part of the salt from the polyion-containing domain to the bathing salt solution.

Despite the effect of counterion condensation to limit the fraction of released counterions, the *osmotic pressure* exerted by a solution of highly charged polyions relative to that of its bathing salt solution (from which it might be separated by a semi-permeable membrane) is still typically enormous, up to 100-fold or more greater than that exerted by the corresponding neutral polymer.[31] When the polyions are sufficiently dilute, this large osmotic pressure is considered to arise from the total concentration of 'free' small ions at those places in the polyion solution where the electric field vanishes.

Electrostatic self-repulsions are expected to contribute to the bending rigidity of DNA or any other semi-flexible polyion, in sufficiently low concentrations of monovalent bathing salt, but are enormously overestimated unless some account is taken of counterion condensation.[32]

Experiments indicate that, in the case of +1-valent counterions, the net interactions between parallel DNA duplexes are repulsive at all center-to-center separations down to ~ 27 Å.[33] However, in the case of +3- and higher valent counterions, *net attractions* between parallel DNA duplexes prevail.[34] These net attractions are believed to be due to (i) the small thickness of the ion atmosphere containing trivalent ions, (ii) local overcharging and (iii) anti-correlation of the instantaneous undulations in local charge of one DNA plus its ion atmosphere with those of the neighboring DNA plus its ion atmosphere. The case of +2-valent counterions is intermediate between these two extremes. Such attractions between DNA molecules are manifested experimentally by aggregation to form compact arrays, which have been observed for all +3- and higher valent counterions so far examined.[35] Attractions have also been observed for certain divalent counterions, including Mn^{2+} and Cu^{2+}, but not for certain others, including Mg^{2+}, Ca^{2+} and putrescine^{2+}, unless the dielectric constant is decreased somewhat from that of water.

9.2 Statistical Mechanics, Configuration Integral and Potential of Mean Force

The chemical potential of a molecule of type j in its standard state of 1 molecule cm^{-3} in a solvent s is given by

$$\mu_j^o = -kT \ln[Q(1_j, N_s, V, T)/Q(N_s, V, T)] \tag{9.1}$$

where $Q(1_j, N_s, V, T)$ is the canonical partition function for one j-molecule plus N_s solvent molecules in a volume, $V = 1.0 \text{ cm}^3$, at temperature T and $Q(N_s, V, T)$ is the corresponding partition function for the N_s solvent molecules alone.[36]

Classical statistical mechanics is assumed to apply in the following sense. Every molecule or ion is assumed to consist of completely rigid subunits, which could be either single atoms or groups of two or more atoms, which interact with other groups in the same or neighboring molecules only via rather soft potentials, so that their rigid-body motions are all classical. For such a classical system, the linear and angular momenta of the rigid subunits always exhibit independent Maxwell–Boltzmann distributions. Consequently, the kinetic energy contribution to the total partition function for any given collection of molecules can always be expressed entirely in terms of the masses and moments of inertia of their constituent rigid subunits and is always the same, independent of the potential energy function. This kinetic energy contribution, which also contains a factor of $1/h$ for every classical spatial or angular coordinate, cancels completely out of all *differences* in thermodynamic properties between any two states of a given system at the same T.[36]

The kinetic energy contribution of the solvent cancels out of the μ_j^o in eqn (9.1), but that of the j-molecule does not. However, in any process, wherein the numbers of each kind of rigid subunit are conserved, such as a non-covalent association, $A + B \rightleftharpoons C$, between two solutes, A and B, to form a solute complex, C, which is held together only by soft van der Waals, H-bond or electrostatic interactions, the rigid subunits of C are precisely the same as those of A and B. Hence their kinetic energy contributions will cancel out of the difference in standard-state chemical potentials between product and reactants, $\Delta G^o = \mu_C^o - \mu_A^o - \mu_B^o$, and also out of the equilibrium constant. This implies that kinetic energies make no contribution to the equilibrium constant, except when covalent forces contribute significantly to the binding.[36]

The classical partition function also contains a factor, $(N_i!)^{-1}$, where N_i is the number of molecules of type i, for every kind of molecule or ion in the system. These factors cancel out of the difference in free energy between any two states of the system, whenever the numbers of each kind of molecule or ion remain unchanged. Likewise, the $(N_s!)^{-1}$ factors and corresponding factors for any other solvent components, cancel out of the standard-state chemical potentials. In classical statistical mechanics, it is customary to omit the kinetic energy contributions and the $(N_i!)^{-1}$ factors from the partition functions in anticipation of their eventual cancellation out of any difference in calculated thermodynamic properties at the same T. One retains, then, only the potential energy contribution to the partition function, namely the configuration integral:

$$Z = \int d\mathbf{R} \exp[-U(\mathbf{R})/kT] \tag{9.2}$$

where $\mathbf{R}$ is an ordered list of all coordinates that position and orient every rigid subunit of every molecule in the system, $U(\mathbf{R})$ is the potential energy function of all those same coordinates and $d\mathbf{R}$ is the product of appropriate volume elements of all those same coordinates. The difference in Helmholtz free energy between any two different states (1 and 2) of the same collection of molecules (with different potential energy functions or total volumes) at the same T is $\Delta A = -kT \ln(Z_2/Z_1)$, from which differences in all other thermodynamic quantities can be obtained by taking appropriate derivatives.

Sometimes one is interested in the *effective* potential energy function for a particular subset of the system coordinates. For example, what is the effective potential energy of a fixed array of ions embedded in a solvent? In this case, one can regard $\mathbf{R}$ as comprising two sub-lists, $\mathbf{R} = (\mathbf{X}, \mathbf{Y})$, where $\mathbf{X}$ contains the coordinates of the fixed ions and $\mathbf{Y}$ contains those of the solvent. The *effective* potential energy function of the fixed coordinates is defined by

$$V(\mathbf{X}) \equiv -kT \ln\left\{\int d\mathbf{Y} \exp[-U(\mathbf{X}, \mathbf{Y})/kT]\right\} \tag{9.3}$$

where the $\mathbf{Y}$ coordinates have been integrated out. $V(\mathbf{X})$ can be regarded as the configurational part of the Helmholtz free energy of the solvent–fixed-ion system. The multi-dimensional probability density of $\mathbf{X}$ is, $P(\mathbf{X}) = \exp[-V(\mathbf{X})/kT]/Z$ and the total configuration integral can be written as $Z = \int d\mathbf{X} \exp[-V(\mathbf{X})/kT]$, which is the same as that in eqn (9.2). In the standard primitive models of electrolyte solutions, $V(\mathbf{X})$ is *assumed* to be the sum of the self-energies and mutual interaction energies of the ions with configuration $\mathbf{X}$ *in a dielectric medium*. The solvent is assumed to provide a dielectric constant, which may vary with position near solution boundaries, but otherwise is not explicitly considered. Henceforth in this chapter, solvent coordinates will be omitted from $\mathbf{R}$ and $U(\mathbf{R})$ will be understood to mean an *effective* potential energy function of the ion coordinates, whose configuration is denoted by $\mathbf{R}$.

A coordinate frame is fixed within the polyion and the positions of small ions of each kind, α, in the polyion frame are denoted by $\mathbf{r}_j^\alpha$, $j=1,\ldots N_\alpha$, where $\mathbf{r}_j^\alpha \equiv (x_j^\alpha, y_j^\alpha, z_j^\alpha)$. The ratio of mean concentrations ($\langle c_\alpha(\mathbf{r})\rangle$) of ions of type α at positions $\mathbf{r}$ and $\mathbf{r}_0$ *in the polyion frame* is found to be

$$\langle c_\alpha(\mathbf{r})\rangle / \langle c_\alpha(\mathbf{r}_o)\rangle = \exp[-W_\alpha(\mathbf{r},\mathbf{r}_o)/kT] \tag{9.4}$$

where

$$\begin{aligned}W_\alpha(\mathbf{r},\mathbf{r}_o) &\equiv V_\alpha(\mathbf{r}) - V_\alpha(\mathbf{r}_o)\\ &= -kT\ln\left\{\int d\mathbf{R}'\exp[-U(\mathbf{R}',\mathbf{r})/kT]\Big/\int d\mathbf{R}''\exp[-U(\mathbf{R}'',\mathbf{r}_o)/kT]\right\}\end{aligned} \tag{9.5}$$

is a *potential of mean force* due to all other ions acting on a single ion of type α at $\mathbf{r}$. In the present case of a primitive model, where solvent is neglected, $\mathbf{R}'$ denotes the coordinates of all ions except one of type α at $\mathbf{r}$, $\mathbf{R}''$ denotes the coordinates of all ions except one of type α at $\mathbf{r}_o$ and $U(\mathbf{R}',\mathbf{r})$ and $U(\mathbf{R}'',\mathbf{r}_o)$ denote the total electrostatic energy of all ions in the dielectric medium, including the one at either $\mathbf{r}$ or $\mathbf{r}_o$. Typically, $\mathbf{r}_o$ is chosen to lie in some region, generally far from the polyion, where $\langle c_\alpha(\mathbf{r}_o)\rangle$ has a known value, c_α^o. Then, $W_\alpha(\mathbf{r},\mathbf{r}_o)$ is the change in free energy when an ion of type α is moved from $\mathbf{r}_o$ to $\mathbf{r}$.

9.3 Ion Atmosphere Theory

9.3.1 The Standard Primitive Model for a Polyanion and Its Small Ion Atmosphere

The polyanion (A^{z_P-}) possesses $z_P \gg 1$ intrinsic negative charges that are fixed in a rigid array either within or upon its surface. The dielectric constant, $\varepsilon(\mathbf{r}) \equiv \varepsilon(x,y,z)$, inside the polyion is typically much lower than that of its surrounding bulk solution and in principle may vary with position.

The small ion hard-core radius, a, is sometimes ignored when treating small ion–polyion or small ion–small ion electrical interactions. However, a realistic radius must be assumed in evaluating the self-energy of an ion interacting with the dielectric medium in which it resides, which is commonly called the Born energy. In an infinite medium of uniform dielectric constant, ε, the self-energy of an ion is given by $(ez_\alpha)^2/(2\varepsilon a)$, where z_α is the ion valence and e is the protonic charge. However, for an ion in water near the surface of a low-dielectric medium, the self-energy cannot be specified so simply and generally depends on the distance of the small ion from the surface, the shape of the surface and the values of the dielectric constant on either side. Most small ions withdraw from an interface between water and a medium of much lower dielectric constant.

Nevertheless, variations in self-energies of the small ions with position in the ion atmosphere are customarily ignored in both NLPB and MC studies.

Effects of the dielectric discontinuity between the interior of the polyion and the solution have been investigated in some detail.[37-40] For a particular grooved model of DNA with dielectric constant 2 inside and 78 outside, the *effective* dielectric constant for Coulombic interactions between small ions in the innermost part of the ion atmosphere of DNA ranged from ~70 for same-side interactions to ~100 for opposite-side interactions.[40] After estimating certain compensating energies, it was cautiously concluded that the net effect of the dielectric discontinuity on the small ion distribution does not appear to be significant.

The value of the dielectric constant of water within ~20 Å of an interface with any other material, including one with a much lower dielectric constant, is to some extent uncertain due to local structure of the water near that interface. Also, saturation of the polarization of water molecules adjacent to a fixed charge diminishes the effective local dielectric constant somewhat. These effects have been assessed,[40-46] and are sometimes treated approximately by incorporating a suitable position-dependent dielectric constant in the neighborhood of the polyion.[47]

Typically, self-energies of the fixed intrinsic charges of the polyion are taken into account, when reckoning its electrostatic free energy and any changes therein due to either a change in conformation or a change in environment on binding to some other species.[48-58]

Effects arising from small ion–small ion correlations due to Coulombic and hard-core repulsions in the inner part of the ion atmosphere are automatically taken into account by MC simulations,[13,59-63] and also by a recently proposed tightly bound ion (TBI) model.[40]

The small ions are treated differently depending on the theoretical method employed and the situation of interest. Whenever MC methods are used, the small ions are treated as *discrete* entities and their different configurations, **R**, around the polyion are randomly sampled and collected in an appropriately weighted manner so as to preserve the expected ratio, $\exp[-(U(\mathbf{R}_2)-U(\mathbf{R}_1))/kT]$, of populations of any two configurations, $\mathbf{R}_1$ and $\mathbf{R}_2$. Various average properties are then computed directly from the collected configurations. Also, in some treatments of closely neighboring DNAs, when the counterions are expected to be localized close to the polyion surface, the counterions are also regarded as discrete and simply assigned fixed positions on the polyion surface.[64-68]

In NLPB treatments of the diffuse ion atmosphere, the concentrations of counterions and co-ions are typically regarded as mean values, $\langle c_\alpha(\mathbf{r})\rangle$, that are continuous smooth functions of position. The local chemical potential of a given kind of small ion with charge ez_α is assumed to be given by

$$\mu_\alpha(\mathbf{r}) = \mu_\alpha^o + kT \ln(\langle c_\alpha(\mathbf{r})\rangle) + V_\alpha(\mathbf{r}) \qquad (9.6)$$

where $V_\alpha(\mathbf{r})$ is the *effective* potential energy experienced by an ion of type α at position **r** due to all *other* charges in the system, including the polyion. μ_α^o is the chemical potential of that ion in its (hypothetical) standard state, where each ion has a concentration of 1 ion cm^{-3}, but experiences only the environment of an infinitely dilute solution (*i.e.* only the local dielectric constant) with no other ions. At equilibrium, $\mu_\alpha(\mathbf{r}) = \mu_\alpha$ has the same value in every region to which the small ions have access, so eqn (9.6) is equivalent to eqn (9.4).

9.3.2 The Non-linear Poisson–Boltzmann Equation

Consider a single rigid polyion of arbitrary shape with a fixed array of intrinsic electronic charges immersed in an enormously larger volume containing small ions in a dielectric medium. The small ions are excluded from the volume of the polyion and the counterions are present in sufficient excess to insure electroneutrality of the overall system. The electrostatic potential at position $\mathbf{r}$ in the polyion frame obeys Poisson's equation,[69] which may be averaged over all configurations of the small ions to yield $\nabla \cdot \varepsilon(\mathbf{r})\nabla\langle\psi(\mathbf{r})\rangle = -4\pi\langle\rho(\mathbf{r})\rangle$, where $\langle\psi(\mathbf{r})\rangle$ and $\langle\rho(\mathbf{r})\rangle$ are the mean electrostatic potential and charge density, respectively.

The mean charge density in the ion atmosphere is, $\langle\rho(\mathbf{r})\rangle = e\sum_\alpha z_\alpha\langle c_\alpha(\mathbf{r})\rangle$. Like the potential of mean force, $W_\alpha(\mathbf{r}, \mathbf{r}_o)$, the electrostatic potential $\langle\psi(\mathbf{r})\rangle$ is also taken to vanish at any point, $\mathbf{r}_o$, sufficiently far from the polyion, that $\langle c_\alpha(\mathbf{r}_o)\rangle = c_\alpha^o$ for all species. The main approximation of NLPB theory is the replacement of $W_\alpha(\mathbf{r}, \mathbf{r}_o)$ by $ez_\alpha\langle\psi(\mathbf{r})\rangle$ in eqn (9.4) to obtain

$$\langle c_\alpha(\mathbf{r})\rangle = c_\alpha^o \exp[-ez_\alpha\langle\psi(\mathbf{r})\rangle/kT] \quad (9.7)$$

Combining eqn (9.7) with the averaged Poisson equation yields the NLPB equation (in cgs electrostatic units):

$$\nabla \cdot \varepsilon(\mathbf{r})\nabla\phi(\mathbf{r}) = -(4\pi e^2/kT)\sum_\alpha z_\alpha c_\alpha^o \exp[-z_\alpha\phi(\mathbf{r})] - (4\pi e/kT)\rho^f(\mathbf{r}) \quad (9.8)$$

where $\phi(\mathbf{r}) = e\langle\psi(\mathbf{r})\rangle/kT$ is the reduced mean electrostatic potential and $\rho^f(\mathbf{r})$ is the density of fixed charge in or on the polyion. For convenience, the angular brackets will henceforth be dropped from $\langle\psi(\mathbf{r})\rangle$, $\langle c_\alpha(\mathbf{r})\rangle$ and $\langle\rho(\mathbf{r})\rangle$.

The validity of eqns (9.7) and (9.8) hinges on the accuracy of replacing $W_\alpha(\mathbf{r}, \mathbf{r}_o)$ by $ez_\alpha\psi(\mathbf{r})$. Even if hard-core forces were negligible and $W_\alpha(\mathbf{r}, \mathbf{r}_o)$ were determined entirely by electrostatic interactions, it would differ in meaning from $ez_\alpha\psi(\mathbf{r})$ in the following way. $W_\alpha(\mathbf{r}, \mathbf{r}_o)$ is the mean electrostatic potential energy at $\mathbf{r}$ (relative to that at $\mathbf{r}_o$) due to all other ions, *when an ion of type α is present at $\mathbf{r}$*, whereas $ez_\alpha\psi(\mathbf{r})$ is the mean electrostatic potential energy at $\mathbf{r}$ (relative to that at $\mathbf{r}_o$) due to all ions, whether or not an ion of type α is present at $\mathbf{r}$. Fixed charges on the polyion should make identical contributions to $W_\alpha(\mathbf{r}, \mathbf{r}_o)$ and $ez_\alpha\psi(\mathbf{r})$. However, excess counterions near the polyion in principle affect these potential energies differently. The effect of a counterion (C) at $\mathbf{r}$ is to oppose close approaches of neighboring counterions,[63] and thereby diminish their repulsive contribution to the mean electrostatic potential energy at $\mathbf{r}$. Hence, for any given value of $c_C(\mathbf{r})$, the inter-counterion repulsions contribute less to $W_C(\mathbf{r}, \mathbf{r}_o)$ than to $ez_C\psi(\mathbf{r})$, so that $W_C(\mathbf{r}, \mathbf{r}_o)$ is overall more negative than $ez_C\psi(\mathbf{r})$ and $c_C(\mathbf{r}) > [c_C(\mathbf{r})]_{NLPB}$, where the subscript indicates the counterion density of NLPB theory. In fact, MC simulations of a polyion with monovalent salt confirm that $[c_C(\mathbf{r})]_{NLPB}$ modestly underestimates $c_C(\mathbf{r})$ in regions very near the polyion surface.[13,14,60–63,70]

The use of eqn (9.7) in the charge density of Poisson's equation eliminates any manifestation of the discrete nature of the counterions. Such discreteness gives rise to damped oscillations of $c_C(\mathbf{r})$ normal to the polyanion surface and for multivalent ions leads to local overcharging and large corrugations of the counterion–counterion pair correlation function along that surface. Therefore, eqn (9.7) precludes such behavior, which is believed to mediate attractions between DNA molecules. Equation (9.7) becomes a much better approximation further from the polyion, where the counterions are more dilute and interspersed with more co-ions. This, too, was confirmed by the MC simulations.

Modifications of eqn (9.8) to incorporate some effects of small ion–small ion correlations[71–75] and the small ion occupied volume[76] have been proposed. Such modifications provide significant improvement over the NLPB eqn (9.8), but in their present state of development appear to be less accurate than either MC simulations or TBI theory.[40,77,78]

9.3.3 Solutions of the NLPB Equation

In the absence of added salt, exact *analytical solutions* of the NLPB equation are available for an infinite uniformly charged cylinder that is contained within a co-axial cylindrical cell of radius $\mathbf{R}_b$, where $-\nabla\phi(\mathbf{R}_b) = 0$ at the outer cell boundary.[2,3] A corresponding solution for the case of two parallel uniformly charged lines is also available.[79]

Nominally analytical solutions for infinite uniformly charged cylinders in infinite salt reservoirs have been developed within the applied mathematics community.[80,81] Numerical evaluations using these more complex solutions may prove more difficult than direct numerical solution of the NLPB equation.

Asymptotic solutions for infinite uniformly charged cylinders in both high and, especially, low concentrations of symmetrical salt have been reported.[82,83] In both the low to moderate-salt and high-salt regimes, relatively simple expressions were obtained for the electrostatic potential at the polyion surface and the polyion–salt preferential interaction coefficient per monomer unit, $\Gamma_{uS}(P)$, which is defined below, in terms of the basic physical parameters.[83]

For infinite uniformly charged cylinders in the presence of salt, a variety of *numerical integration* schemes have been reported, for example the fourth-order Runge–Kutta algorithm used by Stigter.[84,85] The availability of symbol manipulation programs, such as Mathematica, to perform the necessary algebra has made expansion methods, such as that of Sugai and Nitta,[86] a less taxing endeavor. Of course, any solution of the NLPB equation must evolve into a solution of the corresponding linear Poisson–Boltzmann (LPB) equation at a sufficient distance from the cylinder that $z_\alpha \phi(\mathbf{r}) \ll 1.0$.

For more intricate shapes, a variety of finite difference and finite element grid methods have been used.[15,47,51,87–89] Regardless of the method used, the constraint of electroneutrality of the polyion plus its ion atmosphere is always imposed. In addition, a second condition is also required. For a single polyion in an infinite salt solution, one typically takes $\phi(\infty) = 0 = \psi(\infty)$ at any position sufficiently far from the polyion. However, for a cell model, wherein fixed numbers of counterions and co-ions accompany a single polyion in every cell, one typically takes $-\nabla \phi(\mathbf{R}_b) = 0$ at the outer cell boundary, which eliminates electrostatic interactions between cells. Mechanical equilibrium is preserved by the equality of the small ion pressures of adjacent cells at their mutual boundaries.

When treating cell models in either the absence or presence of added salt, one initially lacks a reference concentration at a place where one can set $\phi = 0$. Protocols for dealing with this circumstance in both the absence[87] and presence[71–73] of added salt have been described. In the presence of a mean concentration, $\bar{c}_s$, of added salt, the condition $(1/V)\int dV c_S(\mathbf{r}) = \bar{c}_s$, is imposed as a requirement for an acceptable solution. In the presence of a constant chemical potential of salt, one imposes instead the Donnan constraint, $c_C(\mathbf{R}_b)c_S(\mathbf{R}_b) = (c_S^o)^2$ where c_S^o is the salt concentration in the dialysis reservoir.

9.4 Connection of NLPB Theory to Thermodynamic Properties

9.4.1 Electrostatic Free Energy

The electrostatic free energy, ΔG^{el}, is the difference in free energy between a fully charged polyion immersed in an infinite salt solution and a uncharged molecule of the same shape and dielectric constant immersed in the same salt solution. It is given by two equivalent integrals over the entire volume containing the polyion and its ion atmosphere[87]

$$\Delta G^{el} = \int dV [\rho^f(\mathbf{r})\psi(\mathbf{r}) - \Delta\Pi(\mathbf{r}) - \mathbf{E}(\mathbf{r}) \cdot \mathbf{D}(\mathbf{r})/2]$$
$$= \int dV [-\rho^m(\mathbf{r})\psi(\mathbf{r}) - \Delta\Pi(\mathbf{r}) + \mathbf{E}(\mathbf{r}) \cdot \mathbf{D}(\mathbf{r})/2]$$
(9.9)

where $\rho^f(\mathbf{r})$ is the density of fixed charges, including both intrinsic charges and specifically bound or positioned small ions, $\rho^m(\mathbf{r}) = e\Sigma_\alpha z_\alpha c_\alpha(\mathbf{r})$ is the distribution of small ion charge at $\mathbf{r}$ in the solution, $\Pi(\mathbf{r}) = kT\Sigma_\alpha(c_\alpha(\mathbf{r}) - c_\alpha^o)$ is the excess pressure of all the small ions at $\mathbf{r}$ above that of the bulk salt solution, $\mathbf{E}(\mathbf{r}) \equiv -\nabla\psi(\mathbf{r})$ is the electric field and $\mathbf{D}(\mathbf{r}) \equiv \varepsilon(\mathbf{r})\mathbf{E}(\mathbf{r})$ is the electric displacement vector. The first line of eqn (9.9) was derived using a theorem from the calculus of variations to determine precisely what quantity is minimized by the solution of the NLPB equation. That quantity was identified as ΔG^{el}.[87] The second line in eqn (9.9) follows from the first by use of the electrostatic identity $\int dV[\rho(\mathbf{r})\psi(\mathbf{r})/2] = \int dV[\mathbf{E}(\mathbf{r}) \cdot \mathbf{D}(\mathbf{r})/2]$, together with $\rho(\mathbf{r}) \equiv \rho^f(\mathbf{r}) + \rho^m(\mathbf{r})$.

Equation (9.9) can also be derived by simpler, but less elegant, arguments for a constant V, T system containing only ions or for a constant P, T system containing ions plus an incompressible solvent (unpublished calculations).

ΔG^{el} in eqn (9.9) is readily evaluated, after $\psi(\mathbf{r})$, $\mathbf{E}(\mathbf{r}) = -\nabla\psi(\mathbf{r})$ and the $c_\alpha(\mathbf{r})$ have been obtained by solving the NLPB equation.

Equations (9.8) and (9.9) were applied to reckon ΔG^{el} and the polyion–salt preferential interaction coefficients for uniformly charged spheres, cylinders and plates, all-atom models of B-DNA,[87,91] different RNAs and complexes of such species with various ionic ligands.[48–58]

9.4.2 Polyanion Activity Coefficient and Polyanion–Salt Preferential Interaction Coefficient

For simplicity, only symmetrical $z{:}z$ salts are considered in this section. The salt $S = XY$ is assumed to dissociate completely to its constituent z-valent ions: $S \to X^{z+} + Y^{z-}$. Throughout this section, it is assumed that activity coefficients of the small ions due to small ion–small ion interactions are 1.0. For bulk salt solution in the absence of polyions the small ion activities are then $a_{X^{z+}} = c_{X^{z+}} = c_S \equiv c_{Y^-} = a_{Y^{z-}}$ and the chemical potential of the salt is

$$\mu_S = \mu^0_{X^{z+}} + \mu^0_{Y^{z-}} + kT \ln(c_S^2) \qquad (9.10)$$

A neutral polyanion component is denoted by $P = AX_N$, where $A \equiv A_P^{zP-}$ is a polyanion with z_P intrinsic negative charges, $X \equiv X^{z+}$ is its z-valent counterion and $N = z_P/z$ is the (positive) number of counterions required to neutralize the polyanion charge. The polyanion component dissociates completely: $P \to A^{zP-} + NX^{z+}$, so that $c_A = c_P$, where c_P is the concentration of the neutral polyanion component. In a solution containing salt and polyanions, the co-ion is $Y \equiv Y^{z-}$, its bulk concentration is $c_Y = c_S$ and the bulk counterion concentration is $c_X = Nc_P + c_S$. The chemical potential of the dilute neutral polyanion component, $P = AX_N$, is given by

$$\mu_P = \mu_P^0 + \mu_P^{id} + \mu_P^{el} = \mu_{A^0}^0 + N\mu_{X^{z+}}^0 + kT \ln(\gamma_P c_P c_X^N) \qquad (9.11)$$

where $\mu_P^{id} = kT \ln(c_P c_X^N)$ is the ideal contribution of the polyanion and counterions, as if all were uncharged, γ_P is the *polyanion activity coefficient*, μ_P^{el} is the total electrostatic free energy of a polyanion, which includes interactions between polyanions, and $\mu_P^0 = \mu_{A^0}^0 + N\mu_{X^{z+}}^0$ is the standard state chemical potential of an *uncharged* polyanion, A^0, plus that of its N charged counterions in pure water.

When the polyanions are sufficiently dilute that $Nc_P \ll c_S$, then interactions between polyions can be ignored. In this limit, μ_P^{el} becomes independent of polyanion concentration, so $\mu_P^{el} = \Delta G_P^{el}$ is now the electrostatic free energy of the charged polyanion, A^{zP-}, relative to that of its uncharged counterpart, A^0.

Under such conditions, it follows from eqn (9.11) that

$$kT \ln \gamma_P = \Delta G_P^{el} \tag{9.12}$$

For such a dilute polyanion component, eqn (9.11) can be rewritten as

$$\mu_P = \mu_P^{\square} + kT \ln c_P \tag{9.13}$$

where

$$\mu_P^{\square} \equiv \mu_{A^0}^0 + N\mu_{Xz+}^0 + NkT \ln c_S + \Delta G_P^{el} \tag{9.14}$$

is that part of the chemical potential of the polyanion component that is independent of polyanion concentration. $\mu_P^{\square}$ can be regarded as the standard state chemical potential of the neutral polyanion component, $P = AX_N$, in the presence of the prevailing salt concentration, c_S. Equations (9.13) and (9.14) are used in the next section to treat the effects of salt on ligand binding equilibria.

The relevant polyion–salt preferential interaction coefficient (PIC) is defined by

$$\begin{aligned}\Gamma_S(P) &\equiv -\left(\partial \mu_P^{\square}/\partial \mu_S\right)_{T,P,c_P^{\infty}} \\ &= -[1/(2kT)]\left[\partial(NkT \ln c_S + \Delta G_P^{el})/\partial \ln c_S\right]_{T,P,c_P^{\infty}}\end{aligned} \tag{9.15}$$

where c_P^{∞} denotes dilute $c_P \to 0$. It was shown from eqn (9.9) that[87]

$$\left(\partial \Delta G_P^{el}/\partial \ln c_S\right)_{T,P,c_P^{\infty}} = -kT \int dV \left\{ \sum_{\alpha = X,Y} [c_\alpha(\mathbf{r}) - c_\alpha] \right\} \tag{9.16}$$

Incorporating the electroneutrality condition in the form $\int dV[c_X(\mathbf{r}) - c_Y(\mathbf{r})] = N$ gives

$$\left(\partial \Delta G_P^{el}/\partial \ln c_S\right)_{T,P,c_P^{\infty}} = -kT\left\{N + 2\int dV[c_Y(\mathbf{r}) - c_Y]\right\} \tag{9.17}$$

and also

$$\Gamma_S(P) = \int dV[c_Y(\mathbf{r}) - c_Y] \tag{9.18}$$

which shows that $\Gamma_S(P)$ is the excess accumulation of co-ions, Y, in the vicinity of the polyanion above that in an equivalent volume of the bathing salt solution. Throughout the ion atmosphere, $c_Y(\mathbf{r}) < c_Y = c_S$, so this excess is negative, corresponding to a deficit of co-ions (and salt). Combining eqns (9.12), (9.17) and (9.18) yields

$$kT(\partial \ln \gamma_P/\partial \ln c_S)_{T,P,c_P^{\infty}} = -kT[N + 2\Gamma_S(P)] \tag{9.19}$$

which was presented previously in a form appropriate for a long polyion with translational symmetry.[12,14,90]

Beginning with the definition in eqn (9.15), an alternative expression for $\Gamma_S(P)$ was derived:

$$\Gamma_S(P) = \lim_{m_P \to 0} (\partial m_S / \partial m_P)_{T, \mu_W, \mu_S} \qquad (9.20)$$

where m_α denotes the molality of species α and the subscript W denotes the solvent water.[92] In the derivation, it was assumed for simplicity that the partial molar volumes of the water, macromolecule and osmolyte (salt in this case) were constant over the range of osmolyte concentrations considered. Equation (9.20) applies to both ionic and non-ionic species. By using Kirkwood–Buff theory, it was shown that eqn (9.20) holds exactly for non-ionic species at any value of m_P,[93] which suggests that it might also be exact for ionic species. In any case, eqn (9.20) is regarded here as a practically exact identity.

Consider a dialysis chamber containing a neutral polyanion component, $P = AX_N$, in equilibrium with an enormous reservoir containing salt at concentration c_S. The two chambers are separated by a membrane that is permeable to water and small ions, but not to polyanions. According to eqn (9.20), $\Gamma_S(P)$ is the change in the number of moles of salt in the dialysis chamber per mole of added polyion. For sufficiently small c_P or m_P, it is entirely expected that this number is precisely the number of moles of excess salt in the ion atmosphere of a single polyion, as indicated in eqn (9.18). For polyions, $\Gamma_S(P)$ is typically negative and is sometimes called the Donnan coefficient. In principle, $\Gamma_S(P)$ can be accessed by dialysis experiments and application of eqn (9.20). In practice, it is easier to measure a somewhat different PIC,[94] from which that in eqns (9.15) and (9.20) is obtained via relations developed for that purpose.[11,95,96]

Two length scales are important in polyelectrolyte theory. The Bjerrum length, $L_B = e^2/\varepsilon k T$, was introduced earlier in connection with CC theory. The Debye length of a salt solution is defined by $1/\lambda$, where $\lambda^2 = 4\pi L_B \sum_\alpha z_\alpha^2 c_\alpha$ and the sum runs over each small ion species present. $1/\lambda$ is the length scale for exponential screening of interactions between charges in ionic solutions.

When $1/\lambda$ is significantly smaller than the distance between intrinsic charges on the polyion, those charges no longer interact significantly with one another and can be regarded as practically isolated univalent ions, around which the mean electrostatic potential obeys the linear Poisson–Boltzmann equation. In this case, the compensating charge in the ion atmosphere has equal contributions from the counterion excess and co-ion deficit, so the integral on the right-hand side of eqn (9.18) gives $-N/2$. In this limit, for polyions of any shape, $(\partial \Delta G_P^{el} / \partial \ln c_S)_{T, P, c_P^\infty} = 0$ and $\Gamma_S(P) = -N/2$. For nucleic acids, which exhibit rather close intrinsic charges, this limit is not attainable except at very high salt concentrations, where the assumptions of NLPB theory are no longer valid.

For a sphere in the limit of sufficiently low salt concentration, $\Gamma_S(P) = -N/2$,[90] which means that all of its ions have been released and half a salt molecule is excluded for each of those. The corresponding result for an infinitely long-cylinder with $\xi \geq 1.0$ is $\Gamma_S(P) = -N/(4\xi)$,[83] which means that not all of its ions have been released.

9.4.3 Osmotic Pressure and Osmotic Coefficient

Consider again the dialysis chamber containing a neutral polyanion component, $P \equiv AX_N$, in equilibrium with an enormous reservoir of salt solution at concentration c_S^0. The osmotic pressure of the polyion solution is that of the small ions in the electric field-free regions at the cell boundaries, $\pi = kT[c_X(\mathbf{R}_b) + c_Y(\mathbf{R}_b)]$. As noted above, $c_X(\mathbf{R}_b)c_Y(\mathbf{R}_b) = c_S^{0^2}$. The excess osmotic pressure of the polyanion solution above that of the salt reservoir is

$$\Delta\pi = kT[c_X(\mathbf{R}_b) + c_Y(\mathbf{R}_b) - 2c_S^0] \tag{9.21}$$

and the osmotic coefficient is the ratio, $\phi_P \equiv \Delta\pi/\pi_P^{id}$, where $\pi^{id} = (N+1)c_P$ is the ideal osmotic pressure expected for an uncharged polymer plus N dissociated neutral ligands, all of which remain in the polyion solution. Both $\Delta\pi$ and ϕ_P are readily evaluated from solutions of the NLPB equation.

DNA solutions were treated by combining (i) the cell-model NLPB result for a uniformly charged cylinder of appropriate radius and linear charge density, $\xi = 4.2$, in zero-salt with (ii) the Donnan condition at the cell boundary.[97] ϕ_P was reckoned for various DNA concentrations in both 2 and 10 mM monovalent salt. The results agreed fairly well with the experimental data for DNA phosphate concentrations exceeding ~ 0.01 M, but significantly underestimated the experimental values at lower DNA concentrations.[31] This discrepancy is due to significant invasion of the polyion chamber by neutral salt and consequent failure of the zero-salt approximation under those conditions.

9.5 Polyion Equilibria and Repulsive Interactions

9.5.1 Effect of Salt Concentration on the Association of a Polycation with a Polyanion

Consider the bimolecular association of a neutral polyanion component, $P \equiv AX_N$, with a neutral polycation component, $L \equiv BY_M$, to form a neutral complex component, $Q \equiv ABX_NY_M$. That is,

$$AX_N + BY_M \rightleftharpoons ABX_NY_M \tag{9.22}$$

X and Y are again constituents of a z-valent symmetrical salt, $S \equiv XY$. Also, $A \equiv A^{z_P-}$ where $z_P = Nz$, $B \equiv B^{z_L+}$ with $z_L = Mz$ and $AB = AB^{z_Q-}$, where $z_Q = (N-M)z > 0$. According to eqn (9.13), in any given salt concentration, $c_S = c_X = c_Y$, the equilibrium constant for the overall reaction (22) is given by

$K = c_{AB}/c_A c_B = \exp\{-[\mu_Q^\square - (\mu_P^\square + \mu_L^\square)]/kT\}$, where $\mu_P^\square$ is given by eqn (9.14) and a corresponding equation applies for $\mu_L^\square$. For the complex component, $\mu_Q^\square = \mu_{A^0 B^0}^0 + N\mu_{X^{z+}}^0 + M\mu_{Y^{z-}}^0 + (N+M)kT \ln c_S + \Delta G_Q^{\text{el}}$, where $\mu_{A^0 B^0}^0$ is the standard state chemical potential of the completely uncharged complex. Combining these relations yields

$$K = \exp\left\{-\left[\mu_{A^0 B^0}^0 - (\mu_{A^0}^0 + \mu_{B^0}^0) + \Delta G_Q^{\text{el}} - (\Delta G_P^{\text{el}} + \Delta G_L^{\text{el}})\right]/kT\right\} \quad (9.23)$$

Calculated values of ΔG_Q^{el}, ΔG_P^{el} and ΔG_L^{el} for various choices of c_S can be inserted into eqn (9.23) to calculate the relative variation in K with salt over any range of c_S for which NLPB theory applies.

By using eqns (9.17) and (9.18), the slope of K with respect to $\ln c_S$ is found to be

$$D(\ln K) \equiv (\partial \ln K/\partial \ln c_S)_{T,P,c_P^\infty} = -2M + 2\Delta\Gamma_S \quad (9.24)$$

where $\Delta\Gamma_S \equiv \Gamma_S(Q) - [\Gamma_S(P) + \Gamma_S(L)]$ is the difference in PICs between the product (complex) and reactants (polyanion and polycation).[14,91-98] Equation (9.24) also follows from eqns (9.15) and (9.20), after accounting for the M salt molecules that are formed in the reaction, but diffuse out of the polyion-containing domain, hence it is valid, even when NLPB theory is not. Equations (9.24) and (9.18) allow $(\partial \ln K/\partial \ln c_S)_{T,P,c_P^\infty}$ to be calculated from solutions of the NLPB equation for each component at any particular salt concentration.

According to eqn (9.24), the variation of the equilibrium constant or relative stability of the product and reactants, with increasing salt concentration is governed entirely by (i) the number of molecules of neutral salt, M, added to the solution, when AX_N and BY_M are placed in the same chamber and fully dissociated and (ii) the difference, $-\Delta\Gamma_S$, between the number of salt molecules excluded from the ion atmosphere of the product and the corresponding number excluded from the ion atmospheres of the reactants. Salt formation and salt exclusion are simply the different ways by which salt can be transferred from each neutral polyion component to the bulk salt solution. Whichever side of the reaction transfers the most salt to the solution is least favored by an increase in salt concentration. This principle is general and can be used to generalize eqn (9.24) to either simpler or more complex reactions. It should be carefully noted that $\Gamma_S(Q)$, $\Gamma_S(P)$ and $\Gamma_S(L)$ generally vary rather strongly with salt concentration, whereas $\Delta\Gamma_S$ is often fairly constant over a broad range of c_S. This is discussed further in the section on CC theory.

The symbol $D(F)$ will be used to indicate the slope, $(\partial F/\partial \ln c_S)_{T,P,c_P^\infty}$, for various choices of the quantity F throughout the remainder of this chapter.

The slope, $D(\ln K)$, in eqn (9.24) was evaluated for three different models of DNA binding reactions.[91] In the first example, P was a cylinder of finite length and 10 Å radius with intrinsic charges located every 1.7 Å along its axis (to provide the same linear charge density as B-DNA), L was a 10 Å sphere with a charge, $M = +2$ and Q was the same cylinder as P, but with two missing (neutralized) charges at its center. For rods longer than 100 Å, it was found that

$D(\ln K) = -1.96$, which is very nearly just $-M$. The various contributions to the binding free energy, enthalpy and entropy were also calculated and compared with results of other theories. In the second and third examples, all-atom models from reported crystal structures of the complexes were used for Q. The structures of P and L were taken to be those of the DNA and ligand, respectively, in the complex. The particular ligands studied were the minor groove-binding antibiotic (and dye), DAPI and the lambda repressor DNA-binding domain (λbd). Charge assignments were provided in the paper. The agreement between calculated and experimental values of $D(\ln K)$ was very good for both the DNA–DAPI and DNA–λbd complexes. Similar calculations were performed for the minor groove binding antibiotics, DAPI, Hoechst 33258 and netropsin.[48] In these calculations, the changes in DNA structure from B-form in solution to those in the crystal structures of the different complexes were taken into account. Agreement of the computed $D(\ln K)$ values with experiment was again very good for DAPI and Hoechst 33258, but in the case of netropsin the theory overestimated the experimental values by ~ 1.3–1.4-fold. This general NLPB protocol was also applied to investigate the binding of the λcI repressor, *Eco*RI endonuclease and selected mutants of those proteins to their respective cognate sequences.[49] All-atom models were again employed and rather good agreement between the predicted and measured slopes, $D(\ln K)$, was obtained in each case.

9.5.2 Effect of Salt Concentration on the Equilibrium Constant for a Conformational Transition

Equilibrium constants for conformational transitions can be treated in an analogous manner. Consider the process $AX_N \rightleftharpoons CX_N$, where A and C are simply different conformations of the same polyanion. Both $P \equiv AX_N$ and $Q \equiv CX_N$ dissociate completely to their constituent ions. In this case, the equilibrium constant is

$$K = \exp\left[-\left(\mu^0_{C^0} - \mu^0_{A^0} + \Delta G^{el}_Q - \Delta G^{el}_P\right)/kT\right] \quad (9.25)$$

and its slope with respect to the salt concentration is

$$D(\ln K) = 2\Delta\Gamma_S \quad (9.26)$$

where $\Delta\Gamma_S = \Gamma_S(Q) - \Gamma_S(P)$.

Extension of the preceding NLPB theory to treat ligand-binding and conformational transitions in non-symmetrical salts and mixed salts is straightforward.[90,91]

The general NLPB protocol was applied to the B $\rightleftharpoons$ Z transition of duplex $[d(CG)_n]_2$, for $n = 6$ (12 bp) to 18 (36 bp).[99] All-atom models were used for both the B and Z forms and the respective electrostatic free energies, ΔG^{el}_B and ΔG^{el}_Z and the slope $D(\Delta G^{el}_Z - \Delta G^{el}_B)$, were reckoned. It was found that $\Delta G^{el}_Z < \Delta G^{el}_B$

over the entire range of monovalent salt concentration from 0.01 to ~ 5.0 M. From the known greater stability of B-DNA in low and moderate salt concentrations, it was inferred that $\mu_{B^0}^0 < \mu_{Z^0}^0$, so neutral B is more stable than neutral Z. It was also found that ΔG_Z^{el} descended more rapidly than ΔG_B^{el} with increasing c_S, thereby inducing the transition at the empirically known ~ 2.5 M. The finding that the electrostatic contribution favors Z-DNA under moderate and high salt conditions contradicted numerous prior findings and also one subsequent finding,[100] all of which found that $\Delta G_Z^{el} > \Delta G_B^{el}$. These alternative estimates were all obtained for less realistic models. Not surprisingly, all of the published theories yielded negative slopes, $D(\Delta G_Z^{el} - \Delta G_B^{el})$. The slope predicted by the Misra–Honig model[99] is only about half of the experimentally measured value, whereas that predicted by the less realistic partially porous cylinder model of Gueron et al.[100] is fairly close to the experimental value. At the high monovalent salt concentrations (1.3–4.0 M) over which the transition occurs, the NLPB theory employed by both groups is well outside its range of validity, although the errors may be comparable for both B and Z and consequently nearly cancel out of $\Delta G_Z^{el} > \Delta G_B^{el}$. It was also shown that preferential accumulation of multivalent counterions in the ion atmosphere in the presence of low to moderate concentrations of monovalent salt, together with a modest decrease in the free energy, $\mu_{Z^0}^0$, of the neutral Z-form upon methylation of the cytosines, enables millimolar concentrations of Mg^{2+} and micromolar concentrations of $[Co(NH_3)_6]^{3+}$ to induce the $B \rightarrow Z$ transition in duplex $[(dm^5C - G)_n]_2$ without invoking site-specific ion binding.[100]

9.5.3 Temperature Dependence of a Conformational Transition

At the midpoint of a conformational transition, $K = 1$ and $\Delta G_0 \equiv \mu_C^0 - \mu_A^0 + \Delta G_Q^{el} - \Delta G_P^{el} = 0$. Applying this constraint, $d\Delta G^0 = 0$, to the variations in ΔG^0 with T and $\ln c_S$ yields

$$(\partial T_M / \partial \ln c_S)_{K=1} = -2(\Delta \Gamma_S / N) R T_M^{0\,2} / \Delta h^0 \qquad (9.27)$$

where T_M is the midpoint transition temperature, R is the gas constant and Δh^0 is the observed molar enthalpy of transition per intrinsic charge. When $\Delta \Gamma_S$ is calculated for uniformly charged cylinder models using reasonable estimates of both ξ and the cylinder radius for duplex and single-strand species, the calculated $(\partial T_M / \partial \ln c_S)_{K=1}$ for melting of poly rA·poly rU agrees fairly well with the experimental values.[101]

An alternative approach, often used in the early days, was simply to substitute for each species its limiting constant value, $\Gamma_S(P)/N = -1/(4\xi_P)$, at $c_S \rightarrow 0$ into eqn (9.27). However, $\Gamma_S(P)$ is known to increase substantially with increasing c_S, especially for $c_S \geq 0.001$ M, so the assumed individual $\Gamma_S(P)$ and $\Gamma_S(Q)$ are blatantly incorrect. Yet the predicted $\Delta \Gamma_S$ is often surprisingly close to the experimental value, which is commonly independent of c_S, at all salt concentrations up to ~ 1.0 M. The origins of the 'miraculous' cancellation of the salt dependences of $\Gamma_S(P)$ and $\Gamma_S(Q)$ are still not clearly understood.[83]

9.5.4 Proton Binding to an Acidic Ligand Adjacent to a Polyion

The change in ln K for the binding of a proton to an acidic ligand, when the ligand goes from its free state in solution (L) to a state where it is bound to a DNA or RNA ($DNA-L$), is readily estimated using the same general NLPB protocol. The electrostatic free energy difference, $\Delta G^{el}_{bind}(L) \equiv \Delta G^{el}_{LH^+} - (\Delta G^{el}_L + \Delta G^{el}_{H^+})$, applies for binding the proton to the free ligand and a corresponding quantity, $\Delta G^{el}_{bind}(DNA - L) \equiv \Delta G^{el}_{DNA-LH^+} - (\Delta G^{el}_{DNA-L} + \Delta G^{el}_{H^+})$, applies for binding the proton to the DNA-ligand complex. The latter free energy difference minus the former, $\Delta\Delta G^{el} \equiv \Delta G^{el}_{bind}(DNA - L) - \Delta G^{el}_{bind}(L)$, is just the difference in free energy of proton binding to the ligand between its bound and free states, provided that only electrostatic forces alter the proton affinity. The electrostatic free energy of the proton drops out, so calculations of ΔG^{el}_j are required for only four species: $j = L$, LH^+, $DNA-L$ and $DNA-LH^+$. Finally, $\Delta \ln K \equiv \ln K(DNA-L) - \ln K(L) = -\Delta\Delta G^{el}/kT$. Calculations of the corresponding slopes, $D(\Delta G^{el}_j)$, allow one to calculate the variation of $\Delta \ln K$ with $\ln c_S$. Because $\Delta\Delta G^{el}$ is also the free energy difference to bind the protonated ligand to DNA minus that to bind the unprotonated ligand to DNA, the same calculation also yields the ratio of equilibrium constants for binding LH^+ and L to DNA.[50] These protocols were applied to the intercalator, DAPP, using the reported all-atom crystal structure of the complex.[50] It was assumed that the structure of the DNA-DAPP complex did not change upon protonation. The theoretical results were in excellent agreement with the experiments.[102] The pH dependence of the binding free energy for λcI repressor associating with its operator DNA was reckoned by an extension of these methods and again rather good agreement with experiment was attained.[103]

For a small ligand with a +1 charge near a highly charged polyanion, the ratio of the free ligand concentration $[c_{H^+}(\mathbf{r})]$ at any point in the ion atmosphere to that in bulk solution ($c^0_{H^+}$) should be identical with that of any other univalent counterion: $c_{H^+}(\mathbf{r}) = c^0_{H^+} \exp[-\phi(\mathbf{r})]$. As a first approximation, an otherwise unperturbed neutral proton-binding site at $\mathbf{r}$ senses the local $c_{H^+}(\mathbf{r})$, so $f_{LH^+}/[f_L c_{H^+}(\mathbf{r})] = K_0$, where f_{LH^+} and f_L are the fractions of ligands (attached to the polyanions) with and without bound protons, respectively, and K_0 is the unperturbed binding constant. The effective binding constant (viewed from the bulk) is expressed in terms of $c^0_{H^+}$ as, $K_{eff} = f_{LH^+}/(f_L c^0_{H^+}) = K_0 \exp[-\phi(\mathbf{r})]$. Finally, the relative change in effective proton *dissociation* constant, $(1/K_{eff})/(1/K_0)$, can be expressed simply as, $\Delta pK_a = -\log_{10}(K_0/K_{eff}) = -\phi(\mathbf{r})/2.303$. For a polyanion with a large negative $\phi(\mathbf{r}) = |e|\psi(\mathbf{r})/kT$, the ΔpK_a can be substantial. For the DAPP example cited above, $\Delta pK_a \simeq 3.0$ in 30 mM monovalent salt.[50,102] An advantage of this simpler, but more approximate, approach is that the NLPB eqn must be solved for only one species, either $DNA-L$ or $DNA-LH^+$, whichever has the neutral ligand, in order to determine $c_{H^+}(\mathbf{r})$. This approach was taken to estimate ΔpK_a

for five different amino acids, each bearing a single carboxyl group and a single protonated amino group, which were tethered to the (non-protonated) exocyclic 2-amino group of guanine in the minor groove.[51] Agreement between the computed and experimental pK_a values was fairly good, especially considering the uncertainties in the actual positions of the ionizable groups. So far, good estimates of the errors associated with the simple local pH approach in comparison with the full NLPB protocol are lacking.

The pH dependence of a hairpin $\rightleftharpoons$ triplex transition[104] was investigated by first determining the ΔpK_a of each of the four cytosines in the sequence on going from free cytosine in solution to cytosine in the hairpin DNA and again on going from free solution to the triplex.[57] The method adopted was more sophisticated than that described in the preceding paragraph, although still less rigorous than the full NLPB protocol. The coupling of the assumed two-state transition to the overall titration curve was predicted by considering all 16 possible titration states of the four cytosines.

9.5.5 Binding a Ligand to Different Conformations of the Same Polyion

Analyses of X-ray crystal structures of many RNAs reveal bidentate chelates, wherein two water molecules of the *inner* hydration shell of Mg^{2+} are substituted by two phosphate groups (see Chapter 1). Such chelates are typically found in non-helical, less solvent-exposed regions. Chelates involving a phosphate group and another RNA ligand are also sometimes found.

In a series of papers, the NLPB protocol was applied to examine the contributions of both site-specifically chelated and diffusely bound Mg^{2+} ions to stabilize fully folded tertiary structures relative to less folded intermediates.[105–108] Each site-specific complex of interest contains a Mg^{2+} ion whose inner coordination shell is partially substituted by two phosphate groups or a phosphate group and another ligand. The high cost of partially desolvating both the Mg^{2+} ion and the RNA binding site precludes significant site-specific chelation in (i) yeast tRNA^{phe} and (ii) the P5 helix and P5 stem loop from the P4–P6 domain of the *Tetrahymena thermophila* group I intron. The increase in Mg^{2+} and phosphate self-energies plus the free energy increase due to loss of proximal Na^+ ions in the ion atmosphere outweigh the negative contributions to the total change in electrostatic free-energy upon site-specific Mg^{2+} chelation. For these molecules, the peaks ascribed to Mg^{2+} ions in the electron density maps were attributed simply to high occupancies of small regions of large negative electrostatic potential by hexahydrated Mg^{2+} ions. For yeast tRNA^{phe} and also for poly(AU), the numbers of thermodynamically bound Mg^{2+} ions and their variations with salt concentration were computed as, $\nu_{Mg^{2+}} = \int dV [c_{Mg^{2+}}(\mathbf{r}) - c^0_{Mg^{2+}}]$, from solutions of the NLPB equation. Acceptable agreement with the experimental binding isotherms was obtained without invoking site-specific Mg^{2+} chelation by the RNA. However, in the case of a 58 nucleotide fragment from *E. coli* rRNA, it was suggested that one of the ~ 10 sites of high electron density attributed to Mg^{2+}

ions is an optimal site of such large negative electrostatic potential and such favorable solvent exposure that site-specific chelation at that site provides a modest attractive *electrostatic* contribution to overall Mg^{2+} binding.[108]

The *total* contributions of any site-specifically chelated and, especially, diffusely bound Mg^{2+} ions in the ion atmosphere to stabilize the compact folded structures of the yeast $tRNA^{phe}$ and the 58 nucleotide rRNA fragment relative to their respective intermediates with well-defined secondary structure elements, but no tertiary structure, were purportedly evaluated by the NLPB protocol.[108] However, that protocol provides no information about the free energy change associated with the three translational and three rotational coordinates that position and orient the bound Mg^{2+} ion plus the remaining water molecules of its inner coordination shell when it goes from its standard state in solution to its bound state. That generally large positive contribution was apparently omitted from the total standard state free energy change of site-specific chelation,[108] because it was thought to be much smaller than the other terms. Depending on how restrictive the binding is, the error incurred by neglecting this term can be arbitrarily large. Consequently, it is uncertain by what amount site-specific chelation of Mg^{2+} to the designated sites stabilizes the folded states.

9.5.6 Absolute Equilibrium Constants

In order to calculate the absolute equilibrium constant for a ligand binding reaction, one must go beyond NLPB theory, which provides only $\Delta G_Q^{el} - (\Delta G_P^{el} + \Delta G_L^{el})$ in eqn (9.23). The following analysis is presented in order to see what is involved. Within the NLPB framework, it is permissible to delete the solvent contribution from eqn (9.1) for the chemical potential, to obtain

$$\mu_j^0 = -kT \ln Z(1_j, V, T) = -kT \ln[V \times 8\pi^2 Z(1_j^{fix}, T)] \quad (9.28)$$

where the second equality is obtained as follows: (i) a molecular coordinate frame is fixed in one of the rigid subunits of j; (ii) the potential energy is expressed entirely in terms of the position and orientation coordinates of all other subunits in that molecular frame; and (iii) the position and orientation of that molecular frame in the laboratory frame are integrated over their allowed ranges, which yields the factor $V \times 8\pi^2$, where $V = 1.0 \, cm^3$ is the standard state volume and $8\pi^2$ is the standard state Euler angle volume. There remains, $Z(1^{fix}, V, T) \equiv \int d\widehat{R} \exp[-U(\widehat{R})/kT]$ where $\widehat{R}$ denotes the list of all coordinates that position and orient all other rigid subunits in the subunit-fixed molecular frame. The quantity $K_0 \equiv \exp\{-[\mu_{A^0B^0}^0 - (\mu_{A^0}^0 + \mu_{B^0}^0)]/kT\}$, pertaining to *uncharged* species in eqn (9.23), can now be written as

$$K_0 = Z(1_{A^0B^0}^{fix}, T) / [V8\pi^2 Z(1_{A^0}^{fix}, T) Z(1_{B^0}^{fix}, T)] \quad (9.29)$$

The molecular frame of the complex is chosen to coincide with that of its A moiety. In order to evaluate $Z(1_{A^0B^0}^{fix}, T)$, we first write it as $[Z(1_{A^0B^0}^{fix}, T)/$

$Z_0(1^{\text{fix}}_{A^0B^0}, T)]Z_0(1^{\text{fix}}_{A^0B^0}, T)$, where $Z_0(1^{\text{fix}}_{A^0B^0}, T)$ is the configuration integral for a more convenient fictitious potential energy function, $U_0 \equiv U_A(\widehat{\mathbf{R}}_A) + U_B(\widehat{\mathbf{R}}_B) + U^0_{AB}(\mathbf{r}_{AB}, \Phi^0_{AB})$, where $\widehat{\mathbf{R}}_A$ denotes coordinates of the other subunits of the A moiety relative to its subunit-fixed molecular frame, $\widehat{\mathbf{R}}_B$ denotes coordinates of the other subunits of the B moiety relative to its *own* subunit-fixed molecular frame, $\mathbf{r}_{AB}$ denotes the position of the subunit-fixed molecular frame of the B moiety in the corresponding frame of the A moiety and Φ_{AB} denotes the Euler coordinates that orient the molecular frame of the B moiety in that of the A moiety. $U_A(\widehat{\mathbf{R}}_A)$ and $U_B(\widehat{\mathbf{R}}_B)$ are the respective potential energy functions for the isolated A and B molecules and $U^0(\mathbf{r}_{AB}, \Phi_{AB})$ provides the only interaction potential between species A and B in this model complex and affects only the orientation and position of the subunit-fixed molecular frame of B relative to that of A in the complex. With this potential, the $d\widehat{\mathbf{R}}_A$, $d\widehat{\mathbf{R}}_B$ and $d\mathbf{r}_{AB}d\Phi_{AB}$ integrals can be performed separately to give, $Z_0(1^{\text{fix}}_{A^0B^0}, T) = Z(1^{\text{fix}}_{A^0}, T) Z(1^{\text{fix}}_{B^0}, T) Z^{\text{rel}}_0(T)$, where $Z^{\text{rel}}_0(T) \equiv \int d\mathbf{r}_{AB} d\Phi_{AB} \exp[-U_0(\mathbf{r}_{AB}, \Phi_{AB}/kT]$ is the configuration integral for the relative coordinates. After some cancellation, eqn (9.29) becomes

$$K_0 = [Z^{\text{rel}}_0(T)/(V \times 8\pi^2)] \exp(-\Delta G^{\text{int}}/kT) \qquad (9.30)$$

where $\Delta G^{\text{int}} \equiv -kT \ln[Z(1^{\text{fix}}_{A^0B^0}, T)/Z_0(1^{\text{fix}}_{A^0B^0}, T)]$ is the free energy change to switch off the fictitious potential $U_0(\mathbf{r}_{AB}, \Phi_{AB})$ and simultaneously to switch on all *non-ionic* interaction potentials between subunits of A and subunits of B. Evaluation of ΔG^{int} generally requires a reversible work switching protocol and may be computationally intensive.

It remains to evaluate $Z^{\text{rel}}_0(T)$. For convenience, we take $U_0 = U_{\min} + (g/2)(\xi^2_x + \xi^2_y + \xi^2_z) + (\alpha/2)(\eta^2_x + \eta^2_y + \eta^2_z)$, where $\xi \equiv \mathbf{r}_{AB} - \mathbf{r}^0_{AB}$ is the vector of displacements of the molecular frame of B from its average position, $\mathbf{r}^0_{AB}$. η_x, η_y, η_z are Cartesian rotations of the subunit-fixed molecular frame of B, which has equilibrium orientation Φ^0_{AB} in the corresponding frame of A, around its own body-fixed x, y and z axes. Use of Cartesian rotations is permissible, when the rotations are sufficiently small, which is typical for most complexes. $\mathbf{r}^0_{AB}$ and Φ^0_{AB} are assumed to be average positions in the actual uncharged complex and $U_{\min}$ is assumed to be the average potential therein (relative to that of separate A^0 and B^0), when $\mathbf{r}_{AB}$ and Φ_{AB} are fixed at those average values. g and α are force and torque constants, respectively. In this case, $d\mathbf{r}_{AB}d\Phi_{AB} \to d\xi d\eta$ and standard integrals yield

$$K_0 = (\sigma^3/V)(\tau^3/8\pi^2)\exp[-U_{\min}/kT]\exp[-\Delta G^{\text{int}}/kT] \qquad (9.31)$$

where $\sigma \equiv (2\pi\langle \xi^2_x \rangle)^{\frac{1}{2}} = (2\pi kT/g)^{\frac{1}{2}}$ and $\tau \equiv (2\pi\langle \eta^2_x \rangle)^{\frac{1}{2}} = (2\pi kT\alpha)^{\frac{1}{2}}$ can be regarded as the typical ranges of translational and rotational displacements, respectively, that are allowed by the potential U_0.

In order to gain some idea of the size of K_0, one can choose the force and torque constants such that $\Delta G^{\text{int}} = 0$, which corresponds to equally strong and restrictive binding by the actual and fictitious potentials. In that case, K_0 is just

$\exp(-U_{min}/kT)$ times the ratio, σ^3/V, of the allowed translational volume in the complex to the standard state volume, $V = 1.0\,\text{cm}^3$, multiplied by the corresponding ratio, $\tau^3/8\pi^2$, of the allowed angular volume in the complex to its standard state angular volume, $8\pi^2$. Each of these ratios is typically far less than 1.0, so their product is generally miniscule in cgs units. For example, if one takes $\sigma = 5 \times 10^{-8}\,\text{cm}$ (5 Å) and $\tau \approx 0.175$ rad (10°), then $(\sigma^3/V)(\tau^3/8\pi^2) = 8.4 \times 10^{-27}\,\text{cm}^3\,\text{molecule}^{-1} = 5.1 \times 10^{-6}\,\text{l}\,\text{mol}^{-1}$. The binding free energy for the uncharged species is $\Delta G_0^0 = -RT\ln K_0 = -RT\ln(5.1 \times 10^{-6}) + U_{min} \approx 7.1 + U_{min}$ kcal mol^{-1} at 293 K. In order to obtain a reasonable total equilibrium constant, the ~ 7.1 kcal mol^{-1} confinement penalty in ΔG_0^0 must be overwhelmed by negative U_{min} and/or ΔG_j^{el} terms in eqn (9.23).

Equations (9.29)–(9.31) demonstrate explicitly the complete absence of all kinetic energy contributions and their mass-dependent factors from the equilibrium constant, as required in classical statistical mechanics. The use of standard quantum mechanical equations for the translational, rotational and vibrational entropic contributions to the equilibrium constant should also satisfy this requirement, whenever the vibrational modes involving the relative binding coordinates fall in the classical limit. However, for biological macromolecules in general, and especially for somewhat floppy RNAs, normal modes determined from the curvatures of the potential surface at its minimum (or minima) are not relevant at ambient temperature, where the molecule samples primarily the far more anharmonic regions of the potential surface at higher energy. This substantially reduces the effective elastic constants that govern the lower frequency normal modes, which make the greatest contribution to the entropy. Different minima with different curvatures may also be encountered. Because the entropic term is so large, even modest relative errors can have a large impact on the equilibrium constant.[54] In view of these hazards, a classical reversible work calculation of ΔG^{int} may provide greater accuracy, whenever classical statistical mechanics applies.

An example where classical statistical mechanics does not apply is a recent study of Mg^{2+} hexahydrate binding to the m15 and m21 sites of two fragments of the $\Delta C209$ mutant of the P4–P6 domain of the *Tetrahymena thermophila* group I intron.[58] Density functional theory was applied to the solvated Mg^{2+} hexahydrate, the solvated RNA binding site and the Mg^{2+} hexahydrate–binding site complex to reckon the change in electronic energy upon binding Mg^{2+} hexahydrate to the site. A significant electronic contribution was found. Account was also taken of (i) changes in solvation self-energies that accompany the binding of Mg^{2+} hexahydrate to the RNA, (ii) the interaction of fixed charges on the RNA with both the empty and occupied binding site and (iii) the interaction of the various species with the monovalent salt via the NLPB protocol. The entropic penalty of binding was calculated using the spectrum of normal mode frequencies computed from the curvatures at the minimum of the potential, using the standard quantum mechanical formulas. Results of these microscopic calculations were compared with results of simpler electrostatic methods, which are much more sensitive to the choice of local dielectric constant. In the absence of direct comparisons to experiment, the accuracy of the calculated absolute equilibrium constants for Mg^{2+} binding to the m15 and m21 sites is uncertain.

9.5.7 Repulsive Forces Between Rod-like Polyanions

NLPB theory predicts only repulsions between parallel cylindrical polyions of like charge in the absence and presence of salt.[79,109,110] Indeed, NLPB theory appears to predict only repulsive forces between any two or more objects bearing intrinsic charges all of the same sign.[111,112]

When two infinite charged lines have a distance of closest approach R and a crossing angle ϕ, the interaction potential is given by $w_{12} = 2\pi\nu^2 \exp(-\lambda R)/(\varepsilon\lambda \sin\phi)$, where ν is the effective linear charge density.[85,113] This expression is valid only at sufficiently great R or sufficiently small ν that the electrostatic potential at a distance R from either line is in the linear Poisson–Boltzmann (LPB) regime. A simple equation and the necessary tabular data to compute the effective value of ν for cylinders with a wide range of actual linear charge densities and radii is available.[84] A simple expression for the second virial coefficient (B_2) was derived from w_{12} and evaluated for infinite-cylinder model DNAs in various concentrations of monovalent salt.[85] At DNA concentrations where the second virial term is modest, so $c_{DNA}B_2 \ll 1.0$, nearest-neighbor DNAs in solution preferentially adopt more or less orthogonal configurations. A correction term for B_2 that takes account of the electrostatic end-effect for DNAs of finite length was presented along with an approximate equation for the third virial coefficient, B_3.[114] Higher order interactions become important when $c_{DNA}^2 B_3 \gtrsim c_{DNA} B_2$. Under such conditions, more parallel orientations of neighboring DNAs become favored.[115] With a sufficient further increase in c_{DNA}, an isotropic to cholesteric phase transition begins.[116–119]

In order to treat electrostatic repulsions between arbitrarily curved DNAs, a charged sphere of 12 Å radius is placed every 31.8 Å along each chain.[120] Each sphere is assumed to exhibit a linear Debye–Hückel (screened Coulomb) potential arising from an effective charge (Z), which is calculated in the following way. First, the NLPB equation is solved for a uniformly charged straight cylinder with a diameter of 24 Å and the linear charge density of DNA in the prevailing concentration(s) of salt(s). Then a chain containing 2001 charged spheres is arranged in a straight line. For each salt concentration, the effective charge is adjusted until the sum of screened Coulomb potentials of all the spheres, evaluated along a line perpendicular to the chain near its center and averaged over several different lines, matches the solution of the NLPB equation for all distances greater than 24 Å. Remarkably accurate matches were obtained for 0.01–1.0 M NaCl and for 0.1 M NaCl + 0.01 M $MgCl_2$.[120] The electrostatic interaction energies between charged spheres, i and j, on the same or different chains are then reckoned using the standard Overbeek screened Coulomb form, $U(r_{ij}) = (Ze)^2 [\exp(ka)/(1+\lambda a)]^2 \exp(-\lambda r_{ij})/r_{ij}$.[121] This interaction energy was employed in simulations of the supercoiling free energies of DNA topoisomers, wherein the inter-helix distance is only very rarely ≤ 50 Å. Fairly good quantitative agreement was obtained with various experimental data, provided that torsion elastic constants measured for unstrained DNAs were employed in the simulations.[122,123]

9.6 Counterion Condensation Theory
9.6.1 Relation to NLPB Theory

An important quantity in polyion theory is the effective charge that enables the LPB (or Debye–Hückel) theory to predict correctly the electrostatic potential in the linear region at large distances from the polyion. This potential is important for estimating certain properties of polyions, especially those concerning intra- and inter-polyion interactions. The main idea of CC theory is to avoid the computational difficulties involved in NLPB theory by incorporating a quasi-chemical counterion binding equilibrium in a self-consistent manner directly into the linear theory, where the linear superposability of screened Coulomb potentials enormously simplifies the calculations. In CC theory, the polyion array of intrinsic charges is embedded directly in the solvent, rather than in a low-dielectric medium. Associated with the polyion is a single global binding site or free volume, wherein the bound counterions are distributed so as to partially to neutralize the same fraction, f, of each intrinsic charge.[6,7] The electrostatic potential energy of the polyion is assumed to be a sum over all screened Coulomb pair potentials, $\frac{1}{2}\Sigma_{i \neq j} \Sigma_j e^2(1-f)^2 \exp(-\lambda r_{ij})/\varepsilon r_{ij}$ and the value of f is determined by minimizing the global free energy of the entire system. An alternative model, wherein a binding site is assigned to each intrinsic charge and nearest-neighbor interactions between sites are treated exactly, was also formulated, but the results were practically the same as those obtained for the free-volume model.[29] The singularity due to the mean electrostatic potential of an infinitely long polyion in the low-salt limit is so great that it overwhelms any differences in local energetics. Not surprisingly, CC theory gives a poor account of the electrostatic potential and counterion distribution near the polyion surface, hence also of the electrostatic free energy, energy and entropy.[90,124] It also enormously underestimates the volume occupied by the condensed counterions at low salt.[125] The original quasi-chemical CC theory also incorrectly predicts that the condensed fraction, f, increases with increasing counterion concentration in the region where $\lambda b \geq 1.0$, where b is the distance between intrinsic charges.[27] Nevertheless, CC theory permitted numerous simple predictions that agree remarkably well with experiment and provided many important insights that likely would not have emerged from NLPB theory alone. Especially for experimentalists, CC theory has been an invaluable aid in thinking about polyionic systems.

It was recently proposed that the condensed counterion charge should be just the difference between the actual polyion charge and that smaller effective charge, for which the solution of the LPB equation would match that of the NLPB equation at large distances.[27] When the NLPB equation for a uniformly charged cylinder in monovalent salt is solved by the method of Sugai and Nitta,[86] the effective linear charge density associated with the linear part of the potential is obtained directly.[27] Simply subtracting this value from the total charge density and dividing by the latter yields the condensed fraction of counterions. It was further suggested that CC theory should be constructed so that it yields a very similar condensed counterion charge.

An extension of CC theory was proposed to extend its range of applicability and improve predictions of certain properties.[27] The polyion array may exhibit any geometry, wherein all charges are electrostatically equivalent. Apart from the positions of the polyion charges, the only other relevant parameters are L_B, λ and $K = [8\pi L_B(\frac{1}{2})(z_C^2 + |z_C/z_d|z_d^2)]^{\frac{1}{2}}$, where z_d is the co-ion valence. This extended CC (ECC) theory includes an initially unknown site-binding constant, β', which has units of volume. Although condensation is treated using a grand partition function (GPF) binding method, instead of the free energy minimization procedure used by Manning,[6,7] the resulting extremum condition for the maximum term in the GPF is identical with that of Manning, provided that β' is interpreted as the free volume. This extremum equation has two unknowns, f and β'. Manning assumed that the free volume or equivalently β', was independent of the molar concentration of counterions, c_C^M, which is what led to the (incorrect) continued adsorption of counterions at large c_C^M. In ECC theory, β' is ascertained by imposing a different condition, namely that any bilinear coupling between fluctuations in local counterion concentration, c_C^M and fluctuations in the number, n_C, of bound counterions must vanish. In NLPB theory it is similarly assumed that fluctuated states involving transfer of counterions from the bulk to the ion atmosphere or *vice versa* make no significant contribution to the partition function. This assumption provides a second relation from which the mean number of bound counterions, n_C and β', are determined.

The condensed fraction of counterions, $f = n_C/N$, is given by[27]

$$f = 1 - \frac{1}{(c_C^M)^{\frac{1}{2}} K L_B |z_C|(-)\partial S_j(\lambda)/\partial \lambda} \quad (9.32)$$

where the interaction sum, $S_j(\lambda) \equiv \frac{1}{2}\Sigma_{i \neq j} \exp(-\lambda r_{ij})/r_{ij}$, runs over all intrinsic charges, except the jth. When eqn (9.32) yields a negative value of f, the alternative solution, $f = 0$, must be invoked. For an infinitely long discretely charged line in the low-salt limit, eqn (9.32) gives precisely Manning's result, $f = 1 - \xi^{-1}$, where $\xi \equiv L_B|z_C|/b$ is the reduced linear charge density. However, f falls below that limiting value at higher salt concentration, when $\lambda b \geq 1.0$, as it should. Moreover, f also declines for a linear array of charges with a finite diameter, d, when $\lambda d \geq 1.0$, as it should. The value of β' varies with array structure and also declines with increasing c_C^M, as expected. The condensed fractions, f_{NLPB} and f_{ECC}, were computed via the respective NLPB and ECC theories for a uniformly charged cylinder with diameter $d = 20$ Å and $\xi = 4.2$, over a range of concentrations of monovalent salt.[27] From $c_C^M = 10^{-6}$ to 10^{-3} M, one find $f_{\text{ECC}} = f_{\text{NLPB}} = 1 - \xi^{-1} = 0.76$ and both f_{ECC} and f_{NLPB} decline at higher salt concentration to ~ 0.45 at $c_C^M = 0.1$ M. Between 0.001 and 0.1 M salt, f_{ECC} exceeds f_{NLPB} by up to 15%, but then crosses below f_{NLPB} near 0.1 M salt. By this criterion, ECC theory works moderately well up to $c_C^M = 0.1$ M.

f_{ECC} was calculated for a double-helical array of charges with diameter $d = 20$ Å, and $\xi = 4.2$ in monovalent salt and the results were very similar to those obtained for the corresponding uniformly charged cylinder. In contrast, for a discretely charged line with $\xi = 4.2$, f_{ECC} declines only slightly from 0.76 to 0.72 at $c_C^M = 0.1$ M. Evidently, a discretely charged line is not a good model for DNA when $c_C^M \geq 0.01$ M.

9.6.2 Thermodynamic and Other Properties

The ECC theory was incorporated into a cell model in equilibrium with a salt reservoir and applied to compute the osmotic coefficient of a discretely charged double-helical model of DNA as a function of polyion concentration in either 2 or 10 mM salt.[126a] Values computed using the ECC cell model (ECCC) theory agreed surprisingly well, although not perfectly, with the experimental values over the entire range from 5×10^{-4} to 0.3 M phosphate groups.[126a,31] At low DNA concentrations, the ECCC theory performed significantly better than the aforementioned zero-salt NLPB approximation.[97]

An extension of ECC theory to treat finite arrays of inequivalent charges of the same or mixed signs was also developed.[126b] The result is the same as eqn (9.32), except for replacement of f by f_j and the incorporation of an appropriate sign and the ratio $(1 - f_i)/(1 - f_j)$ into each term of $S_j(\lambda)$. The f_j are calculated either by matrix inversion or iteration. Although this theory yields unphysical oscillations of the f_j (about the mean value) with position j on a uniform linear lattice, reasonable average condensed fractions for the entire molecule are obtained. MC simulations of *integral* counterion exchanges between fully occupied and vacant sites for any given number of condensed ions completely suppress the oscillations, which arise only when different *fractional* occupations of the sites are allowed. For a double-helical array of 72 charges with $d = 20$ Å and $\xi = 4.2$, an iterative protocol involving simulated counterion interchanges predicts an ~ 10 bp electrostatic end-effect in 0.0018 M monovalent salt. This agrees well with that obtained by MC simulations of a discretely charged line of the same length and total charge located at the center of an impenetrable cylinder.[127]

At present there is no good way to calculate $\Gamma_S(P)$ from the number of condensed counterions, n_p, obtained via ECC theory. NLPB calculations indicate that $\Gamma_S(P)$ for a uniformly charged cylinder model increases substantially (becomes less negative) with decreasing salt concentration and approaches a limiting value, $\Gamma_S(P)/N = -1/(4\xi)$.[83,90] Nevertheless, the observed $\Delta\Gamma_S$ for binding small linear oligocations to DNA are typically independent of c_S over a wide range. In principle, ECC theory gives the residual *net* valence of the polyion, P, plus its condensed counterions, namely $|z_P^{eff}| = N - |z|n_P$ and for the corresponding complex, $Q = P \cdot L$, it gives $|z_Q^{eff}| = N - M - |z|n_Q$, where M is the ligand valence. If, under fairly low-salt conditions, ECC theory predicts nearly the same values for $|z_P^{eff}|$ and $|z_Q^{eff}|$, then one might expect to find fairly similar electrostatic potentials and co-ion distributions. In that case, eqn (9.18) would give, $\Gamma_S(Q) \cong \Gamma_S(P)$. If the oligocation, L, condenses no small ions, then its ion atmosphere is linear, $D(\Delta G_L^{el}) = 0$ and eqn (9.18) gives $\Gamma_S(L) = -M/2$. Insertion of these results into eqn (9.24) yields $D(\ln K) = -M$. This prediction agrees extremely well with results obtained via the NLPB protocol for binding small doubly charged ligands to DNA models of sufficient length (≥ 50 Å) with $\xi = 4.2$ in 0.1 M monovalent salt.[91] However, for oligocations of arbitrary shape and mixed valence, this simple result does not apply.[91]

9.6.3 Electrostatic Contribution to the Bending Rigidity and Persistence Length

The increase in electrostatic free energy upon bending a polyion contributes to its bending rigidity, κ and persistence length, P, which for an intrinsically straight rod is given by $P = (\kappa_0 + \kappa_{el})/kT = P_0 + P_{el}$, where the subscripts '0' and 'el' denote the intrinsic and electrostatic contributions, respectively. Provided that $P_0 \gg P_{el}$, the shift in the condensation equilibrium due to thermally excited bending does not significantly affect κ_{el}, which then applies for bending at constant fraction, f, of condensed counterions.[29] For monovalent counterions, the electrostatic persistence length of a discretely charged line is

$$P_{el} = (1-f)^2 L_B / (4\lambda^2 b^2) \qquad (9.33)$$

where b is the mean axial charge spacing.[32,128] For DNA in 0.01 M monovalent salt at ambient T, the predicted value is $P_{el} \cong 30$ Å, which is barely significant compared with the intrinsic persistence length, $P_0 = 480 \pm 20$ Å.[129,130] With further increases in salt concentration, P_{el} declines to truly insignificant values. Because $P_{el} \propto c_S^{-1}$, the rise in P_{el} with decreasing salt below 0.01 M is very substantial. The expression $P = P_0 + P_{el}$ agrees fairly well with numerous experimental data over the range 5×10^{-4} to 1.0 M salt.[129,130] At higher c_S, DNA undergoes a shift toward alternative sub-conformations within the B-family, which are more flexible for some sequences.[131a] Numerical predictions derived from NLPB theory for bending a uniformly charged cylinder do not match the observed variation of persistence length with c_S as well as eqn (9.33).[131b–d] This circumstance remains an unresolved puzzle.

9.6.4 Effect of Oligocations to Bend DNA

Ions of sufficient valence cause permanent bends that diminish the persistence length of DNA, but not necessarily its bending rigidity. Ions may induce bends either by shifting a prevailing slow equilibrium between differently curved sub-conformations within the B-family or by directly bending the considerably stiffer DNA at fixed sub-conformation.[132]

A single tethered univalent cation above the floor of the major groove of a 14 bp DNA caused no detectable net bend.[133] Similar placement of two adjacent charges introduced a small net bend.

Trivalent counterions were observed to reduce substantially the persistence length of DNA. This effect was attributed to local bending of the DNA toward the trivalent ion.[134] Ion-induced curvature was predicted to decrease both the persistence length and the effective contour length without decreasing the bending rigidity.

The wrapping of a 500 Å segment of a uniformly charged, semi-flexible line with $\xi = 4.2$ and $P = 500$ Å around a uniformly charged sphere with the radius and charge of a nucleosome core particle was investigated by mapping the minimum free energy structures as a function of the screening parameter, λ.[135] All charges of the line and oppositely charged sphere were assumed to interact via screened Coulomb potentials and no account was taken of counterion condensation, so DNA–DNA self-repulsions were substantially overestimated. With increasing monovalent salt concentration from the zero-salt limit, the stable structure of this model DNA–nucleosome complex evolved stepwise from a center-bound planar bend to an end-bound planar $\sim 180°$ wrap, then to an end-bound non-planar (helical) $\sim 180°$ wrap, then to a centered full ~ 1.5 turn helical wrap, typical of DNA on nucleosomes, then finally at high salt to a center-bound planar bend of small amplitude.[135] The salt concentrations where the first and last changes in structure are predicted to occur, namely 4 and 900 mM, correspond roughly to the experimental values for release of DNA from nucleosomes at low and high salt.

9.7 Attractions Between Like-charged Filaments and Condensed Phase Formation

9.7.1 Simulations and Analyses

Early theoretical indications that attractive forces could arise between like-charged polyions in solution under certain conditions came largely from integral eqn methods and were reviewed by Schmitz.[17] Greater insight has come from more recent simulations of counterions and in some cases also co-ions, interacting with rigid,[136–138] semi-flexible,[136,139,140] or flexible[136,141] linear polyanions. In order to keep the counterion charges from coalescing with the polyion charges, some non-electrostatic repulsive potential is always employed. For two parallel uniformly charged rods in the presence of divalent counterions, but no salt, short range attractive interactions are predicted at $T \approx 300$ K, when the linear charge density is sufficiently great ($\xi \geq 2.1$).[137] At $T = 0$ K, all counterions are associated with the two rods and exhibit two interlocking lattice structures in the space between them. Although such periodic order is substantially attenuated with increasing T, the residual order gives rise to the attractive force at finite T. The attractive force is maximal at $T = 0$ K and decreases with increasing T. The rod radii in these Brownian dynamics simulations were less than half that of DNA, so the detailed results are not directly transferable. The residual liquid-like order and acoustic phonon dynamics of condensed layers of Mg^{2+}, Sr^{2+} or Ba^{2+} ions between f-actin filaments were observed by X-ray scattering methods.[142]

An analytical theory of the interactions between two parallel uniformly charged rods was based on fluctuations in the local number of condensed counterions about the mean value.[143] This theory yields a general expression for the total free energy as a function of rod separation in terms of various parameters, including ξ and the counterion valence. A critical feature of this theory is a complete summation of all fluctuating moments of the multipole expansion, not just the lowest few. For divalent counterions, the predictions of this theory agree quantitatively with the above simulations for temperatures in the ambient range.[143] This theory was extended to treat bundles containing arbitrary numbers, $N \geq 2$, of rod-like polyions on a square lattice and yields a general expression for the free energy in terms of various parameters, including N and the lattice constant.[144] The free energy of the N-rod bundle differs substantially from the sum of pairwise interactions of the constituent rods. The negative free energy of the full N-rod bundle is essentially proportional to N, so there appears to be no upper limit to the equilibrium bundle size. It was suggested that the observed cessation of aggregate growth might be due to kinetic rather than equilibrium reasons. Experiments suggest that both factors influence aggregate size.[145]

Langevin molecular dynamics (MD) simulations of the counterion-induced self-aggregation of a single semi-flexible polyanion were performed for +2-, +3- and +4-valent counterions.[139] Divalent ions induced no condensed structures, whereas +3- and +4-valent counterions induced both toroidal and rod-like structures over a range of parameters. Increasing polyion stiffness favors toroids over rods, presumably due to the high energetic cost of the required large bends at the ends of a rod. A subsequent Langevin MD simulation examined the self-aggregation of a single polyanion over a wider range of charge, stiffness and length parameters.[140] This study revealed a number of intermediate structures and mechanisms by which they inter-converted and evolved toward the final structure. Again, increasing bending stiffness was found to favor toroids.

The variation of attractive interactions with the relative orientation of two charged rods was studied as a function of mono- and multivalent counterion concentration by Langevin MD simulations.[138] With increasing concentration of multivalent ions (in the absence of monovalent salt), the stable state evolved stepwise from two isolated rods to two associated rods with perpendicular orientation and counterions accumulated near the crossing point, then to two associated parallel rods with many longitudinally dispersed counterions and finally at sufficiently high concentration back to two isolated rods. The prediction that perpendicular orientations are more stable over some range of counterion concentration suggests that network aggregates might form below the threshold concentration for the appearance of bundles or toroids.[138]

A greatly simplified model of the interaction potential between linkers and rigid rods was incorporated into a more general theory, wherein reversible adsorption of linkers from solution onto one or two adjacent charged rods was allowed.[146] It was assumed in advance that a system containing rods and linkers plus some background electrostatic screening might be able to form an ordinary nematic phase, a denser hexagonal bundle phase at higher linker concentration and a cubatic cross-linked network phase, in addition to an isotropic phase. Phase diagrams as a function of linker and rod concentration were constructed. Interestingly, three different kinds of phase diagram emerged. At low linker concentration and moderate rod concentration, an isotropic to nematic transition was always predicted. For one range of potential parameters, only a bundle phase was predicted to occur with increasing linker concentration and for another range of those same parameters only the cross-linked cubatic phase was predicted to occur with increasing linker concentration. Within a fairly narrow range of potential parameters and a limited region of rod concentration both cubatic and bundle phases were predicted to be encountered with increasing linker concentration. It was found that shorter linkers cannot adequately stabilize isolated junctions and cross-linked networks, but in sufficient numbers are able to stabilize bundles of parallel rods. Longer linkers stabilize the cross-linked networks relative to bundles at all linker concentrations.[146] This theory may explain the different kinds of aggregates formed upon adding 30 µM of either spermidine or a homologue with 5-, ..., 8-carbon chains in place of its 4-carbon chain to ϕ29 DNA in 0.001 M monovalent salt.[147] These polyamines are denoted by sph$n$, with $n = 4$–8. Both sph4 and sph5 formed dense aggregates that readily passed through a 0.45 µm filter without significant loss and gave similar static and dynamic light scattering results, which were consistent with toroids of diameter 1800 Å, as seen in electron micrographs (EM) of the same sample. The sph6 formed aggregates that filtered slowly and much of the DNA did not pass through the filter. The sph7 and sph8 aggregates effectively blocked the filter so nothing passed through it. The formation of large cross-linked flocs instead of dense aggregates by sph7 and sph8 would be entirely consistent with the theoretical predictions. The same theory is relevant to both the bundle and crosslinked network phases formed by f-actin filaments.[146]

The likelihood of observing both a cross-linked state and a bundled state might be significantly enhanced in a system where the rods are semi-flexible rather than rigid. In that case, the preference of two rods for high-angle crossings at low concentrations of multivalent ions does not prevent them from achieving repeated high-angle crossings. By bending to form large loops between each crossing point, two rods can achieve multiple mutual crossings at moderately high angles.[148] If each multivalent ion is regarded as a rod cross-linker that can slide along the pair of rods, then an interesting circumstance emerges.[148] These mobile linkers prefer the crossing zones, where the two rods are closest and within such a zone they experience an effective short-range attraction when the crossing angle is high. However, more distant linkers at the next crossing point experience a repulsive barrier, due to either a bending or a low-angle penalty, when they try to move toward the linkers at the first crossing point. An effective potential with these properties was formulated for the interaction potential between linkers and the statistical mechanics of linkers on the two-rod system was analyzed. A broad transition from a state of isolated linkers separated by large loops of the oppositely bent rods to a state of parallel rods bridged by many randomly located linkers was predicted to occur with increasing numbers of linkers. This picture was confirmed by Langevin MD simulations.[148]

When each pair of semi-flexible rods has ~ 10 or more crossing points, one would expect aggregation to form clusters, wherein any given rod may exchange partners with another rod between one crossing and the next. Cluster formation is driven by the enormous configurational degeneracy of the multi-chain clusters due to multiple partner exchanges, which easily overwhelms the translational entropy lost upon aggregation. Such a configurational entropy drives the aggregation of DNAs and collagens of sufficient length in their mid- and upper melting regions,[149] and also gives a reasonable account of the DNA aggregation transitions driven by various $+2$- and $+3$-valent ions in the lower to mid-melting regions.[150] In this scenario, the first aggregates formed would be lightly crosslinked, but increasing numbers of linkers would favor densely crosslinked bundles. Such a scenario might prevail, when $\phi 29$ DNA interacts with spermidine in 0.001 M NaCl. Tenuous and somewhat anisotropically extended structures, consistent with lightly crosslinked aggregates, prevail in 25 µM spermidine, whereas dense aggregates, primarily toroids, are formed in 30 µM spermidine.[147]

9.7.2 The Tightly Bound Ion (TBI) Model

An important modification of the NLPB method was proposed to account explicitly for correlations among the most tightly bound ions in the inner part of the ion atmosphere and thereby to mediate attractions between like-charged polyions.[40,77,78] In general terms, the protocol is first to solve the NLPB equation and calculate the local ion concentrations. The excess counterion concentration above the neutral salt value at a given point is used to define the radius, a_{ws}, of the spherical volume per excess counterion at that point. When a_{ws} is sufficiently small that the electrostatic interaction between two counter-ions separated by $2a_{ws}$ exceeds $1.3\,kT$, an ion at that position is judged to be highly correlated with its neighbors in an electrostatic sense. An ion may also become highly correlated with its neighbors due to hard-core repulsions, when $2a_{ws}$ is smaller than the hard-core diameter plus thermal displacements, which are estimated as $0.2a_{ws}$ according to the Lindemann melting criterion. All regions wherein excess counterions are sufficiently dense that they are highly correlated by either criterion are identified as tightly bound ion regions and are treated separately. These regions, which typically cover much or all of the molecule, are divided into as many cells (N) as there are intrinsic charges and assigned so that each cell contains a single intrinsic charge. Each cell is allowed to contain 0, 1 or 2 counterions, but no co-ions and all 3^N possible cell-occupation configurations are treated individually. For each occupation configuration, the free energy of interaction of all the counterions and phosphates within the tightly bound layer is evaluated by summing the free energies associated with intra- and inter-cell interactions. The former are evaluated according to $\Delta G_1 = -kT \sum_i \ln\langle\exp(-u_{ii}/kT)\rangle$, where the average is taken over all positions of one or both ions in the cell and the sum runs over all cells. The latter are evaluated according to, $\Delta G_{12} = -kT \sum_{j>i} \sum_i \ln\langle\exp(-u_{ij}/kT\rangle$, where the average is taken over positions of all ions in the ith and jth cells. Excluded volume interactions and also Coulomb interactions are included in the u_{ii} and u_{ij}. For each cell-occupation configuration, ions are fixed at their mean positions in each cell and the NLPB equation is solved with and without the diffuse ions in the space exterior to the tightly bound region. These solutions are used to calculate the electrostatic free energy of interaction between the diffuse ions and the fixed charges, which now include the tightly bound ions, and also the free energy of the diffuse-ion solution. The electrostatic free energies of each occupation configuration are incorporated into a total partition function ratio, which yields ΔG^{el} for the polyion.

TBI theory was employed to calculate ΔG^{el}_{ds} and ΔG^{el}_{ss} for discretely charged helical models of double-stranded (ds) and single-stranded (ss) DNAs, which contain 5–13 nucleotides per strand, as a function of NaCl or $MgCl_2$ concentration at 37°C.[77] By subtracting the computed $\Delta\Delta G^{el}$ for the ds $\rightleftharpoons$ ss transition from the empirical value of ΔG^0 in 1M NaCl, the non-electrostatic contribution was estimated and assumed to remain invariant to changes in salt concentration. For NaCl, the predicted values of ΔG^{el}_{ds}, ΔG^{el}_{ss} and ΔG^0 over a wide range of salt concentration agreed fairly well with the values predicted by NLPB theory and the predicted ΔG^0 also agreed fairly well with the experimental values. For $MgCl_2$, the TBI theory predicted significantly lower values of all three quantities than did NLPB theory, as expected, and the predicted ΔG^0 values agreed not quite as well with experiment as in the case of NaCl, perhaps due to modest structural changes induced by Mg^{2+} in either the DNA or its hydration shell.[151] Similar remarks apply to the calculated variation of the melting temperature, T_m, with NaCl or $MgCl_2$ concentration. Again, agreement with the experimental data is better for NaCl than for $MgCl_2$. A simplified set of fitted analytical relations was proposed to predict the variations of ΔG^0, ΔS^0 and $1/T_m$ with DNA length and concentration of either NaCl or $MgCl_2$ from the measured values in 1.0 M solutions of either salt. These are potentially valuable tools for predicting the T_m of oligonucleotides of different length under various ionic conditions.

TBI theory was applied to two parallel model DNA helices (12 bp) at various separations over a range of NaCl or $MgCl_2$ concentrations.[78] The azimuthal orientations of the two DNAs (about their helix-axes) were such that the major groove of one was juxtaposed with the minor groove of the other and *vice versa*. In NaCl, the TBI theory agreed well with NLPB theory and no attraction was predicted at any salt concentration. However, in 0.001, 0.01 and 0.1 M $MgCl_2$, TBI theory predicted attractive interactions with minima at separations of 32, 28 and 27 Å, respectively. The largest well depth, $\sim 3kT$, was found for 0.1 M $MgCl_2$. Because the attractive well extends only over ~ 15 Å in the separation dimension and less than a duplex diameter in the other two and because other azimuthal orientations are likely repulsive, it is most probable that an estimated binding constant would be too small to measure, in agreement with many experiments, wherein side-by-side aggregation of dilute short DNAs in $MgCl_2$ has not been observed.

The TBI results for two DNAs of long, but unspecified, length were used to estimate the osmotic pressures of hexagonal arrays of parallel DNAs as a function of interhelix distance.[78] The total free energy was computed simply as the sum of six independent DNA–DNA pair interactions for particular concentrations of various divalent salts that do or do not induce aggregation.[78] Although overall quantitative agreement with the experimental osmotic compression data is remarkably good in many respects, it is far from perfect in certain regards. In addition, these calculations contain assumptions that may introduce potentially substantial changes, when removed. Particular problems include the following: (i) it is not geometrically possible for all pairs of rods in the array to have simultaneously optimal azimuthal orientations with one another; (ii) the neglect of multi-rod correlations is known to introduce a substantial error;[144] (iii) due to the Donnan effect, the prevailing salt concentrations inside the arrays are far lower than those in the bathing salt solutions, for which the osmotic pressures were calculated. Nevertheless, these calculations show that the TBI method holds considerable promise for treating accurately the correlations of +2- and higher valent ions that lead to mutual attractions between DNAs.

9.7.3 Effects of Helical Symmetry, DNA Shape and Fixed-counterion Models

Because the intrinsic charges on DNA are helically arranged, the counterions in the innermost part of the ion atmosphere must also be organized to some extent in a related helical fashion. The permanent helical structure of the negative and positive charges significantly alters the interaction between two duplexes from that predicted for uniformly charged rods in an orientation-dependent manner.[64,65] For parallel duplexes, the relevant orientation of each DNA is its azimuthal angle (ζ) for rotation around its helix-axis. (ζ is the angle between a vector normal to the helix-axis, which defines the orientation of the central base-pair of the DNA and the plane containing both the helix-axis and an inter-axis line connecting that DNA to its neighbor.) For non-parallel duplexes, the crossing angle (χ) is also important. (χ is the angle between the helix-axis of one DNA and the plane containing both the helix-axis of the other DNA and the unique joint normal to both DNAs.) An expression for the electrostatic energy as a function of the separation distance, R, ζ_1, ζ_2 and χ was derived from the LPB equation.[64] Attractions between helices are not expected to arise under any conditions in NLPB theory and presumably also not in LPB theory under conditions where it is valid, unless some account is taken of correlated fluctuations in the counterion distribution. Such fluctuations were effected simply by fixing discrete counterions at particular points in regions of large negative electrostatic potential. It was proposed that, in dense arrays, practically all of the condensed counterions would become so localized and the electrostatic potential in the region outside the localized counterions could be adequately described by LPB theory.[64] This proposal is justified to a limited extent by the results of certain experiments,[152] and also by simulations of model DNAs with helical shapes and charge distributions,[153] all of which indicate some adsorption of small counterions into the minor groove in a structured manner. Once ions of both signs are part of the fixed-charge array, then attractions are in principle possible and are predicted to occur for certain ranges of ζ_1, ζ_2, χ and R for ~ 0.7 monovalent salt.[64,65] So far, neither experiments nor simulations have provided significant evidence of the predicted attractive interactions in ≤ 1.0 M monovalent salt.

This fixed-counterion theory of electrostatic interactions between double-helical DNA molecules was applied to analyze numerous properties of dense DNA arrays, including (i) the exponentially decaying short-range repulsive forces in osmotically compressed arrays;[65] (ii) the relative stabilities of A- and B-helices under different conditions of counterion adsorption into the minor groove;[154] (iii) azimuthal alignment of neighboring DNAs and the origins of the in-register helix pitch, 10.0 bp/turn, in dense fibers of B-DNA, which differs significantly from the 10.4 bp/turn in solution;[65,155] (iv) the stability and pitch of a cholesteric phase that appears upon expansion of osmotically compressed arrays;[64] (v) possible unidentified array geometries or phases, with distinctly different azimuthal orientations of nearest neighbors;[67] and (vi) melting in DNA aggregates.[66] A lengthy review of this fixed-counterion model and its predictions has recently appeared.[163] Unambiguous experimental support for the many interesting predictions of this model is still lacking in most cases. For example, very short-range repulsive forces in osmotically compressed arrays are found not only for DNA, but also for other non-charged and rather weakly charged polymers and assigned to hydration forces. For DNA, there remains the question of whether such forces originate primarily from hydration forces or instead from electrostatic forces, as predicted by the fixed-counterion model.

Numerous DNA fiber diffraction data were examined for evidence of the predicted correlations between the azimuthal angles of neighboring DNAs at interhelical distances from ~ 25 to ~ 40 Å and features consistent with that possibility were identified.[156] However, the conclusion that azimuthal orientations of neighboring molecules are significantly locked together or pinned by electrostatic forces at such large interhelix distances (~ 33–40 Å) appears to be inconsistent with three other findings. (i) Simulations of two parallel DNAs with helical shapes and charge distributions and optimal azimuthal alignment for attraction predict no significant forces beyond 32 Å.[157] (ii) Solid-state NMR studies of dense DNAs, which are locally aligned in randomly oriented domains, were performed over hydration levels ranging from 0 up to ~ 33 D_2O/phosphate using ^{31}P or added ^2H labels at the 8-positions of adenines and guanines.[158–160] The results indicate the onset of large amplitude rotations of the DNAs about their local helix-axes on a 10^{-7}–10^{-4} s time-scale, when the hydration levels were ≥ 20 H_2O (or D_2O) per phosphate. The amplitude and probably also the rate of this motion increased substantially with further hydration up to ~ 33 D_2O/phosphate. The pinning of azimuthal orientations of these DNAs at lower hydration levels (≤ 12 H_2O/phosphate) is evidently released already by ~ 20 H$_2$O/phosphate and additional water most likely simply reduces the frictional drag resisting the local spinning/twisting motions. The interhelix distances for 20 and 33 H$_2$O/phosphate in hexagonal geometry are estimated to be 28.5 and 32 Å, respectively. (iii) Diffraction studies with the X-ray beam parallel to the fiber-axis and also solid-state ^{31}P NMR studies of osmotically compressed non-oriented samples both indicated an abrupt loss of bond-orientational order in a direction perpendicular to the fiber axis and an abrupt onset of large amplitude azimuthal rotations at an interhelical spacing of ~ 33 Å.^{161} A subsequent study found no sign of angular correlations at an interhelical spacing of ~ 36 Å.^{162} The findings (ii) and (iii) above are difficult to reconcile with pinned azimuthal orientations of adjacent helices at interhelix distances of 40 Å (60 H$_2$O/phosphate). Resolution of the paradox posed by these apparently contradictory findings awaits further investigation.

References

1. P. Debye and E. Hückel, *Phys. Z.*, 1923, **24**, 185.
2. T. Alfrey, Jr., P. W. Berg and H. Morawetz, *J. Polym. Sci. Part A.*, 1951, **7**, 543.
3. R. M. Fuoss, A. Katchalsky and S. Lifson, *Proc. Natl. Acad. Sci. USA*, 1951, **37**, 579.
4. S. Lifson and A. Katchalsky, *J. Polym. Sci.*, 1953, **13**, 43.
5. F. Oosawa, *J. Polm. Sci.*, 1957, **23**, 421.
6. G. S. Manning, *Biophys. Chem.*, 1977, **7**, 95.
7. G. S. Manning, *Q. Revs. Biophys.*, 1978, **11**, 179.
8. S. A. Rice and M. Nagasawa, *Polyelectrolyte Solutions*, Academic Press, New York, 1961.
9. F. Oosawa, *Polyelectrolytes*, Marcel Dekker, New York, 1971.
10. A. Katchalsky, *Pure Appl. Chem.*, 1971, **26**, 326.
11. H. Eisenberg, *Biological Macromolecules and Polyelectrolytes in Solution*, Clarendon Press, Oxford, 1976.
12. C. F. Anderson and M. T. Record, Jr., *Annu. Rev. Phys. Chem.*, 1982, **33**, 191.
13. C. F. Anderson and M. T. Record, Jr., *Annu. Rev. Biophys. Biophys. Chem.*, 1990, **19**, 423.
14. C. F. Anderson and M. T. Record, Jr., *Annu. Rev. Phys. Chem.*, 1995, **46**, 657.
15. K. A. Sharp and B. Honig, *Annu. Rev. Biophys. Biophys. Chem.*, 1990, **19**, 301.
16. B. Jayaram and D. L. Beveridge, *Annu. Rev. Biophys. Biomol. Struct.*, 1996, **25**, 367.
17. K. S. Schmitz, *Macroions in Solution and Colloidal Suspension*, VCH, New York, 1993.
18. S. A. Allison and S. Mazur, *Biopolymers*, 1998, **46**, 359.
19. S. A. Allison, *Macromolecules*, 1998, **31**, 4164.

20. S. Mazur, C. Y. Chen and S. A. Allison, *J. Phys. Chem. B*, 2001, **105**, 1100.
21. S. A. Allison, C. Y. Chen and D. Stigter, *Biophys. J.*, 2001, **81**, 2558.
22. S. A. Allison, *Biophys. Chem.*, 2001, **93**, 197.
23. S. A. Allison, Z. Li, D. Reed and N. C. Stellwagen, *Electrophoresis*, 2002, **23**, 2678.
24. S. A. Allison, *J. Colloid Interface Sci.*, 2005, **288**, 616.
25. M. Fixman, *J. Chem. Phys.*, 1979, **70**, 4995.
26. M. T. Record, Jr., T. M. Lohman and P. L. de Haaseth, *J. Mol. Biol.*, 1976, **107**, 145.
27. J. M. Schurr and B. S. Fujimoto, *Biophys. Chem.*, 2002, **101–102**, 425.
28. J. W. Klein and B. R. Ware, *J. Chem. Phys.*, 1984, **80**, 1334.
29. P. J. Heath and J. M. Schurr, *Macromolecules*, 1992, **25**, 4149.
30. C. Tanford, *Physical Chemistry of Macromolecules*, Wiley, New York, 1961.
31. E. Raspaud, M. da Conceicao and M. Livolant, *Phys. Rev. Lett.*, 2000, **84**, 2533.
32. T. Odijk and A. C. Houwaart, *J. Polym. Sci., Polym. Phys. Ed.*, 1978, **16**, 627.
33. D. C. Rau, B. Lee and V. A. Parsegian, *Proc. Natl. Acad. Sci. USA*, 1984, **81**, 2621.
34. D. C. Rau and V. A. Parsegian, *Biophys. J.*, 1992, **61**, 246.
35. V. A. Bloomfield, *Biopolymers*, 1997, **44**, 269.
36. J. B. Clendenning and J. M. Schurr, *Biopolymers*, 1994, **34**, 849.
37. J. Skolnick and M. Fixman, *Macromolecules*, 1978, **11**, 867.
38. J. Conrad, M. T. Troll and B. M. Zimm, *Biopolymers*, 1988, **27**, 1711.
39. M. T. Troll, D. Roitman, J. Conrad and B. M. Zimm, *Macromolecules*, 1986, **19**, 1186.
40. Z.-J. Tan and S.-J. Chen, *J. Chem. Phys.*, 2005, **122**, 044903.
41. D. Stigter, *Biopolymers*, 1998, **46**, 503.
42. M. A. Young, B. Jayaram and D. L. Beveridge, *J. Phys. Chem. B.*, 1998, **102**, 7666.
43. J. Ramstein and R. Lavery, *Proc. Natl. Acad. Sci. USA*, 1988, **85**, 7231.
44. G. Lamm and G. R. Pack, *J. Phys. Chem. B.*, 1997, **101**, 959.
45. G. Lamm and G. R. Pack, *Int. J. Quantum Chem.*, 1997, **65**, 1087.
46. J. R. C. van der Maarel, *Biophys. J.*, 1999, **76**, 2673.
47. G. R. Pack, G. A. Garrett, L. Wong and G. Lamm, *Biophys. J.*, 1993, **65**, 1363.
48. V. K. Misra, J. L. Hecht, K. A. Sharp, R. A. Friedman and B. Honig, *J. Mol. Biol.*, 1994, **238**, 245.
49. V. K. Misra, J. L. Hecht, K. A. Sharp, R. A. Friedman and B. Honig, *J. Mol. Biol.*, 1995, **238**, 264.
50. V. K. Misra and B. Honig, *Proc. Natl. Acad. Sci. USA*, 1995, **92**, 4691.
51. S. Hanlon, L. Wong and G. R. Pack, *Biophys. J.*, 1996, **72**, 291.

52. V. K. Misra, J. L. Hecht, A.-S. Yang and B. Honig, *Biophys. J.*, 1998, **75**, 2262.
53. D. Bashford and D. A. Case, *Annu. Rev. Phys. Chem.*, 2000, **51**, 129.
54. V. Tsui and D. A. Case, *J. Phys. Chem. B.*, 2001, **205**, 11314.
55. J. Srinivasan, M. W. Trevathan, P. Beroza and D. A. Case, *Theor. Chim. Acta*, 2002, **101**, 426.
56. V. K. Misra and D. E. Draper, *Biopolymers*, 1998, **48**, 113.
57. A. S. Petrov, G. Lamm and G. R. Pack, *Biophys. J.*, 2004, **87**, 3954.
58. A. S. Petrov, G. Lamm and G. R. Pack, *Biopolymers*, 2005, **77**, 137.
59. D. Bratko and V. Vlachy, *Chem. Phys. Lett.*, 1982, **90**, 434.
60. V. Vlachy and D. A. McQuarrie, *J. Chem. Phys.*, 1985, **83**, 1927.
61. M. Le Bret and B. M. Zimm, *Biopolymers*, 1984, **23**, 271.
62. R. Bacquet and P. J. Rossky, *J. Phys. Chem.*, 1984, **88**, 2660.
63. J. Piñero, L. B. Bhuiyan, J. Rescic and V. Vlachy, *Acta Chim. Slov.*, 2006, **53**, 316.
64. A. A. Kornyshev and S. Leikin, *Phys. Rev. E*, 2000, **62**, 2576.
65. A. A. Kornyshev and S. Leikin, *Biophys. J.*, 1998, **75**, 2513.
66. A. G. Cherstvy and A. A. Kornyshev, *J. Phys. Chem. B*, 2005, **109**, 13024.
67. H. M. Harrais, A. A. Kornyshev, C. N. Likos, H. Löwen and G. Suttman, *Phys. Rev. Lett.*, 2002, **89**, 018303.
68. D. J. Lee, A. Wynveen and A. A. Kornyshev, *Phys. Rev. E*, 2004, **70**, 051913.
69. J. D. Jackson, *Classical Electrodynamics*, Wiley, New York, 1962.
70. G. R. Pack, L. Wong and G. Lamm, *Biopolymers*, 1999, **49**, 575.
71. C. W. Outhwaite, *J. Chem. Soc. Faraday Trans. 2*, 1986, **82**, 789.
72. L. B. Bhuiyan and C. W. Outhwaite, *Philos. Mag. B*, 1994, **69**, 1051.
73. L. B. Bhuyian, C. W. Outhwaite and D. Bratko, *Chem. Phys. Lett.*, 1992, **193**, 203.
74. T. Das, D. Bratko, L. B. Bhuyian and C. W. Outhwaite, *J. Phys. Chem.*, 1995, **99**, 410.
75. T. Das, D. Bartko, L. B. Bhuyian and C. W. Outhwaite, *J. Chem. Phys.*, 1997, **107**, 9197.
76. I. Borukhov, *J. Polym. Sci. Part B.*, 2004, **42**, 3598.
77. Z.-J. Tan and S.-J. Chen, *Biophys. J.*, 2006, **90**, 1175.
78. Z.-J. Tan and S.-J. Chen, *Biophys. J.*, 2006, **91**, 518.
79. T. Ohnishi, N. Imai and F. Oosawa, *J. Phys. Soc. Jpn.*, 1960, **15**, 896.
80. J. S. McCaskill and E. D. Feckerell, *J. Chem. Soc. Faraday Trans. 2*, 1988, **84**, 161.
81. C. A. Tracy and H. Widom, *Physica A*, 1997, **244**, 402.
82. I. A. Shkel, O. V. Tsodikov and M. T. Record, Jr., *J. Phys. Chem. B*, 2000, **104**, 5161.
83. I. A. Shkel, O. V. Tsodikov and M. T. Record, Jr., *Proc. Natl. Acad. Sci. USA*, 2002, **99**, 2597.
84. D. Stigter, *J. Colloid Interface Sci.*, 1975, **53**, 296.
85. D. Stigter, *Biopolymers*, 1977, **16**, 1435.
86. S. Sugai and K. Nitta, *Biopolymers*, 1973, **12**, 1363.

87. K. A. Sharp and B. Honig, *J. Phys. Chem. B*, 1990, **94**, 7684.
88. M. E. Davis, J. D. Madura, B. A. Luty and J. A. McCammon, *Comput. Phys. Commun.*, 1991, **62**, 187.
89. J. D. Madura, J. M. Briggs, R. C. Wade, M. E. Davis, B. A. Luty, A. Ilin, J. Antosiewicz, M. K. Gilson, B. Bagheri, L. R. Scott and J. A. McCanmon, *Comput. Phys. Commun.*, 1995, **91**, 57.
90. K. A. Sharp, *Biopolymers*, 1995, **36**, 227.
91. K. A. Sharp, R. Friedman, V. Misra, J. Hecht and B. Honig, *Biopolymers*, 1995, **36**, 245.
92. J. M. Schurr, D. P. Rangel and S. R. Aragon, *Biophys. J.*, 2005, **89**, 2258.
93. P. E. Smith, *Biophys. J.*, 2006, **91**, 849.
94. E. S. Courtenay, M. W. Capp, C. F. Anderson and M. T. Record, Jr., *Biochemistry*, 2000, **39**, 4455.
95. C. F. Anderson, D. J. Felitsky, J. Hong and M. T. Record, Jr., *Biophys. Chem.*, 2002, **101–102**, 497.
96. C. F. Anderson, E. S. Courtenay and M. T. Record, Jr., *J. Phys. Chem. B*, 2002, **106**, 418.
97. P. L. Hansen, R. Podgornik and V. A. Parsegian, *Phys. Rev. E*, 2001, **64**, 64.
98. C. F. Anderson and M. T. Record, Jr., *J. Phys. Chem.*, 1993, **97**, 7116.
99. V. Misra and B. Honig, *Biochemistry*, 1996, **35**, 1115.
100. M. Gueron, J.-Ph. Demaret and M. Filoche, *Biophys. J.*, 2000, **78**, 1070.
101. J. Bond, C. F. Anderson and M. T. Record, Jr., *Biophys. J.*, 1994, **67**, 825.
102. R. L. Jones and W. D. Wilson, *Biopolymers*, 1981, **20**, 141.
103. V. K. Misra, J. L. Hecht, A.-S. Yang and B. Honig, *Biophys. J.*, 1998, **75**, 2262.
104. M. J. P. van Dongen, J. F. Doreleijers, G. A. van der Maarel, J. H. van Boom, C. W. Hilbers and S. J. Wijmenga, *Nat. Struct. Biol.*, 1999, **6**, 854.
105. V. K. Misra and D. E. Draper, *J. Mol. Biol.*, 1999, **294**, 1135.
106. V. K. Misra and D. E. Draper, *J. Mol. Biol.*, 2000, **299**, 813.
107. V. K. Misra and D. E. Draper, *J. Mol. Biol.*, 2002, **317**, 507.
108. V. K. Misra and D. E. Draper, *Biopolymers*, 2003, **69**, 118.
109. V. A. Bloomfield, *Biopolymers*, 1991, **31**, 1471.
110. R. Podgornik and V. A. Parsegian, *Biophys. J.*, 1994, **66**, 962.
111. D. Andelman, in *Handbook of Biological Physics*, ed. R. Lipowsky and E. Sackmann, Elsevier, New York, 1995, Vol. **1**, Chap. 12, 603–640.
112. G. Telléz and E. Trizac, *J. Chem. Phys.*, 2003, **118**, 3362.
113. S. L. Brenner and V. A. Parsegian, *Biophys. J.*, 1974, **14**, 327.
114. T. Odijk, *J. Chem. Phys.*, 1990, **93**, 5172.
115. A. N. Naimushin, B. S. Fujimoto and J. M. Schurr, *Macromolecules*, 1999, **32**, 8210.
116. K. Merchant and R. L. Rill, *Biophys. J.*, 1997, **73**, 3154.
117. T. E. Strzelecka and R. L. Rill, *Bipolymers*, 1990, **30**, 803.
118. T. E. Strzelecka and R. L. Rill, *Biopolymers*, 1990, **30**, 57.
119. A. Leforestier and F. Livolant, *Biophys. J.*, 1993, **65**, 56.
120. J. J. Delrow, J. A. Gebe and J. M. Schurr, *Biopolymers*, 1997, **42**, 455.

121. E. J. W. Verwey and J. Th. C. Overbeek, *Theory of the Stability of Lyophobic Colloids*, Elsevier, Amsterdam, 1948.
122. B. S. Fujimoto, G. P. Brewood and J. M. Schurr, *Biophys. J.*, 2006, **91**, 4166.
123. C. A. Sucato, D. P. Rangel, D. Aspleaf, B. S. Fujimoto and J. M. Schurr, *Biophys. J.*, 2004, **86**, 3079.
124. D. Stigter, *Biophys. J.*, 1995, **69**, 380.
125. B. M. Zimm and M. LeBret, *J. Biomol. Struct. Dyn.*, 1983, **1**, 461.
126. (a) J. M. Schurr and B. S. Fujimoto, *J. Phys. Chem. B*, 2003, **107**, 4451; (b) J. M. Schurr, *Biophys. J.*, 2001, **80**, 308a.
127. M. C. Olmsted, C. F. Anderson and M. T. Record, Jr., *Proc. Natl. Acad. Sci. USA*, 1989, **86**, 7766.
128. J. M. Schurr and S. A. Allison, *Biopolymers*, 1981, **20**, 251.
129. G. Maret and G. Weill, *Biopolymers*, 1983, **22**, 2727.
130. M. D. Wang, H. Yin, R. Landick, G. Gelles and S. M. Block, *Biophys. J.*, 1997, **72**, 1335.
131. (a) U.-S. Kim, B. S. Fujimoto, C. E. Furlong, J. A. Sundstrom, R. Humbert, D. C. Teller and J. M. Schurr, *Biopolymers*, 1993, **33**, 1725; (b) M. Le Bret, *C.R. Acad. Sci.*, 1981, **292**, 291; (c) M. Fixman, *J. Chem. Phys.*, 1982, **76**, 6346; (d) J. M. Schurr and K. S. Schmitz, *Annu. Rev. Phys. Chem.*, 1986, **37**, 271.
132. A. N. Naimushin, B. S. Fujimoto and J. M. Schurr, *Biophys. J.*, 2000, **78**, 1498.
133. S. L. Williams, L. K. Parkhurst and L. J. Parkhurst, *Nucleic Acids Res.*, 2006, **34**, 1028.
134. I. Rouzina and V. A. Bloomfield, *Biophys. J.*, 1998, **74**, 3152.
135. K.-K. Kunze and R. R. Netz, *Phys. Rev. Lett.*, 2000, **85**, 4389.
136. M. J. Stevens, *Phys. Rev. Lett.*, 1999, **82**, 101.
137. N. Grønbach-Jensen, R. J. Mashl, R. F. Bruinsma and W. M. Gelbart, *Phys. Rev. Lett.*, 1997, **78**, 2477.
138. K.-C. Lee, I. Borukhov, W. M. Gelbart, A. J. Liu and M. J. Stevens, *Phys. Rev. Lett.*, 2004, **93**, 128101.
139. M. J. Stevens, *Biophys. J.*, 2001, **80**, 130.
140. Z. Ou and M. Muthukumar, *J. Chem. Phys.*, 2005, **123**, 074905.
141. J. Klos and T. Pakula, *J. Chem. Phys.*, 2004, **120**, 2496.
142. T. E. Angelini, R. Golestanian, R. H. Coridan, J. C. Butler, A. Beraud, M. Krisch, H. Sinn, K. S. Schweizer and G. C. L. Wong, *Proc. Natl. Acad. Sci. USA*, 2006, **103**, 7962.
143. B.-Y. Ha and A. J. Liu, *Phys. Rev. Lett.*, 1997, **79**, 1289.
144. B.-Y. Ha and A. J. Liu, *Phys. Rev. Lett.*, 1998, **81**, 1011.
145. N. V. Hud and I. D. Vilfan, *Annu. Rev. Biophys. Biomol. Struct.*, 2005, **34**, 295.
146. I. Borukhov, R. F. Bruinsma, W. M. Gelbart and A. J. Liu, *Proc. Natl. Acad. Sci. USA*, 2005, **102**, 3673.
147. S. A. Allison, J. C. Herr and J. M. Schurr, *Biopolymers*, 1981, **20**, 469.

148. I. Borukhov, R. F. Bruinsma, W. M. Gelbart and A. J. Liu, *Phys. Rev. Lett.*, 2001, **86**, 2182.
149. J. H. Shibata and J. M. Schurr, *Biopolymers*, 1981, **20**, 525.
150. J. G. Duguid and V. A. Bloomfield, *Biophys. J.*, 1995, **69**, 2642.
151. B. S. Fujimoto, J. Miller, N. S. Ribeiro and J. M. Schurr, *Biophys. J.*, 1994, **67**, 304.
152. L. McFail-Isom, C. C. Sines and L. D. Williams, *Curr. Opin. Struct. Biol.*, 1999, **9**, 298.
153. E. Allahyarov, H. Löwen and G. Gompper, *Phys. Rev. E*, 2003, **68**, 061903.
154. A. A. Kornyshev and S. Leikin, *Proc. Natl. Acad. Sci. USA*, 1998, **95**, 13579.
155. A. A. Kornyshev and A. Wynveen, *Phys. Rev. E*, 2004, **69**, 041905.
156. A. A. Kornyshev, D. J. Lee, S. Leikein, A. Wynveen and S. B. Zimmerman, *Phys. Rev. Lett.*, 2004, **95**, 148102.
157. E. Allahyarov, G. Gompper and H. Löwen, *Phys. Rev. E*, 2004, **69**, 041904.
158. M. T. Mai, D. Wemmer and O. Jardetzky, *J. Am. Chem. Soc.*, 1983, **103**, 7149.
159. R. Brandes, R. R. Vold, R. L. Vold and D. R. Kearns, *Biochemistry*, 1986, **25**, 7744.
160. H. Shindo, Y. Hiyama, S. Roy, J. S. Cohen and D. A. Torchia, *Bull Chem. Soc. Jpn.*, 1987, **60**, 1631.
161. R. Podgornik, H. H. Strey, K. Gawrisch, D. C. Rau, A. Rupprecht and V. A. Parsegian, *Proc. Natl. Acad. Sci. USA*, 1996, **93**, 4261–4266.
162. H. H. Strey, J. Wang, R. Podgornik, A. Rupprecht, L. Yu, V. A. Parsegian and E. B. Sirota, *Phys. Rev. Lett.*, 2000, **84**, 3105.
163. A. A. Kornyshev, D. J. Lee, S. Leikin and A. Wynveen, *Rev. Mod. Phys.*, 2007, **79**, 943.

CHAPTER 10
Metal Ion–Nucleic Acid Interactions in Disease and Medicine

ANA M. PIZARRO AND PETER J. SADLER

Department of Chemistry, University of Warwick, Gibbet Hill Road, Coventry, CV4 7AL, UK

10.1 Introduction

Metal ions play important roles in natural biological processes, including the activity of RNA and DNA. Metal ions interact directly with RNA and DNA and they are components of the proteins and enzymes that are involved in their biosynthesis and their processing. Recent X-ray crystallographic studies are now revealing the roles of metal ions in these processes at the atomic level.

Less well understood, despite enormous interest, are the interactions of nonessential, potentially carcinogenic and mutagenic metal ions with DNA. There is often a temptation to label every compound of particular metals as carcinogenic, but in reality it is probably only certain compounds or species. We know that the biochemistry of a metal is determined by its oxidation state, the number and nature of the bound ligands and the stereochemistry of the complex. Different complexes can have different redox potentials, different ligand affinities and ligand exchange rates.

Finally, of much current interest, following the clinical success especially of platinum anticancer drugs, are the potential applications of metal complexes as therapeutic (and diagnostic) agents. The mechanism of action of metal complexes as drugs can be quite distinct from that of purely organic drugs. Drugs with new mechanisms of action are much sought after by pharmaceutical companies.

We review here some of the reported studies of metal–polynucleotide interactions which relate to the above areas. In some areas there is still a surprising lack of knowledge and in others exciting new developments. DNA and RNA continue to prove to be important targets in medicine.

10.2 Mutagenesis and Carcinogenesis

Mutations are heritable changes in genetic material (either DNA or RNA) involving changes in covalent bonds. Damaged DNA can exist for periods long enough to lead to misincorporation of sequences by DNA polymerase. The major forms of damage are oxidation, hydrolysis and alkylation. Exposure of biological systems to certain metal ions can lead to DNA damaging processes. Carcinogenesis is a complex process, caused by mutation of the genetic material of normal cells, which upsets the normal balance between cell proliferation and cell death.

Many cancer-related international agencies have classified some elements (*e.g.* chromium, nickel, arsenic, iron, copper, cadmium) as carcinogens on the basis of evidence from epidemiological data and experimental studies in animals. It is usually certain specific species of elements which are active in this way and not every species. Unfortunately, little is known about these exact species and the mechanism of action of most of carcinogens. We review the best studied carcinogens, chromium and arsenic. For those metals with redox properties, such as nickel, iron and copper, the formation of reactive oxygen species appears to be the cause of the carcinogenicity through catalysed oxidative DNA damage.

10.2.1 Chromium

Chromium is potentially an essential element, yet can cause serious damage to DNA. It exists in two major oxidation states at physiological pH, Cr^{III} and Cr^{VI}, the latter as chromate, CrO_4^{2-}, which can be absorbed through the skin and can localize in various cellular areas. Currently, Cr^{VI} compounds are recognized human carcinogens.

Cellular uptake of Cr^{VI}, followed by its reduction to Cr^{III} with the formation of reactive $Cr^{V/IV}$ intermediates, is currently a generally accepted cause of Cr^{VI}-induced genotoxicity and carcinogenicity. This assumption (reviewed in 2001, 2003 and 2005)[1–3] is based on Wetterhahn's uptake-reduction model,[4] which includes: (i) the active transport of Cr^{VI} into cells (through anion channels for soluble chromates or through phagocytosis for insoluble compounds such as $PbCrO_4$); (ii) the intracellular reduction of Cr^{VI} with the formation of potentially DNA-damaging $Cr^{V/IV}$ intermediates (which can be stabilized by intracellular ligands) and organic radicals; (iii) the formation of kinetically inert Cr^{III} complexes, including highly genotoxic DNA–Cr^{III}–protein and DNA–Cr^{III}–DNA crosslinks, as a result of such reduction.[4]

The redox reactions of chromium may explain Cr^{VI}-induced genotoxicity. Such reactions can involve all the oxidation states from Cr^{VI} to Cr^{II} and are complicated by ligand-exchange processes, reactions between Cr species and formation of organic radicals (capable of initiating radical chain reactions with O_2).[5] Owing to this complexity, the mechanisms of Cr reactions under biologically relevant conditions are generally poorly understood. Levina and Lay have compiled an extensive review of all these redox mechanisms and the possible biological implications of the different intermediates.[3]

Reactive Cr^V intermediates and their implication in DNA damage are a focus for studies of the chemical mechanisms involved in Cr^{VI} toxicity.[1,2,6] It is likely that the first Cr^V complexes formed in biological systems upon the addition of Cr^{VI} are those of strong reductants, such as glutathione (GSH) or ascorbate.[1,2] Such complexes have been implicated as the likely DNA-damaging species in the Cr^{VI} + reductant + O_2 systems or Cr^{III} + H_2O_2 systems (as Cr^{VI} itself does not react with isolated DNA).[2,7,8]

Although the mechanisms of Cr^{VI} reduction to Cr^{III} under biologically relevant conditions is poorly understood, the existing mechanistic evidence points to Cr^V as the main, if not ultimate, reactive species in Cr^{VI}-induced genotoxicity.[1-3,6,9]

Formation of DNA lesions, including strand breaks, abasic sites and Cr^{III}–DNA crosslinks in Cr^{VI}-exposed cells, is believed to be the primary cause of Cr^{VI}-induced carcinogenicity.

10.2.2 Arsenic

Chronic exposure to arsenic compounds in occupational and environmental settings has been linked to increased prevalences of cancers of the skin, lung, liver, kidney and bladder.[10] In humans and many other species, arsenic compounds undergo complex cellular metabolism. Arsenite ($As(OH)_3$ at pH 7) is methylated in the liver and other tissues to monomethylarsonic acid ($[CH_3AsO_3]^{2-}$) and dimethylarsinic acid ($[(CH_3)_2AsO_2]^-$), that contain pentavalent arsenic.[11] The pathway for conversion of As into these metabolites can be summarized as $As^{III} \rightarrow MeAs^V \rightarrow MeAs^{III} \rightarrow Me_2As^V \rightarrow Me_2As^{III}$, where alternating steps of oxidative methylation and the reduction of As^V to As^{III} yield a series of intermediates and products.[12] Although methylation of arsenic has sometimes been described as a detoxification process,[13] studies indicating that trivalent methylated arsenicals may be genotoxic *in vitro* have been published.[14]

The molecular and cellular mechanisms by which arsenic acts as a carcinogen are controversial. A study by Mass *et al.* showed that the trivalent methylated arsenicals $MeAs^{III}$ and Me_2As^{III} are directly genotoxic.[14] Supercoiled ΦX174 DNA was nicked by Me_2As^{III} (150 μM) and $MeAs^{III}$ (30 mM). Neither arsenite, arsenate, $MeAs^V$ nor Me_2As^V were able to nick supercoiled DNA. Me_2As^{III} and $MeAs^{III}$ also damaged human lymphocyte DNA and were 386 and 77 times more potent, respectively, than arsenite.[14] However, it is generally accepted that arsenic does not interact directly with DNA.[15]

Reviews on the possible mechanisms of toxicity and carcinogenesis of arsenic and the evidence for oxidative stress being a major contributor to arsenic carcinogenesis have been published.[16–19] Shi *et al.* have also summarized the literature data on arsenic-mediated generation of reactive oxygen species (ROS) and reactive nitrogen species (RNS) in various biological systems and discussed the role of ROS and RNS in arsenic-induced DNA damage.[18]

10.2.3 Other Metals

Metals such as nickel, cadmium, iron and copper have also been considered to be human carcinogens based on growing evidence from epidemiological studies and *in vitro* model system investigations.[20–28] However, their molecular mechanisms of carcinogenesis are not well understood and there is no convincing evidence for direct mutagenic metal–DNA adduct formation. For example, two mechanisms have been discussed that play an important role in cadmium mutagenicity, (i) induction of ROS and (ii) inhibition of DNA repair,[29] whereas it has been observed that cadmium binds only weakly to DNA.[30] Current hypotheses on the mechanisms of nickel,[31–33] iron[34] and copper[35] carcinogenesis have recently been reviewed.

Carcinogenesis of these metals has been related to their redox activity. They are believed to induce formation of ROS, such as $O_2^{\bullet-}$ and $^{\bullet}OH$ radicals, which react rapidly with DNA and produce DNA damage by breaking strands or modifying the bases and/or deoxyribose leading to carcinogenesis.

10.3 Antiviral Drugs

Although there are currently no metal-based antiviral drugs in the clinic, efforts have been devoted to the development of this type of drug and the importance of metal–nucleic acid interactions has also been apparent in investigations of the mechanism of action of some antiviral compounds.

Aoki and Kimura have synthesized lipophilic Zn^{II}–cyclen complexes that might be effective carriers of the antiviral drug AZT.[36] AZT, 3′-azido-3′-deoxythymidine (Zidovudine), belongs to a family of nucleoside drugs that together with 2′,3′-dideoxycytidine (ddC) and 2′,3′-dideoxyadenosine (ddA), are currently used in the treatment of AIDS.[37,38] These drugs enter cells primarily by simple passive diffusion[39] and then undergo phosphorylation to the nucleotide triphosphate derivatives to act as HIV reverse transcriptase inhibitors.[40,41]

Kimura and co-workers have exploited the idea of creating lipophilic carriers for these nucleotide phosphates, on the basis that this would greatly increase the efficacy of the drug.[42] Lipophilic Zn^{II}–cyclen derivatives such as

Zn^{II}–hexadecylcyclen complex **1**, were found to be selective and effective carriers of dT and U.

![Structure 1: Zn^{2+} cyclen complex with H$_2$O and hexadecyl chain]

1

The same workers have also attempted to achieve anti-HIV activity using a different approach. This involves preventing the formation of key complexes of viral RNA with proteins, which again is based on the highly selective interaction of Zn^{II}–cyclens with uridine (U) and thymidine (T) nucleotides.[43]

The TAR-RNA of HIV-1 is recognized by the HIV-1 regulatory protein Tat. The binding of Tat to TAR-RNA up-regulates HIV-1 mRNA transcription.[44–46] The TAR (*trans*-activation responsive) element comprising the first 59 nucleotides of the HIV-1 primary transcript adopts a hairpin structure with a uracil-rich bulge (UUU or UCU) located four base pairs below a six-nucleotide loop (Figure 10.1). The UUU bulge is the Tat binding site and the loop is a homing site for cellular proteins.[47] The binding of the Tat protein (Figure 10.2) to the TAR-RNA is a crucial event in the transcription of viral DNA, because it stabilizes the active complex between the polymerase, the DNA template and the nascent mRNA. When the Tat protein is not bound to the TAR-RNA, premature abortion of transcription takes place.[46] These findings showed that the TAR-RNA and the Tat protein are potential targets for anti-HIV agents.

Zn^{II} complexes of cyclen interact selectively with uridine (U) and thymidine (T) nucleotides. The unique base-sequence recognition ability of bis(Zn^{II}–cyclen),

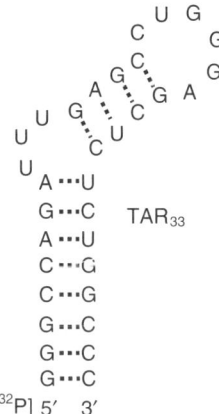

Figure 10.1 A TAR model (TAR$_{33}$), containing residues 17–43 of HIV-1 mRNA and three additional GC pairs.

```
M E P V D P R L E P W K H P G S Q P K T A C T N C T Y C K K
C C F H C Q V C F I T K A L T I S Y G R K K R⁵² R⁵³ Q⁵⁴ R R R P
P Q G S Q T H Q V S L S K Q P T S Q S R G D P T G P K E
```

Figure 10.2 The Tat protein. The arginine-rich region is written in bold letters. Residues Arg52 and Arg53 are responsible for the selective binding to the bulge region of HIV-1 mRNA.

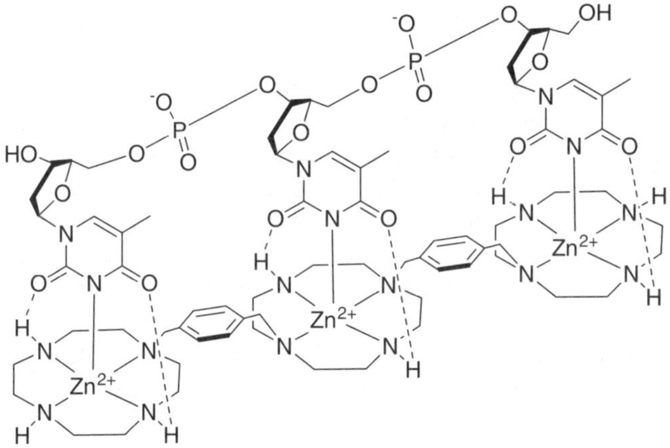

Figure 10.3 1:1 3–d(TpTpT) adduct.[437] Hydrogen bonds are indicated by dashed lines.

2, and tris(Zn^{II}–cyclen), **3**, for TAR-RNA (see Figure 10.3)[43] gives rise to the most potent inhibitors of TAR-Tat binding ever reported. Footprinting analysis using micrococcal nuclease (ribo- and deoxyribonuclease) indicate that **2** and **3** strongly protect the loop region UUU in the TAR model sequence, with IC_{50} values (drug concentration required for 50% nuclease inhibition) of 3 μM and 25 nM, respectively.[43]

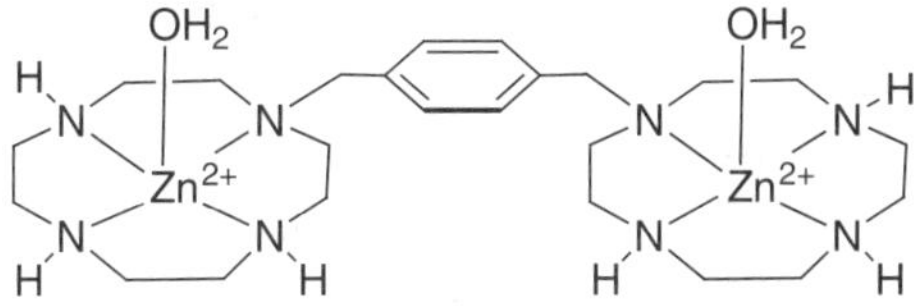

2

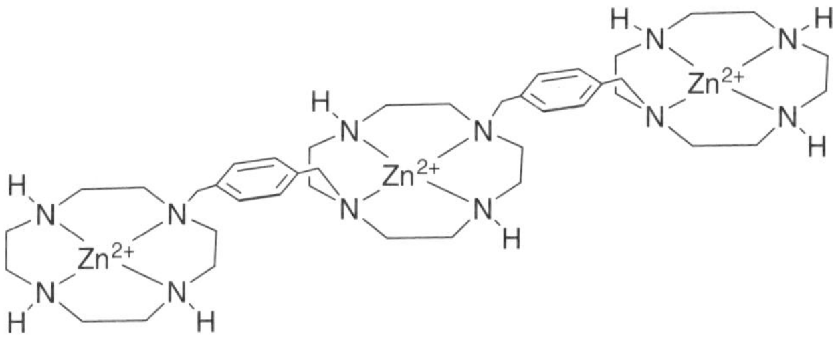

3

The potential medical applications of the interactions of nucleic acids with both zinc complexes and, more generally, non-metallated macrocyclic polyamines, have been recently reviewed.[48,49]

Natile and co-workers have synthesized complexes that hybridize two types of drugs, one antitumour active (cisplatin-like) and the other bearing antiviral properties (*e.g.* acyclovir (ACV), penciclovir (PEN), famciclovir (FAM)), in search of a synergistic effect of both agents. The resulting compound contains a *cis*-{Pt(NH$_3$)$_2$} moiety, a labile leaving halide and a guanine analogue as an antiviral drug acting as a ligand on the metal centre (see Chart 10.1).[50–52] These complexes exhibit antiviral activities but appear to have lost some of the antitumour properties of the parent cisplatin.

They have also synthesized antiviral-Pt derivatives with non-labile ligands[53–58] (Chart 10.2), obtaining cationic species that appear to keep or improve, depending on the virus, the antiviral properties of acyclovir or penciclovir themselves and with a cytotoxicity lower than that of cisplatin.[54,55]

There is increasing interest in this use of known anticancer (Pt-based) complexes as antiviral or virucidal and microbicidal agents.[59–61]

Morrison and co-workers have developed a class of octahedral Rh^{III} complexes that are photoactivated by UVA light and possess both anticancer and antiviral activities (see Section 10.4.4 for anticancer properties).[62–64]

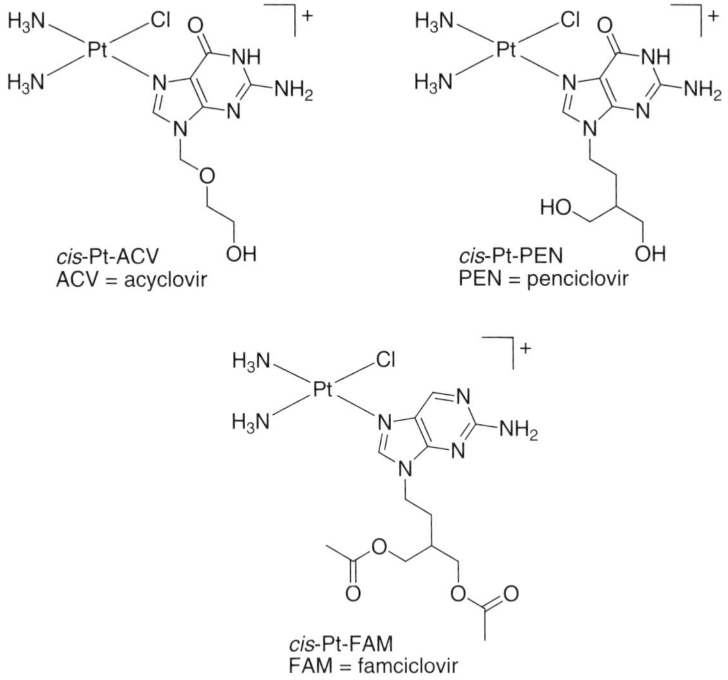

Chart 10.1

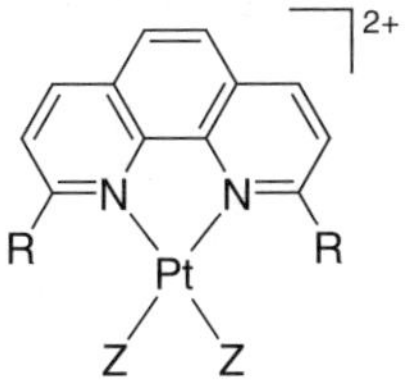

R = CH₃ or H
Z = acyclovir or penciclovir

Chart 10.2

Complex **4**, *cis*-dichloro(dipyrido-[3,2-*a*:2′,3′-*c*]-phenazine)(1,10-phenanthroline) rhodium(III) chloride, DPPZPHEN, can act as a photobinding agent toward nucleic acids in an hypoxic environment and as a photonuclease. DPPZPHEN is phototoxic to tumour cells and to the alphavirus Sindbis (SINV) and renders the viral genome non-infectious.

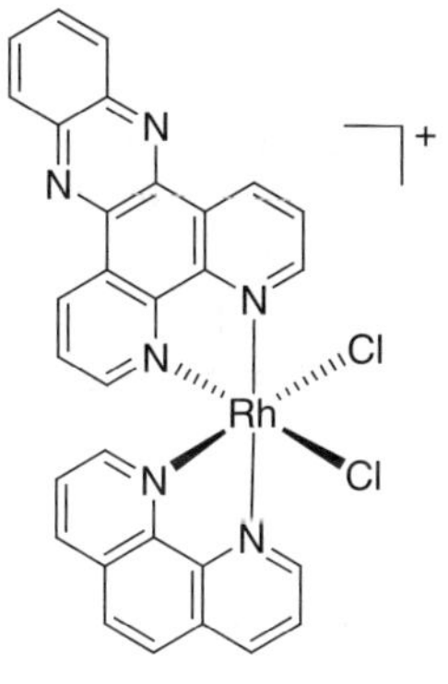

DPPZPHEN

4

Analysis of Sindbis virus, following irradiation with the metal complex, confirmed that the viral genome is a primary target for the photoactivated complex and is rendered non-infectious by this treatment. Model studies with calf thymus and supercoiled plasmid DNA indicate that the complex can both bind to and nick polynucleotides. In its antiviral activity, DPPZPHEN can penetrate the Sindbis viral capsid, showing the ability to migrate through the viral glycoprotein, lipid and capsid layers before inactivating the viral genome.

The metal complex is able to reach the RNA core and react with the nucleic acid, despite the lipid bilayer and glycoprotein and nucleocapsid shells shielding the RNA.[62]

10.4 Metal-containing Anticancer Agents

Drugs can be classified according to the type of DNA association: coordinative binding agents, groove binders, intercalators and – most recently classified – phosphodiester backbone binders. The non-coordinative interactions rely on electrostatic interactions, shape and size molecular recognition and hydrogen bonding. To date, the only DNA-targeting metal-containing anticancer drugs in clinical use are platinum-based drugs and these are believed to interact primarily with the nucleic acid through coordination bond formation.

We will highlight here those metals that have reached clinical trials (see Table 10.1) and those that are promising and currently under preclinical investigation. For example, there are six platinum complexes in the clinic today and at least 10 others in clinical trials. There are two ruthenium complexes in trials (NAMI-A and KP1019), but trials of two titanium complexes (budotitane and titanocene dichloride) have now been abandoned, whereas trials of gallium complexes (gallium tris-maltolate and gallium tris-8-hydroxyquinolinolate KP46) are still in progress. Organometallic complexes of ruthenium and complexes of rhodium, gold and iron are undergoing preclinical investigations for potential trials.

10.4.1 Platinum Drugs

Since the discovery of the antineoplastic activity of cisplatin (*cis*-diamminedichloridoplatinum(II)) at the end of the 1960s, thousands of platinum complexes have been designed, synthesized and tested for anticancer activity. However, only three – cisplatin, carboplatin and oxaliplatin – have been approved worldwide for use in the clinic (approved by the FDA in 1978, 1989 and 2002, respectively), whereas another three – nedaplatin, lobaplatin and heptaplatin – have been approved for clinical use only in Japan, China and South Korea, respectively (Chart 10.3). Several other platinum complexes are still in trials (Chart 10.4). Progress with some appears to be slow (see Table 10.1) as a consequence of low efficiency and/or no advantages over cisplatin and carboplatin and/or unacceptable toxicity issues.

Current research on platinum complexes is directed partly towards the development of complexes that kill cancer cells with inherent low response to current chemotherapy. *trans*-Platinum complexes and multinuclear platinum complexes are promising candidates as they are likely to give rise to unusual DNA adducts and hence different activity profiles compared with cisplatin and its analogues. Another, maybe even more important feature to consider in the design of novel platinum complexes, is the capability of certain drugs to be activated only under selected circumstances, as a solution for the severe side-effects of currently-used platinum complexes. Examples are photoactivatable platinum complexes, which are inert until activated by certain wavelengths of light.

Understanding the individual steps involved in binding to the ultimate target of the drug, DNA, including the binding kinetics, the nature of different DNA adducts and their recognition by nuclear proteins and other intracellular

Table 10.1 Metal anticancer compounds which have entered clinical trials or been approved as drugs.

Metal	Complex name	Clinical state
Ti^{IV}	Budotitane[a]	Phase I trials in late 1990s[414]
	Titanocene dichloride[b]	Phase II trials 2000[238,239]
Ga^{III}	Gallium nitrate	Phase II clinical trials[415,416]
	KP46 (or FFC11)[c]	Phase I trials completed[417]
Pt^{II}	Cisplatin[d]	Clinical use worldwide since 1978
	Carboplatin[d]	Clinical use worldwide since 1989
	Oxaliplatin[d]	First approval in France (second-line treatment in 1996, first-line treatment in 1998)
		Clinical use in USA since 2002 as second-line treatment (accelerated approval)
		Full approval by FDA in 2004
	Nedaplatin (254-S)[d]	Clinical use in Japan recently
	Lobaplatin[d]	Clinical trials worldwide
		Approved in China in 2003
	Heptaplatin[d]	Approved in South Korea in 1999
	Picoplatin (ZD0473)[e]	Entered clinical trials in 1997[418–420]
	Aroplatin$^+$ (L-NDDP)[e]	Trials in dogs in 1989 and cats in 1999[421,422]
		In phase II in 2005 and 2006.[423,424]
	Enloplatin[e]	Phase II in 1997[425]
	Miboplatin (DWA2114R)[e]	Trials in Japan since the 1990s[129,426]
	Sebriplatin (or Cl-973 or NK-121)[e]	Trials in Japan since the 1990s[129,426]
	Zeniplatin[e]	Phase II in the 1990s[427]
	BBR3464[f]	Phase II in USA[428,429]
Pt^{IV}	Iproplatin (or CHIP or JM9)[g]	Phase III.[145,430]
	Tetraplatin (or ormaplatin)[g]	Trials abandoned in Phase I. Severe neurotoxicity[147]
	Satraplatin (or JM216)[g]	Phase III trials[149,150]
Ru^{III}	NAMI-A[h]	Antimetastatic agent. Completed phase I in 2006[279]
	KP1019[h]	Completed phase I in 2005[298]

[a]Complex **17**.
[b]Complex **16**.
[c]In Chart 10.13.
[d]In Chart 10.3.
[e]In Chart 10.4.
[f]In Chart 10.8.
[g]In Chart 10.5.
[h]In Chart 10.12.

processes as cellular responses to nucleic acid attack, will improve the design and further development of platinum-based drugs.

In this section, we explore the nature of cisplatin–DNA adducts as the cause of the mechanism of action of cisplatin and the cellular consequences of such adduct formation. We also review current developments with other promising platinum drugs and their possible future use in chemotherapy.

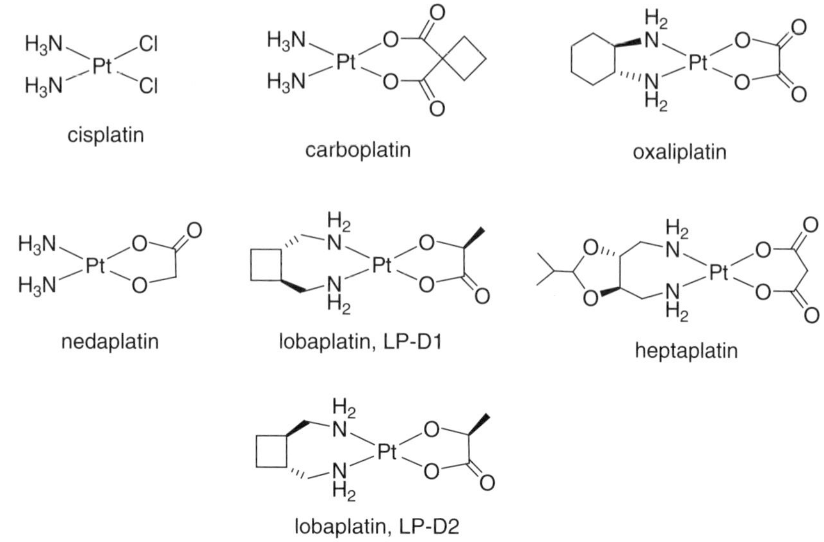

cisplatin

carboplatin

oxaliplatin

nedaplatin

lobaplatin, LP-D1

heptaplatin

lobaplatin, LP-D2

Chart 10.3

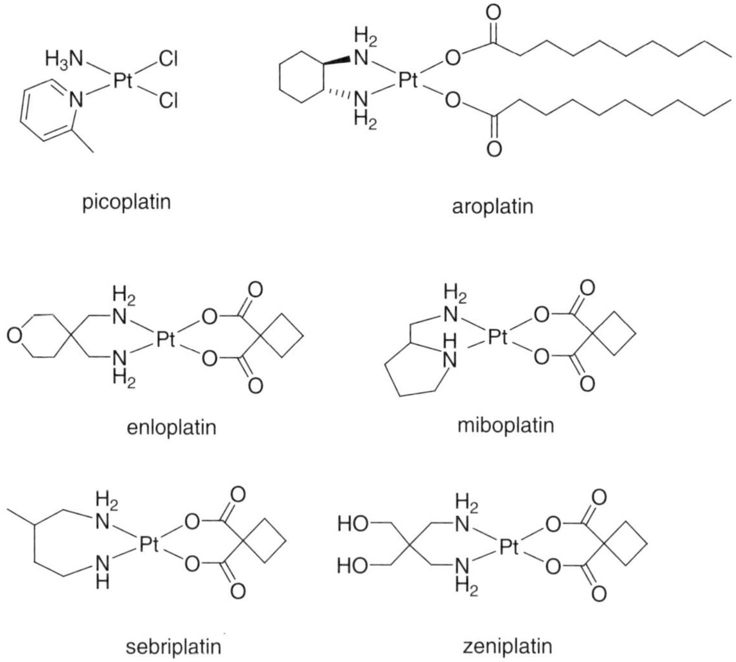

picoplatin

aroplatin

enloplatin

miboplatin

sebriplatin

zeniplatin

Chart 10.4

10.4.1.1 Cisplatin

In the mid-1960s, Rosenberg and co-workers were investigating the effect of electric fields on the growth of *Escherichia coli* using platinum electrodes in a solution containing chloride and ammonium salts, among other constituents, when they observed an unexpected event: the bacteria became long filaments and did not replicate. Cell division in *E. coli* was thus found to be inhibited.[65] Extensive analysis of this unpredicted observation led them to the conclusion that the bacterial replicative cycle was inhibited by platinum ammine complexes formed by electrolysis at the platinum electrodes and this led to the subsequent use of *cis*-[PtCl$_2$(NH$_3$)$_2$] as an anticancer drug.

Today, 30 years after its approval as a chemotherapeutic agent by the US Food and Drug Administration (FDA), cisplatin is still one of the world's best selling anticancer drugs. It is responsible for the cure of over 90% of cases of testicular cancer and it plays an important role in some cancer treatments such as ovarian, head and neck cancer, bladder cancer, cervical cancer, melanoma and lymphomas.

Regardless of the achievements of cisplatin, the drug is efficient in only a limited range of cancers and some tumours develop acquired immunity during treatment. Additionally, cisplatin can cause severe side-effects. Nausea and vomiting, bone marrow suppression and kidney toxicity are among the most common side-effects. An understanding of the molecular basis of these effects may aid the design of platinum analogues that overcome toxicity while maintaining the effectiveness of cisplatin.

Evidence from both pre-clinical studies and clinical investigations has strongly implicated DNA as the biological target for cisplatin through the formation of irreversible adducts *via* the process of ligand exchange.[66,67] In this section, we review the fate of cisplatin from its aquation leading to its reaction with nuclear DNA, we describe in detail the nature of cisplatin–DNA adduct formation and relate this interaction to the induction of cell death.

Aquation. Any drug able to hydrolyse will be susceptible to aquation as soon as it enters body fluids. In the case of cisplatin, the high concentration of chloride ions in blood plasma (*ca.* 100 mM) preserves its stability towards hydrolysis.[68] However, the significantly lower intracellular chloride concentration (*ca.* 4–23 mM) facilitates rapid hydrolysis of the chloride ligands of cisplatin, leading to activated cationic species capable of reacting mono- and bifunctionally.

The aquation of cisplatin has been extensively investigated and reviewed.[69–81] The two hydrolysis steps, to *cis*-[PtCl(H$_2$O)(NH$_3$)$_2$]$^+$ and further to *cis*-[Pt(H$_2$O)$_2$(NH$_3$)$_2$]$^{2+}$ (see Scheme 10.1) follow first-order kinetics and have been investigated in the absence and presence of nucleic acid and oligonucleotides.

The first aquation step, considered the rate-limiting step for DNA interaction,[82–84] is remarkably faster than the second step under different conditions of concentration, temperature and ionic strength. However, regarding the second aquation in the presence of DNA, there is controversy in the literature. On one hand, it has been suggested that cisplatin undergoes the second aquation during

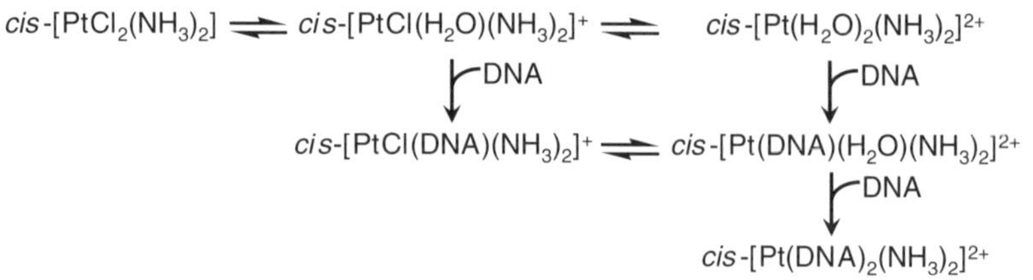

Scheme 10.1

cis-[PtCl$_2$(NH$_3$)$_2$] ⇌ cis-[PtCl(H$_2$O)(NH$_3$)$_2$]$^+$ ⇌ cis-[Pt(H$_2$O)$_2$(NH$_3$)$_2$]$^{2+}$

↓ DNA ↓ DNA

cis-[PtCl(DNA)(NH$_3$)$_2$]$^+$ ⇌ cis-[Pt(DNA)(H$_2$O)(NH$_3$)$_2$]$^{2+}$

↓ DNA

cis-[Pt(DNA)$_2$(NH$_3$)$_2$]$^{2+}$

pre-association with the nucleic acid and is already transformed into the diaqua product prior to binding to DNA.[74] The second and perhaps most popular theory is that the monoaqua complex attacks DNA. Then, the remaining platinum–chloride bond of the cis-[PtCl(DNA)(NH$_3$)$_2$] adduct undergoes hydrolysis (second hydrolysis step) and closure to the bifunctional adduct cis-[Pt(DNA)$_2$(NH$_3$)$_2$] takes place (see Scheme 10.1). Reactions between the 14-mer 5′-d(AATTGGTACCAATT)-3′ and cisplatin and cis-[PtCl(NH$_3$)$_2$(OH$_2$)]$^+$ have been investigated and are consistent with the latter theory. Reactions between the oligonucleotide and the diaqua species were also considered, but accounted for only about 1% of the platination products. This result was primarily due to kinetic incompatibility; the second aquation is slower than the platination of DNA by the monoaqua species.[76,81] In this chapter we focus on the cisplatin interaction with DNA, the kind of DNA–cisplatin adducts formed and which adducts are the critical lesions.

Cisplatin–DNA adducts. From Rosenberg and co-workers' early studies with *E. coli*, it was suspected that any agent which inhibits cell division should interfere with DNA replication, while the filamentous growth suggested that RNA and proteins were performing normally. Thus, already in the 1970s the implication of DNA in the cytotoxic properties of cisplatin was well established,[85–88] and nuclear DNA has since been thought to be the ultimate target of cisplatin and related metal-based antineoplastic agents.[66,89–93]

Once in body fluids, although activated cisplatin can interact with various biomolecules,[94] its antitumour activity derives from its capability to form bifunctional DNA crosslinks. The predominant adducts formed by cisplatin with DNA are bifunctional GG intrastrand crosslinks, where the chlorido ligands have been replaced by the nitrogen N7 of adjacent guanine (G) residues.[95] Chelation by DNA triggers a specific distortion (bending, kinking) of the duplex. This particular kink in DNA is considered to be the critical lesion due to the chain of events that the bimolecular recognition eventually provokes in the cell.

The most electron-dense and accessible sites in DNA for electrophilic attack of platinum are nitrogen N7 atoms of purines. They are exposed in the major groove of the double helix and not involved in base-pair hydrogen bonding. The major cisplatin–DNA adducts – comprising around 90% of all the adducts –

both *in vitro* and *in vivo* were indeed found to be: 1,2-intrastrand d(GpG) crosslinks (between adjacent guanines) in *ca.* 50–65% and 1,2-intrastrand d(ApG) crosslinks in *ca.* 25% (between adenine and adjacent guanine from 5′ to 3′). In addition, 1,3-intrastrand d(GpNpG) crosslinks (intrastrand adducts between purines separated by one or more intervening bases) were present in less than 10%, together with interstrand crosslinks, monofunctional adducts to guanine residues and DNA–Pt–protein crosslinks (see Figure 10.4).[89–90,95–98]

In recent years, however, a discrepancy has emerged regarding the scarce interaction of cisplatin or its aquated species with d(GpA) binding sequences.[99,100] Kozelka and co-workers have shown that cisplatin forms both d(GpA) and d(ApG) adducts with a 17-mer hairpin in a ratio *ca.* 1:2 favouring, as would be expected, d(ApG).[99] The authors discussed possible reasons why the bifunctional d(GpA) adducts had not been identified in previous studies.

The significant prevalence of intrastrand GG crosslinks over any other adduct has been examined in the context of contributions of hydrogen bonding, the electrostatics in the vicinity of the binding site and the steric effects of adduct conformation. It has been argued that hydrogen bonding between the NH_3 of the DNA–cisplatin complex and O6 of a nearby guanine – not possible for adenine – has a large impact on the adduct configuration, being essential for stability and eventual cytotoxic activity.[101–103] However, it has also been

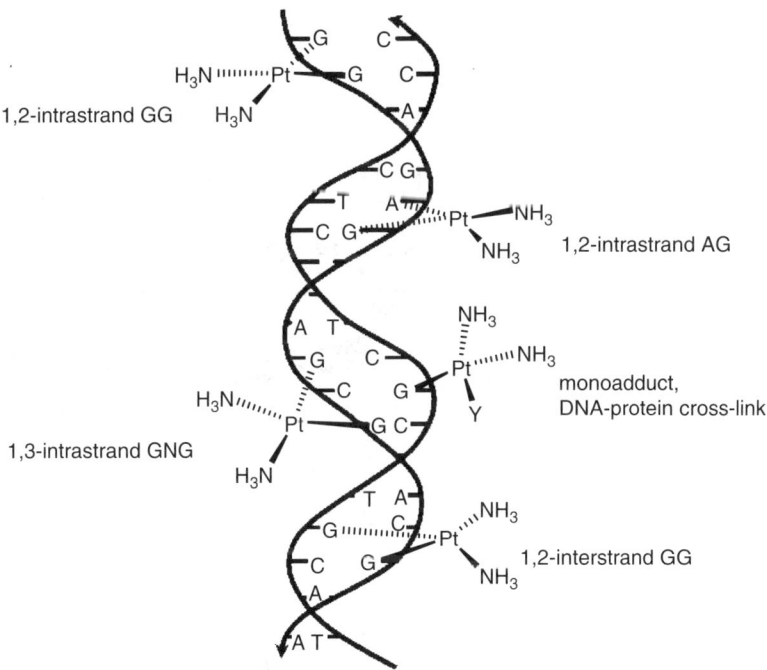

Figure 10.4 Different types of DNA–cisplatin adducts for a given random sequence. The 1,2-intrastrand GG cross-link (top left) is believed to be the critical lesion. Y = H_2O/OH, Cl, protein.

proposed that it might be the small size of the NH_3 group, rather than its hydrogen bonding ability, that is important.[104]

The different possible conformations of the GG intrastrand crosslink depend on the orientation of the exocyclic oxygen O6 with respect to the platinum coordination plane (see Figure 10.5). Various studies, recently reviewed by Marzilli and Natile,[105] on the binding of cisplatin to guanine bases using the cis-[Pt(NH$_3$)G$_2$] model, where G is a detached or tethered guanine base, have shown that the head-to-tail (HT) conformation is the most stable.[56,106–109] This is the conformation occurring in the minor GG crosslinks between different strands. In platinated DNA, however, the major conformation is head-to-head (HH), occurring in the most common 1,2-intrastrand d(GpG) crosslinks, but seldom happening in the majority of the investigated cis-[Pt(NH$_3$)G$_2$] models.[53,106,110]

Lippard and co-workers published the first crystal structure of an adduct of cisplatin with a dinucleotide, cis-[Pt(NH$_3$)$_2${d(pGpG)}].[108] The nucleobases are in an *anti* conformation and the two O6 oxygen atoms are on the same side of the platinum coordination plane, *i.e.* head-to-head orientation. The same group published the first crystal structure of the cisplatin fragment coordinated to a double-stranded oligonucleotide (a dodecamer), cis-[Pt(NH$_3$)$_2$ d(CCTCTGGTCTCC)·d(GGAGACCAGAGG)] (see Figure 10.6, left).[111] They observed rolling of the adjacent guanines towards one another by 49°;

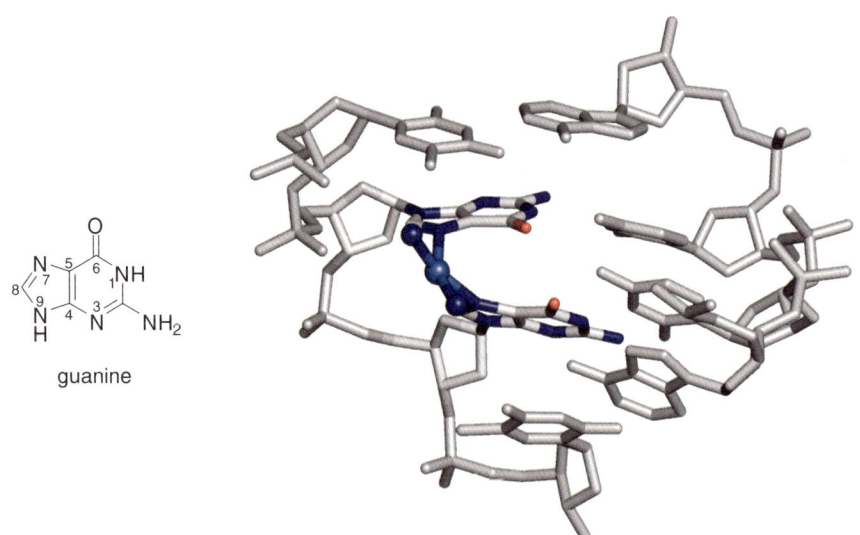

Figure 10.5 Guanine base (left) and cis-{Pt(NH$_3$)$_2$}–GG (right) in a head-to-head conformation as part of the adduct formed from the reaction of cisplatin with the duplex 5′-d(CCTCTGGTCTCC)·d(GGAGACCAGAGG). Atom colours: Pt, light blue; N coordinated to Pt (and other guanine nitrogens), deep blue; C6O in cross-linked guanines, red; other atoms, grey for clarity. Coordinates taken from the Protein Data Bank, 1AIO.[111]

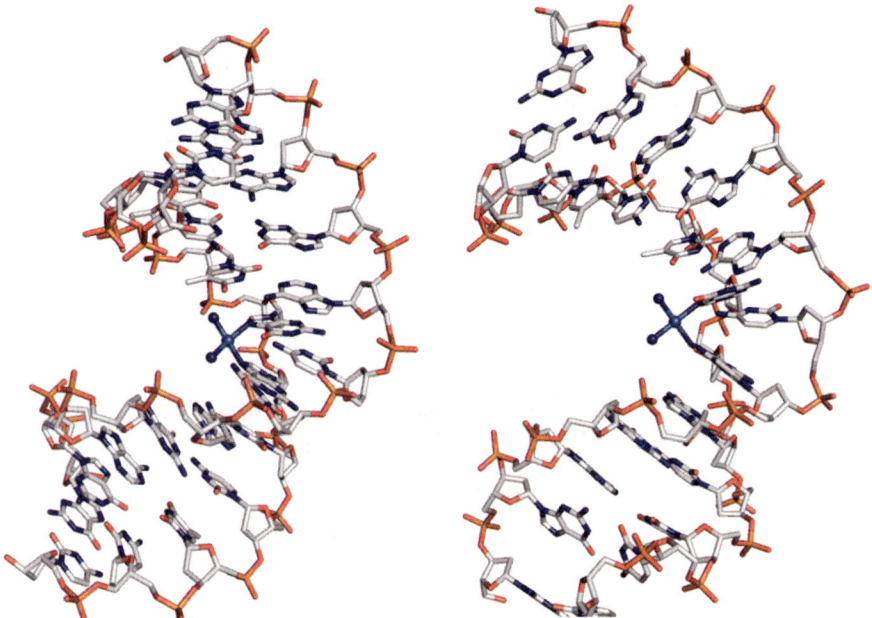

Figure 10.6 X-ray structure[111] (left) and NMR solution structure[112] (right) of the adduct formed by interaction of cisplatin and the 12-mer d(CCTCTG-GTCTCC)·d(GGAGACCAGAGG). Pt, light blue; N, deep blue; O, red; P, orange; C, grey. PDB: 1AIO and 1A84, respectively.

however, the overall bend in the duplex could not be quantified. Three years after this crystal structure, the same group published the NMR structure of the same adduct in solution.[112] They found similar features, although more pronounced in solution (Figuer 10.6, right). The rolling of the guanines towards one another was 61°, leading to an overall helix bend angle of 78°. The NMR study also gave the unwinding of the helix at the site of platination, which was 25°. Common to both structures is that the coordination of the cis-$\{Pt(NH_3)_2\}^{2+}$ fragment to G6 and G7 bends the duplex significantly towards the major groove without disrupting the Watson–Crick hydrogen bonding. As a result, the minor groove opposite the platinum adduct is widened and flattened, affording a geometry with characteristics of A- and B-type DNA.

Cellular response to DNA damage. Once cisplatin has reached and attacked DNA, the response of the cell towards repair of the damage is crucial for the anticancer activity of cisplatin. Cisplatin–DNA adducts are primarily removed by nucleotide excision repair (NER), which consists of recognition, incision, excision, repair synthesis and ligation of DNA.[67] If this repair mechanism fails, the probability of cancer cell survival decreases.

To answer why cisplatin–DNA damage is not repaired, we must go back to the kink that the GG intrastrand crosslink induces in duplex DNA. Lippard and co-workers argue that the GG crosslink, which helps stabilize an opened and flattened minor groove, may be an essential structural feature in the recognition and binding of the HMG-domain proteins (see Figure 10.7). In the ternary adduct HMG1-DNA-Pt a phenylalanine residue of protein helix II intercalates between the bases in the platinated GG crosslink.[115] This can lead to repair shielding; the damage persists and ultimately provokes apoptosis.[113–115] A second hypothesis suggests that cisplatin-damaged DNA hijacks HMG-domain

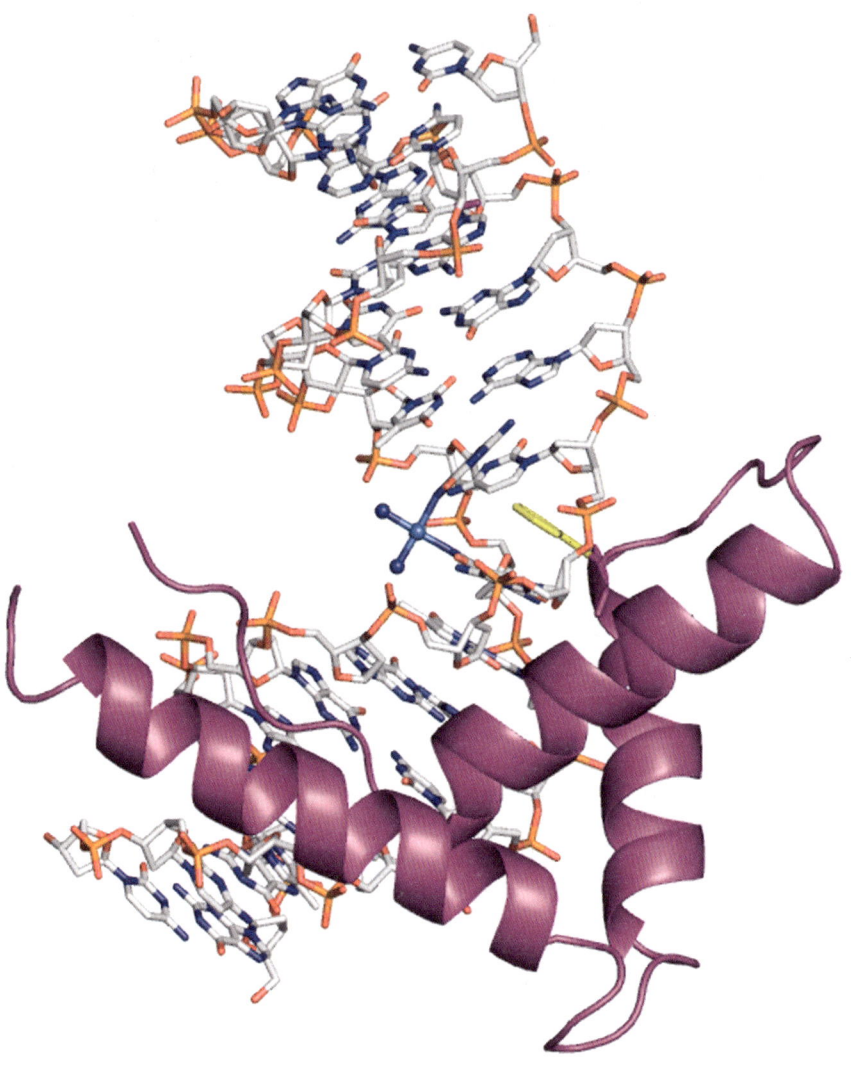

Figure 10.7 Adduct formed between the non-sequence-specific domain A of HMG1 and cisplatin-modified DNA. Pt, light blue; N, deep blue; O, red; P, orange; C, grey. HMG is a purple ribbon with the intercalating phenylalanine highlighted in yellow. Coordinates taken from the Protein Data Bank, 1CKT.[115]

proteins and other nuclear proteins away from their natural binding sites, leading to cellular stress and eventually cell death. These two hypotheses are not mutually exclusive.[116]

The HMG domain is an 80-amino acid structure-specific DNA binding motif. HMGB1 is an architectural protein important in maintaining chromatin structure. It increases transcription activation and functions as an extracellular signalling agent during inflammation, cell differentiation, cell migration and tumour metastases.[117–119]

However, this is not the only protein which provides the cellular response to cisplatin-damaged DNA. A number of proteins have also been implicated in Pt–DNA adduct recognition. The TATA-binding protein (TBP) also binds to cisplatin-modified DNA and appears to have the highest specificity for Pt-damaged DNA of all known platinated DNA-binding proteins.[120,121] Other proteins can contribute to enhance the binding, such as p53,[122] sex-determining region Y protein (SRY) and human upstream binding factor (hUBP).[67,116,123]

10.4.1.2 DNA Adduct Formation of Cisplatin-like Drugs in Clinical Use

There are six platinum-based drugs in clinical use today, *i.e.* cisplatin, carboplatin and oxaliplatin (*l*-OHP) worldwide, and nedaplatin, (254-S), lobaplatin, (D-19466) and heptaplatin (SKI 2053R) in Japan, China and South Korea, respectively.

Carboplatin was developed in the search for a drug less toxic than cisplatin. It was thought that a platinum analogue with the same carrier ligands as cisplatin would form the same DNA lesions, maintaining cisplatin antitumour activity. At the same time, by having less labile ligands, this analogue would react with nucleophiles to a lesser extent than cisplatin, being less toxic. Thus, it was found that carboplatin, where the ligand susceptible to nucleophilic substitution is cyclobutanedicarboxylate, is less toxic than cisplatin, with no significant renal or neural toxicity, and at the same time is as effective as or even more active than cisplatin in some tumours. Due to the presence of the same carrier ligands (NH_3), carboplatin and cisplatin have been reported to form eventually the same DNA adducts and the differences between these two drugs is due to the much slower DNA binding kinetics of carboplatin.[124–126] The same scenario is expected for nedaplatin where the carrier ligands are also ammonia groups.

Oxaliplatin, (*trans*-1R,2R-diaminocyclohexane)oxalatoplatinum(II), first synthesized by Kidani *et al.* in 1976,[127] is the latest worldwide approved platinum-based anticancer agent. It has shown a broad spectrum of antitumour activity[128] with a partial or a non-cross-resistance with cisplatin in a wide range of human tumours *in vitro* and *in vivo*.[128–131]

Oxaliplatin also has similar DNA-binding properties to its congener, leading to the same type of DNA–Pt crosslinks, although the amount of the different types of adduct varies in comparison to cisplatin. Several studies show that oxaliplatin is less reactive and forms fewer adducts with cellular DNA than cisplatin. Therefore, the reason why oxaliplatin adducts lead to fatal lesions in cisplatin-resistant cell lines and differences in mutagenicity in comparison with cisplatin and carboplatin might be explained by the different discrimination of cisplatin and oxaliplatin adducts by cellular proteins.[132] There are two X-ray crystal structures of oligonucleotide–oxaliplatin adducts in the literature for two different dodecamers: a 2.4 Å crystal structure of a 1,2-d(GpG) intrastrand crosslink in a DNA dodecamer duplex formed by reaction with oxaliplatin,[133] and a related NMR solution structure.[134]

The crystal structure is comparable to that of a cisplatin adduct published by Takahara et al.[111] The oligonucleotide sequence, 5′-d(CCTCTGGTCTCC)·(GGAGACCAGAGG), was identical and so are the crystal structures of the two adducts. Therefore, the differences in cytotoxic activity of the two platinum complexes are not clear on the basis of these crystal structures. Wu et al. made an attempt to explain these differences based on the solution structure with a similar dodecamer, with the sequence 5′-d(CCTCAGGCCTCC)·(GGAGACCAGAGG).[134] In addition, they reported more recently the solution structures of the above DNA dodecamer duplex with and without a cisplatin 1,2-d(GG) intrastrand crosslink and a detailed comparison with the 1,2-d(GG) intrastrand crosslink from the reaction of the same oligonucleotide with oxaliplatin.[135] For example, when comparing average solution structures, the platinum–oligonucleotide adduct from reaction of the oligonucleotide with oxaliplatin bends the DNA by 31° (overall bend angle), whereas the {Pt(NH$_3$)$_2$(–GG–)} adduct bends the DNA by 22°. The different response of cellular proteins to these structurally different DNA adducts may be responsible for the differences in cytotoxicity and tumour range between oxaliplatin and cisplatin.

There appear to be few published studies on DNA adducts formed by lobaplatin or heptaplatin. The former has been reported to form DNA adducts slower than cisplatin and oxaliplatin after 24 h of incubation but to a higher extent after incubation for 3 days.[136] The formation of adducts in vitro by calf thymus DNA has also been investigated using ^{32}P-post-labelling. The major DNA–Pt crosslinks formed in vitro in drug-treated cells are Pt–GG and Pt–AG intrastrand crosslinks. The authors found that the reactivity of lobaplatin towards DNA in vitro showed large differences compared with cisplatin, which forms more GG and AG adducts. This difference was smaller in cells, suggesting enhancement of adduct formation by intracellular mechanisms.[137]

10.4.1.3 Platinum(IV) Complexes

The first evidence of a platinum(IV) complex showing anticancer activity appeared in the very early days in Rosenberg's laboratory during the time of cisplatin discovery. In the experiments in which cell division of E. coli was inhibited, it was thought that cis-[Pt^{IV}Cl$_4$(NH$_3$)$_2$] was the active species in solution, the agent that could provoke the filamentous elongation of the bacteria.[65,138–141]

iproplatin   tetraplatin   satraplatin JM216

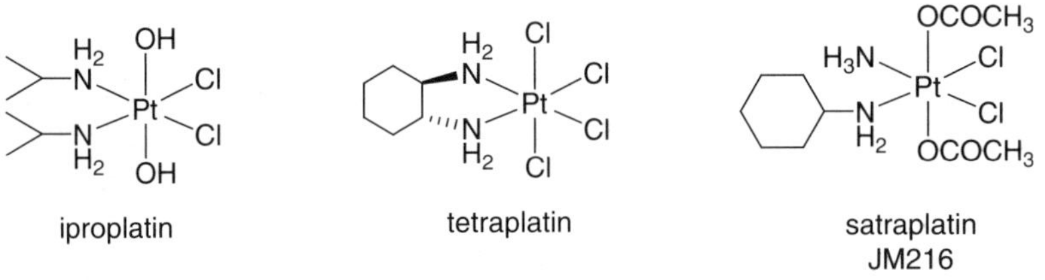

Chart 10.5

However, soon after this finding, efforts regarding the clinical development of a new platinum-based anticancer drug were focused on the reduced form cis-[Pt^{II}Cl$_2$(NH$_3$)$_2$], cisplatin, due to the better dose–response in screening against the leukaemia L1210 tumours in mice by the National Cancer Institute. Despite the fact that a number of platinum(IV) drugs have since entered clinical trials (Table 10.1, Chart 10.5), evidence for their direct interaction with the biological target, nucleic acid, *in vivo*, has yet to be reported.

The oxidation state of platinum in complexes has a major effect on its geometry and reactivity. Octahedral platinum(IV) complexes have been found to be advantageous in medicinal chemistry because they are less susceptible to substitution reactions and so undergo fewer reactions on route to the tumour. This results in fewer side-effects and reduced drug loss due to deactivation compared with square-planar platinum(II) complexes. In other words, these complexes can have reduced toxicity and higher activity, *i.e.* a higher therapeutic index. Another advantage of these complexes is their higher aqueous solubility. Precisely this feature of platinum(IV) complexes has been exploited by Tobe and co-workers, who selected iproplatin (CHIP, JM9, cis,trans,cis-[PtCl$_2$(OH)$_2$(isopropylamine)$_2$]) from a range of platinum(IV) complexes for its high aqueous solubility.[142] Iproplatin entered phase I and II clinical trials,[143,144] and even phase III,[145] but was ultimately found to be less active than cisplatin and so has not been registered for clinical use.[146]

Tetraplatin ([PtCl$_4$(D,L-cyclohexane-1,2-diamine)], also called ormaplatin) also showed great promise in preclinical studies but caused severe neurotoxicity in treated patients and the trials were subsequently abandoned at phase I.[147]

JM216 (satraplatin, cis,trans,cis-[PtCl$_2$(OAc)$_2$(NH$_3$)cyclohexylamine]) is a rationally designed, lipophilic drug, which can be administered orally[148] and circumvent the poor gastrointestinal absorption of cisplatin or carboplatin. It is currently in phase III trials.[149,150]

Mechanism of action of platinum(IV) complexes. Despite studies reporting *in vitro* binding of platinum(IV) species to nucleobase models in the absence of reductants,[151–153] evidence that platinum(IV) species can enter cells,[154] and that they can bind to DNA more slowly but at similar sites to their platinum(II) counterparts,[155–158] it is widely believed that reduction to platinum(II) is essential for *in vivo* activation and effective antitumour activity of platinum(IV) complexes.[159–164] For this reason, platinum(IV) complexes are often considered to be pro-drugs. Additionally, catalytic amounts of platinum(II) complexes play a role in the reduction of platinum(IV).[165–167]

The effect of axial and carrier ligands on the reduction and cytotoxicity of Pt^{IV} complexes with chlorido ligands has been investigated. The complexes undergoing the fastest reduction bind most readily to DNA.[168] There is no clear correlation between reduction rates and cytotoxicity, although a correlation is observed for *trans*-[PtX$_2$Cl$_2$(en)] (where en is ethylenediamine) complexes differing only in the axial ligands X (cytotoxicity increases in the order OH < OCOCH$_3$ < Cl < OCOCF$_3$).[169] Kratochwil and Bednarski found no relationship between reduction potentials and cell growth inhibitory activity for *cis*-dichloro- and *cis*-diiodo(en)platinum(IV) complexes with different axial ligands.[170] Hambley *et al.* also found no clear and general correlation between reduction potential and cytotoxicity for a number of organometallic platinum(IV) complexes, containing anionic polyfluoroaryl ligands, ethylenediamine or ammonia ligands and different axial ligands.[171] The *in vitro* biological behaviour of Pt^{IV} complexes can be related to how readily and rapidly they are reduced, as reviewed by Hall and Hambley.[172]

Platinum(IV) complexes would not be expected to interact with nucleic acids due to their slow substitution kinetics in comparison with their reduction rates. Since substitution is slow, Pt^{IV} would be reduced before undergoing substitution, making it mechanistically unlikely that the Pt^{IV} complex survives the number of reductants in the body.[172] However, Choi *et al.* have observed such substitution *in vitro*.[167] The reaction of some platinum(IV) complexes with 5'-GMP appears to go through substitution on Pt^{IV} followed by reduction. They also observed the previously reported catalytic effect of platinum(II). In a more detailed study, the mechanism of action of tetraplatin, [Pt^{IV}Cl$_4$(dach)], was investigated, suggesting that chloride substitution in the Pt^{IV} complex by N7 of 5'-dGMP is followed by cyclization *via* nucleophilic attack of a phosphate oxygen at C8 of 5'-dGMP. In the following step, an internal two-electron transfer produces a cyclic phosphodiester intermediate and [Pt^{II}Cl$_2$(dach)] (dach = (*trans-R,R*)-1,2-diaminocyclohexane), which eventually leads to the [Pt^{II}Cl(dach)(GMP)] adduct. The cyclic intermediate reacts with water to give 8-oxo-dGMP. Recently, it was reported that oxidation of 3'-dGMP and 5'-d(GTTTT)-3' by tetraplatin leads to Pt^{II}–G and cyclic guanosine products. The interaction with N7 of cGMP, 9-Mxan (9-methylxanthine), 5'-d(TTGTT)-3' and 5'-d(TTTTG)-3' is observed, but no electron transfer is seen in these cases.[153,173–175]

Finally, a class of platinum(IV) pro-drugs which is not activated by chemical substitution or reduction but by light is described later in this chapter.

10.4.1.4 trans-*Platinum Complexes*

Transplatin (*trans*-diamminedichloridoplatinum(II)) was found not to be effective in blocking cell division in bacteria during the early stages of platinum investigations by Rosenberg.[141] Other *trans*-platinum analogues were also found not to be cytotoxic in comparison with their *cis* analogues.[176,177] *trans*-Platinum complexes were therefore not developed as potential chemotherapeutics. In the early 1970s, one of the structure–activity relationships established that "the requirement of *cis* leaving groups is a necessary (but insufficient) criterion for observing activity".[177]

The reason for the poor cytotoxicity of *trans*-diam(m)ine complexes could be related at least in part to their rapid reactivity and hence deactivation on their way to the intracellular target. The possibility of slowing this reactivity – by using, for example, bulky ligands that somehow impede or retard access of incoming nucleophiles to the axial coordination sites used in associative substitution pathways – has been investigated by Farrell and colleagues. Almost 20 years after *trans*-Pt was discarded on the basis of structure–activity relationships, Farrell and co-workers discovered the antineoplastic activity of new *trans*-platinum complexes different from those synthesized in the early days.[178,179] Their findings for transplatin analogues with aromatic N-donor heterocycles substituting ammonia provided inspiration for further exploration by many other research groups. The first *trans*-platinum complexes with reported cytotoxic activity *in vitro* were *trans*-[PtCl$_2$(NH$_3$)(quinoline)] (**5**, Chart 10.6) and *trans*-[PtCl$_2$(NH$_3$)(thiazole)] (**6**, Chart 10.6). Their mode of binding to DNA has also been elucidated.[180–183] DNA binding of *trans*-platinum complexes in comparison with cisplatin is listed in Table 10.2.

Soon after Farrell and colleagues' work, Natile and co-workers also exploited the *trans*-Pt geometry, finding complexes with imino ligands that exhibited potency towards *in vitro* tumour cell growth, often being active towards cisplatin-resistant tumour cells.[184,185] They found that the complex *trans*-[PtCl$_2$(*E*-iminoether)$_2$] (**7**, Chart 10.6) formed essentially monofunctional adducts at guanine residues on DNA.[186] They also found that these adducts are not recognized by HMGB1 proteins but readily crosslink proteins, which markedly enhances their efficiency in terminating DNA polymerization by DNA polymerases *in vitro* and inhibits removal of this adduct from DNA by nuclear excision repair proteins (NER). They suggested that DNA–Pt–protein ternary crosslinks are responsible for the antitumour activity of this drug.[187]

Navarro-Ranninger and co-workers have developed a series of branched-chain am(m)ine *trans*-Pt complexes[188,189] with remarkable cytotoxicity.[190,191] Particularly, the DNA binding properties of the complex *trans*-[PtCl$_2$(dimethylamine)(isopropylamine)] (**8**, Chart 10.7), for which cytotoxic activity has been shown to implicate apoptosis, involve mainly interstrand crosslinks between guanine and its complementary cytosine.[191,192] Curiously, a comparison of the *in vivo* activity of

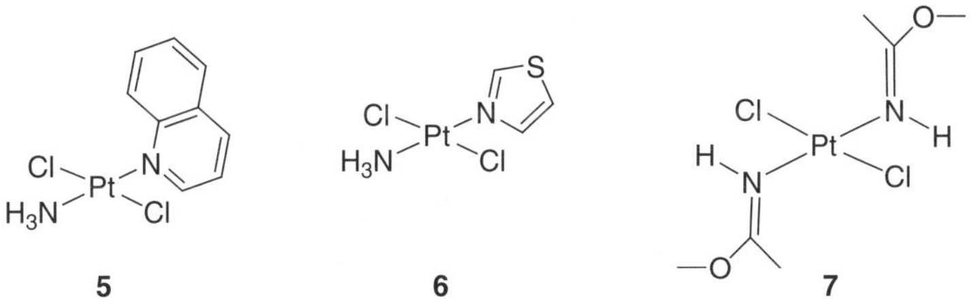

Chart 10.6

Table 10.2 Some DNA-binding preferences of *trans* diam(m)ine platinum(II) complexes in comparison with cisplatin.

Ligands in trans positions	Type of DNA adduct			DNA bending[a]	DNA unwinding angle (adduct)	Ref.
	Intra (%)	ICL (%)	Mono (%)			
Cisplatin, NH₃ vs Cl	GG + AG (~90), GNG	GG (~6)	G	32–34° tmag (1,2), 40–45° tmig (ICL)	13° (1,2), 76–79° (ICL)	97,180,431–434
Transplatin, NH₃ vs NH₃	28	GC (~12), 1,3-GNG	G, C	20° tmig (ICL)	12° (ICL)	97,180,431,435
NH₃ vs quinoline	GG, AG	GG (30)	G		17°	180,181
NH₃ vs thiazole	1,3-GG (20–40)	GG (30–40)	G (30–40)	40° tmig (1,3), 22° tmig (ICL), 34° tmag (mono)	15° (1,3), 20° (ICL), 12° (mono)	181–183
E-iminoether vs E-iminoether			G	21° tmig		186,187
Dimethylamine vs isopropylamine		GC				191,192
NH₃ vs methylbutylamine		40	36		8°	436
NH₃ vs sec-butylamine		50	36		8°	436
Isopropylamine vs 3-hydroxymethylpyridine	GG (69)	10–12	G, A (20)		16°	194
Isopropylamine vs 4-hydroxymethylpyridine	GG (64)	10–12	G, A (25)		16°	194
4-Picoline vs piperidine	GG (92)	3	G (5)		13°	202
4-Picoline vs piperazine	GG (76)	6	G (18)		16°	202

[a] tmig = toward minor groove; tmag = toward major groove

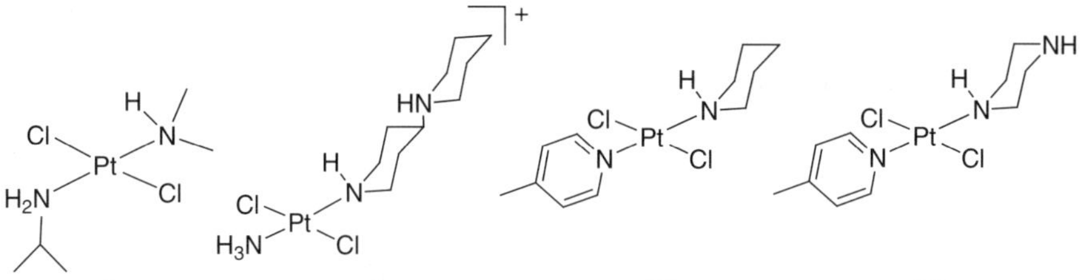

Chart 10.7

the Pt^{II}/Pt^{IV} pair *trans*-[PtCl$_2$(OH)$_2$(dimethylamine)(isopropylamine)] and *trans*-[PtCl$_2$(dimethylamine)(isopropylamine)] showed that only the Pt^{IV} analogue was able to inhibit the growth of CH1 human ovarian carcinoma xenografts in mice. Moreover, binding studies with serum albumin indicated that the platinum(II) complex possesses a much higher reactivity towards albumin than the platinum(IV) complex, suggesting that the lack of *in vivo* antitumour activity shown by the former might be related to its extracellular inactivation before reaching the tumour site because of its high rate of binding to plasma proteins.[193]

In contrast to other *trans*-Pt^{II} complexes, complexes with aromatic ligands *trans* to isopropylamine such as *trans*-[PtCl$_2$(isopropylamine)(3-hydroxymethylpyridine] and *trans*-[PtCl$_2$(isopropylamine)(4-hydroxymethylpyridine)] have been found to form largely intrastrand crosslinks.[194]

Finally, Gibson and co-workers have recently developed platinum(II) complexes with planar and non-planar cyclic ligands in *trans* geometry which are highly cytotoxic towards a number of cancer cell lines.[195–200] Some of the results observed in DNA binding experiments are striking. The complex *trans*-[PtCl$_2$(NH$_3$)(pip-pip)]$^+$ (where pip-pip is 4-piperidinopiperidine) (**9**, Chart 10.7) binds to DNA extremely rapidly. The $t_{1/2}$ for binding to calf thymus DNA (CT-DNA) is 13 min at 37 °C[201] and without the complex undergoing initial aquation;[199] the complexes *trans*-[PtCl$_2$(4-picoline)(piperidine)] and *trans*-[PtCl$_2$(4-picoline)(piperazine)] (**10** and **11**, respectively, Chart 10.7) form a large number of intrastrand crosslinks in double-helical DNA, 92 and 76%, respectively, after 48 h.[202]

In summary, substitution of one or more ammine ligands in transplatin by aromatic N-donor heterocyclic ligands, branched aliphatic amines or imino ligands has led to compounds with significant inhibitory potency towards *in vitro* tumour cell growth. These *trans* complexes are often active towards cisplatin-resistant tumour cells and in some cases endowed with significant activity also *in vivo*. The DNA binding characteristics of these *trans*-platinum complexes are different from those of the parent cisplatin, perhaps explaining the lack of cross-resistance; the perturbations of DNA structure are also different from those induced by transplatin itself. A summary of reported DNA binding studies is given in Table 10.2.

Unfortunately, no *trans*-platinum complex has yet entered clinical trials. Future work will involve understanding why certain *trans*-platinum complexes are active and sometimes more active than their *cis*-platinum counterparts, often active in cisplatin-resistant cell lines, and understanding the nature of the crucial lesions on DNA.

10.4.1.5 Polyplatinum Complexes

Compounds based on the *cis*-[PtX$_2$(amine)$_2$] geometry produce very similar types of adduct on target DNA – as exemplified by cisplatin, carboplatin and oxaliplatin. It is therefore not surprising that they induce similar biological consequences. This consideration led to the hypothesis that development of platinum compounds structurally different to cisplatin may form different types of DNA–Pt adducts and leading to a different spectrum of clinical activity maybe complementary to cisplatin. This turned out to be true for *trans*-platinum complexes and is also the case for multinuclear platinum complexes.

In the late 1980s, Farrell's group began a fruitful investigation of linear multiplatinum complexes, where the platinum centres are separated by aliphatic chains and the chloride ligands are *cis* or *trans* in the ends of the complex. These polyplatinum complexes are generally highly positively charged, which leads to a better initial electrostatic recognition of DNA in comparison with monomeric species. Such electrostatic recognition can be observed by NMR as chemical shift changes of the complex in the presence of DNA immediately upon mixing.[203] The pre-association step might be crucial before direct coordination to DNA nucleobases. Another critical feature of these multinuclear complexes is flexibility. Farrell and co-workers observed that steric hindrance within the diamine linker results in diminished cytotoxicity perhaps due to reduction in DNA binding affinity.

The most successful of this type of complex so far is [{*trans*-PtCl(NH$_3$)$_2$}$_2$-μ{*trans*-Pt(NH$_3$)$_2$(NH$_2$(CH$_2$)$_6$NH$_2$)$_2$}]$^{4+}$, BBR3464 (Chart 10.8), which was first reported in 1993[204] and successfully completed phase I clinical trials.[205] There is extensive literature reporting preclinical *in vitro* and *in vivo* investigations of BBR3464.[206–209] The cellular uptake is enhanced relative to CDDP, being high and rapid.[210,211] Hydrolysis of BBR3464 and DNA adduct formation are well documented.[212,213] In this chapter, we focus on the DNA binding mode of BBR3464.

As is general for polynuclear platinum compounds, BBR3464 shows rapid binding to DNA with a $t_{1/2}$ of *ca.* 40 min, probably due to its high charge and

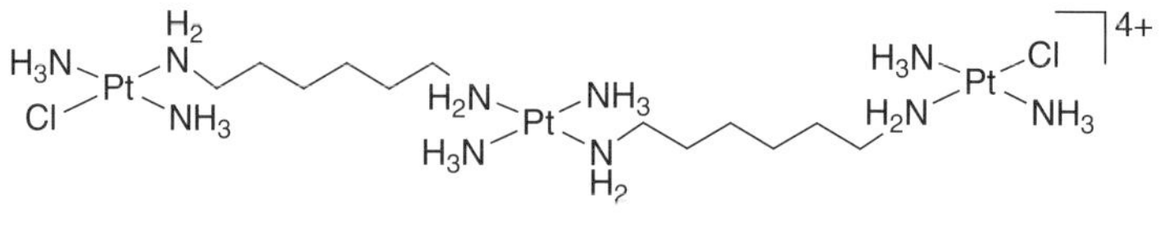

BBR3464

Chart 10.8

charge-mediated pre-association. The unwinding angle of 14° for supercoiled plasmid DNA is indicative of bifunctional DNA binding, with *ca.* 20% of the DNA adducts being interstrand crosslinks, through binding to guanine bases.[214]

BBR3464 forms monofunctional adducts with guanine residues and gives rise to 1,4-interstrand crosslinks on the self-complementary DNA strand of the 12-mer duplex [d(ATATGTACATAT)]$_2$ and octamer duplex [d(ATGTA-CAT)]$_2$. The two platinum atoms are coordinated in the major groove at the N7 positions of guanines that are four base pairs apart on opposite DNA strands. The central tetraamine linker unit of [*trans*-H$_2$N(CH$_2$)$_6$NH$_2$Pt(NH$_3$)$_2$NH$_2$(CH$_2$)$_6$NH$_2$] is located in the minor groove.[213,215]

The most relevant feature of this special DNA binding is the lack of severe DNA distortions such as a kink or significant unwinding of the helices, which are characteristic of DNA adducts of mononuclear platinum complexes. One of the direct consequences of this mild conformational change in the DNA is that these adducts appear to be poor substrates for recognition by proteins that bind to rigidly bent DNA, such as those containing the HMG domain, in contrast to cisplatin adducts. This might be due to the central platinum unit lying non-covalently on the minor groove in the place in which the phenylalanine intercalates in the case of the ternary adduct formed by HMG and DNA adducts of cisplatin.[115] The flexibility of the DNA–Pt adduct would also confer modest attraction for nucleotide-excision repair proteins. Therefore, it seems that the recognition by HMG domain proteins of DNA–BBR3464 adducts is not a crucial step in the mechanism of cytotoxicity of BBR3464, as claimed for cisplatin. This would support the original hypothesis that polynuclear platinum compounds coordinating and modifying DNA in a different way than the mononuclear complex represent a novel class of platinum anticancer drugs acting by a different mechanism than cisplatin and its analogues.

Farrell's group has recently synthesized a new generation of multinuclear complexes. The latest example of these polynuclear rule-breaking complexes is the trinuclear non-leaving group-containing TriplatinNC, of formula [{*trans*-Pt(NH$_3$)$_2$(NH$_2$(CH$_2$)$_6$(NH$_3^+$)}$_2$-μ-{*trans*-Pt(NH$_3$)$_2$(NH$_2$(CH$_2$)$_6$NH$_2$)$_2$}](NO$_3$)$_8$. Illustrated in Chart 10.9, with a positive charge of +8, the complex has shown micromolar activity against a human ovarian cancer cell line,[216] being the first example of a non-coordinating platinum compound with cytotoxicity equivalent to cisplatin. Moreover, TriplatinNC displays *in vivo* antitumour activity in 2008 ovarian carcinoma xenografts.

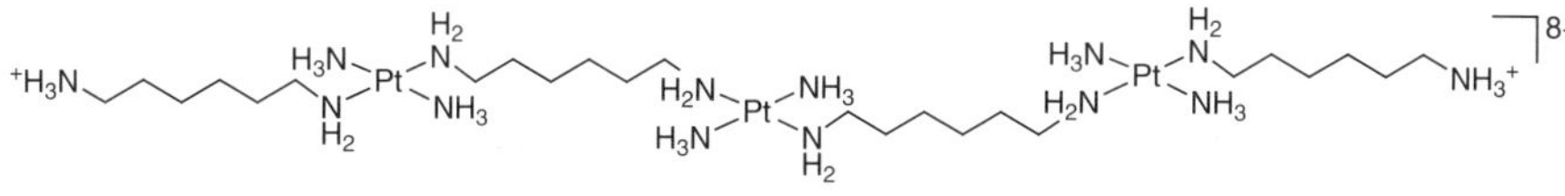

TriplatinNC

Chart 10.9

The DNA binding mode of triplatinNC has been classified as the third non-covalent binding mode of a drug to nucleic acid, after intercalators and groove binders. In this new type of interaction, the drugs exclusively utilize backbone functional groups.[217] The structure of this platinum–DNA complex at 1.2 Å resolution shows that TriplatinNC has selectivity for phosphate oxygens. The square-planar tetraam(m)ine platinum(II) coordination unit forms bidentate NH···O···HN (amine···phosphate···ammine) complexes with OP atoms through hydrogen bonding interactions called by the authors 'phosphate clamps'. A phosphate clamp is defined as a tetracyclic structure with a single OP, which accepts two hydrogen bonds, one from each of two am(m)ine ligands of a single Pt^{II} centre.

Following an inherently different line, Reedijk and co-workers have developed a series of polyplatinum complexes linked with rigid ligands that show high cytotoxicity.[218–220] Their aim was to develop complexes that minimize the distortion in DNA upon formation of 1,2-d(GG) intrastrand crosslinks in the search for different activity profiles compared with cisplatin. They investigated the interaction of complexes such as $[(cis-\{Pt-(NH_3)_2\})_2(\mu-OH)(\mu-pz)](NO_3)_2$, where pz is pyrazolido (**12**), with a double-stranded DNA decamer, d(CTCTGGTCTC)·d(GAGACCAGAG), by NMR spectroscopy and molecular dynamic simulations.[221] Pt^{II}–pz–Pt^{II} crosslinks the duplex at two adjacent guanines and induces relatively minor structural perturbations in the adduct, without destabilizing the duplex and without changing the directionality of the helix axis. The DNA appears to accommodate the platinum complex without significant bending or base-pair disruption, although it unwinds the duplex by 15°, as does cisplatin.

12

A comprehensive overview of multi-nuclear platinum drugs has recently been published by Wheate and Collins.[222,223]

10.4.1.6 Photoactivatable Platinum(IV) Complexes

Because anticancer agents stop replication of rapidly dividing cells, *e.g.* cells of the bone marrow, gastrointestinal tract and skin, these cells will be particularly vulnerable to chemotherapy. Additionally, the therapeutic index (*i.e.* correlation between toxic dose and curative dose) of active platinum(II) complexes is relatively narrow, so cancer cells only need to acquire a low level of resistance to a toxic agent to unbalance the ratio and the treatment will eventually fail. In other words, classical platinum drugs have severe toxicity and the development of resistance can prevent the success of chemotherapy. There is, therefore, a clinical need to develop an anticancer drug with significantly reduced toxicity without diminishing efficiency. The groups of Sadler and Bednarski are developing a type of platinum-based anticancer drug that would be ideal candidates to satisfy this need.

Reduction to Pt^{II} is an essential requirement for Pt^{IV} complexes to exert their anticancer activity. When and where this reduction takes place controls the activation of the pro-drug and hence its toxicity. Many clinically useful antitumour agents are pro-drugs and indeed their success is due to this fact. The applicability of this concept to platinum drugs becomes feasible with Pt^{IV} photoactivatable complexes.

There are two non-metal-based photosensitzers in use currently in anticancer therapy: Photofrin (porfirmer sodium) since 1995 and Foscan (temoporfin) since 2001. Photochemical reduction of Pt^{IV} to Pt^{II} can occur by complex mechanisms. With appropriate choice of ligands, hexacoordinated platinum(IV) photolabile complexes can be prepared. Then, the light-sensitive Pt^{IV} pro-drugs can be photoactivated to antitumour Pt^{II} agents directly at the site of the tumour with laser light. These complexes represent a new area for drug development.

Desirable features for an ideal photosensitzer include the following. First, stability under physiological conditions is essential with a plasma half-life long enough to allow accumulation, but at the same time the elimination time should be sufficiently short to reduce light toxicity after treatment. Second, a wavelength of light to activate the drug in the red and near-infrared region (600–1000 nm, the so-called 'phototherapeutic window') would provide increased depth of penetration into tissue. An ideal compound would also have minimal dark toxicity. It would also be convenient if the photosensitzer is oxygen-independent as cancer cells are often hypoxic. Thus, a major advantage of photoactivated platinum compounds over currently used photosensitzers is that they are independent of oxygen.

There are two types of reported platinum(IV) photoactivatable anticancer complexes: diiodo complexes and diazido complexes. The first photoactivated Pt^{IV} drugs contained iodide as a reducing ligand with general formula *trans,cis*-$[Pt(X)_2I_2(en)]$, where $X = Cl$, OH, acetate or methylsulfonate.[224,225] 1,2-Diaminoethane (en), a bidentate chelating ligand, was chosen as the non-leaving ligand to avoid photoisomerizations (Chart 10.10). The diiodido complexes gave rise to intense iodido-to-Pt^{IV} ligand-to-metal charge-transfer bands and were photolabile at *ca.* $\lambda = 400$ nm.

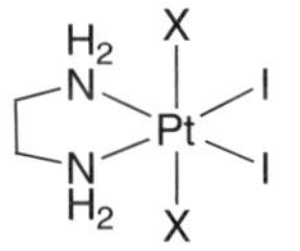

X = Cl, OH, acetate,
or methylsulfonate

Chart 10.10

Pt–DNA binding modes ranged from irreversible platination when X = Cl or SO$_2$CH$_3$ (not affected by light), to no platination over 6 h in the case of X = OH and to 25% platination only after light exposure in the case of X = OAc.[224,225] The results of the DNA platination studies were confirmed by NMR studies in which 5′-GMP was used as a model.[226] Moreover, the complexes with X = Cl, OH or SO$_2$CH$_3$ were found to be cytotoxic in a human bladder cancer cell line, but the differences between light and dark toxicities were not as great as had been expected.[225] Thus, the cytotoxicity and DNA binding studies involving irradiation of the diiodo–Pt^{IV} complexes were promising but these complexes were not suitable as photoactivated drugs because they are too easily reduced by thiols in and outside of cells, causing a facile reduction of the diiodo–Pt^{IV} complexes to cytotoxic Pt^{II} species before light activation. Nevertheless, these studies showed that the development of photoactivatable Pt^{IV} complexes is feasible.

The diazido complexes developed by Sadler and co-workers, however, are stable enough to reduction and exhibit the main features required for a photoactivated drug: non-toxicity in the dark and retention of cisplatin-like activity when irradiated. These azide–Pt^{IV} complexes have the general formula [Pt^{IV}(N$_3$)$_2$(OH)$_2$(am)$_2$], where only the hydroxo ligands are maintained *trans* to one another and the am(m)ines can be *cis* and chelated as *e.g.* ethylenediamine (Chart 10.11) or *trans*. One possible mechanism is that upon irradiation the azides leave in the form of azidyl radicals (•N$_3$) (Figure 10.8), which combine to form molecular nitrogen (N$_2$), preventing reoxidation of the platinum centre. This fast decomposition may contribute to the efficiency of photoredox reactions involving complexes containing azide ligands.

Stability in solution and towards potential targets such as the guanine bases of RNA and DNA in the dark has been investigated by NMR. The *cis* complexes **13** and **14** have been shown to be stable towards hydrolysis for 90 days, inert to reaction with nucleobases such as 5′-GMP and d(GpC) in the dark and, most

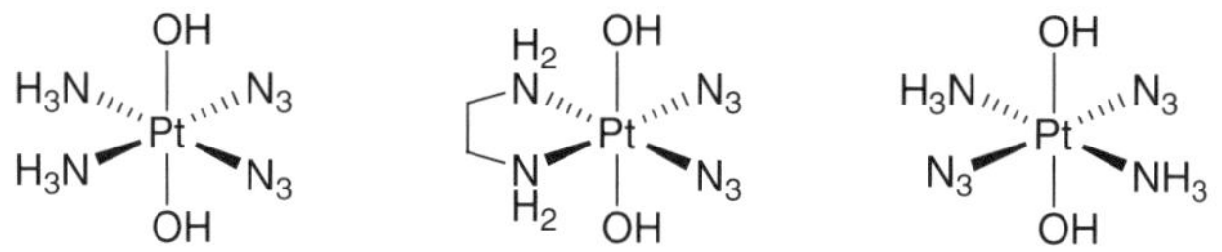

Chart 10.11

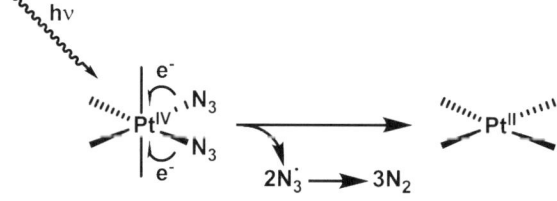

Figure 10.8 A possible mechanism for the photoreduction of a Pt^{IV}–diazido complex.

significantly, react only very slowly with GSH over a period of several weeks.[227,228] Reaction of the ^{15}N-labelled complexes, **13** and **14**, have been followed in aqueous solution in the presence of 5'-GMP and d(GpG), by 1D ^{1}H, 2D [^{1}H,^{15}N] HSQC and 2D [^{1}H,^{15}N] HSQC-TOCSY NMR spectroscopy. The complexes are cytotoxic toward 5637 human bladder cancer cells under irradiation. Nucleotide crosslinking induced by photoreactions of **13** and **14** was observed. The formation of bifunctional intrastrand GG adducts, such as those formed by cisplatin on DNA, has been demonstrated with light of various wavelengths.[227–229]

The all-*trans* complex *trans,trans,trans*-[Pt^{IV}(N$_3$)$_2$(OH)$_2$(^{15}NH$_3$)$_2$], **15**, has comparable cytotoxicity to cisplatin towards human HaCaT keratinocytes and lacks toxicity to cells in the dark. Additionally, it has a much higher aqueous solubility (>20 mM) than the *cis* isomer **13** and more favourable absorption spectrum, with an LMCT band more intense and shifted to longer wavelength, *i.e.* more penetrating. Photoreaction and nucleotide interactions have been also investigated in the presence of 5'-GMP by 1D ^{1}H and 2D [^{1}H,^{15}N] HSQC NMR techniques,[230] showing stability in the dark and reaction only upon irradiation.

In summary, the Pt^{IV}–diazido complexes **13–15** are stable under biological conditions, but are reduced and bind to nucleotides upon radiation with light of various wavelengths. These complexes have been shown to be phototoxic towards several different cell lines, including a cisplatin-resistant cell lines, while remaining inactive in the dark. So far, investigations into the mechanism of action of these compounds indicate that they exert their cytotoxic effect in a different way to cisplatin. This new class of photoactivatable Pt^{IV}–azide complexes has the potential to avoid the side-effects associated with cisplatin, while having the advantage of being independent of oxygen. Importantly, the use of photoactivation can apparently lead to the formation of platinated DNA lesions which are not readily formed by conventional Pt^{II} anticancer complexes, especially by *trans* diam(m)ine complexes. This may lead to novel mechanisms of action and therefore to clinically useful drugs.[231] A highly active complex from this class containing pyridine *trans* to ammonia has been reported recently.[232]

10.4.2 Titanium Agents

In the late 1970s, Köpf and Köpf-Maier investigated the antitumour activity of various series of metallocenes, and in 1979 reported in detail the antitumour activity of titanocene dichloride (**16**).[233]

16   **17**

Titanocene dichloride is active against a diverse range of human carcinomas, including gastrointestinal and breast carcinomas,[234] but not against head and neck cancers. It went successfully through phase I clinical trials,[235–237] and reached phase II. However, the efficacy rates of titanocene dichloride in phase II clinical trials in patients with metastatic renal cell carcinoma and metastatic breast cancer were too low to support further evaluation of the drug.[238,239] The trials were unsuccessful due to formulation problems as a result of rapid hydrolysis.[240,241] No matter how disappointing these results were, titanocene dichloride still has the merit of being the first reported metallocene derivative with anticancer activity[233] and the only organometallic complex to have entered clinical trials to date. However, titanocene dichloride is not the first non-Pt metal-based anticancer drug, or the first titanium-based drug to have reached clinical trials. In 1986, the bis(β-diketonato)Ti^{IV} complex [Ti^{IV}(bzac)$_2$(OEt)$_2$] (Hbzac = 1,3-diphenylpropane-1,3-dione) (budotitane), 17 (shown to exist as the major *cis,cis,trans* isomer in the solid state[242]), developed by Keppler and Schmaehl in the 1980s,[243] entered phase I clinical trials in Germany for the treatment of colon cancer.

10.4.2.1 Budotitane Mode of Action

Although budotitane entered phase I clinical trials over 20 years ago, neither its cellular nor molecular mechanism of action or its metabolism in mammals have been elucidated. Thus, it is not known whether budotitane acts by establishing coordinative bonds to biologically important macromolecules or by targeting and structurally disrupting DNA or RNA strands *via* intercalation. In general, the interaction of Ti^{IV} derivatives with DNA is poorly understood. When budotitane was studied for growth inhibition in a cell proliferation assay, no metal–DNA crosslink was seen, although dose-dependent inhibition was observed; then DNA–budotitane intercalation was suggested.[244] Also, the latter mechanism of action might be supported by the experimental finding that the antitumour activity of bis(β-diketonato)metal complexes increases when bulky groups and planar aromatic systems, such as phenyl groups, are introduced into the diketonato ligand.[245,246] On the other hand, budotitane shows a strong tendency towards hydrolysis, which presents challenges for formulation prior to administration. Such analytical problems resulted in the cessation of its clinical development.

10.4.2.2 Titanocene Mode of Action

The first chemical reaction that titanocene dichloride undergoes in body fluids is hydrolysis. Ti–Cl aquation is fast and results from its phase I clinical studies show 70–80% titanium bound to proteins in blood.[236,237] Moreover, Sadler and co-workers found that Ti^{IV} binds strongly to human serum transferrin and suggested that this protein could be involved in the delivery of Ti^{IV} into cancer cells.[247] They proposed a transferrin-mediated mechanism for the uptake of the drug *via* cleavage of the two Ti–Cp bonds and binding to the two specific iron sites of human transferrin (hTF), leading the adduct Ti$_2$–hTF. They also observed that ATP facilitates the release of Ti^{IV} from transferrin adducts (as might happen inside cells).[248]

Most biological studies of metal–DNA interactions for titanocenes and metallocenes suggest that nuclear DNA is a potential intracellular target of Cp$_2$TiCl$_2$ and other metallocene dihalides. The hydrolysed forms of the drug appear to interact with DNA.[241,249–253] Köpf-Maier and Krahl[254] observed the presence of titanium near phosphorus-rich areas in the nuclear chromatin of human carcinoma cells after treatment with the drug. This appeared to be an indication of the interaction of titanium species with nucleic acids, especially with DNA.[255,256] Other early reports confirmed the formation of Ti–DNA adducts by ICP-AES in tumour cells treated with Cp$_2$TiCl$_2$.[257] Moreover, lack of cross-resistance between titanocene and cisplatin was observed.[251]

In general, metallocenes exhibit high affinity for the phosphate groups of DNA, binding to a phosphate oxygen atom, OP, with the variation of additional binding to a base nitrogen atom, N7, in a chelate fashion for {Cp$_2$Ti}$^{2+}$ and {Cp$_2$Mo}$^{2+}$. This binding may disturb DNA base pairing and damage the secondary structure of DNA. The role of phosphate binding under physiological conditions seems to be potentially important for hydrogen bonding and the binding between metal ion and phosphate might be the first step in metal–DNA interactions. The binding of the drug to nucleotides in aqueous solution has been investigated in detail by NMR spectroscopy. This binding is weak at neutral pH, but stronger at low pH with the formation of phosphate-bound adducts in acidic solution.[258]

Among all the metallocene dichlorides (Cp$_2$MCl$_2$) investigated so far, those with M=Ti, V and Mo show notable antitumour activity, although titanocene is the most promising of the series.[259–266] Their binding to DNA has been confirmed.[267–270] Metallocenes based on iron, germanium, tin, niobium, zirconium and hafnium have also shown anticancer activity.[250,261,266,271]

10.4.3 Ruthenium Drugs

In the 1970s Clarke and co-workers reported pentaammine(purine)-ruthenium(III) complexes capable of inhibiting DNA and protein synthesis in human nasopharyngeal carcinoma cells *in vitro*,[272] which subsequently triggered interest in ruthenium complexes as potential anticancer pharmaceuticals.[273] During the following decade, Mestroni *et al.* developed hexaco-ordinated Ru^{II} complexes with dimethylsulfoxide and chloride ligands, particularly *cis*- and *trans*-RuCl$_2$(dimethylsulfoxide)$_4$, which exhibit anticancer activity *in vitro* and *in vivo*.[274] The complexes were shown to interact both *in vitro* and *in vivo* with DNA, their most likely target. Today, there are two Ru-based anticancer drugs in clinical trials: NAMI-A, developed by Mestroni and

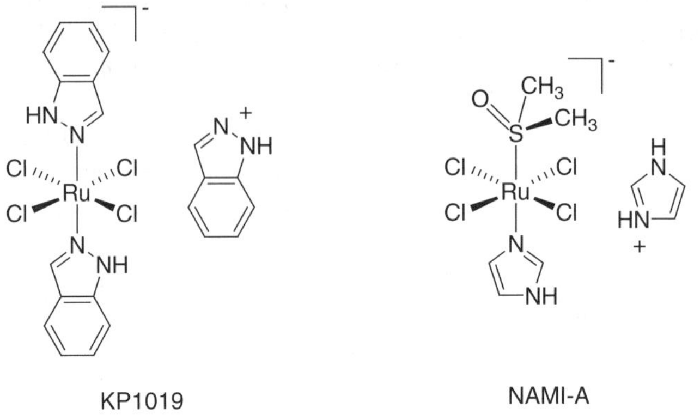

Chart 10.12

co-workers in Trieste, and KP1019, developed by Keppler and co-workers in Vienna (Chart 10.12).

NAMI-A has a relatively low toxicity towards cancer cells, but is particularly effective against solid tumour metastases both in experimental mouse tumours and against human tumours engrafted in the nude mouse,[275–278] being scarcely effective in solid tumours. The drug has now completed phase I clinical trials.[279] However, in contrast to the other metal-based anticancer drugs reviewed here, the antimetastatic activity of NAMI-A is thought to be due to the combined effects on the control of angiogenesis (possibly because it interferes with NO metabolism *in vivo*)[280,281] and anti-invasive properties towards tumour cells and on blood vessels and not to its interaction with nucleic acids, although it does interact with DNA *in vitro*.[282] There are several reviews on the preclinical development of NAMI-A and the different approaches concerning the mechanisms of action of this type of drug by Sava and co-workers.[283–285]

Indazolium *trans*-[tetrachlorobis(1*H*-indazole)ruthenate(III)] (KP1019 or FFC14A) exhibited cytotoxic activity against colon carcinomas in rats in preclinical investigations.[286–288] The preclinical development of KP1019 has also been recently reviewed.[289,290]

10.4.3.1 KP1019 Mode of Action

Octahedral ruthenium(III) complexes are relatively inert towards ligand substitution. The reduction from ruthenium(III) to ruthenium(II) as an activation process prior to DNA binding was first suggested in the late 1970s by Clarke and co-workers.[272,291] Correlations between the Ru^{III} reduction potential and increased antiproliferative activity for the colon carcinoma cell line SW480 in the series [Ru^{III}Cl$_{6-n}$X$_n$]$^{(3-n)-}$, where $n = 0$–4 for X = 1*H*-indazole, and $n = 2$ for X = 1*H*-imidazole or 1*H*-1,2,4-triazole,[292–294] confirm this proposal. KP1019 can therefore be considered a pro-drug, with a reduction potential accessible to biological reductants,[293] which undergoes activation *in vivo* through reduction from Ru^{III} to Ru^{II} and then interacts with biomolecules after labilization of the Ru^{II}–Cl bonds.

DNA has been proposed as one of the biological targets of KP1019 and it has also been reported that the drug triggers apoptosis.[295] However, the cellular mechanism for the activation of apoptosis is not understood. Competitive studies of the binding of KP1019 to DNA nucleoside 5′-monophosphates in buffered solution by capillary electrophoresis have shown a preference for guanosine and adenosine 5′-monophosphates (5′-GMP and 5′-AMP).[296] The X-ray structure of the product from reaction of the model DNA base 9-methyladenine with KP1019 is *mer,trans*-[RuCl$_3$(1*H*-ind)$_2$(9-methyladenine)],[297] which contains a monofunctional N7-bound purine. Further studies with DNA have indicated that KP1019 is able to unwind DNA,[298] which the authors interpreted as strong formation of monofunctional adducts with poor closure to bifunctional adducts.

An antineoplastic agent containing gallium(III) has also been synthesized that is currently in clinical trials: tris(8-quinolinolato)gallium(III) (KP46)[299] (Chart 10.13). The structurally similar complex tris(3-hydroxy-2-methyl-4*H*-pyran-4-onato-O3,O4)gallium(III) (gallium maltolate) is also under clinical development.[300] However, mechanism-of-action studies on Ga^{III} have so far been centred on uptake *via* transferrin binding and ribonucleotide reductase inhibition. Further work is required to investigate the possibility that RNA–Ga^{III} or DNA–Ga^{III} interactions can occur.

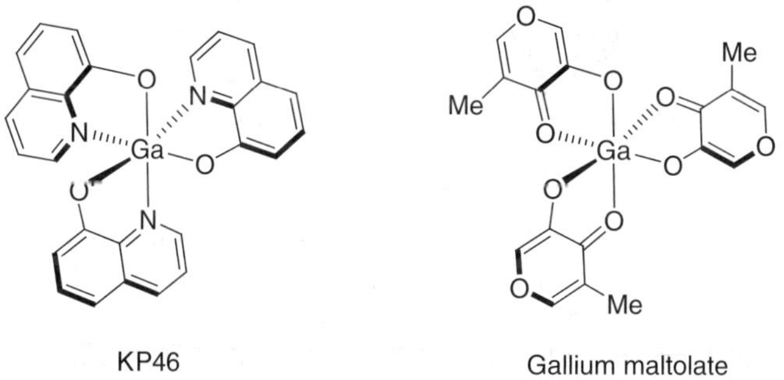

KP46

Gallium maltolate

Chart 10.13

10.4.3.2 Other Coordination Ruthenium Compounds. Polypyridyl Compounds

Luminescent ruthenium(II) coordination compounds with polypyridyl ligands have been extensively investigated by Barton's group in the last two decades.[301] Complexes of the type [Ru(phen)$_3$]$^{2+}$ (phen = 1,10-phenanthroline) (**18**) and analogues,[302–305] and [Ru(bpy)$_2$(dppz)]$^{2+}$ (bpy = 2,2′-bipyridine, dppz = dipyrido-[3,2-*a*:2′,3′-*c*]-phenazine)[306] (**19**) have been developed as (non-radioactive) chiral-discriminating probes for polynucleotide structures due to their shape and the sensitivity of their excited-state properties. [Ru(bpy)$_2$(tactp)]$^{2+}$ (where tactp is 4,5,9,18-tetraazachryseno[9,10-b]triphenylene) (**20**) is luminescent in aqueous solution at micromolar concentrations and contains a bulky intercalating ligand to promote selectivity in targeting single base mismatches in DNA. The complex appears to bind preferentially to CC mismatches.[307]

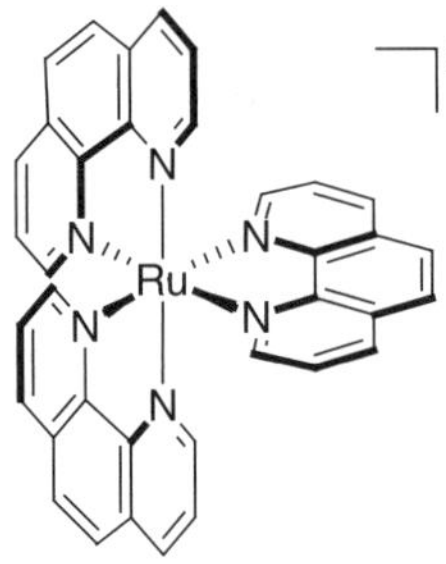

18

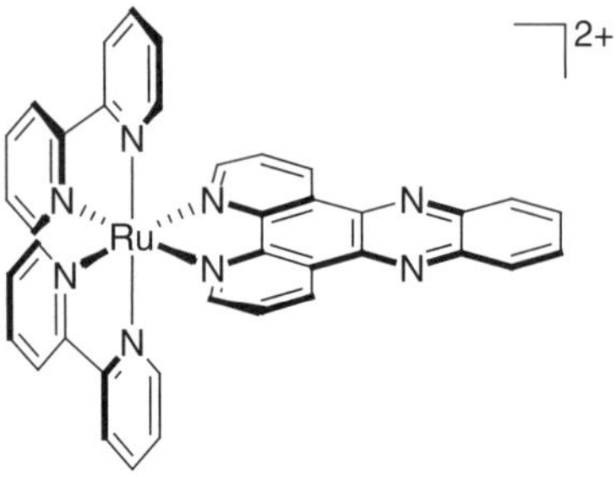

19

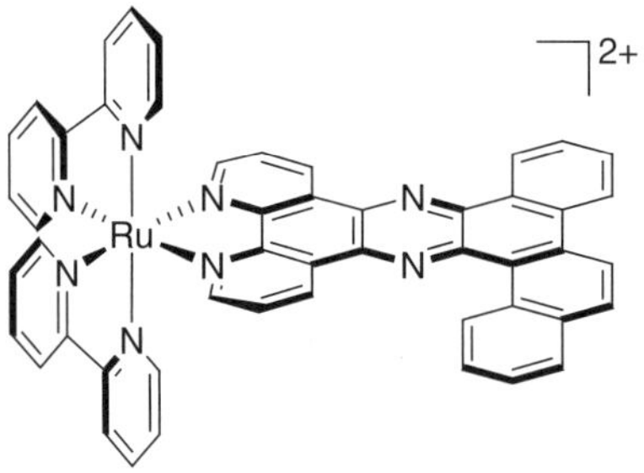

20

Reedijk *et al.* have investigated polypyridyl ruthenium complexes for antitumour activity. The Ru^{III}–trichloro complex *mer*-[Ru(terpy)Cl$_3$] (terpy = 2,2′: 6′,2″-terpyridine) (**21**) is exceptionally cytotoxic in comparison with analogous mono- and dichloro complexes of Ru^{II} such as [Ru(terpy)(bpy)Cl]Cl and *cis*-[Ru(bpy)$_2$Cl$_2$] (bpy=2,2′-bipyridine). In a cell-free medium, the trichlorido complex coordinates to DNA preferentially at guanine residues forming interstrand crosslinks.[308]

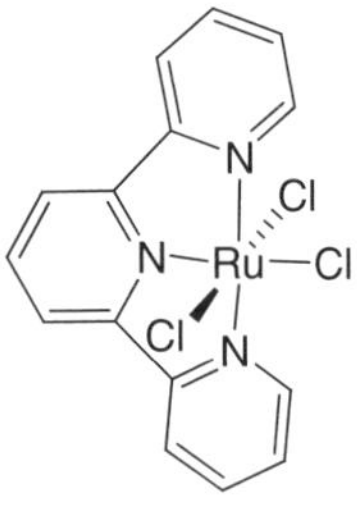

21

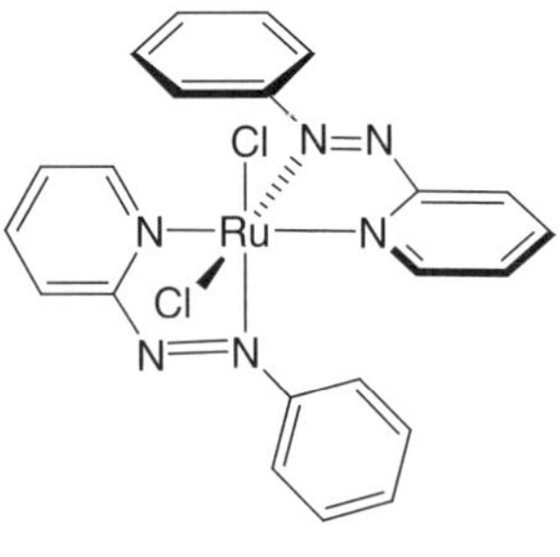

22

The Ru^{II} complex α-[Ru(azpy)$_2$Cl$_2$] (azpy = 2-phenylazopyridine, α denotes the coordinating pairs Cl, N(py) and N(azo) as *cis,trans,cis*, respectively) (**22**) exhibits high cytotoxicity in a series of tumour cell lines in comparison with the β (*cis,cis,cis*) and γ (*trans,cis,cis*) isomers.[309] The activity of these complexes has also been compared to their methylated-azpy Ru^{II} analogues, showing only slight differences in activity.[310] Water-soluble bis(2-phenylazopyridine)ruthenium(II) analogues containing leaving groups other than chlorides such as nitrates,[311] or *O,O*-chelating ligands (*e.g.* cyclobutane-1,1-dicarboxylato, oxalato, malonato) have also been investigated. All the bis(2-phenylazopyridine)ruthenium(II) derivatives exhibit promising cytotoxicity against A2780 human ovarian carcinoma cells ($IC_{50} = 0.9$–$10\,\mu M$).[312] Recently, analogous polypyridyl Ru^{II} compounds without labile groups in the inner-coordination sphere have been reported. Surprisingly, these complexes show low to moderate cytotoxicity, which may imply a different mechanism of action.[313]

10.4.3.3 Ruthenium Arene Complexes

Recently, Sadler and co-workers have explored the potential of Ru^{II}(arene) complexes [(η^6-arene)Ru(X)(Y)(Z)] as antitumour drugs.[314,318] These complexes possess characteristic 'piano-stool' structures (Chart 10.14, where X–Y is a neutral chelating ligand and Z is monoanionic). In these complexes, ruthenium is already in its lower oxidation state, which may be important for the cytotoxicity of the drug *in vivo*. The arene ligand, binding as an η^6 electron donor and a π-acceptor, confers stability to the +2 oxidation state. The presence of a chelating ligand, X–Y, seems to provide additional stability to the whole structure. The ligand Z is the 'leaving group', such as a halide, and allows activation of the molecule: if labile, it can provide a coordination site for biomolecules. Small variations of the arene and the 'legs of the stool' afford versatility to the molecule and the possibility of fine tuning their pharmacological properties.

In general, Ru(arene) complexes show promising cytotoxic activity against human ovarian cancer cell lines, some of them as potent as cisplatin and carboplatin. Sadler and co-workers have established some structure–activity relationships.[314–318] As an example, when the linker is ethylenediamine and the leaving group is chloride, the cytotoxicity against A2780 human ovarian cancer cells increases with increasing size of the coordinated arene. However, substitution of chloride by other halides has only a small effect on the cytotoxicity of

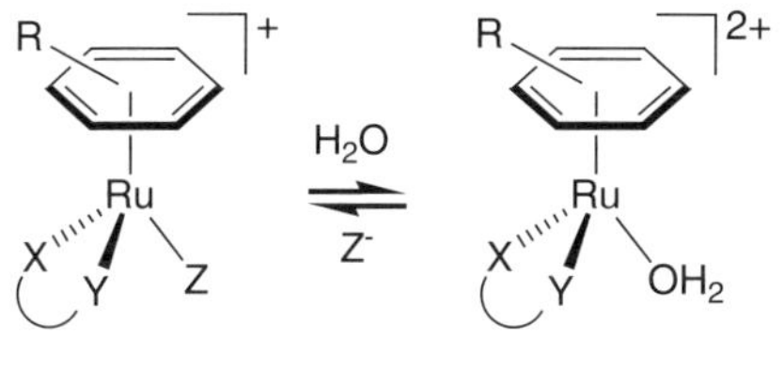

Chart 10.14

ethylenediamine complexes. Additionally, substitution of the chelating ligand for labile ligands such as other chlorides, acetonitrile or isonicotinamide, led to less active complexes, perhaps due to inactivation of a too-reactive complex before reaching the target. Changing the nature of the donors in the chelating ligand (*e.g.* from N to O) has also been investigated but the structure–activity relationships become more complicated.[319]

Ruthenium(II) complexes often exhibit good solubility in aqueous solution, a feature of great importance for a convenient administration of potential drugs. Additionally, activation through hydrolysis of the Ru–Cl bond may be important for the mechanism of activation of this type of antitumour agent; their chemical behaviour in aqueous media has been extensively investigated.[316,317,320–323] In general, arene–ruthenium complexes that hydrolyse also exhibit cancer cell cytotoxicity, whereas those which do not undergo aquation exhibit little or no activity.

Once the Ru(arene) complex is activated and the aqua species, $[(\eta^6\text{-arene})\text{Ru}(X)(Y)(OH_2)]^{2+}$ (Chart 10.14) has been formed, ruthenium is a potential centre for nucleophilic attack by biomolecules. The binding of ruthenium(II) arene complexes to nucleobases and DNA in particular is of special interest, as this may be the ultimate target for this type of agent. A number of studies have confirmed this postulate[324,325] and investigated in detail such interactions.[314,317,326–332]

Ruthenium arene complexes of formula $[(\eta^6\text{-arene})\text{Ru}(\text{en})\text{Cl}](PF_6)$ (en = 1,2-diaminoethane, arene = biphenyl, dihydroanthracene, tetrahydroanthracene, *p*-cymene or benzene) have been found to interact with natural DNA in a cell-free medium and bind preferentially to guanine residues in double helical DNA. In addition to coordination to guanine, non-covalent, hydrophobic interactions between the arene ligand and DNA can occur. This may include arene-base stacking and arene intercalation and minor groove binding.[324] The nature of the arene can influence the level of repair synthesis of ruthenium-damaged DNA. When the arene is tetrahydroanthracene, the level of repair is lower than with *p*-cymene as arene or for cisplatin. This observation provides additional support for a mechanism underlying antitumour activity of Ru^{II} arene compounds different from that of cisplatin.[325]

The interaction between $[(\eta^6\text{-arene})\text{Ru}(\text{en})\text{Cl}]^+$ complexes (and aqua adducts) and guanine, thymine, cytosine and adenine derivatives has been investigated.[328] The reactivity of the various binding sites of nucleobases towards Ru^{II} at neutral pH decreases in the order G(N7) > T(N3) > C(N3) > A(N7), A(N1) (with binding to cytosine being very weak and almost none to adenine derivatives). This base selectivity appears to be enhanced by the ethylenediamine NH_2 groups, which form hydrogen bonds with exocyclic oxygens (*e.g.* C6O of G) but are non-bonding and repulsive towards exocyclic amino groups of the nucleobases (*e.g.* C6NH$_2$ of A) (Chart 10.15). This pattern of selectivity is found for 5′-mononucleotides.[328] It also appears that the phosphate group might play a role in the binding of these complexes to mononucleotides as the initial binding site before transfer of ruthenium to the base. Reactions with nucleotides appear to proceed *via* aquation of $[(\eta^6\text{-arene})$

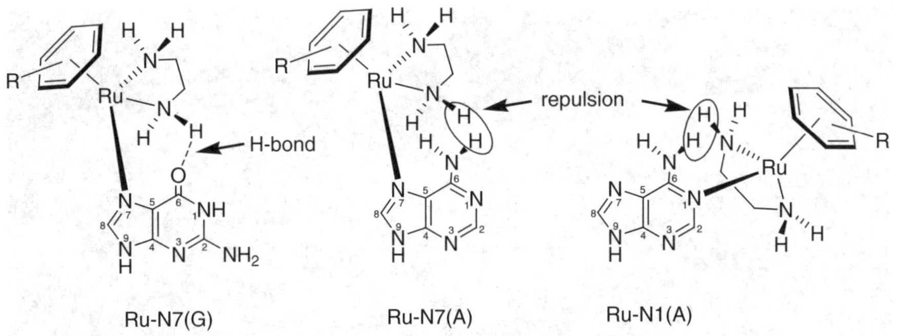

Chart 10.15

Ru(en)Cl]$^+$ followed by rapid binding to the 5′-phosphate group and then rearrangement to give N7-, N1- or N3-bound products.

Since the intracellular tripeptide glutathione (γ-L-glutamyl-L-cysteinylglycine, GSH) is present at millimolar concentrations in cells and can inactivate cisplatin, reactions of Ru^{II}–arene complexes with guanine derivatives have been studied in the presence of excess GSH. These studies have revealed a new mechanism for DNA ruthenation involving oxidation of coordinated S-bound glutathione to the sulfenate which weakens the Ru–S bond and facilitates guanine coordination.[333]

Nucleoside selectivity is affected by exchanging the H-bond donor *N,N*-chelating ligand, such as ethylenediamine (en), for an H-bond acceptor *O,O*-chelating ligand, such as acetylacetonate (acac). The affinity of acac-type complexes for adenosine (N1- and N7-bound) can be greater than for guanosine and there is little binding to cytidine or thymidine.[322,323] This work illustrates how the specific interactions in the more restricted octahedral sites of organometallic Ru^{II} complexes can alter biomolecular recognition properties in comparison with squareplanar complexes such as cisplatin. Recently, Liu *et al.* studied the interaction of the monofunctional fragment {(η^6-biphenyl)Ru(en)}$^{2+}$ with the 14-mer d(ATACATGGTACATA)·d(TATG^{18}TACCATGTAT) by HPLC-ESI-MS and 2D NOESY NMR.[331] Ruthenation occurs at N7 of every guanine residue of the 14-mer and at one site (G^{18}), two different conformers were identified. In one conformer, intercalation of the arene between G^{18} and adjacent T is observed. In the second conformer, the arene is non-intercalated but stacked on a tilted adjacent thymine, lying on the surface of the major groove (Figure 10.9).

10.4.3.4 Osmium Arene Complexes

Recently, Sadler and co-workers investigated the chemical and biological activity of analogous half-sandwich osmium arene complexes.[334–336] Osmium, the heavier congener of ruthenium and a third-row transition metal, commonly exhibits slower kinetics than ruthenium and is often considered to be relatively

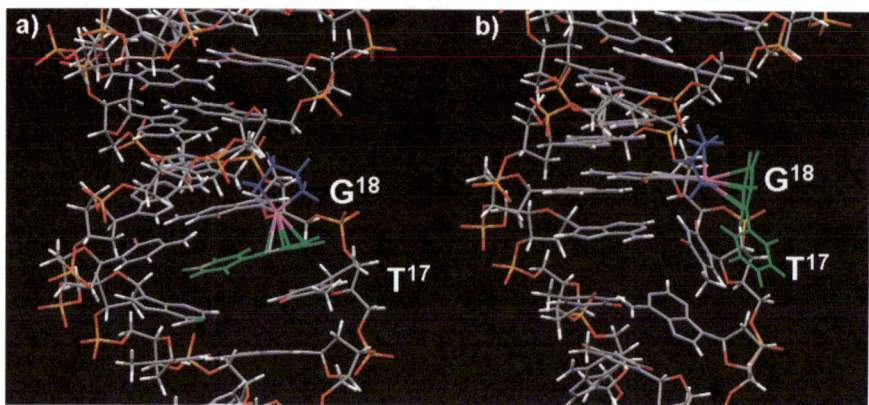

Figure 10.9 Molecular models of two conformers of duplex 14-mer d(ATA-CATGGTACATA)·d(TATG^{18}TACCATGTAT) ruthenated at N7 of G^{18} with monofunctional fragment {(η^6-biphenyl)Ru(en)}$^{2+}$: (a) showing the intercalation of the arene between G^{18} and T^{17}; (b) in which the arene is non-intercalated but stacked on a tilted T17. Colour code: ruthenium, purple; en, blue; biphenyl, green. Adapted from ref. 331.

inert. However, it is apparent that it is possible to tune the biochemical reactivity of the Os^{II}–arene complexes through understanding their aqueous solution chemistry. Thus, active Os^{II}–arene complexes have been designed. Peacock *et al.* found that the chelating ligand plays a major role in the stability, nucleobase binding and cancer cell toxicity.[336] They synthesized complexes containing a series of *N,O*-chelates and found that the choice of the type of N- and O-donors is crucial. Stability (towards the formation of inactive hydroxo-bridged dimers) and therefore activity of the osmium complexes is provided by the more acidic chelating oxygen donors and the introduction of a π-acceptor nitrogen such as pyridine.

Control over both the thermodynamics and kinetics of osmium(II) and ruthenium(II) arene derivatives in aqueous solution can be tuned by variation of the arene, chelating ligand and leaving group.

10.4.3.5 *Other Organometallic Ruthenium Compounds*

Ruthenium(II)–arene complexes with phosphine ligands which possess antitumour and DNA-binding properties have been reported by Dyson and co-workers.[337–339] These include bifunctional Ru^{II}–arene complexes of general formula [Ru(η^6-arene)Cl$_2$(pta)] (pta = 1,3,5-triaza-7-phosphatricyclo[3.3.1.1] decane), *e.g.* **23**. These so-called RAPTA compounds show pH-dependent DNA-damaging properties. The active agent is believed to have a protonated pta ligand, perhaps formed at the acidic pH in some cancer cells.[340] On binding to nucleic acids, loss of the arene ligand can occur.[338] Related phosphine complexes are also being studied in which the labile chlorides are replaced by other chelating ligands to produce more stable complexes.[341]

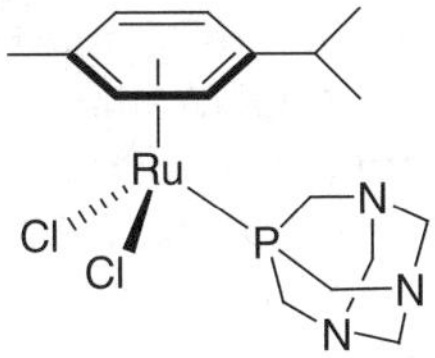

RAPTA-C

23

10.4.4 Rhodium Complexes

Since the 1970s, dirhodium carboxylate complexes have been reported to exert activity against a wide variety of cancers. Their inhibition of DNA replication, *in vitro* transcription and protein synthesis has also been reported.[342–348]

Dunbar and co-workers have developed a series of these antitumour agents, which consists of a paddlewheel structure with a metal–metal-bonded core and at least two bridging ligands (Chart 10.16; for example, $Rh_2(\mu\text{-}O_2CR)_4L_2$, where R=Me, Et, Pr, CF_3; L=solvent).[349–351] The interactions of these complexes with nucleic acid fragments have been investigated in detail. 1D and 2D NMR studies have revealed guanine N7/O6 coordination of d(pGpG) to the equatorial sites of the dirhodium core spanning the Rh–Rh bond to form $[Rh_2(OAc)_2\{d(pGpG)\}]$ in a similar manner to the platinated crosslink *cis*-$[Pt(NH_3)_2\{d(pGpG)\}]$ adduct. Gel shift assays have also shown that the dirhodium compounds form stable Rh–DNA intrastrand crosslinks and a variety of other adducts with double-stranded DNA in aqueous buffer at neutral pH. Thus, DNA is a potential biological target for these compounds.[352,353]

Recently, a new class of dirhodium(II) carboxylate complexes with intercalating groups has been developed jointly by Dunbar, Turro and co-workers as potential antitumour agents for photochemotherapy.[354]

Turro's group had previously investigated the excited-state properties and reactivity of quinonediimine–Rh^{III} complexes such as $[Rh(phi)_2(phen)]^{3+}$ (phi=9, 10-phenanthrenequinonediimine, phen=1,10-phenanthroline), **24**, and related complexes.[355,356] The photophysical properties and DNA binding of quinonediimine–Rh^{III} and related Zn^{II} and Ru^{II} complexes have also been reported.

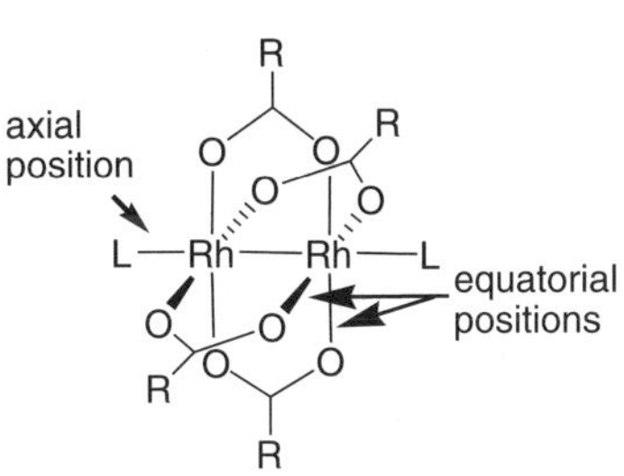

Chart 10.16

Complex **24** and its ruthenium dication analogue [Ru(phi)$_2$(phen)]$^{2+}$ inhibit transcription *via* stabilization of the DNA duplex.[357] Studies of the ligand-loss photochemistry and photoinduced DNA binding of Ru^{II}–bipyridyl–am(m)ine complexes such as *cis*-[Ru(bpy)$_2$(am)$_2$]$^{2+}$ [bpy = 2,2′-bipyridine and (am)$_2$ = (NH$_3$)$_2$ or ethylenediamine] suggest that the *cis*-[Ru(bpy)$_2$(NH$_3$)$_2$]$^{2+}$ complex may be useful photoactivated analogues of cisplatin.[358] Photoactivatable 1,12-diazaperylene (DAP)–Ru^{II} complexes can intercalate into DNA *via* the DAP ligand. Additionally, the complex [Ru(bpy)$_2$(DAP)]$^{2+}$ photocleaves plasmid DNA upon irradiation with visible light.[359]

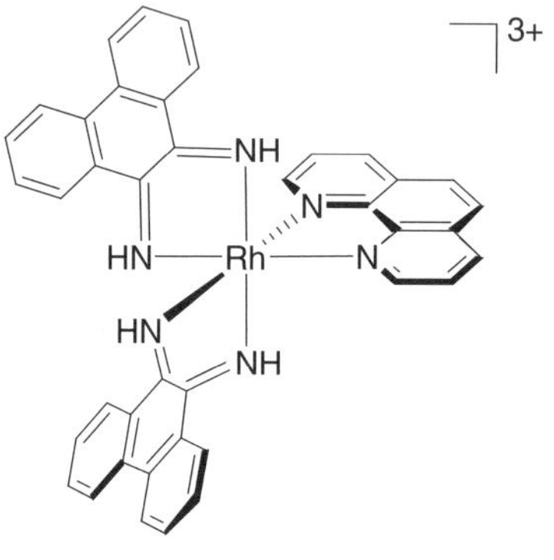

[Rh(phi)$_2$(phen)]$^{3+}$

24

In the last few years, the groups of Turro and Dunbar have reported the chemical reactivity and cytotoxic activity of dirhodium–carboxylate complexes with intercalating ligands such as dppz and phen (dppz = dipyrido-[3,2-a:2′,3′-c]-phenazine; phen = 1,10-phenanthroline).[354,360,361] They observed that DNA photocleavage by *cis*-[Rh$_2$(μ-O$_2$CCH$_3$)$_2$(η^1-O$_2$CCH$_3$)(CH$_3$OH)(dppz)]$^+$ with visible light proceeds through the generation of a reactive oxygen species upon irradiation with $\lambda_{irr} > 395$ nm both in air and in deoxygenated solution.[354,360] The analogous *cis*-[Rh$_2$(μ-O$_2$CCH$_3$)$_2$(dppz)(bpy)]$^{2+}$ complex **25**, where bpy = 2,2′-bipyridine, does not intercalate between the DNA bases and DNA photocleavage with visible light proceeds *via* both oxygen-dependent and -independent mechanisms. Its light toxicity is similar and its dark toxicity is lower than the PDT agent hematoporphyrin.[362]

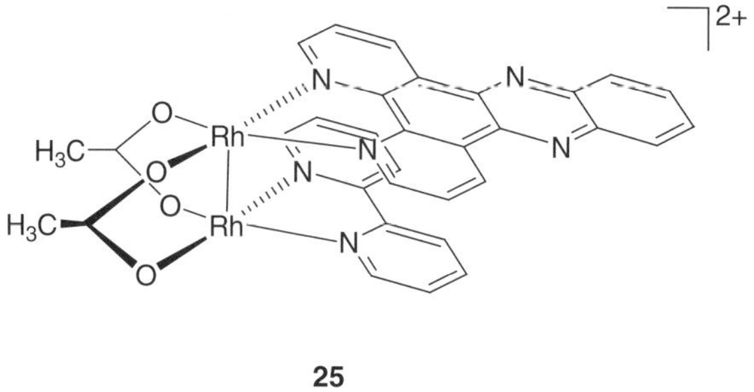

25

The benefits of photodynamic therapy (PDT) have already been discussed in Section 10.4.1.6 on Pt^{IV} photoactivatable drugs.

There is also much interest in photoactivatable Rh and Ru complexes bearing heterocyclic aromatic ligands. Dirhodium complexes with intercalating bidentate nitrogen chelates are one example. Morrison and co-workers' bipyridyl-rhodium(III) complexes are also promising candidates for photodynamic therapy.[62] Photocleavage of DNA by metal complexes has been recently reviewed by Loganathan and Morrison,[63] and by Boerner and Zaleski.[363]

In addition to antiviral activity (see Section 10.3), octahedral bis(bipyridyl) rhodium(III) dichloride complexes, such as cis-dichlorobis(1,10-phenanthroline)rhodium(III) chloride, cis-$[Rh(phen)_2Cl_2]^+$ (BISPHEN) (26), also exhibit antitumour properties when irradiated by light of certain wavelengths. Despite being substitution-inert in the dark, complex 26 undergoes photo-aquation reactions upon irradiation to generate electrophilic species that bind strongly to nucleic acids, primarily to guanine,[364–370] to form crosslinks in RNA,[369] and inactivate naked viral DNA.[367] Additionally, the complex shows minimal dark interaction with DNA, due to the absence of a ligand capable of intercalation. Unfortunately, BISPHEN lacks the hydrophobicity needed for uptake into mammalian cells. Complexes containing methylated analogues of BISPHEN may have improved cell uptake.[64] The related 1,10-phenanthroline complex cis-dichloro(dipyrido-[3,2-a:2',3'-c]-phenazine)(1,10-phenanthroline) rhodium(III) chloride (DPPZPHEN) (27) is capable of penetrating tumour cell membranes and displays phototoxic activity towards tumour cells. When irradiated with light of $\lambda_{irr} > 330$ nm, DPPZPHEN binds strongly to calf thymus DNA. It can also photolytically nick DNA. More importantly, its photoactivation is oxygen independent and does not produce reactive oxygen species (ROS).[62]

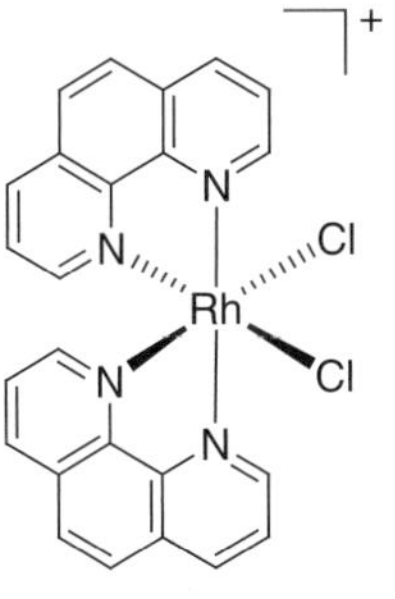

BISPHEN
26

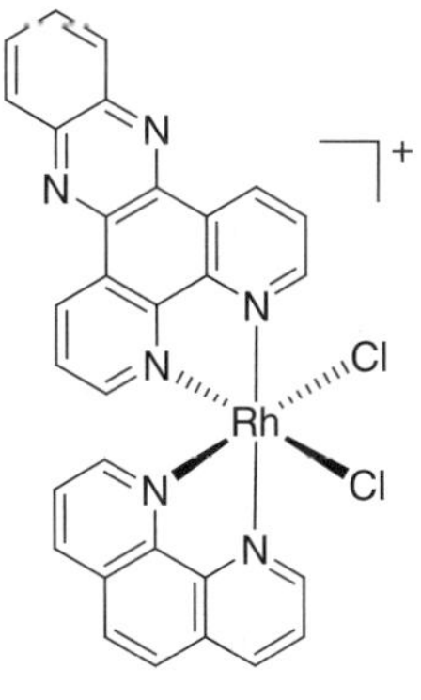

DPPZPHEN
27

Finally, rhodium(III) complexes such as $[Rh(phi)(Me_2trien)]^{3+}$, complex 28 in Chart 10.17, intercalate into DNA. Water molecules mediate interactions between the DNA major groove and the ancillary ligands of this rhodium complex.[371]

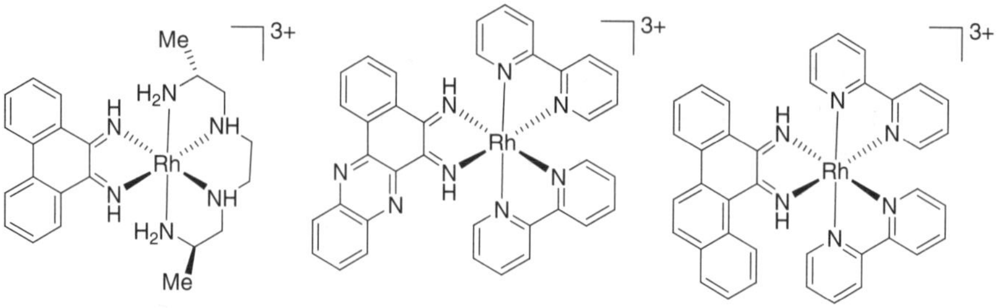

[Rh(phi)(Me₂trien)]³⁺
28

Δ-[Rh(bpy)₂phzi]³⁺
29

Δ-[Rh(bpy)₂chrysi]³⁺
30

Chart 10.17

Some complexes can bind to DNA by both coordinative and intercalative modes, *e.g.* the mixed Ru and Pt complexes synthesized by Brewer and co-workers.[372] These heterobimetallic complexes contain Pt and Ru centres joined by a linker. Platinum offers coordinative binding through ligand substitution and octahedral ruthenium chelated by terpyridine or extended pyridyl ligands acts as an intercalator and possesses useful optical properties.[372] In a similar direction, van der Schilden *et al.* reported a heterodinuclear Ru–Pt complex in which the Pt unit is capable of interacting with DNA by π–π stacking or coordination through chloride substitution and the ruthenium unit can bind to DNA by electrostatic or surface binding or partial intercalation modes.[373]

Accurate DNA replication is the basis of maintaining the fidelity of the genome. Non-complementary base pairs or mismatches within DNA occur naturally during its synthesis *via* polymerase errors, genotoxic chemicals or UV or ionizing radiation.[374–376] Deficiencies in mismatch repair (MMR) are correlated with cancerous transformations. Targeting DNA mismatches may therefore provide a cell-selective strategy for chemotherapeutic design.

In order to probe DNA mismatches, Barton and co-workers have designed bulky Rh^{III} and Ru^{II} complexes containing intercalating ligands that exhibit high DNA sequence selectivity or structural specificity, in search of effective DNA probing.[307,377–379] This is achieved with metal complexes containing sterically demanding intercalating heterocyclic aromatic imine ligands targeted to destabilized mismatched sites in double-helical DNA. The ligands are too wide to insert into matched B-form DNA and bind instead to the enhanced intercalation volume and preferential base stacking offered by CC, CA or CT mismatch sites. Recognition is therefore associated with the local helix instability of a mispair. Examples of these complexes are Δ-[Rh(bpy)₂phzi]³⁺ (bpy = 2,2′-bipyridine, phzi = benzo[*a*]phenazine-5,6-quinone diimine), **29**, and Δ-[Rh(bpy)₂chrysi]³⁺ (chrysi = chrysene-5,6-quinonediimine), **30** (Chart 10.17). Experiments based on the luminescence and photochemical reactivity of these metal complexes confirm DNA photocleavage at the mismatch location. Upon photoactivation at their target site, complexes **29** and **30** trigger direct DNA backbone scission neighbouring the mismatch sites within the duplex with efficiency at concentrations in the nanomolar range. Complexes **29** and **30** demonstrate the correlation between photocleavage and mismatch repair (MMR) deficiency in photocleavage assays of genomic DNA in a series of cell lines, some proficient in MMR and others completely or partially deficient in repair.[379,380]

A strategy for the design of a bifunctional fluorescent probe for mismatches has involved conjugating a fluorophore (Oregon Green) to compound **30**. The rhodium-fluorophore complex binds selectively, photocleaves and shows enhanced fluorescence with mismatched DNA.[381]

The 1.1 Å resolution crystal structure of Δ-[Rh(bpy)$_2$chrysi]$^{3+}$, **30**, bound in mismatched and matched sites in the oligonucleotide 5'-d(CGGAAA-TTCCCG)$_2$-3'.[382] The structure reveals site-specific ligand insertion from the minor groove at the **AC** mismatch site with ejection of both mismatched bases and no change in rise. At the matched site, however, intercalation occurs via the major groove and is typified by a doubling of the rise and no ejection of bases. Furthermore, the large width of the major groove does not sterically hinder the ancillary bipyridine ligands of the complex and can accommodate both the Λ and Δ isomers, whereas the narrow width of the minor groove can only house the Δ enantiomer within the right-handed B-DNA helix. These different groove orientations also lead to distinct photochemical strand cleavage reactions. Mass spectral analysis reveals different photocleavage products associated with the two binding modes in the crystal, with only products characteristic of mismatch binding in solution.[382] The structure illustrates how the recognition of thermodynamically destabilized mismatched sites in DNA might occur and permits a direct comparison with intercalation at a matched site.

10.4.5 Gold Complexes

Gold(III) complexes are isostructural and isoelectronic with Pt^{II} and have attracted interest as possible improved antitumour drugs compared with the parent complex cisplatin. The high oxidation state of Au^{III} gives rise to strongly oxidizing complexes, which makes their synthesis challenging and their pharmacology complicated due to instability under physiological conditions. Recently, however, Au^{III} complexes more stable towards reduction have been synthesized. Gold(III) complexes reported to exhibit significant antitumour activity against different types of anticancer cell lines in the micromolar range include organogold(III),[383–388] gold(III) polyamines,[389,390] gold(III) polypyridines,[386,389] gold(III) porphyrins,[391] gold(III) dithiocarbamate complexes[392–394] and heterocyclic gold(III) complexes.[395–397]

31, **32**, **33**

Heterocyclic chlorogold(III) compounds **31** and **32**, for example, are reported to bind to DNA and this may be the mechanism for their cytotoxic action. However, a lack of stability in physiological media has led to the halting of the preclinical development of these complexes.[395–397]

The gold(III) complex containing dipeptide glycylhistidine, **33**, is stable for days in a physiological phosphate buffer at pH 7.4, as judged by CD and ^{1}H NMR spectroscopies. CD, atomic absorption and DNA melting data suggest that the complex coordinates to calf thymus DNA.[390]

Au^{III} *meso*-tetraarylporphyrins exhibit remarkable solution stability towards reduction by glutathione (GSH) under physiological conditions and potent activity against a panel of human cancer cell lines *in vitro*. Che *et al.* examined the interaction of the gold(III) porphyrin **34**, where X = H, with duplex DNA by UV–visible absorption titrations. By confocal microscopy, the gold(III) porphyrins were found to induce apoptosis of cancer cells.[391]

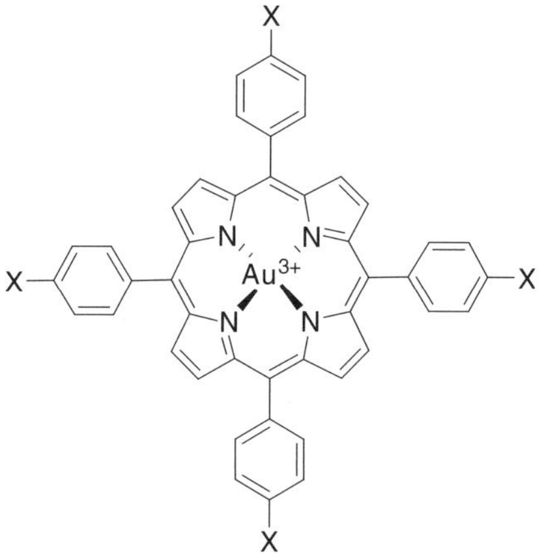

X = H, Me, OMe, Br, Cl

34

Dinuclear gold(III) compounds containing an extended planar system consisting of a common Au$_2$O$_2$ core and bidentate bipyridyl ligands bearing a variety of substituents (analogues of compound **35**) exhibit antiproliferative activity in the A2780 human ovarian carcinoma cell line, with IC_{50} values in the range *ca.* 2–30 µM. Significant differences are observed in reactions of these complexes with calf thymus DNA. Whereas compound **35** interacts weakly with Cl⁻ DNA, most likely through electrostatic interactions, compound **36** binds tightly to the DNA double helix as shown by circular dichroism.[398]

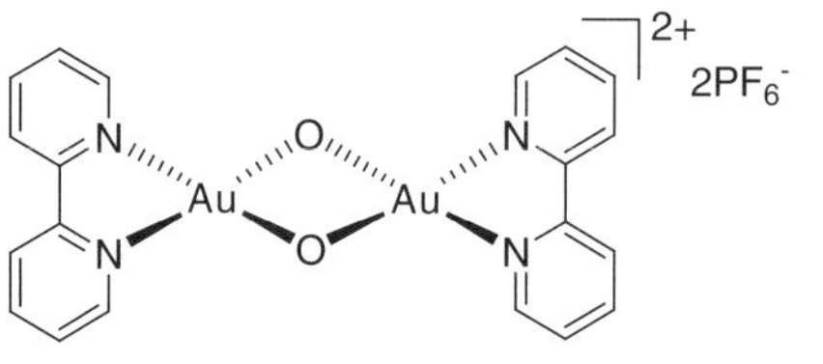

35 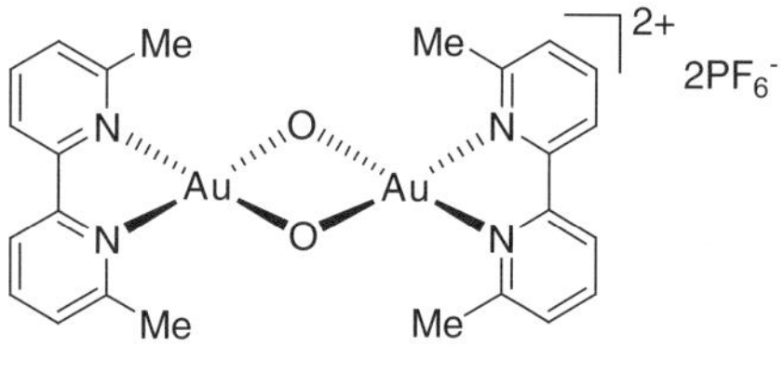   **36**

In general, the role of DNA (or RNA) as potential targets for gold(III) antitumour complexes has yet to be fully elucidated.

Gold(I) has in general a low affinity for oxygen and nitrogen ligands and DNA is unlikely to be the primary target for gold(I) compounds.[399,400] Interference in mitochondrial function, formation of DNA–protein crosslinks in tumour cells, inhibition of protein synthesis and pathways of oxidative phosphorylation are most likely their main intracellular targets.[400–406]

10.4.6 Iron Helicates

Hannon and co-workers are developing a class of metal-based supramolecular assemblies for B-DNA non-covalent recognition *via* major groove sequence-selective targeting and binding.[407,408] For example, the iron complex of formula $[Fe_2L_3]^{4+}$ (L = $C_{25}H_{20}N_4$) (Figure 10.10), is a triple-helical tetracationic cylinder 19 Å in length and 11 Å in diameter, comparable to that of zinc fingers or helix-turn-helix structures. The cylinder is too big to fit in the minor groove, but just the right size and shape to lie along the major groove. The interaction of

Figure 10.10 X-ray structure of triple-helical tetracationic cylinder, $[Fe_2L_3]^{4+}$ (L = $C_{25}H_{20}N_4$). Fe is orange, N is blue. Adapted from ref. 408.

this complex with a duplex palindromic oligonucleotide 5'-d(GAC-GGCCGTC)$_2$ has been studied by NMR and other methods, such as linear dichroism and atomic force microscopy, and the binding to the major groove has been confirmed.[407] Only one enantiomer (ΛΛ) appears to adapt to the shape and size of the groove.[409] The interaction with DNA duplexes (spanning five or more bases) is very strong (binding constant $> 10^7$ M^{-1} in 20 mM NaCl), inducing DNA bending (40–60°) and dramatic intramolecular DNA coiling.[410,411] An X-ray structure shows that the complex is bound not to a DNA duplex but to a three-way junction formed from the palindromic hexanucleotide 5'-d(CGTACG)-3', which defines a triangular-shaped hydrophobic binding site. Both the structures of the helicate and the junction undergo little conformational change upon binding (Figure 10.11). There appears to be selectivity for central TA sequences.[408]

A similar dinuclear ruthenium analogue with two bridging ligands instead of three of formula [Ru$_2$Cl$_4$L$_2$], where L is a dinucleating bisazopyridine ligand with a di(4-phenyl)methane spacer (Chart 10.18) has also been reported. The synthesis of this complex produces three isomeric helicates, which show different but high cytotoxicity.[412]

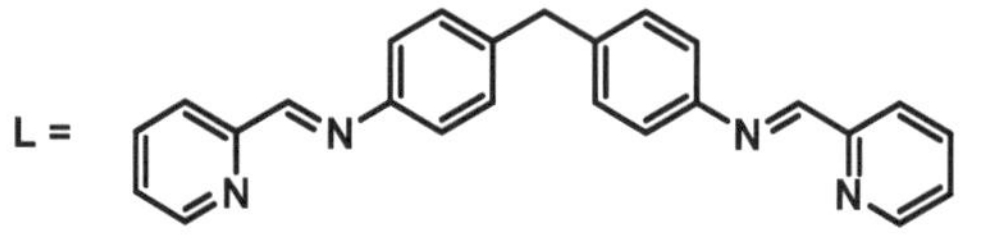

Chart 10.18

The dinuclear dicationic double-bridged copper(I) analogue [Cu$_2$L'$_2$]$^{2+}$, where L' is a tetramethyl derivative of L, binds to DNA more strongly than the dicationic Ru spherical monomer [Ru(phen)$_3$]$^{2+}$, suggesting that charge might not be crucial for strong binding. On the other hand, the copper dication helicate binds more weakly than the tetracation helicate. It seems to bind in a different orientation from that of the iron cylinder and does not coil DNA. However, the copper derivative cleaves nucleic acid in the presence of peroxide suggesting a possible role as an artificial nuclease.[413]

10.5 Conclusions and Perspectives

Interference in the natural biochemistry of DNA by metal ions can lead to problems of carcinogenicity but can also be utilized in therapy. This often involves not only metal coordination to DNA but also the generation of redox, free-radical, reactions which can modify DNA. Although there is a tendency to classify particular metals as carcinogenic in general, it is particular species which exhibit such reactivity. In the case of chromium, for example, Cr^V species formed by reduction of Cr^{VI} (CrO$_4^-$) appear to be especially damaging.

The clinical success of platinum anticancer drugs (cisplatin, carboplatin, oxaliplatin) with their target as DNA has stimulated interest in designing complexes of other metal ions which can cause specific structural alterations in

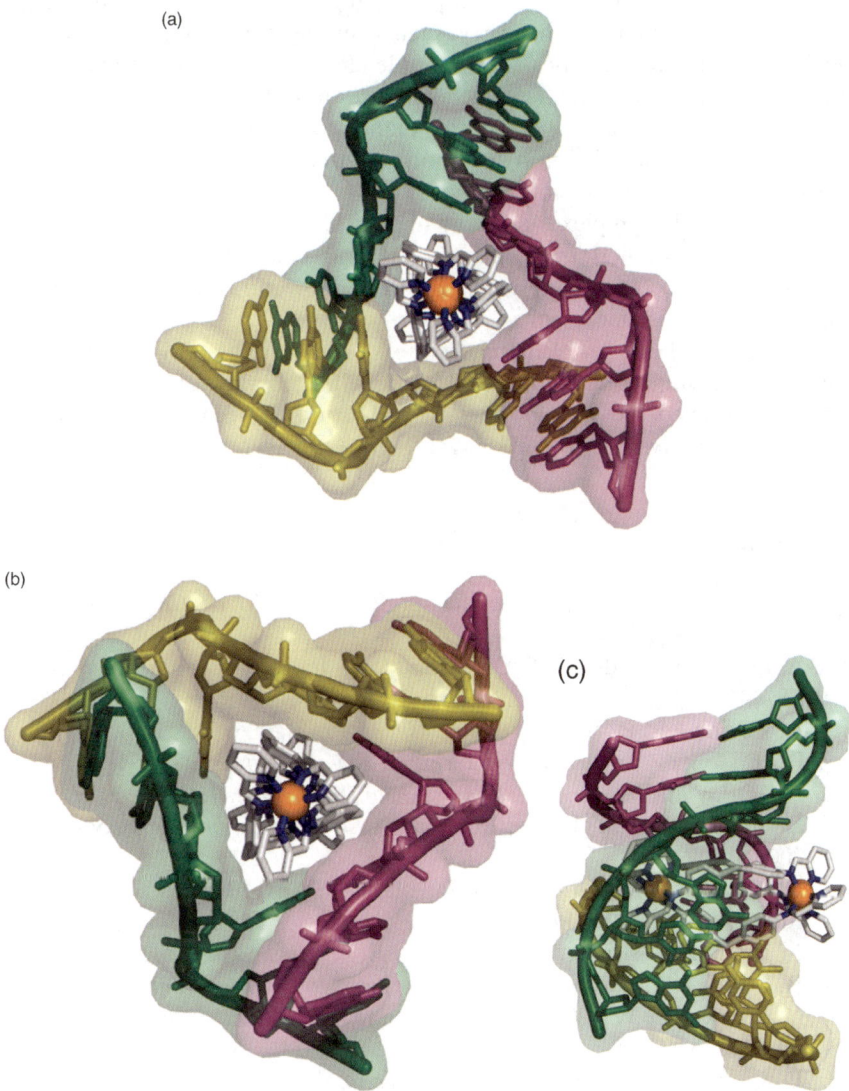

Figure 10.11 Three-dimensional structure of the $[Fe_2L_3]^{4+}$–DNA complex (PDB 2ET0). The helicate is shown with stick representation (N, deep blue; C, grey), the DNA is shown with both stick and surface representations (palindromic oligonucleotides in purple, yellow, and green), whereas the Fe^{2+} ions (orange) are shown as spheres. (a) Major groove side view; (b) Minor groove side view, (c) Lateral view. One of the Fe^{2+} ions and three terminal phenyl rings project from the major groove side of the junction. Adapted from ref. 408.

DNA and influence downstream cell processing. This is presenting interesting challenges in coordination chemistry, such as in the control of redox potentials and ligand substitution rates, for example. Recent work on organometallic ruthenium(II) and osmium(II) arene complexes has illustrated how this can be achieved for these pseudo-octahedral 'piano-stool' complexes and the roles which the ligands can play in hydrophobic and H-bonding interactions and in controlling ligand substitution reactions.

Ruthenium(III) complexes are now in clinical trials and promising data on titanium(IV), gallium(III), rhodium(II/III) and gold(III) complexes suggest that other complexes will soon reach the clinic. However, DNA may not be the only target for these metal ions.

Although a difficult problem, on account of the flexibility of DNA and RNA structures, we can expect computational modelling to play a bigger part in the design of targeted metal complexes in the future. It seems likely that such approaches will give rise to truly novel therapeutic agents.

References

1. R. Codd, C. T. Dillon, A. Levina and P. A. Lay, *Coord. Chem. Rev.*, 2001, **216–217**, 537.
2. A. Levina, R. Codd, C. T. Dillon and P. A. Lay, *Prog. Inorg. Chem.*, 2003, **51**, 145.
3. A. Levina and P. A. Lay, *Coord. Chem. Rev.*, 2005, **249**, 281.
4. P. H. Connett and K. E. Wetterhahn, *Struct. Bond.*, 1983, **54**, 93.
5. P. A. Lay and A. Levina, *J. Am. Chem. Soc.*, 1998, **120**, 6704.
6. K. D. Sugden and D. M. Stearns, *J. Environ. Pathol. Toxicol. Oncol.*, 2000, **19**, 215.
7. D. I. Pattison, M. J. Davies, A. Levina, N. E. Dixon and P. A. Lay, *Chem. Res. Toxicol.*, 2001, **14**, 500.
8. L. Joudah, S. Moghaddas and R. N. Bose, *Chem. Commun.*, 2002, 1742.
9. R. Codd, J. A. Irwin and P. A. Lay, *Curr. Opin. Chem. Biol.*, 2003, **7**, 213.
10. C. J. Chen, C. W. Chen, M. M. Wu and T. L. Kuo, *Br. J. Cancer*, 1992, **66**, 888.
11. D. J. Thomas, M. Styblo and S. Lin, *Toxicol. Appl. Pharmacol.*, 2001, **176**, 127.
12. W. R. Cullen, B. C. McBride and J. Reglinski, *J. Inorg. Biochem.*, 1984, **21**, 45.
13. T. W. Gebel, *Int. J. Hyg. Environ. Health*, 2002, **205**, 505.
14. M. J. Mass, A. Tennant, B. C. Roop, W. R. Cullen, M. Styblo, D. J. Thomas and A. D. Kligerman, *Chem. Res. Toxicol.*, 2001, **14**, 355.
15. K. Yamanaka, K. Kato, M. Mizoi, Y. An, F. Takabayashi, M. Nakano, M. Hoshino and S. Okada, *Toxicol. Appl. Pharmacol.*, 2004, **198**, 385.
16. E. M. Kenyon and M. F. Hughes, *Toxicology*, 2001, **160**, 227.
17. M. F. Hughes, *Toxicol. Lett.*, 2002, **133**, 1.
18. H. Shi, X. Shi and K. J. Liu, *Mol. Cell. Biochem.*, 2004, **255**, 67.

19. M. F. Hughes and K. T. Kitchin, in *Oxidative Stress, Disease and Cancer*, K. K. Singh ed., Imperial College Press, London, 2006, p. 825.
20. P. Grandjean, *IARC Sci. Publ.*, 1984, **53**, 469.
21. K. Salnikow, W. G. An, G. Melillo, M. V. Blagosklonny and M. Costa, *Carcinogenesis*, 1999, **20**, 1819.
22. S.-H. Lee, Y.-H. Shiao, S. Y. Plisov and K. S. Kasprzak, *Biochem. Biophys. Res. Commun.*, 1999, **258**, 592.
23. Z. Zhong, W. Troll, K. L. Koenig and K. Frenkel, *Cancer Res.*, 1990, **50**, 7564.
24. J. A. Knight, M. R. Plowman, S. M. Hopfer and F. W. Sunderman Jr., *Ann. Clin. Lab. Sci.*, 1991, **21**, 275.
25. E. Denkhaus and K. Salnikow, *Crit. Rev. Oncol. Hematol.*, 2002, **42**, 35.
26. A. R. Oller, M. Costa and G. Oberdorster, *Toxicol. Appl. Pharmacol.*, 1997, **143**, 152.
27. F. W. Sunderman Jr., *Ann. Clin. Lab. Sci.*, 1973, **3**, 156.
28. M. P. Waalkes, *Mutat. Res. Fundam. Mol. Mech. Mutagen.*, 2003, **533**, 107.
29. M. Filipic, T. Fatur and M. Vudrag, *Hum. Exp. Toxicol.*, 2006, **25**, 67.
30. M. P. Waalkes and L. A. Poirier, *Toxicol. Appl. Pharmacol.*, 1984, **75**, 539.
31. M. Costa, *Biol. Chem.*, 2002, **383**, 961.
32. K. S. Kasprzak, F. W. Sunderman and K. Salnikow, *Mutat. Res. Genet. Toxicol. Environ. Mutagen.*, 2003, **533**, 67.
33. H. Lu, X. Shi, M. Costa and C. Huang, *Mol. Cell. Biochem.*, 2005, **279**, 45.
34. X. Huang, *Mutat. Res. Fundam. Mol. Mech. Mutagen.*, 2003, **533**, 153.
35. T. Theophanides and J. Anastassopoulou, *Crit. Rev. Oncol. Hematol.*, 2002, **42**, 57.
36. S. Aoki and E. Kimura, *Rev. Mol. Biotechnol.*, 2002, **90**, 129.
37. D. M. Huryn and M. Okabe, *Chem. Rev.*, 1992, **92**, 1745.
38. E. De Clercq, *J. Med. Chem.*, 1995, **38**, 2491.
39. T. P. Zimmerman, W. B. Mahony and K. L. Prus, *J. Biol. Chem.*, 1987, **262**, 5748.
40. H. Mitsuya and S. Broder, *Nature*, 1987, **325**, 773.
41. J. Bourdais, R. Biondi, S. Sarfati, C. Guerreiro, I. Lascu, J. Janin and M. Veron, *J. Biol. Chem.*, 1996, **271**, 7887.
42. S. Aoki, Y. Honda and E. Kimura, *J. Am. Chem. Soc.*, 1998, **120**, 10018.
43. E. Kikuta, S. Aoki and E. Kimura, *J. Am. Chem. Soc.*, 2001, **123**, 7911.
44. C. Dingwall, I. Ernberg, M. J. Gait, S. M. Green, S. Heaphy, J. Karn, A. D. Lowe, M. Singh and M. A. Skinner, *EMBO J.*, 1990, **9**, 4145.
45. U. Delling, L. S. Reid, R. W. Barnett, M.Y.-X. Ma, S. Climie, M. Summer-Smith and N. Sonnenberg, *J. Virol.*, 1992, **65**, 7012.
46. P. A. Sharp and R. A. Marciniak, *Cell*, 1989, **59**, 229.
47. K. A. Jones and B. M. Peterlin, *Annu. Rev. Biochem.*, 1994, **63**, 717.

48. S. Aoki and E. Kimura, *Chem. Rev.*, 2004, **104**, 769.
49. F. Liang, S. Wan, Z. Li, X. Xiong, L. Yang, X. Zhou and C. Wu, *Curr. Med. Chem.*, 2006, **13**, 763.
50. L. Cavallo, R. Cini, J. Kobe, L. G. Marzilli and G. Natile, *J. Chem. Soc., Dalton Trans.*, 1991, 1867.
51. M. Coluccia, A. Boccarelli, C. Cermelli, M. Portolani and G. Natile, *Met.-Based Drugs*, 1995, **2**, 249.
52. Z. Balcarova, J. Kasparkova, A. Zakovska, O. Novakova, M. F. Sivo, G. Natile and V. Brabec, *Mol. Pharmacol.*, 1998, **53**, 846.
53. R. Cini, S. Grabner, N. Bukovec, L. Cerasino and G. Natile, *Eur. J. Inorg. Chem.*, 2000, 1601.
54. P. Papadia, N. Margiotta, A. Bergamo, G. Sava and G. Natile, *J. Med. Chem.*, 2005, **48**, 3364.
55. N. Margiotta, A. Bergamo, G. Sava, G. Padovano, E. De Clercq and G. Natile, *J. Inorg. Biochem.*, 2004, **98**, 1385.
56. S. Grabner, J. Plavec, N. Bukovec, D. Di Leo, R. Cini and G. Natile, *J. Chem. Soc., Dalton Trans.*, 1998, 1447.
57. N. Margiotta, F. P. Fanizzi, J. Kobe and G. Natile, *Eur. J. Inorg. Chem.*, 2001, 1303.
58. L. Cerasino, F. P. Intini, J. Kobe, E. de Clercq and G. Natile, *Inorg. Chim. Acta*, 2003, **344**, 174.
59. V. Maheshwari, D. Bhattacharyya, F. R. Fronczek, P. A. Marzilli and L. G. Marzilli, *Inorg. Chem.*, 2006, **45**, 7182.
60. D. A. Sartori, B. Miller, U. Bierbach and N. Farrell, *J. Biol. Inorg. Chem.*, 2000, **5**, 575.
61. N. Farrell and U. Bierbach, US Pat., 6 001 872, 1999.
62. E. L. Menon, R. Perera, M. Navarro, R. J. Kuhn and H. Morrison, *Inorg. Chem.*, 2004, **43**, 5373.
63. D. Loganathan and H. Morrison, *Curr. Opin. Drug Discov. Dev.*, 2005, **8**, 478.
64. D. Loganathan and H. Morrison, *Photochem. Photobiol.*, 2006, **82**, 237.
65. B. Rosenberg, L. Van Camp and T. Krigas, *Nature*, 1965, **205**, 698.
66. B. Lippert, in *Cisplatin: Chemistry and Biochemistry of a Leading Anticancer Drug*, B. Lippert ed., Verlag Helvetica Chimica Acta, Zurich, 1999, p. 71.
67. E. R. Jamieson and S. J. Lippard, *Chem. Rev.*, 1999, **99**, 2467.
68. M. Jennerwein and P. A. Andrews, *Drug Metab. Dispos.*, 1995, **23**, 178.
69. C. J. Boreham, J. A. Broomhead and D. P. Fairlie, *Aust. J. Chem.*, 1981, **34**, 659.
70. S. E. Miller and D. A. House, *Inorg. Chim. Acta*, 1989, **166**, 189.
71. S. E. Miller and D. A. House, *Inorg. Chim. Acta*, 1990, **173**, 53.
72. S. J. Berners-Price, T. A. Frenkiel, U. Frey, J. D. Ranford and P. J. Sadler, *J. Chem. Soc., Chem. Commun.*, 1992, 789.
73. M. S. Davies, S. J. Berners-Price and T. W. Hambley, *J. Am. Chem. Soc.*, 1998, **120**, 11380.
74. F. Legendre, V. Bas, J. Kozelka and J. C. Chottard, *Chem. Eur. J.*, 2000, **6**, 2002.

75. J. Vinje, E. Sletten and J. Kozelka, *Chem. Eur. J.*, 2005, **11**, 3863.
76. M. S. Davies, S. J. Berners-Price and T. W. Hambley, *Inorg. Chem.*, 2000, **39**, 5603.
77. M. S. Davies, S. J. Berners-Price and T. W. Hambley, *J. Inorg. Biochem.*, 2000, **79**, 167.
78. J. Kozelka, F. Legendre, F. Reeder and J.-C. Chottard, *Coord. Chem. Rev.*, 1999, **190–192**, 61.
79. R. B. Martin, in *Cisplatin: Chemistry and Biochemistry of a Leading Anticancer Drug*, B. Lippert ed., Verlag Helvetica Chimica Acta, Zurich, 1999, p. 183.
80. S. J. Berners-Price and T. G. Appleton, in *Platinum-Based Drugs in Cancer Therapy*, L. R. Kelland and N. Farrell ed., Humana Press, Totowa, N, 2000, p. 3.
81. T. W. Hambley, *J. Chem. Soc., Dalton Trans.*, 2001, 2711.
82. D. P. Bancroft, C. A. Lepre and S. J. Lippard, *J. Am. Chem. Soc.*, 1990, **112**, 6860.
83. N. P. Johnson, J. D. Hoeschele and R. O. Rahn, *Chem. Biol. Interact.*, 1980, **30**, 151.
84. K. J. Barnham, S. J. Berners-Price, T. A. Frenkiel, U. Frey and P. J. Sadler, *Angew. Chem. Int. Ed. Engl.*, 1995, **34**, 1874.
85. H. C. Harder and B. Rosenberg, *Int. J. Cancer*, 1970, **6**, 207.
86. J. A. Howle and G. R. Gale, *Biochem. Pharmacol.*, 1970, **19**, 2757.
87. J. J. Roberts and J. M. Pascoe, *Nature*, 1972, **235**, 282.
88. J. M. Pascoe and J. J. Roberts, *Biochem. Pharmacol.*, 1974, **23**, 1345.
89. A. Eastman, *Biochemistry*, 1983, **22**, 3927.
90. A. Eastman, *Biochemistry*, 1986, **25**, 3912.
91. A. L. Pinto and S. J. Lippard, *Biochim. Biophys. Acta, Rev. Cancer*, 1985, **780**, 167.
92. A. Eastman, in *Cisplatin: Chemistry and Biochemistry of a Leading Anticancer Drug*, B. Lippert ed., Verlag Helvetica Chimica Acta, Zurich, 1999, p. 111.
93. U.-M. Ohndorf and S. J. Lippard, *DNA Damage Recognit.*, 2006, 239.
94. J. Reedijk, *Chem. Rev.*, 1999, **99**, 2499.
95. J. P. Caradonna, S. J. Lippard, M. J. Gait and M. Singh, *J. Am. Chem. Soc.*, 1982, **104**, 5793.
96. J. Filipski, K. W. Kohn and W. M. Bonner, *Chem. Biol. Interact.*, 1980, **32**, 321.
97. A. M. J. Fichtinger-Schepman, J. L. Van der Veer, J. H. J. Den Hartog, P. H. M. Lohman and J. Reedijk, *Biochemistry*, 1985, **24**, 707.
98. A. M. Fichtinger-Schepman, A. T. van Oosterom, P. H. Lohman and F. Berends, *Cancer Res.*, 1987, **47**, 3000.
99. V. Monjardet-Bas, S. Bombard, J.-C. Chottard and J. Kozelka, *Chem. Eur. J.*, 2003, **9**, 4739.
100. R. Gupta, J. L. Beck, M. M. Sheil and S. F. Ralph, *J. Inorg. Biochem.*, 2005, **99**, 552.
101. T. W. Hambley, *Drug Des. Deliv.*, 1988, **3**, 153.

102. A. F. Struik, C. T. M. Zuiderwijk, J. H. Van Boom, L. I. Elding and J. Reedijk, *J. Inorg. Biochem.*, 1991, **44**, 249.
103. M.-H. Baik, R. A. Friesner and S. J. Lippard, *J. Am. Chem. Soc.*, 2003, **125**, 14082.
104. S. T. Sullivan, A. Ciccarese, F. P. Fanizzi and L. G. Marzilli, *J. Am. Chem. Soc.*, 2001, **123**, 9345.
105. G. Natile and L. G. Marzilli, *Coord. Chem. Rev.*, 2006, **250**, 1315.
106. H. Schoellhorn, G. Raudaschl-Sieber, G. Mueller, U. Thewalt and B. Lippert, *J. Am. Chem. Soc.*, 1985, **107**, 5932.
107. B. Lippert, *Prog. Inorg. Chem.*, 1989, **37**, 1.
108. S. E. Sherman, D. Gibson, A. H. J. Wang and S. J. Lippard, *Science*, 1985, **230**, 412.
109. A. Sinur and S. Grabner, *Acta Crystallogr. Sect. C.*, 1995, **C51**, 1769.
110. B. Lippert, G. Raudaschl, C. J. L. Lock and P. Pilon, *Inorg. Chim. Acta*, 1984, **93**, 43.
111. P. M. Takahara, A. C. Rosenzweig, C. A. Frederick and S. J. Lippard, *Nature*, 1995, **377**, 649.
112. A. Gelasco and S. J. Lippard, *Biochemistry*, 1998, **37**, 9230.
113. S. J. Brown, P. J. Kellett and S. J. Lippard, *Science*, 1993, **261**, 603.
114. J.-C. Huang, D. B. Zamble, J. T. Reardon, S. J. Lippard and A. Sancar, *Proc. Natl. Acad. Sci. USA*, 1994, **91**, 10394.
115. U.-M. Ohndorf, M. A. Rould, Q. He, C. O. Pabo and S. J. Lippard, *Nature*, 1999, **399**, 708.
116. M. Kartalou and J. M. Essigmann, *Mutat. Res.*, 2001, **478**, 1.
117. P. Scaffidi, T. Misteli and M. E. Bianchi, *Nature*, 2002, **418**, 191.
118. V. Wunderlich and M. Bottger, *J. Cancer Res. Clin. Oncol.*, 1997, **123**, 133.
119. J. O. Thomas, *Biochem. Soc. Trans.*, 2001, **29**, 395.
120. M. Wei, S. M. Cohen, A. P. Silverman and S. J. Lippard, *J. Biol. Chem.*, 2001, **276**, 38774.
121. S. M. Cohen, E. R. Jamieson and S. J. Lippard, *Biochemistry*, 2000, **39**, 8259.
122. T. Imamura, H. Izumi, G. Nagatani, T. Ise, M. Nomoto, Y. Iwamoto and K. Kohno, *J. Biol. Chem.*, 2001, **276**, 7534.
123. J. Zlatanova, J. Yaneva and S. H. Leuba, *FASEB J.*, 1998, **12**, 791.
124. R. J. Knox, F. Friedlos, D. A. Lydall and J. J. Roberts, *Cancer Res.*, 1986, **46**, 1972.
125. W. I. Sundquist, S. J. Lippard and B. D. Stollar, *Proc. Natl. Acad. Sci. USA*, 1987, **84**, 8225.
126. F. A. Blommaert, H. C. M. van Dijk-Knijnenburg, F. J. Dijt, L. den Engelse, R. A. Baan, F. Berends and A. M. J. Fichtinger-Schepman, *Biochemistry*, 1995, **34**, 8474.
127. Y. Kidani, K. Inagaki and S. Tsukagoshi, *Gann*, 1976, **67**, 921.
128. O. Rixe, W. Ortuzar, M. Alvarez, R. Parker, E. Reed, K. Paull and T. Fojo, *Biochem. Pharmacol.*, 1996, **52**, 1855.
129. R. B. Weiss and M. C. Christian, *Drugs*, 1993, **46**, 360.

130. E. Raymond, S. G. Chaney, A. Taamma and E. Cvitkovic, *Ann. Oncol.*, 1998, **9**, 1053.
131. S. G. Chaney and A. Vaisman, *J. Inorg. Biochem.*, 1999, **77**, 71.
132. S. G. Chaney, S. L. Campbell, E. Bassett and Y. Wu, *Crit. Rev. Oncol. Hematol.*, 2005, **53**, 3.
133. B. Spingler, D. A. Whittington and S. J. Lippard, *Inorg. Chem.*, 2001, **40**, 5596.
134. Y. Wu, P. Pradhan, J. Havener, G. Boysen, J. A. Swenberg, S. L. Campbell and S. G. Chaney, *J. Mol. Biol.*, 2004, **341**, 1251.
135. Y. Wu, D. Bhattacharyya, C. L. King, I. Baskerville-Abraham, S.-H. Huh, G. Boysen, J. A. Swenberg, B. Temple, S. L. Campbell and S. G. Chaney, *Biochemistry*, 2007, **46**, 6477.
136. C. Kloft, C. Eickhoff, K. Schulze-Forster, H. R. Maurer, W. Schunack and U. Jaehde, *Pharm. Res.*, 1999, **16**, 470.
137. C. P. Saris, P. J. M. van de Vaart, R. C. Rietbroek and F. A. Blommaert, *Carcinogenesis*, 1996, **17**, 2763.
138. B. Rosenberg, L. Van Camp, E. B. Grimley and A. J. Thomson, *J. Biol. Chem.*, 1967, **242**, 1347.
139. B. Rosenberg, E. C. Renshaw, L. Van Camp, J. Hartwick and J. Drobnik, *J. Bacteriol.*, 1967, **93**, 716.
140. B. Rosenberg, L. Van Camp, J. E. Trosko and V. H. Mansour, *Nature*, 1969, **222**, 385.
141. B. Rosenberg, *Plat. Met. Rev.*, 1971, **15**, 42.
142. P. D. Braddock, T. A. Connors, M. Jones, A. R. Khokhar, D. H. Melzack and M. L. Tobe, *Chem. Biol. Interact.*, 1975, **11**, 145.
143. V. H. Bramwell, D. Crowther, S. O'Malley, R. Swindell, R. Johnson, E. H. Cooper, N. Thatcher and A. Howell, *Cancer Treat. Rep.*, 1985, **69**, 409.
144. P. J. Creaven, L. Pendyala and S. Madajewicz, *Drugs Exp. Clin. Res.*, 1986, **12**, 287.
145. H. Anderson, J. Wagstaff, D. Crowther, R. Swindell, M. J. Lind, J. McGregor, M. S. Timms, D. Brown and P. Palmer, *Eur. J. Cancer Clin. Oncol.*, 1988, **24**, 1471.
146. P. D. Bonomi, D. M. Finkelstein, J. C. Ruckdeschel, R. H. Blum, M. D. Green, B. Mason, R. Hahn, D. C. Tormey, J. Harris and R. Comis, *J. Clin. Oncol.*, 1989, **7**, 1602.
147. R. J. Schilder, F. P. LaCreta, R. P. Perez, S. W. Johnson, J. M. Brennan, A. Rogatko, S. Nash, C. McAleer, T. C. Hamilton and D. Roby, *Cancer Res.*, 1994, **54**, 709.
148. L. R. Kelland, G. Abel, M. J. McKeage, M. Jones, P. M. Goddard, M. Valenti, B. A. Murrer and K. R. Harrap, *Cancer Res.*, 1993, **53**, 2581.
149. H. Choy, *Expert Rev. Anticancer Ther.*, 2006, **6**, 973.
150. C. N. Sternberg, P. Whelan, J. Hetherington, B. Paluchowska, P. H. T. J. Slee, K. Vekemans, P. Van Erps, C. Theodore, O. Koriakine, T. Oliver, D. Lebwohl, M. Debois, A. Zurlo and L. Collette, *Oncology*, 2005, **68**, 2.

151. J. L. Van der Veer, G. J. Ligtvoet and J. Reedijk, *J. Inorg. Biochem.*, 1987, **29**, 217.
152. R. M. Roat and J. Reedijk, *J. Inorg. Biochem.*, 1993, **52**, 263.
153. S. Choi, R. B. Cooley, A. S. Hakemian, Y. C. Larrabee, R. C. Bunt, S. D. Maupas, J. G. Muller and C. J. Burrows, *J. Am. Chem. Soc.*, 2004, **126**, 591.
154. S. G. Chaney, G. R. Gibbons, S. D. Wyrick and P. Podhasky, *Cancer Res.*, 1991, **51**, 969.
155. V. Brabec, O. Vrana and V. Kleinwachter, *Stud. Biophys.*, 1986, **114**, 199.
156. O. Vrana, V. Brabec and V. Kleinwachter, *Anticancer Drug Des.*, 1986, **1**, 95.
157. O. Novakova, O. Vrana, V. I. Kiseleva and V. Brabec, *Eur. J. Biochem.*, 1995, **228**, 616.
158. M. Defais, M. Germanier and N. P. Johnson, *Chem. Biol. Interact.*, 1990, **74**, 343.
159. E. Rotondo, V. Fimiani, A. Cavallaro and T. Ainis, *Tumori*, 1983, **69**, 31.
160. E. E. Blatter, J. F. Vollano, B. S. Krishnan and J. C. Dabrowiak, *Biochemistry*, 1984, **23**, 4817.
161. J. L. Van der Veer, A. R. Peters and J. Reedijk, *J. Inorg. Biochem.*, 1986, **26**, 137.
162. A. Eastman, *Biochem. Pharmacol.*, 1987, **36**, 4177.
163. L. Pendyala, J. W. Cowens, G. B. Chheda, S. P. Dutta and P. J. Creaven, *Cancer Res.*, 1988, **48**, 3533.
164. S. G. Chaney, S. Wyrick and G. K. Till, *Cancer Res.*, 1990, **50**, 4539.
165. R. M. Roat, M. J. Jerardi, C. B. Kopay, D. C. Heath, J. A. Clark, J. A. Demars, J. M. Weaver, E. Bezemer and J. Reedijk, *J. Chem. Soc., Dalton Trans.*, 1997, 3615.
166. E. L. Weaver and R. N. Bose, *J. Inorg. Biochem.*, 2003, **95**, 231.
167. S. Choi, S. Mahalingaiah, S. Delaney, N. R. Neale and S. Masood, *Inorg. Chem.*, 1999, **38**, 1800.
168. L. T. Ellis, H. M. Er and T. W. Hambley, *Aust. J. Chem.*, 1995, **48**, 793.
169. S. Choi, C. Filotto, M. Bisanzo, S. Delaney, D. Lagasee, J. L. Whitworth, A. Jusko, C. Li, N. A. Wood, J. Willingham, A. Schwenker and K. Spaulding, *Inorg. Chem.*, 1998, **37**, 2500.
170. N. A. Kratochwil and P. J. Bednarski, *Arch. Pharm. (Weinheim)*, 1999, **332**, 279.
171. T. W. Hambley, A. R. Battle, G. B. Deacon, E. T. Lawrenz, G. D. Fallon, B. M. Gatehouse, L. K. Webster and S. Rainone, *J. Inorg. Biochem.*, 1999, **77**, 3.
172. M. D. Hall and T. W. Hambley, *Coord. Chem. Rev.*, 2002, **232**, 49.
173. S. Choi, L. Orbai, E. J. Padgett, S. Delaney and A. S. Hakemian, *Inorg. Chem.*, 2001, **40**, 5481.
174. S. Choi, R. B. Cooley, A. Voutchkova, C. H. Leung, L. Vastag and D. E. Knowles, *J. Am. Chem. Soc.*, 2005, **127**, 1773.
175. S. Choi, L. Vastag, C.-H. Leung, A. M. Beard, D. E. Knowles and J. A. Larrabee, *Inorg. Chem.*, 2006, **45**, 10108.

176. M. J. Cleare and J. D. Hoeschele, *Bioinorg. Chem.*, 1973, **2**, 187.
177. M. J. Cleare and J. D. Hoeschele, *Plat. Met. Rev.*, 1973, **17**, 2.
178. N. Farrell, T. T. B. Ha, J. P. Souchard, F. L. Wimmer, S. Cros and N. P. Johnson, *J. Med. Chem.*, 1989, **32**, 2240.
179. N. Farrell, L. R. Kelland, J. D. Roberts and M. Van Beusichem, *Cancer Res.*, 1992, **52**, 5065.
180. A. Zakovska, O. Novakova, Z. Balcarova, U. Bierbach, N. Farrell and V. Brabec, *Eur. J. Biochem.*, 1998, **254**, 547.
181. V. Brabec, K. Neplechova, J. Kasparkova and N. Farrell, *J. Biol. Inorg. Chem.*, 2000, **5**, 364.
182. J. Kasparkova, O. Novakova, N. Farrell and V. Brabec, *Biochemistry*, 2003, **42**, 792.
183. V. Marini, P. Christofis, O. Novakova, J. Kasparkova, N. Farrell and V. Brabec, *Nucleic Acids Res.*, 2005, **33**, 5819.
184. M. Coluccia, A. Nassi, F. Loseto, A. Boccarelli, M. A. Mariggio, D. Giordano, F. P. Intini, P. Caputo and G. Natile, *J. Med. Chem.*, 1993, **36**, 510.
185. M. Coluccia, A. Boccarelli, M. A. Mariggio, N. Cardellicchio, P. Caputo, F. P. Intini and G. Natile, *Chem. Biol. Interact.*, 1995, **98**, 251.
186. V. Brabec, O. Vrana, O. Novakova, V. Kleinwaechter, F. P. Intini, M. Coluccia and G. Natile, *Nucleic Acids Res.*, 1996, **24**, 336.
187. O. Novakova, J. Kasparkova, J. Malina, G. Natile and V. Brabec, *Nucleic Acids Res.*, 2003, **31**, 6450.
188. E. I. Montero, S. Diaz, A. M. Gonzalez-Vadillo, J. M. Perez, C. Alonso and C. Navarro-Ranninger, *J. Med. Chem.*, 1999, **42**, 4264.
189. J. M. Perez, M. A. Fuertes, C. Alonso and C. Navarro-Ranninger, *Crit. Rev. Oncol. Hematol.*, 2000, **35**, 109.
190. J. M. Perez, E. I. Montero, A. M. Gonzalez, A. Alvarez-Valdes, C. Alonso and C. Navarro-Ranninger, *J. Inorg. Biochem.*, 1999, **77**, 37.
191. E. I. Montero, J. M. Perez, A. Schwartz, M. A. Fuertes, J. M. Malinge, C. Alonso, M. Leng and C. Navarro-Ranninger, *ChemBioChem*, 2002, **3**, 61.
192. J. M. Perez, E. I. Montero, A. M. Gonzalez, X. Solans, M. Font-Bardia, M. A. Fuertes, C. Alonso and C. Navarro-Ranninger, *J. Med. Chem.*, 2000, **43**, 2411.
193. J. M. Perez, L. R. Kelland, E. I. Montero, F. E. Boxall, M. A. Fuertes, C. Alonso and C. Navarro-Ranninger, *Mol. Pharmacol.*, 2003, **63**, 933.
194. F. J. Ramos-Lima, O. Vrana, A. G. Quiroga, C. Navarro-Ranninger, A. Halamikova, H. Rybnickova, L. Hejmalova and V. Brabec, *J. Med. Chem.*, 2006, **49**, 2640.
195. Y. Najajreh, J. M. Perez, C. Navarro-Ranninger and D. Gibson, *J. Med. Chem.*, 2002, **45**, 5189.
196. E. Khazanov, Y. Barenholz, D. Gibson and Y. Najajreh, *J. Med. Chem.*, 2002, **45**, 5196.
197. J. Kasparkova, O. Novakova, V. Marini, Y. Najajreh, D. Gibson, J.-M. Perez and V. Brabec, *J. Biol. Chem.*, 2003, **278**, 47516.

198. J. Kasparkova, V. Marini, Y. Najajreh, D. Gibson and V. Brabec, *Biochemistry*, 2003, **42**, 6321.
199. Y. Najajreh, D. Prilutski, Y. Ardeli-Tzaraf, J. M. Perez, E. Khazanov, Y. Barenholz, J. Kasparkova, V. Brabec and D. Gibson, *Angew. Chem. Int. Ed.*, 2005, **44**, 2885.
200. Y. Najajreh, E. Khazanov, S. Jawbry, Y. Ardeli-Tzaraf, J. M. Perez, J. Kasparkova, V. Brabec, Y. Barenholz and D. Gibson, *J. Med. Chem.*, 2006, **49**, 4665.
201. Y. Najajreh, Y. Ardeli-Tzaraf, J. Kasparkova, P. Heringova, D. Prilutski, L. Balter, S. Jawbry, E. Khazanov, J. M. Perez, Y. Barenholz, V. Brabec and D. Gibson, *J. Med. Chem.*, 2006, **49**, 4674.
202. Y. Najajreh, J. Kasparkova, V. Marini, D. Gibson and V. Brabec, *J. Biol. Inorg. Chem.*, 2005, **10**, 722.
203. J. W. Cox, S. J. Berners-Price, M. S. Davies, Y. Qu and N. Farrell, *J. Am. Chem. Soc.*, 2001, **123**, 1316.
204. N. Farrell, Y. Qu and J. D. Hoeschele, US Pat., 5380897, 1993.
205. C. Sessa, G. Capri, L. Gianni, F. Peccatori, G. Grasselli, J. Bauer, M. Zucchetti, L. Vigano, A. Gatti, C. Minoia, P. Liati, S. Van den Bosch, A. Bernareggi, G. Camboni and S. Marsoni, *Ann. Oncol.*, 2000, **11**, 977.
206. N. Farrell and S. Spinelli, in *Uses of Inorganic Chemistry in Medicine*, N. P. Farrell ed., Royal Society of Chemistry, Cambridge, 1999, p. 124.
207. P. Perego, C. Caserini, L. Gatti, N. Carenini, S. Romanelli, R. Supino, D. Colangelo, I. Viano, R. Leone, S. Spinelli, G. Pezzoni, C. Manzotti, N. Farrell and F. Zunino, *Mol. Pharmacol.*, 1999, **55**, 528.
208. G. Pratesi, P. Perego, D. Polizzi, S. C. Righetti, R. Supino, C. Caserini, C. Manzotti, F. C. Giuliani, G. Pezzoni, S. Tognella, S. Spinelli, N. Farrell and F. Zunino, *Br. J. Cancer*, 1999, **80**, 1912.
209. C. Manzotti, G. Pratesi, E. Menta, R. D. Domenico, E. Cavalletti, H. H. Fiebig, L. R. Kelland, N. Farrell, D. Polizzi, R. Supino, G. Pezzoni and F. Zunino, *Clin. Cancer Res.*, 2000, **6**, 2626.
210. J. D. Roberts, J. Peroutka and N. Farrell, *J. Inorg. Biochem.*, 1999, **77**, 51.
211. P. Perego, L. Gatti, C. Caserini, R. Supino, D. Colangelo, R. Leone, S. Spinelli, N. Farrell and F. Zunino, *J. Inorg. Biochem.*, 1999, **77**, 59.
212. M. S. Davies, D. S. Thomas, A. Hegmans, S. J. Berners-Price and N. Farrell, *Inorg. Chem.*, 2002, **41**, 1101.
213. A. Hegmans, S. J. Berners-Price, M. S. Davies, D. S. Thomas, A. S. Humphreys and N. Farrell, *J. Am. Chem. Soc.*, 2004, **126**, 2166.
214. V. Brabec, J. Kasparkova, O. Vrana, O. Novakova, J. W. Cox, Y. Qu and N. Farrell, *Biochemistry*, 1999, **38**, 6781.
215. Y. Qu, N. J. Scarsdale, M.-C. Tran and N. P. Farrell, *J. Biol. Inorg. Chem.*, 2003, **8**, 19.
216. A. L. Harris, X. Yang, A. Hegmans, L. Povirk, J. J. Ryan, L. Kelland and N. P. Farrell, *Inorg. Chem.*, 2005, **44**, 9598.
217. S. Komeda, T. Moulaei, K. K. Woods, M. Chikuma, N. P. Farrell and L. D. Williams, *J. Am. Chem. Soc.*, 2006, **128**, 16092.

218. S. Komeda, M. Lutz, A. L. Spek, M. Chikuma and J. Reedijk, *Inorg. Chem.*, 2000, **39**, 4230.
219. S. Komeda, G. V. Kalayda, M. Lutz, A. L. Spek, Y. Yamanaka, T. Sato, M. Chikuma and J. Reedijk, *J. Med. Chem.*, 2003, **46**, 1210.
220. S. Komeda, H. Yamane, M. Chikuma and J. Reedijk, *Eur. J. Inorg. Chem.*, 2004, 4828.
221. S. Teletchea, S. Komeda, J.-M. Teuben, M.-A. Elizondo-Riojas, J. Reedijk and J. Kozelka, *Chem. Eur. J.*, 2006, **12**, 3741.
222. N. J. Wheate and J. G. Collins, *Coord. Chem. Rev.*, 2003, **241**, 133.
223. N. J. Wheate and J. G. Collins, *Curr. Med. Chem. Anti Cancer Agents*, 2005, **5**, 267.
224. N. A. Kratochwil, P. J. Bednarski, H. Mrozek, A. Vogler and J. K. Nagle, *Anticancer Drug Des.*, 1996, **11**, 155.
225. N. A. Kratochwil, M. Zabel, K.-J. Range and P. J. Bednarski, *J. Med. Chem.*, 1996, **39**, 2499.
226. N. A. Kratochwil, J. A. Parkinson, P. J. Bednarski and P. J. Sadler, *Angew. Chem. Int. Ed.*, 1999, **38**, 1460.
227. P. Müller, B. Schroder, J. A. Parkinson, N. A. Kratochwil, R. A. Coxall, A. Parkin, S. Parsons and P. J. Sadler, *Angew. Chem. Int. Ed.*, 2003, **42**, 335.
228. P. Müller, *PhD Thesis*, University of Edinburgh, 2002.
229. J. Kasparkova, F. S. Mackay, V. Brabec and P. J. Sadler, *J. Biol. Inorg. Chem.*, 2003, **8**, 741.
230. F. S. Mackay, J. A. Woods, H. Moseley, J. Ferguson, A. Dawson, S. Parsons and P. J. Sadler, *Chem. Eur. J.*, 2006, **12**, 3155.
231. P. J. Bednarski, F. S. Mackay and P. J. Sadler, *Anti Cancer Agents Med. Chem.*, 2007, **7**, 75.
232. F. S. Mackay, J. A. Woods, P. Heringova, J. Kasparkova, A. M. Pizarro, S. A. Moggach, S. Parsons, V. Brabec and P. J. Sadler, *Proc. Natl. Acad. Sci. USA*, 2007, **104**, 20743.
233. H. Köpf and P. Köpf-Maier, *Angew. Chem.*, 1979, **91**, 509.
234. P. Köpf-Maier, *Eur. J. Clin. Pharmacol.*, 1994, **47**, 1.
235. M. Hartmann and B. K. Keppler, *Comments Inorg. Chem.*, 1995, **16**, 339.
236. A. Korfel, M. E. Scheulen, H.-J. Schmoll, O. Grundel, A. Harstrick, M. Knoche, L. M. Fels, M. Skorzec, F. Bach, J. Baumgart, G. Sass, S. Seeber, E. Thiel and W. E. Berdel, *Clin. Cancer Res.*, 1998, **4**, 2701.
237. C. V. Christodoulou, D. R. Ferry, D. W. Fyfe, A. Young, J. Doran, T. M. T. Sheehan, A. Eliopoulos, K. Hale, J. Baumgart, G. Sass and D. J. Kerr, *J. Clin. Oncol.*, 1998, **16**, 2761.
238. G. Luemmen, H. Sperling, H. Luboldt, T. Otto and H. Ruebben, *Cancer Chemother. Pharmacol.*, 1998, **42**, 415.
239. N. Krögera, U. R. Kleebergb, K. Mrossc, L. Edlerd, G. Saße and D. K. Hossfeld, *Onkologie*, 2000, **23**, 60.
240. J. H. Toney and T. J. Marks, *J. Am. Chem. Soc.*, 1985, **107**, 947.
241. M. M. Harding and G. Mokdsi, *Curr. Med. Chem.*, 2000, **7**, 1289.

242. E. Dubler, R. Buschmann and H. W. Schmalle, *J. Inorg. Biochem.*, 2003, **95**, 97.
243. B. K. Keppler and D. Schmaehl, *Arzneim.-Forsch.*, 1986, **36**, 1822.
244. S. Fruehauf and W. J. Zeller, *Cancer Res.*, 1991, **51**, 2943.
245. B. K. Keppler and M. E. Heim, *Drugs Future*, 1988, **13**, 637.
246. B. K. Keppler, M. R. Berger and M. E. Heim, *Cancer Treat. Rev.*, 1990, **17**, 261.
247. H. Sun, H. Li, R. A. Weir and P. J. Sadler, *Angew. Chem. Int. Ed.*, 1998, **37**, 1577.
248. M. Guo, H. Sun, H. J. McArdle, L. Gambling and P. J. Sadler, *Biochemistry*, 2000, **39**, 10023.
249. P. Köpf-Maier, *J. Struct. Biol.*, 1990, **105**, 35.
250. P. Köpf-Maier and H. Köpf, in *Metal Compounds in Cancer Therapy*, S. P. Fricker ed., Chapman and Hall, London, 1994, p. 109.
251. C. V. Christodoulou, A. G. Eliopoulos, L. S. Young, L. Hodgkins, D. R. Ferry and D. J. Kerr, *Br. J. Cancer*, 1998, **77**, 2088.
252. L. Y. Kuo, A. H. Liu and T. J. Marks, *Met. Ions Biol. Syst.*, 1996, **33**, 53.
253. P. Yang and M. Guo, *Coord. Chem. Rev.*, 1999, **185–186**, 189.
254. P. Koepf-Maier and D. Krahl, *Cult. Tech., Symp. Prenatal Dev.*, 5th, 1981, 509.
255. P. Köpf-Maier and D. Krahl, *Chem.–Biol. Interact.*, 1983, **44**, 317.
256. P. Köpf-Maier and R. Martin, *Virchows Arch. B*, 1989, **57**, 213.
257. M. L. McLaughlin, J. M. Cronan Jr., T. R. Schaller and R. D. Snelling, *J. Am. Chem. Soc.*, 1990, **112**, 8949.
258. M. Guo, Z. Guo and P. J. Sadler, *J. Biol. Inorg. Chem.*, 2001, **6**, 698.
259. P. Köpf-Maier, A. Moormann and H. Koepf, *Eur. J. Cancer Clin. Oncol.*, 1985, **21**, 853.
260. P. Köpf-Maier, *J. Cancer Res. Clin. Oncol.*, 1987, **113**, 342.
261. P. Köpf-Maier and H. Köpf, *Chem. Rev.*, 1987, **87**, 1137.
262. P. Köpf-Maier and H. Köpf, *Struct. Bonding*, 1988, **70**, 103.
263. M. S. Murthy, L. N. Rao, L. Y. Kuo, J. H. Toney and T. J. Marks, *Inorg. Chim. Acta*, 1988, **152**, 117.
264. P. Köpf-Maier, *Cancer Chemother. Pharmacol.*, 1989, **23**, 225.
265. V. J. Moebus, R. Stein, D. G. Kieback, I. B. Runnebaum, G. Sass and R. Kreienberg, *Anticancer Res.*, 1997, **17**, 815.
266. F. Caruso and M. Rossi, *Met. Ions Biol. Syst.*, 2004, **42**, 353.
267. P. Köpf-Maier and H. Köpf, *Naturwissenschaften*, 1980, **67**, 415.
268. P. Köpf-Maier, W. Wagner and E. Liss, *J. Cancer Res. Clin. Oncol.*, 1981, **102**, 21.
269. J. Aubrecht, R. K. Narla, P. Ghosh, J. Stanek and F. M. Uckun, *Toxicol. Appl. Pharmacol.*, 1999, **154**, 228.
270. M. M. Harding, G. J. Harden and L. D. Field, *FEBS Lett.*, 1993, **322**, 291.
271. P. Ghosh, O. J. D'Cruz, R. K. Narla and F. M. Uckun, *Clin. Cancer Res.*, 2000, **6**, 1536.
272. A. D. Kelman, M. J. Clarke, S. D. Edmonds and H. J. Peresie, *J. Clin. Hematol. Oncol.*, 1977, **7**, 274.

273. M. J. Clarke, *Met. Ions Biol. Syst.*, 1980, **11**, 231.
274. G. Mestroni, E. Alessio, M. Calligaris, W. M. Attia, F. Quadrifoglio, S. Cauci, G. Sava, S. Zorzet and S. Pacor, *Prog. Clin. Biochem. Med.*, 1989, **10**, 71.
275. G. Sava, A. Bergamo, S. Zorzet, B. Gava, C. Casarsa, M. Cocchietto, A. Furlani, V. Scarcia, B. Serli, E. Iengo, E. Alessio and G. Mestroni, *Eur. J. Cancer*, 2002, **38**, 427.
276. A. Bergamo, R. Gagliardi, V. Scarcia, A. Furlani, E. Alessio, G. Mestroni and G. Sava, *J. Pharmacol. Exp. Ther.*, 1999, **289**, 559.
277. G. Sava, I. Capozzi, K. Clerici, G. Gagliardi, E. Alessio and G. Mestroni, *Clin. Exp. Metastasis*, 1998, **16**, 371.
278. G. Sava, S. Zorzet, C. Turrin, F. Vita, M. Soranzo, G. Zabucchi, M. Cocchietto, A. Bergamo, S. DiGiovine, G. Pezzoni, L. Sartor and S. Garbisa, *Clin. Cancer Res.*, 2003, **9**, 1898.
279. J. M. Rademaker-Lakhai, D. Van Den Bongard, D. Pluim, J. H. Beijnen and J. H. M. Schellens, *Clin. Cancer Res.*, 2004, **10**, 3717.
280. A. Vacca, M. Bruno, A. Boccarelli, M. Coluccia, D. Ribatti, A. Bergamo, S. Garbisa, L. Sartor and G. Sava, *Br. J. Cancer*, 2002, **86**, 993.
281. L. Morbidelli, S. Donnini, S. Filippi, L. Messori, F. Piccioli, P. Orioli, G. Sava and M. Ziche, *Br. J. Cancer*, 2003, **88**, 1484.
282. G. Sava, E. Alessio, A. Bergamo and G. Mestroni, *Top. Biol. Inorg. Chem.*, 1999, **1**, 143.
283. P. J. Dyson and G. Sava, *Dalton Trans.*, 2006, 1929.
284. E. Alessio, G. Mestroni, A. Bergamo and G. Sava, *Curr. Top. Med. Chem.*, 2004, **4**, 1525.
285. E. Alessio, G. Mestroni, A. Bergamo and G. Sava, *Met. Ions Biol. Syst.*, 2004, **42**, 323.
286. F. T. Garzon, M. R. Berger, B. K. Keppler and D. Schmaehl, *Cancer Chemother. Pharmacol.*, 1987, **19**, 347.
287. M. R. Berger, F. T. Garzon, B. K. Keppler and D. Schmaehl, *Anticancer Res.*, 1989, **9**, 761.
288. M. H. Seelig, M. R. Berger and B. K. Keppler, *J. Cancer Res. Clin. Oncol.*, 1992, **118**, 195.
289. M. Galanski, V. B. Arion, M. A. Jakupec and B. K. Keppler, *Curr. Pharm. Des.*, 2003, **9**, 2078.
290. C. G. Hartinger, S. Zorbas-Seifried, A. Jakupec Michael, B. Kynast, H. Zorbas and K. Keppler Bernhard, *J. Inorg. Biochem.*, 2006, **100**, 891.
291. M. J. Clarke, S. Bitler, D. Rennert, M. Buchbinder and A. D. Kelman, *J. Inorg. Biochem.*, 1980, **12**, 79.
292. V. B. Arion, E. Reisner, M. Fremuth, M. A. Jakupec, B. K. Keppler, V. Y. Kukushkin and A. J. L. Pombeiro, *Inorg. Chem.*, 2003, **42**, 6024.
293. E. Reisner, V. B. Arion, M. F. C. Guedes da Silva, R. Lichtenecker, A. Eichinger, B. K. Keppler, V. Y. Kukushkin and A. J. L. Pombeiro, *Inorg. Chem.*, 2004, **43**, 7083.
294. M. A. Jakupec, E. Reisner, A. Eichinger, M. Pongratz, V. B. Arion, M. Galanski, C. G. Hartinger and B. K. Keppler, *J. Med. Chem.*, 2005, **48**, 2831.

295. S. Kapitza, M. Pongratz, M. A. Jakupec, P. Heffeter, W. Berger, L. Lackinger, B. K. Keppler and B. Marian, *J. Cancer Res. Clin. Oncol.*, 2005, **131**, 101.
296. A. Kung, T. Pieper and B. K. Keppler, *J. Chromatogr. B*, 2001, **759**, 81.
297. A. Egger, V. B. Arion, E. Reisner, B. Cebrian-Losantos, S. Shova, G. Trettenhahn and B. K. Keppler, *Inorg. Chem.*, 2005, **44**, 122.
298. C. G. Hartinger, S. Zorbas-Seifried, M. A. Jakupec, B. Kynast, H. Zorbas and B. K. Keppler, *J. Inorg. Biochem.*, 2006, **100**, 891.
299. A. V. Rudnev, L. S. Foteeva, C. Kowol, R. Berger, M. A. Jakupec, V. B. Arion, A. R. Timerbaev and B. K. Keppler, *J. Inorg. Biochem.*, 2006, **100**, 1819.
300. L. R. Bernstein, T. Tanner, C. Godfrey and B. Noll, *Met.-Based Drugs*, 2000, **7**, 33.
301. J. K. Barton, *Science*, 1986, **233**, 727.
302. J. K. Barton, A. Danishefsky and J. Goldberg, *J. Am. Chem. Soc.*, 1984, **106**, 2172.
303. C. V. Kumar, J. K. Barton and N. J. Turro, *J. Am. Chem. Soc.*, 1985, **107**, 5518.
304. J. K. Barton and E. Lolis, *J. Am. Chem. Soc.*, 1985, **107**, 708.
305. J. K. Barton, J. M. Goldberg, C. V. Kumar and N. J. Turro, *J. Am. Chem. Soc.*, 1986, **108**, 2081.
306. A. E. Friedman, J. C. Chambron, J. P. Sauvage, N. J. Turro and J. K. Barton, *J. Am. Chem. Soc.*, 1990, **112**, 4960.
307. E. Rueba, J. R. Hart and J. K. Barton, *Inorg. Chem.*, 2004, **43**, 4570.
308. O. Novakova, J. Kasparkova, O. Vrana, P. M. van Vliet, J. Reedijk and V. Brabec, *Biochemistry*, 1995, **34**, 12369.
309. A. H. Velders, H. Kooijman, A. L. Spek, J. G. Haasnoot, D. De Vos and J. Reedijk, *Inorg. Chem.*, 2000, **39**, 2966.
310. A. C. G. Hotze, S. E. Caspers, D. De Vos, H. Kooijman, A. L. Spek, A. Flamigni, M. Bacac, G. Sava, J. G. Haasnoot and J. Reedijk, *J. Biol. Inorg. Chem.*, 2004, **9**, 354.
311. A. C. G. Hotze, A. H. Velders, F. Ugozzoli, M. Biagini-Cingi, A. M. Manotti-Lanfredi, J. G. Haasnoot and J. Reedijk, *Inorg. Chem.*, 2000, **39**, 3838.
312. A. C. G. Hotze, M. Bacac, A. H. Velders, B. A. J. Jansen, H. Kooijman, A. L. Spek, J. G. Haasnoot and J. Reedijk, *J. Med. Chem.*, 2003, **46**, 1743.
313. A. C. G. Hotze, E. P. L. van der Geer, H. Kooijman, A. L. Spek, J. G. Haasnoot and J. Reedijk, *Eur. J. Inorg. Chem.*, 2005, 2648.
314. R. E. Morris, R. E. Aird, P. d. S. Murdoch, H. Chen, J. Cummings, N. D. Hughes, S. Parsons, A. Parkin, G. Boyd, D. I. Jodrell and P. J. Sadler, *J. Med. Chem.*, 2001, **44**, 3616.
315. A. Habtemariam, M. Melchart, R. Fernandez, S. Parsons, I. D. H. Oswald, A. Parkin, F. P. A. Fabbiani, J. E. Davidson, A. Dawson, R. E. Aird, D. I. Jodrell and P. J. Sadler, *J. Med. Chem.*, 2006, **49**, 6858.
316. S. J. Dougan, M. Melchart, A. Habtemariam, S. Parsons and P. J. Sadler, *Inorg. Chem.*, 2006, **45**, 10882.

317. F. Wang, A. Habtemariam, E. P. L. van der Geer, R. Fernandez, M. Melchart, R. J. Deeth, R. Aird, S. Guichard, F. P. A. Fabbiani, P. Lozano-Casal, I. D. H. Oswald, D. I. Jodrell, S. Parsons and P. J. Sadler, *Proc. Natl. Acad. Sci. USA*, 2005, **102**, 18269.
318. R. E. Aird, J. Cummings, A. A. Ritchie, M. Muir, R. E. Morris, H. Chen, P. J. Sadler and D. I. Jodrell, *Br. J. Cancer*, 2002, **86**, 1652.
319. M. Melchart and P. J. Sadler, in *Bioorganometallics*, G. Jaouen ed., Viley-VCH Verlag GmbH & Co. KGaA, Weinheim, 2006, p. 39.
320. F. Wang, H. Chen, J. A. Parkinson, P. d. S. Murdoch and P. J. Sadler, *Inorg. Chem.*, 2002, **41**, 4509.
321. F. Wang, H. Chen, S. Parsons, I. D. H. Oswald, J. E. Davidson and P. J. Sadler, *Chem. Eur. J.*, 2003, **9**, 5810.
322. R. Fernandez, M. Melchart, A. Habtemariam, S. Parsons and P. J. Sadler, *Chem. Eur. J.*, 2004, **10**, 5173.
323. M. Melchart, A. Habtemariam, S. Parsons, S. A. Moggach and P. J. Sadler, *Inorg. Chim. Acta*, 2006, **359**, 3020.
324. O. Novakova, H. Chen, O. Vrana, A. Rodger, P. J. Sadler and V. Brabec, *Biochemistry*, 2003, **42**, 11544.
325. O. Novakova, J. Kasparkova, V. Bursova, C. Hofr, M. Vojtiskova, H. Chen, P. J. Sadler and V. Brabec, *Chem. Biol.*, 2005, **12**, 121.
326. S. J. Dougan and P. J. Sadler, *Chimia*, 2007, **61**, 704.
327. H. Chen, J. A. Parkinson, S. Parsons, R. A. Coxall, R. O. Gould and P. J. Sadler, *J. Am. Chem. Soc.*, 2002, **124**, 3064.
328. H. Chen, J. A. Parkinson, R. E. Morris and P. J. Sadler, *J. Am. Chem. Soc.*, 2003, **125**, 173.
329. H. Chen, J. A. Parkinson, O. Novakova, J. Bella, F. Wang, A. Dawson, R. Gould, S. Parsons, V. Brabec and P. J. Sadler, *Proc. Natl. Acad. Sci. USA*, 2003, **100**, 14623.
330. F. Wang, J. Bella, J. A. Parkinson and P. J. Sadler, *J. Biol. Inorg. Chem.*, 2005, **10**, 147.
331. H.-K. Liu, S. J. Berners-Price, F. Wang, J. A. Parkinson, J. Xu, J. Bella and P. J. Sadler, *Angew. Chem. Int. Ed.*, 2006, **45**, 8153.
332. H.-K. Liu, F. Wang, J. A. Parkinson, J. Bella and P. J. Sadler, *Chem. Eur. J.*, 2006, **12**, 6151.
333. F. Wang, J. Xu, A. Habtemariam, J. Bella and P. J. Sadler, *J. Am. Chem. Soc.*, 2005, **127**, 17734.
334. A. F. A. Peacock, A. Habtemariam, R. Fernandez, V. Walland, F. P. A. Fabbiani, S. Parsons, R. E. Aird, D. I. Jodrell and P. J. Sadler, *J. Am. Chem. Soc.*, 2006, **128**, 1739.
335. A. F. A. Peacock, M. Melchart, R. J. Deeth, A. Habtemariam, S. Parsons and P. J. Sadler, *Chem. Eur. J.*, 2007, **13**, 2601.
336. A. F. A. Peacock, S. Parsons and P. J. Sadler, *J. Am. Chem. Soc.*, 2007, **129**, 3348.
337. C. S. Allardyce and P. J. Dyson, *J. Cluster Sci.*, 2001, **12**, 563.
338. A. Dorcier, P. J. Dyson, C. Gossens, U. Rothlisberger, R. Scopelliti and I. Tavernelli, *Organometallics*, 2005, **24**, 2114.

339. C. Scolaro, A. Bergamo, L. Brescacin, R. Delfino, M. Cocchietto, G. Laurenczy, T. J. Geldbach, G. Sava and P. J. Dyson, *J. Med. Chem.*, 2005, **48**, 4161.
340. C. S. Allardyce, P. J. Dyson, D. J. Ellis and S. L. Heath, *Chem. Commun.*, 2001, 1396.
341. W. H. Ang, E. Daldini, C. Scolaro, R. Scopelliti, L. Juillerat-Jeannerat and P. J. Dyson, *Inorg. Chem.*, 2006, **45**, 9006.
342. A. Erck, L. Rainen, J. Whileyman, I. M. Chang, A. P. Kimball and J. Bear, *Proc. Soc. Exp. Biol. Med.*, 1974, **145**, 1278.
343. J. L. Bear, H. B. Gray Jr., L. Rainen, I. M. Chang, R. Howard, G. Serio and A. P. Kimball, *Cancer Chemother. Rep., Part 1*, 1975, **59**, 611.
344. S. H. Lee, D. L. Chao, J. L. Bear and A. P. Kimball, *Cancer Chemother. Rep. Part 1*, 1975, **59**, 661.
345. A. Erck, E. Sherwood, J. L. Bear and A. P. Kimball, *Cancer Res.*, 1976, **36**, 2204.
346. R. A. Howard, E. Sherwood, A. Erck, A. P. Kimball and J. L. Bear, *J. Med. Chem.*, 1977, **20**, 943.
347. R. A. Howard, A. P. Kimball and J. L. Bear, *Cancer Res.*, 1979, **39**, 2568.
348. P. N. Rao, M. L. Smith, S. Pathak, R. A. Howard and J. L. Bear, *J. Natl. Cancer Inst.*, 1980, **64**, 905.
349. H. T. Chifotides, K. M. Koshlap, L. M. Perez and K. R. Dunbar, *J. Am. Chem. Soc.*, 2003, **125**, 10703.
350. H. T. Chifotides, K. M. Koshlap, L. M. Perez and K. R. Dunbar, *J. Am. Chem. Soc.*, 2003, **125**, 10714.
351. H. T. Chifotides, J. S. Hess, A. M. Angeles-Boza, J. R. Galan-Mascaros, K. Sorasaenee and K. R. Dunbar, *Dalton Trans.*, 2003, 4426.
352. S. U. Dunham, H. T. Chifotides, S. Mikulski, A. E. Burr and K. R. Dunbar, *Biochemistry*, 2005, **44**, 996.
353. H. T. Chifotides and K. R. Dunbar, *Acc. Chem. Res.*, 2005, **38**, 146.
354. A. M. Angeles-Boza, P. M. Bradley, P. K. L. Fu, S. E. Wicke, J. Bacsa, K. R. Dunbar and C. Turro, *Inorg. Chem.*, 2004, **43**, 8510.
355. C. Turro, A. Evenzahav, S. H. Bossmann, J. K. Barton and N. J. Turro, *Inorg. Chim. Acta*, 1996, **243**, 101.
356. C. Turro, D. B. Hall, W. Chen, H. Zuilhof, J. K. Barton and N. J. Turro, *J Phys. Chem. A*, 1998, **102**, 5708.
357. P. K. L. Fu, P. M. Bradley and C. Turro, *Inorg. Chem.*, 2003, **42**, 878.
358. T. N. Singh and C. Turro, *Inorg. Chem.*, 2004, **43**, 7260.
359. A. Chouai, S. E. Wicke, C. Turro, J. Bacsa, K. R. Dunbar, D. Wang and R. P. Thummel, *Inorg. Chem.*, 2005, **44**, 5996.
360. P. M. Bradley, A. M. Angeles-Boza, K. R. Dunbar and C. Turro, *Inorg. Chem.*, 2004, **43**, 2450.
361. A. M. Angeles-Boza, H. T. Chifotides, J. D. Aguirre, A. Chouai, P. K. L. Fu, K. R. Dunbar and C. Turro, *J. Med. Chem.*, 2006, **49**, 6841.
362. A. M. Angeles-Boza, P. M. Bradley, P. K. L. Fu, M. Shatruk, M. G. Hilfiger, K. R. Dunbar and C. Turro, *Inorg. Chem.*, 2005, **44**, 7262.

363. L. J. K. Boerner and J. M. Zaleski, *Curr. Opin. Chem. Biol.*, 2005, **9**, 135.
364. R. E. Mahnken, M. Bina, R. M. Deibel, K. Luebke and H. Morrison, *Photochem. Photobiol.*, 1989, **49**, 519.
365. R. E. Mahnken, M. A. Billadeau, E. P. Nikonowicz and H. Morrison, *J. Am. Chem. Soc.*, 1992, **114**, 9253.
366. M. A. Billadeau, K. V. Wood and H. Morrison, *Inorg. Chem.*, 1994, **33**, 5780.
367. T. Mohammad, I. Tessman, H. Morrison, M. A. Kennedy and S. W. Simmonds, *Photochem. Photobiol.*, 1994, **59**, 189.
368. H. L. Harmon and H. Morrison, *Inorg. Chem.*, 1995, **34**, 4937.
369. T. Mohammad, C. Chen, P. Guo and H. Morrison, *Bioorg. Med. Chem. Lett.*, 1999, **9**, 1703.
370. H. Morrison and H. Harmon, *Photochem. Photobiol.*, 2000, **72**, 731.
371. C. L. Kielkopf, K. E. Erkkila, B. P. Hudson, J. K. Barton and D. C. Rees, *Nat. Struct. Biol.*, 2000, **7**, 117.
372. R. L. Williams, H. N. Toft, B. Winkel and K. J. Brewer, *Inorg. Chem.*, 2003, **42**, 4394.
373. K. van der Schilden, F. Garcia, H. Kooijman, A. L. Spek, J. G. Haasnoot and J. Reedijk, *Angew. Chem. Int. Ed.*, 2004, **43**, 5668.
374. T. A. Kunkel, *Cancer Cell*, 2003, **3**, 105.
375. J. H. J. Hoeijmakers, *Nature*, 2001, **411**, 366.
376. L. J. Marnett, *Carcinogenesis*, 2000, **21**, 361.
377. B. A. Jackson, V. Y. Alekseyev and J. K. Barton, *Biochemistry*, 1999, **38**, 4655.
378. B. A. Jackson and J. K. Barton, *J. Am. Chem. Soc.*, 1997, **119**, 12986.
379. H. Junicke, J. R. Hart, J. Kisko, O. Glebov, I. R. Kirsch and J. K. Barton, *Proc. Natl. Acad. Sci. USA*, 2003, **100**, 3737.
380. J. R. Hart, O. Glebov, R. J. Ernst, I. R. Kirsch and J. Barton, *Proc. Natl. Acad. Sci. USA*, 2006, **103**, 15359.
381. B. M. Zeglis and J. K. Barton, *J. Am. Chem. Soc.*, 2006, **128**, 5654.
382. V. C. Pierre, J. T. Kaiser and J. K. Barton, *Proc. Natl. Acad. Sci. USA*, 2007, **104**, 429.
383. R. G. Buckley, A. M. Elsome, S. P. Fricker, G. R. Henderson, B. R. C. Theobald, R. V. Parish, B. P. Howe and L. R. Kelland, *J. Med. Chem.*, 1996, **39**, 5208.
384. R. V. Parish, B. P. Howe, J. P. Wright, J. Mack, R. G. Pritchard, R. G. Buckley, A. M. Elsome and S. P. Fricker, *Inorg. Chem.*, 1996, **35**, 1659.
385. R. V. Parish, *Met.-Based Drugs*, 1999, **6**, 271.
386. G. Marcon, S. Carotti, M. Coronnello, L. Messori, E. Mini, P. Orioli, T. Mazzei, M. A. Cinellu and G. Minghetti, *J. Med. Chem.*, 2002, **45**, 1672.
387. L. Messori, G. Marcon, M. A. Cinellu, M. Coronnello, E. Mini, C. Gabbiani and P. Orioli, *Bioorg. Med. Chem.*, 2004, **12**, 6039.
388. M. Coronnello, E. Mini, B. Caciagli, M. A. Cinellu, A. Bindoli, C. Gabbiani and L. Messori, *J. Med. Chem.*, 2005, **48**, 6761.

389. L. Messori, F. Abbate, G. Marcon, P. Orioli, M. Fontani, E. Mini, T. Mazzei, S. Carotti, T. O'Connell and P. Zanello, *J. Med. Chem.*, 2000, **43**, 3541.
390. S. Carotti, G. Marcon, M. Marussich, T. Mazzei, L. Messori, E. Mini and P. Orioli, *Chem.–Biol. Interact.*, 2000, **125**, 29.
391. C.-M. Che, R. W.-Y. Sun, W.-Y. Yu, C.-B. Ko, N. Zhu and H. Sun, *Chem. Commun.*, 2003, **1**, 1718.
392. L. Giovagnini, L. Ronconi, D. Aldinucci, D. Lorenzon, S. Sitran and D. Fregona, *J. Med. Chem.*, 2005, **48**, 1588.
393. L. Ronconi, L. Giovagnini, C. Marzano, F. Bettio, R. Graziani, G. Pilloni and D. Fregona, *Inorg. Chem.*, 2005, **44**, 1867.
394. L. Ronconi, C. Marzano, P. Zanello, M. Corsini, G. Miolo, C. Macca, A. Trevisan and D. Fregona, *J. Med. Chem.*, 2006, **49**, 1648.
395. P. Calamai, S. Carotti, A. Guerri, L. Messori, E. Mini, P. Orioli and G. P. Speroni, *J. Inorg. Biochem.*, 1997, **66**, 103.
396. P. Calamai, S. Carotti, A. Guerri, T. Mazzei, L. Messori, E. Mini, P. Orioli and G. P. Speroni, *Anticancer Drug Des.*, 1998, **13**, 67.
397. P. Calamai, A. Guerri, L. Messori, P. Orioli and G. P. Speroni, *Inorg. Chim. Acta*, 1999, **285**, 309.
398. A. Casini, M. A. Cinellu, G. Minghetti, C. Gabbiani, M. Coronnello, E. Mini and L. Messori, *J. Med. Chem.*, 2006, **49**, 5524.
399. C. K. Mirabelli, C. M. Sung, J. P. Zimmerman, D. T. Hill, S. Mong and S. T. Crooke, *Biochem. Pharmacol.*, 1986, **35**, 1427.
400. M. J. McKeage, L. Maharaj and S. J. Berners-Price, *Coord. Chem. Rev.*, 2002, **232**, 127.
401. G. F. Rush, P. F. Smith, D. W. Alberts, C. K. Mirabelli, R. M. Snyder, S. T. Crooke, J. Sowinski, H. B. Jones and P. J. Bugelski, *Toxicol. Appl. Pharmacol.*, 1987, **90**, 377.
402. G. D. Hoke, G. E. Bossard, J. V. McArdle, B. D. Jensen, C. K. Mirabelli and G. F. Rush, *J. Biol. Chem.*, 1988, **263**, 11203.
403. G. D. Hoke, R. A. Macia, P. C. Meunier, P. J. Bugelski, C. K. Mirabelli, G. F. Rush and W. D. Matthews, *Toxicol. Appl. Pharmacol.*, 1989, **100**, 293.
404. Y. Dong, S. J. Berners-Price, D. R. Thorburn, T. Antalis, J. Dickinson, T. Hurst, L. Qiu, S. K. Khoo and P. G. Parsons, *Biochem. Pharmacol.*, 1997, **53**, 1673.
405. M. P. Rigobello, G. Scutari, R. Boscolo and A. Bindoli, *Br. J. Pharmacol.*, 2002, **136**, 1162.
406. M. P. Rigobello, L. Messori, G. Marcon, M. Agostina Cinellu, M. Bragadin, A. Folda, G. Scutari and A. Bindoli, *J. Inorg. Biochem.*, 2004, **98**, 1634.
407. M. J. Hannon, V. Moreno, M. J. Prieto, E. Moldrheim, E. Sletten, I. Meistermann, C. J. Isaac, K. J. Sanders and A. Rodger, *Angew. Chem. Int. Ed.*, 2001, **40**, 880.
408. A. Oleksi, A. G. Blanco, R. Boer, I. Uson, J. Aymami, A. Rodger, M. J. Hannon and M. Coll, *Angew. Chem. Int. Ed.*, 2006, **45**, 1227.

409. E. Moldrheim, M. J. Hannon, I. Meistermann, A. Rodger and E. Sletten, *J. Biol. Inorg. Chem.*, 2002, **7**, 770.
410. I. Meistermann, V. Moreno, M. J. Prieto, E. Moldrheim, E. Sletten, S. Khalid, P. M. Rodger, J. C. Peberdy, C. J. Isaac, A. Rodger and M. J. Hannon, *Proc. Natl. Acad. Sci. USA*, 2002, **99**, 5069.
411. S. Khalid, M. J. Hannon, A. Rodger and P. M. Rodger, *Chem. Eur. J.*, 2006, **12**, 3493.
412. A. C. G. Hotze, B. M. Kariuki and M. J. Hannon, *Angew. Chem. Int. Ed.*, 2006, **45**, 4839.
413. L. J. Childs, J. Malina, B. E. Rolfsnes, M. Pascu, M. J. Prieto, M. J. Broome, P. M. Rodger, E. Sletten, V. Moreno, A. Rodger and M. J. Hannon, *Chem. Eur. J.*, 2006, **12**, 4919.
414. T. Schilling, K. B. Keppler, M. E. Heim, G. Niebch, H. Dietzfelbinger, J. Rastetter and A. R. Hanauske, *Invest. New Drugs*, 1996, **13**, 327.
415. W. M. Stadler, T. M. Kuzel, D. Raghavan, E. Levine, N. J. Vogelzang, B. Roth and F. A. Dorr, *Eur. J. Cancer*, 1997, **33**, S23.
416. C. R. Chitambar, *Oncology*, 2004, **18**, 39.
417. R. D. Hofheinz, C. Dittrich, M. A. Jakupec, A. Drescher, U. Jaehde, M. Gneist, N. Graf von Keyserlingk, B. K. Keppler and A. Hochhaus, *Int. J. Clin. Pharmacol. Ther.*, 2005, **43**, 590.
418. D. R. Newell, J. Silvester, C. McDowell and S. S. Burtles, *Eur. J. Cancer*, 2004, **40**, 899.
419. K. T. Flaherty, J. P. Stevenson, M. Redlinger, K. M. Algazy, B. Giantonio and P. J. O'Dwyer, *Cancer Chemother. Pharmacol.*, 2004, **53**, 404.
420. M. E. Gore, R. J. Atkinson, H. Thomas, H. Cure, D. Rischin, P. Beale, P Bougnoux, L. Dirix and W. M. Smit, *Eur. J. Cancer*, 2002, **38**, 2416.
421. R. Perez-Soler, J. Lautersztain, L. C. Stephens, K. Wright and A. R. Khokhar, *Cancer Chemother. Pharmacol.*, 1989, **24**, 1.
422. L. E. Fox, K. Toshach, M. Calderwood-Mays, A. R. Khokhar, P. Kubilis, R. Perez-Soler and E. G. MacEwen, *Am. J. Vet. Res.*, 1999, **60**, 257.
423. C. Lu, R. Perez-Soler, B. Piperdi, G. L. Walsh, S. G. Swisher, W. R. Smythe, H. J. Shin, J. Y. Ro, L. Feng, M. Truong, A. Yalamanchili, G. Lopez-Berestein, W. K. Hong, A. R. Khokhar and D. M. Shin, *J. Clin. Oncol.*, 2005, **23**, 3495.
424. T. Dragovich, D. Mendelson, S. Kurtin, K. Richardson, D. Hoff and A. Hoos, *Cancer Chemother. Pharmacol.*, 2006, **58**, 759.
425. A. P. Kudelka, Z. H. Siddik, D. Tresukosol, C. L. Edwards, R. S. Freedman, T. L. Madden, R. Rastogi, M. Hord, E. E. Kim, C. Tornos, R. Mante and J. J. Kavanagh, *Anticancer Drugs*, 1997, **8**, 649.
426. Y. Nakai, *Gan To Kagaku Ryoho*, 1992, **19**, 2157.
427. I. Olver, M. Green, W. Peters, A. Zimet, G. Toner, J. Bishop, W. Ketelbey, R. Rastogi and M. Birkhofer, *Am. J. Clin. Oncol.*, 1995, **18**, 56.

428. D. I. Jodrell, T. R. J. Evans, W. Steward, D. Cameron, J. Prendiville, C. Aschele, C. Noberasco, M. Lind, J. Carmichael, N. Dobbs, G. Camboni, B. Gatti and F. De Braud, *Eur. J. Cancer*, 2004, **40**, 1872.
429. T. A. Hensing, N. H. Hanna, H. H. Gillenwater, M. G. Camboni, C. Allievi and M. A. Socinski, *Anticancer Drugs*, 2006, **17**, 697.
430. R. P. Castleberry, A. B. Cantor, A. A. Green, V. Joshi, R. L. Berkow, G. R. Buchanan, B. Leventhal, D. H. Mahoney, E. I. Smith and F. A. Hayes, *J. Clin. Oncol.*, 1994, **12**, 1616.
431. V. Brabec and M. Leng, *Proc. Natl. Acad. Sci. USA*, 1993, **90**, 5345.
432. S. F. Bellon and S. J. Lippard, *Biophys. Chem.*, 1990, **35**, 179.
433. S. F. Bellon, J. H. Coleman and S. J. Lippard, *Biochemistry*, 1991, **30**, 8026.
434. K. Stehlikova, H. Kostrhunova, J. Kasparkova and V. Brabec, *Nucleic Acids Res.*, 2002, **30**, 2894.
435. F. Paquet, M. Boudvillain, G. Lancelot and M. Leng, *Nucleic Acids Res.*, 1999, **27**, 4261.
436. R. Prokop, J. Kasparkova, O. Novakova, V. Marini, A. M. Pizarro, C. Navarro-Ranninger and V. Brabec, *Biochem. Pharmacol.*, 2004, **67**, 1097.
437. E. Kimura, M. Kikuchi, H. Kitamura and T. Koike, *Chem. Eur. J.*, 1999, **5**, 3113.

Subject Index

ξ *see* linear charge density
α$_x$ *see* proportionality constant
β' *see* site-binding constant
β value *see* Tanford β value
χ *see* crossing angle
ΔG *see* free energy
ΔG^{el} *see* electrostatic free energy
η *see* absolute hardness
κ *see* bending rigidity
ζ *see* azimuthal angle; charge density

A1061U RNA 199
A1088U RNA 199
A2780 human ovarian carcinoma cell line 394
absolute equilibrium constants 329–31
absolute hardness (η) 263–5
abundance of elements 262–3, 264
accuracy, 3D structures 13, 15, 16
acidic ligands 327–8
acid–base catalysis 295–8
acid–base equilibria 43–4, 62–5
acquired immune deficiency syndrome (AIDS) 353–5
active sites 260–99
acyclovir 356, 357
adenine 44, 49
adenosine diphosphate (ADP) 8–11
adenosine trisphosphate (ATP) 8–11
ADP *see* adenosine diphosphate
affinities *see* binding affinities
A-form DNA 75, 76, 84–5

AIDS (acquired immune deficiency syndrome) 353–5
allosteric activation 266
all-*trans* platinum(IV) complexes 379
alternative backbone conformations 107
amino groups, exocyclic 45–8
ammine ligands 372–3
ammonium cations 97–100, 132, 135–8
anionic nucleobases 50
anomalous SAXS (ASAX) 169–70
anticancer agents 39, 40
  gold complexes 393–5
  metal-containing 358–96
  osmium complexes 387–8
  platinum complexes 358–79, 392
  rhodium complexes 389–93
  ruthenium complexes 381–7
  titanium complexes 379–81
anti-Hammond behaviour 236
antiviral drugs 353–7
AOCN *see* average observed coordination number
approved drugs 359, 367–8
aquation, cisplatin 361–2
arene–ruthenium complexes 385–7, 388
argininimide (ARG) 190, 191
aromatic N-donor heterocyclic ligands 373
aroplatin 359, 360
arsenic 352–3
artificial M-triplets 61–2
ASAX *see* anomalous SAXS

atom numbering 77
ATP *see* adenosine triphosphate
A-tracts
　ammonium cation and minor groove 98, 99, 100
　definition 76, 79
　divalent cations and minor groove 92
　monovalent cation size selection and minor groove 103–4
　single-atom substitutions 111
AT-rich tracts *see* A-tracts
attractive forces 337–44
average observed coordination number (AOCN) 5, 7
azimuthal angle (ζ) 342–3
Azoarcus 184, 185, 188, 210, 213, 282–3

Bacillus subtilis 231, 232, 286, 288
backbone cleavage 66
backbone conformations 107, 111, 336–7, 362
base pairs 43, 64–5, 77, 222–3
base selectivity 386
BBR3464 374, 375
beet western yellows virus (BWYV) 198, 199, 205, 206
bending, axial 111, 336–7, 362
bending rigidity (κ) 336
beta value *see* Tanford β value
B-form DNA
　A-form transition 75, 76, 106
　groove width/divalent cation binding 85
　helicity inversion 59
　iron helicates 395
　magnesium cation binding 82, 83
bicycles 15, 17
bidentate chelation 8, 9, 12–14, 30
bifunctional DNA crosslinks 362–4
binding
　affinities
　　cations to DNA 156, 158
　　G-quadruplexes and ammonium ions 138
　　metal ions 263–5
　hierarchies 174

models
　cations to DNA 77–9
　RNA
　　diffuse ion association models 204–5
　　direct binding methods 197–200
　　folding/metal ion binding 192–207
　　general considerations 194–7
　　Hill model 200–4
　　protonation link to magnesium cation binding/RNA folding 205–7
sites
　A-form major groove divalent cation binding 84–5
　G-quadruplexes 132
　monovalent/divalent ions competition on RNA 186–7
　multivalent cations 225
　tetraloop–tetraloop receptor 183, 184, 185
BISPHEN 391
Bjerrum length 322
bond strength 40
branched-chain am(m)ine *trans*-Pt complexes 371, 372
broadening *see* paramagnetic broadening
8-bromoguanosine gels 118, 119, 145
budotitane 380
bundle phase 338, 339
BWYV *see* beet western yellows virus

cadmium cations 160, 280, 353
caesium cations 101, 102, 104, 105
calcium cations
　d(TG$_4$T) crystal structure 125
　alternative backbone conformations 107
　coordination chemistry 5, 7
　G-quadruplex coordination 128–9
　G-quadruplex stability 140
　minor groove localisation 90–1
　nucleic acid complexes 1–32
　sequence-specific binding 82, 83
capillary electrophoresis 104
carbon-bonded H atoms 65–6
carboplatin 358, 359, 360, 367, 374

Subject Index 419

carcinogenesis 351–3
catalysis 260–99
catalytic (C) domain 284
cationic base modifications 111
cation-induced RNA folding kinetics 226–53
cation probes, NMR-active 160–3
cations
   see also individual cations
   exchange, G-quadruplex direct observation 134–8
   non-natural base studies 110–11
   nucleic acid complexes 1–32
   physical properties 5, 7
   position determinations 24–5
   sequence-specific interactions 75–112
CC *see* counterion condensation theory
cellular DNA damage responses 365–7
chameleon-like behaviour 54–6
charge density (ζ) 182, 183, 236–7, 264, 265
charge undulations 310
chelation 4–24, 52
   chelated ions environment 192
   ring size 11, 13, 14, 15
chemical numbering 77
chemical potential 313
chemical shift effects 159, 160, 161
chemotherapy *see* anticancer agents
chlorido ligands 370
chromium, mutagenesis 351–2
cisplatin 359, 360, 374, 379
   aquation 361–2
   discovery 358
   DNA adducts 359, 362–8
   guanine chelation 52
   history 41
   classification
      cation interactions 155
      metal-containing anticancer agents 358, 359
      metal ligand exchange kinetics 263–6
clinical trials 359
cobalt ions 172
compensating charges 309

computational methods 26, 272–3
condensed counterion charge 307–8, 311, 333–7
condensed fraction 311
condensed phase formation 337–44
configuration integral 313–15
conformational deviants *see* non-motif RNA
conformational transition
   ligand binding to polyions 328–9
   RNA spectroscopy 158
   salt concentration 325–6
   temperature dependence 326–7
$Co(NH_3)_6^{3+}$ 161, 162
conservation 18, 20
controls (experimental) 279
conventional metal binding sites 44–5, 46
coordination bond formation
   di-nuclear metal complexes 52–3
   chameleon-like behaviour 54–6
   consequences 57–68
   DNA condensation 59
   Group I/II cation–nucleic acid complexes 4–24
   helicity inversion 59
   history 41–2
   irreversible alterations 66–8
   multinuclear metal complexes 52–3
   nucleobases as ligands 42–56
   subtle effects 62–4
   transition metal ions–nucleobases 39–69
   types of interactions 40
coordination number 264
copper 353, 396
Coulombic force *see* electrostatics
Coulomb's law 184
counterion condensation (CC) theory 307–8, 311, 333–7
counterions
   definition 309
   RNA folding kinetics 223–53
   valence 310, 311, 312–13
crossing angle (χ) 342, 343
crossing zones 339–40

crosslinks
  cisplatin–DNA adducts 362–4, 365–6
  duplex DNA 57–9
  polyplatinum complexes 376
  trans-Pt^{II}(NH$_3$)$_2$ 57–8
crystal structure *see* X-ray crystal structure
cyclic hexanuclear purine complex 49, 51
cytosine 44

d(A$_n$T$_n$) motif 88–90
DAPI 325, 327
dark stability 378, 391
database methods 272–3
daunomycin 125
3-deazaadenine 111
Debye length 322
Debye–Huckel theory *see* linear Poisson–Boltzmann theory
density functional theory 331
deprotonation 44, 45–50, 51
d(G$_3$T$_4$G$_3$) 131, 132
d(G$_3$T$_4$G$_4$) 132, 137, 138
d(G$_4$T$_4$G$_3$) 142
diagonal loops 122
cis-diamminedichloridoplatinum(II) *see* cisplatin
diazido photoactivable platinum(IV) complexes 377–8, 379
Dickerson–Drew dodecamer (DDD)
  divalent cations and minor groove 90–1
  monovalent cation localisation in major groove 93–6
  non-natural base studies for fine structure 111
  sequence-specific cation localisation 108–10
  water/caesium cation minor groove co-occupation 101, 102
diffuse ion environment 155, 192, 204–5
difluorotoluene 111
diiodo photoactivable platinum(IV) complexes 377–8
dinuclear gold(III) compounds 394–5

di-nuclear metal complexes, coordination bond formation 52–3
direct binding methods 197–200
direct cation coordination 119
direct observation 134–8
dirhodium–carboxylate complexes 389, 390
disease aspects 350–98
dispersal in conformational space 22–3
distortion 57–8
  see also bending
divalent cations
  G-quadruplexes
    binding 140
    coordination 128–9
    structures 120
  group II intron requirement 289–90
  localisation
    major groove binding 79–85
    minor groove binding 76, 77–9
    sequence-specific interactions 79–92
  minor groove width modulation 108
  physico–chemical properties 182–5, 263–5
  RNA folding 246
  small ribozyme active site independence 261, 295, 298
DNA
  see also Dickerson–Drew dodecamer; duplex dodecamers
  adducts 359, 362–8, 375
  bending 111, 336–7, 362
  binding 372
  condensation 59
  damage responses 365–7
  distortion 57–9
  duplex 43–4, 57–9
  fine structure 106–11
  irreversible sequence modifications 66–7
  polyanion models 307–44
docked/undocked forms 232, 233, 240, 241
dodecamers *see* duplex dodecamers
Donnan effect 310, 311, 312

Subject Index

donor sites 42–3
double chain reversal loops 121, 122
dppz ligand 390
DPPZPHEN 357, 391
drug resistance 376–9
d(TG$_4$T) 123–5, 127, 128
d(T$_n$A$_n$) motif 88–90
duplex DNA 43–4, 57–9
duplex dodecamers
   see also Dickerson–Drew dodecamer
   cisplatin adducts 364, 365
   ruthenated 387, 388
   sequence-specific metal ion binding 82–8, 90–6, 106–10
dynamical restriction 187–92

early transition states 238, 240–1
ECC theory *see* extended CC theory
edge wise loops 122
effective potential energy function 314–15
electron nuclear double resonance (ENDOR) 163, 164–6, 167
electron paramagnetic resonance (EPR) 163–7, 271
electron spin-echo envelope modulation (ESEEM) 163, 164–7
electrostatic free energy (ΔG^{el}) 310–11, 319–20, 327
electrostatics
   contribution, bending rigidity/persistence length 336
   environment, magnesium cation-mediated RNA folding 251–3
   potential energy 317
   repulsion 223
   tertiary interactions 28, 29
endonucleolytic cleavage 292, 295–8
ENDOR *see* electron nuclear double resonance
energetics 23–4
   see also electrostatics; thermodynamics
enloplatin 359, 360
EPR *see* electron paramagnetic spectroscopy

equilibrium properties
   definitions 310–13
   polyanion absolute equilibrium constants 329–31
   polyanion models 307–44
   polyion equilibria 323–32
equilibrium thermal unfolding 204–5
Escherichia coli 235, 286, 288, 361, 362
ESEEM *see* electron spin-echo envelope modulation
europium ions 171, 172–3
EXAFS *see* extended X-ray absorption fine strcture
exocyclic groups 45–8, 50, 51, 67
extended X-ray absorption fine structure (EXAFS) 168–9
extended CC theory 334, 335
extended ribonuclease P RNA 232, 233

famciclovir 356
fiber diffraction studies 75–6, 102
final state 226–35
fine structure of DNA 106–11
   see also extended X-ray absorption fine structure
fixed-counterion models 342–4
fluorescence resonance energy transfer (FRET) 231, 233, 234
fluorescence titration data 238, 239
fluorescent probes 393
folding 2–4, 227
   intermediates 28
   kinetics 221–53
   study techniques 187
   thermodynamics 180–213
   topologies 121–3, 142, 143
   unfolding 207–12, 227, 241–6
force–extension curves 242–5
four-way RNA junction constructs 233
frameshifting pseudoknots 206–7, 211
free energy (ΔG)
   phi-value analysis 238
   polyanion absolute equilibrium constants 329–31
   polyanion equilibrium properties 310–11

polyanion models 319–20
RNA folding, ionic strength effects 226–35
RNA magnesium cation binding 192–200
fully-ionised electrolytes 307
see also non-linear Poisson–Boltzmann equation
functional roles 19, 20
functions derivation 31–2

gallium maltolate 383
GC-rich tracts see G-tracts
gels 118, 119, 145
gene pairing 141
general acid–base catalysis 269–70, 295–8
gene regulation 118, 120
generic DNA 79
glmS catalytic riboswitch 184, 185, 212
globular structures 2, 3
glutathione 394
GO6 carbonyl groups 120
gold complexes 393–5
GPF see grand partition function
G-quadruplex structures 118–47
 X-ray crystallography 123–9
 cation movement direct observation 134–8
 folding toplogies 121–3
 melting temperatures 139
 NMR-active probes 162, 164, 165
 NMR studies 130–4
 non-biological applications 145–7
 polymorphism 141–5
 stability 139–41
G-quartets
 cation interactions, non-biological applications 145–7
 definition 118–19
 manganese cation EPR spectroscopy 165
 solid-state NMR studies of cation coordination 133–4
 structure 121
 syn–anti–anti–syn pattern 142, 143

grand partition function (GPF) 334
groove width 106–10, 112
Group I cations 4–7, 24–5
group I intron
 active site structures 282
 Hammond behaviour 236
 non-productive folding intermediates 228–9, 230
 P4–P6 domain 183–5, 238–9, 244–5, 247–8, 252
 RNA tertiary structure 224
 transition state model 287
 two-metal ion mechanism 285
Group II cations 6, 7–24, 25
group II intron 285, 287, 288–93
G-tracts 76, 79, 105, 106
guanine
 see also G-quadruplex structures; G-quartets
 chelation, cisplatin 52
 cisplatin binding 364
 divalent cation localisation 80, 81, 82, 83
 hemideprotonated base pair 65
 protonated/deprotonated existence ranges 44
 redox chemistry 67–8
 ruthenium arene complexes 386
guanosine
 gels 119, 145
 telomeres 119
 Tetrahymena thermophila group I ribozyme active site 278, 280–3
G-wires 146

hairpin ribozyme 240, 241
hammerhead ribozymes
 allosteric activation by metal ions 266
 spectroscopic studies 156–7, 164, 165, 168, 169
 structure elucidation 181
Hammond behaviour 235–7
hard metal ions 263
hardness see absolute hardness
hard/soft Lewis acid–base theory 183
HDV see hepatitis delta virus

heavy atom isotope effects 275–6
helical structure
   axial bending 111
   folded nucleic acids 2, 3
   inversion, coordinative bond formation 59
   salt conditions 75
   symmetry, polyanion models 342–4
hemideprotonated guanine base pair 65
hepatitis delta virus (HDV) 206, 269–70, 295–7
heptaplatin 359, 360, 368
heterobimetallic complexes 392
heterocyclic chlorogold(III) compounds 394
hexaaqua magnesium cation complexes 7
HHRz see hammerhead ribozymes
hierarchies
   binding 174
   G-quadruplex stabilisation by cations 140
   RNA folding 3–4, 185–6, 222–3
Hill model 189, 200–4, 286
history
   coordinative bond formation 41–2
   polyanion models 308
   RNA structure elucidation 180–1
HIV-1 TAR RNA 189, 190–1, 209, 211, 353–5
HMG-domain 366–7, 375
Holoarcula marismortui 2, 6
Hoogsteen-paired guanine residues 118, 119
hopping behaviour 242–3, 246
HQS see 8-hydroxyquinolone-5-sulfonic acid
human bladder cancer cell line 378, 379
human immunodeficiency virus (HIV-1) 353–5
human ovarian carcinoma 394
human telomere quadruplex 126, 127, 141–5
Hybrid-1/Hybrid-2 structures 144
hydrated magnesium cations 225

hydration
   enthalpy 264, 265
   measurements 75, 104–6
   shells 182–3
hydrodynamic collapse 187–92
hydrogen 65–6, 131–2
hydrolytic processes 67
hydroxyl groups 280–3
8-hydroxyquinolone-5-sulfonic acid (HQS) 197–8
HYSCORE 164–5

I see intermediate state
imino tautomer structure 45, 46–8
independently stable RNA subdomains 247, 248
indicator dyes 197–8
indirect catalytic contributions 269–70
indirect chelates 52
indirect detection 159–60
induced intramolecularity 268, 269
infinite uniformly charged cylinders 318–19
infrared spectroscopy 173–4
initial (RNA) state 226–35, 246–53
inner hydration shell 225
inner sphere binding see coordination bond formation
intermediate (I) state 192, 195–7, 200, 201, 226–35
intermediate coordination geometry 123–5
Internet-based databases 273
interstrand crosslinks 57–8
intramolecularity, induced 268, 269
intrastrand crosslinks 57–9, 362–4, 365–6, 376
ion atmosphere 309–10, 315–19
ion channels 146
ion hydration shells 182, 183
ionic radius 119, 120, 264, 265
ionic strength (ζ) 156, 226–35
   see also counterions
iproplatin 369
iron, mutagenesis 353
iron helicates 395–6, 397

irreversible bond alterations 66–8
Irving–Williams series 265
isoguanosine ionophore 146
isotopes
  see also individual isotopes
  effects 275–6
  isotopic hydrogen exchange 65–6

JM216 (satraplatin) 369

keto tautomeric form 46
kinetics 25–6, 221–53, 313
kinetic traps 222–3
kinking see bending
kink-turn (KT) motif 191
Kirkwood–Buff theory 322
kissing hairpins 210, 211
KP46 383
KP1019 358, 382–3
K_T see tautomer equilibrium constant
KY see kink-turn motif

L-21 group I intron, *Tetrahymena* 210
L-21 *ScaI* ribozyme
  RNA folding 229, 230
  secondary structure 229, 244, 245
lambda repressor DNA-binding domain (λbd) 325
Langevin molecular dynamics simulations 338
lanthanide luminescence 170–3, 274–5
large electric charges 307–44
large monovalent cations 103–4
large ribozymes
  group I intron
    active site structures 282
    Hammond behaviour 236
    nonproductive folding intermediates 228–9, 230
    P4–P6 domain 183–5, 238–9, 244–5, 247–8, 252
    RNA tertiary structure 224
    transition state model 287
    two-metal ion mechanism 285
  group II intron 285, 287, 288–93
  precise positioning of metal cations 292

Rnase P 284–8
spliceosome 285, 287, 293–5
transition state coordination environment 283–95, 298
two-metal ion mechanistic paradigm 261, 283–95
large RNA assemblies 2
large RNA folding transition states 238
laser optical tweezers 207–12, 241–6
L coordination site 132, 138
lead cations 140
leaving group stabilisation 267, 268
Lewis acidity 263, 264, 265
like-charged filaments 337–44
linear charge density (ξ) 311–12
linear multiplatinum complexes 374
linear Poisson–Boltzmann (LPB) theory 332, 333
linear polyions 307–8, 311, 333–5
linear REFER 227
linkers (Langevin MD) 338–9
lipophilic Zn^{II}–cyclen complexes 353–5
lobaplatin 359, 360, 368
localised cation interactions 155
loop topologies 121–3, 145
LPB see linear Poisson–Boltzmann theory
LRET see luminescence resonance energy transfer
luminescence resonance energy transfer (LRET) 171, 172
luminescent ruthenium(II) coordination compounds 383–4

M_A see metal A
macrochelates 52
magnesium-25 isotope 160–1
magnesium 262–3, 264
magnesium cations
  alternative backbone conformations 107
  coordination chemistry 7–24, 25
  Dickerson–Drew dodecamer crystal structures 108
  divalent cations and cation minor groove 90–1

Subject Index

hydration measurements 104–5
lanthanide luminescence studies 171, 172, 274
major groove of B-form cation binding 82
Nature 262–3
nucleic acid complexes 1–32
P4–P6 domain folding 247, 248
RNA
  folding 240, 241
    rate scenarios 230–5
    thermodynamics 180–213
  interaction statistical analysis 28–32
  large ribozyme transition states 286–8
  motif chelation 20–2, 30, 31–2
  nucleotide/RNA chelation 7–24, 25
  structure stability 225–6
  Tetrahymena thermophila 247, 248, 278, 281, 328–9, 331
  unfolding 241–6
sequence-specific cation binding 84–5
major groove
  divalent cation localisation 79–85
  donor site locations 42–3
  iron helicates 395–6, 397
  monovalent cation localisation 92–6
  narrowing 106–10
  sequence-specific cation interactions 76, 77–9
manganese cations
  ENDOR/ESEEM studies 164–6, 167
  NMR probe 80, 81
  sequence-specific major groove cation binding 84
  sequence-specific minor groove cation binding 86–90
  Tetrahymena thermophila group I ribozyme active site studies 281
  thrombin-binding aptamer interaction 141
Maxwell–Boltzmann distributions 313
M_B *see* metal B
M-box magnesium cation-sensing riboswitch 184, 185
MC *see* Monte Carlo methods

MD *see* molecular dynamics
mean electrostatic potential energy 317
mechanical unfolding studies, RNA 207–12, 241–6
mechanism of action, platinum(IV) complexes 369–70
mechanisms of action
  ribozyme reactions and metal ions 266–70
  two-metal ion paradigm 283–95
medical applications 353–98
  anticancer agents 358–96
  antiviral drugs 353–7
  gold complexes 393–5
  iron helicates 395–6, 397
  osmium arene complexes 387–8
  platinum(IV) complexes 368–70
  platinum drugs 358–79
  polyplatinum complexes 374–6
  rhodium complexes 389–93
  ruthenium drugs 381–7
  titanium agents 379–81
  trans-platinum complexes 370–4
melting temperatures
  8-bromoguanosine gels 119
  G-quadruplexes 139
mercury cations, organometallic nuclebase complexes 50–1
metal C (M_C) 278, 280 3
metal A (M_A) 278, 279
metal B (M_B) 278, 280
metal ion rescue experiments 262, 276–83
metallocenes 379–81
metal migration 54
metal-stabilized rare nucleobase tautomers 47, 48
metal–nucleobase quartets 61–2
miboplatin 359, 360
minor groove
  donor site locations 42–3
  monovalent cation localisation 96–102
  monovalent cation size selection 103–4
  narrowing 106–10
  sequence-specific cation interactions 76, 77–9, 85–92
  width in G-tract DNA 106–10

MIR *see* multiple isomorphous replacement
mismatch repair (MMR) 392, 393
Misra–Honig model 326
mixed-nucleobase/mixed-metal complexes 44–5, 46
MMR *see* mismatch repair
modelling 307–44
   molecular dynamics 95–6, 108–9, 309, 338
   Monte Carlo methods 308, 316, 318
molecular dynamics (MD) 95–6, 108–9, 309, 338
monodentate chelation 8, 9, 11, 30
monovalent cations
   G-quadruplexes 123–8, 130–3
   localisation
      MD simulations 95–6
      major groove 92–6
      minor groove 96–102
   physical properties, RNA folding 182–5
   physicochemical properties 263–5
   RNA folding 247–51
   small ribozyme catalysis 261
Monte Carlo (MC) methods 308, 316, 318
motifs *see* RNA motifs
M-quartets 61–2
multinuclear metal complexes 52–3
multiple isomorphous replacement (MIR) 51
multiple parallel folding pathways 246–53
multistranded nucleic acids 59–60
multivalent cations 225, 235–7
mutagenesis 351–3

N *see* native state
NAIM *see* nucleotide analogue interference mapping
NAMI-A 358, 382
nanowires 146
narrowing, groove width 106–10, 112
native (N, folded) state 2–4, 195–7, 200, 201, 227

natural systems 59–60, 350
nedaplatin 359, 360
negative electrostatic potential 181–2
Neidle crystal structure 143–4
NER *see* nucleotide excision repair
net attractions 312–13
net energetic stability 226–35
neutral nucleobases 44–5, 46
nickel 353
nitrogen-15 isotope 97–100, 132, 135–8
NLPB *see* non-linear Poisson–Boltzmann equation
NMR *see* nuclear magnetic resonance
NOESY spectra 97–8, 99, 100, 103
non-biological applications 145–7
non-bridging phosphoryl oxygens 268–9, 278, 281–3
non-linear Poisson–Boltzmann (NLPB) equation
   counterion condensation theory relationship 333–5
   polyanion models 307, 308, 317–23, 340–2
   RNA folding 193, 194, 199, 212, 240, 241
   solvent–RNA interactions 273
   two binding-mode formalism 26
non-linear REFER 227
non-motif RNA 20–3
non-natural base studies 110–11
nonproductive RNA folding intermediates 228, 229, 230, 246
non-specific cation associations 25
non-zero nuclear spin 160–3
nuclear magnetic resonance (NMR)
   cation probes 160–3
   cisplatin–DNA adduct 365
   divalent cations and major groove 79–81
   divalent cations and minor groove 88–90
   G-quadruplex structures 125, 130–4
   metal interaction indirect detection 159–60
   monovalent cations and minor groove 96–101
   ribozymal catalytic metal ion–ligand interactions 272

Subject Index

nuclear Overhauser effect *see* NOESY spectra
nucleic acid folding *see* folded nucleic acids
nucleobase carbon-bonded H atoms 65–6
nucleobases
  see also guanine
  acid–base equilibria 43–4
  as ligands 42–3
  organometallic complexes 50–1
  pK_a shifted values due to metal coordination 64
  redox chemistry 67–8
  transition metal ion coordinative bond formation 39–69
nucleobase triplets 60, 63
nucleophile activation 267–8
nucleotide analogue interference mapping (NAIM) 273–4, 281
nucleotide excision repair (NER) 365, 371, 375
nucleotide phosphoroamidates 110–11
NzExHSQC spectra 136, 137

Oceanobacillus iheyensis 292
octahedral platinum(IV) complexes 369
off-pathway RNA intermediates 228, 229, 230, 246
oligocations 336–7
oligonucleotide-bound paramagnetic ions 164
OP *see* oxygen–phosphate
optical methods 170–3, 274–5, 393
organic solvents 146
organometallic nucleobase complexes 50–1
osmium arene complexes 387–8
osmotic pressure 312, 323
oxaliplatin 358, 359, 360, 367–8, 374
oxidation state 350, 353, 369
oxygen 118–19, 120, 268–9, 278, 281–3
  non-bridging phosphoryl oxygens 268–9, 278, 281–3
  oxygen-17 isotope 134
  reactive oxygen species 68

oxygen–phosphate (OP) 6, 11
Oxytricha nova 125, 142, 163

P3–P7 domain 230
P4 helix 288
P4–P6 domain
  force–extension curves 244, 245
  ion binding sites 183, 184, 185
  magnesium cation-induced folding 238, 239, 247, 248, 252
  mechanical unfolding map 244, 245
P5abc RNA 242–6
p see persistence length
paddlewheel structures 389
pairing of genes 141
paleo-magnesium ions 12, 18–20
parallel/antiparallel strand orientations 121–3, 142, 144
parallel RNA folding pathways 222–3
parallel uniformly charged rods 337–40
paramagnetic broadening 80, 81, 86, 87, 160
partition function 314
pathway partitioning 246–53
pea enation mosaic virus-1 (PEMV-1) 207
Pearson hardness 263, 264
PEMV-1 *see* pea enation mosaic virus-1
penciclovir 356, 357
pentaammine(purine)ruthenium(III) complexes 381
peptidyl transfer centres 19
periodic order 337
persistence length (P) 336
pH 43–4, 48–50, 62–5, 295–8
phen ligand 390
phi-value analysis 238, 239
phosphates 1–32, 181–2
phosphorothioate substitutions 160, 168
phosphoryl transfer reactions
  indirect catalytic contributions 269–70
  induced intramolecularity 268, 269
  Lewis acid stabilisation of leaving group by metal ions 267
  non-bridging phosphoryl oxygens coordination 268–9, 278, 281–3
  nucleophile activation 267–8
  ribozyme catalysis 260–99

phosphotransesterification 277–83
photoactivable drugs 356–7, 376–9, 391
photosensitisers 377
physical properties
  cations 5, 7
  metal ions 263–6
  RNA
    folded polyanion nature 181–2
    hierarchical folding 185–6
    ion competition for sites 186–7
    monovalent *versus* di(multi)valent ions 182–5
piano-stool structures 385
PIC *see* preferential interaction coefficient
picoplatin 359, 360
pK_a values 64
planar/non-planar cyclic ligands 373
plastic RNA transition state 235–46
platinum
  (II) complexes 372, 373
  (IV) complexes 368–70, 376–9
  drugs 356–7, 358–79
  heterobimetallic complexes 392
  ion coordination 49, 50–1, 57–8
Poisson–Boltzmann equation 212, 240, 241
polyamines 236–7, 339
polyanion activity coefficient 320–3
polyanion models 307–44
  compensating charges 309
  configuration integral 313–15
  counterion condensation theory 307–8, 311, 333–7
  definitions/terminology 309–13
  helical symmetry effects 342–4
  ion atmosphere theory 315–19
  like-charged filament attractions 337–44
  non-linear Poisson–Boltzmann theory 307, 308, 317–23
  osmotic pressure 323
  polyion equilibria 323–32
  simulations 337–40
  statistical mechanics 313–15
  thermodynamic properties 319–23, 335–6
  tightly bound ion model 340–2

polyanion–salt preferential interaction coefficient 320–3
polycations 323–5
polyelectrolyte models 76
polyion equilibria 323–32
polymorphic G-quadruplex structures 121–3, 141–5
polyplatinum complexes 374–6
polypyridyl ruthenium compounds 383–5
porfirmer sodium 377
positional RNA transition aspects 235–46
positioning of metal cations 292
potassium-39 isotope 130
potassium cations
  coordination chemistry 5–7
  DNA helical structure 108
  G-quadruplexes 120
    $d(G_4T_4G_4)$ crystal structure 125–8
    polymorphism 141–5
    proton NMR 131, 132
    stability studies 139–40
  magnesium cation-mediated RNA folding 251–3
  major groove localisation 92–3
  nucleic acid complexes 1–32
  RNA sites competition 186–7
potential energy function 314–15
potential of mean force 313–15
precise positioning of metal cations 292
preferential interaction coefficient (PIC) 320–3
probes, NMR-active 160–3
pro-drugs 369–70, 377
productive RNA intermediates 246
proportionality constant (α_x) 235
protonation
  acidic ligand to adjacent polyion 327–8
  exocyclic amino groups 45–8
  nucleobases 44
  purine bases 54
  RNA folding/magnesium cation binding link 205–7
proton NMR studies 131, 132
pseudoknot structures 207, 230, 235
  see also beet western yellows virus

Subject Index

quadrupolar NMR studies 130–3
quartets 61–2
quasi-globular structures 2, 3

Raman spectroscopy 173–4
random binding model 21, 30–2
RAPTA compounds 388–9
rare nucleobase tautomers 46–8
rate–equilibrium free energy
    relationship (REFER) 226–7
reaction coordinates 27–8, 241–6
redox chemistry 67–8, 350, 353, 370
REFER *see* rate–equilibrium free
    energy relationship
repulsive forces 311, 323–5, 332
residence times 134–7
resistance to drugs 376–9
resonance line broadening 80, 81, 86,
    87, 160
Resonance Raman spectroscopy 173–4
reverse second step splicing 291–2
rhodium(III) photoactivable
    complexes 356–7, 391
rhodium complexes 389–93
ribonuclease P RNA 231, 232, 233
riboswitches 184, 185, 212, 213
ribozymes
  see also large ribozymes
  catalytic metal ion–ligand interactions
    270–83
  glmS 184, 185, 212
  hairpin 240, 241
  hammerhead 156–7, 164, 165, 168,
    169, 181
  HDV 206, 269–70, 295–7
  metal ion mechanisms of action
    266–70
  reaction kinetics, metal ion rescue
    experiments 262
  RNA catalysis 260–99
rigidity *see* bending rigidity
rigid ligands 376
rips *see* unfolding
RNA
  see also ribozymes
  assemblies 1–32

catalysis and metal ions 260–99
$Co(NH_3)_6^{3+}$ NMR 161–2
folding
  hierarchical 3–4, 222–3
  hydrodynamic collapse/dynamical
    restriction 187–92
  kinetics 221–53
  metal ion binding models 158,
    192–207
  physical properties 181–7
  reaction coordinates 27–8
  simplified mechanism 28, 29
  single-molecule unfolding
    mechanical force studies 207–12
  thermodynamics 180–213
irreversible sequence modifications
    66–7
LRET studies 171, 172
magnesium-25 NMR 161
polyanion models 307–44
tertiary structure 224–6
unfolding 207–12, 227, 241–6
RNase P 284–8
rod-like polyions 307, 332
23S-rRNA^{HM}
  Holoarcula marismortui 2, 6
  magnesium chelation 13, 15, 17
  paleo-magnesium ions 12, 18–20
  secondary structure 18
23S-rRNA^{TT} 2, 12, 18–20
rubidium-87 isotope 130
rubidium cations 101
$r(UG_4U)$ 128–9
Runge–Kutta algorithm 318
ruthenium drugs 381–7, 388–9

salt conditions
  conformational transition
    equilibrium effects 325–6
  Debye length of solution 322
  DNA helical structure 75, 112
  ion atmosphere in solution 309–10
  polyanion–salt preferential
    interaction coefficient 320–3
  polycation associations with
    polyanions 323–5

satraplatin 369
SAXS *see* small-angle X-ray scattering
scattering shells 168
scissile *pro*-S_P non-bridging oxygen 278, 281–3
SDSL *see* site-directed spin labelling
sebriplatin 359, 360
secondary structure
  G-quadruplexes 118–47
  RNA
    L-21 ribosome 244, 245
    definition 186
    elements stability 223
    folding hierarchy 3, 29, 222
    23S-rRNA^{HM} 18
    Tetrahymena ribosome 243
second step splicing 291–2
self-assembled reversible component amplification process 146
self-energies 316
sensing sites 184, 185, 212, 213
sequence-specific cation localisation 75–112
  divalent 79–92
  groove width 106–10
  monovalent 92–102
sequence structure 66–7, 202, 243
simulations
  molecular dynamics 95–6, 108–9, 309, 338
  Monte Carlo (MC) methods 308, 316, 318
  polyanion models 337–40
Sindbis virus (SINV) 357
single atom A-tract substitutions 111
single molecule FRET (smFRET) 202–4, 247–51
single molecule unfolding studies 207–12, 241–6
SINV *see* Sindbis virus
site-binding constant (β') 334
site-bound cation interactions 25, 155
site-directed spin labelling (SDSL) 164
site-specifically chelated magnesium cations 328–9
size selection 103–4

small-angle X-ray scattering (SAXS) 166–8, 169–70
small ion atmosphere 315–17
small ribozyme catalysis 238, 240–1, 261, 292, 295–8
snRNAs, spliceosomal 293
sodium-23 isotope 100, 130, 133–4
sodium cations
  coordination chemistry 5, 6
  Dickerson–Drew dodecamer crystal structures 108–9, 110
  G-quadruplexes
    X-ray crystallography 123–5
    proton NMR 131, 132
    stability studies 139–40
    structures 120
  magnesium cation comparison in nucleic acid chelation 16
  magnesium cation-mediated RNA folding 251–3
  major groove localisation 92–3
  minor groove 100–1
  nucleic acid complexes 1–32
  RNA folding conformational heterogeneity in initial state 247–51
  RNA sites competition 186–7
soft metal ions 263
solid-state NMR studies 120, 133–4
solution-state NMR studies 120, 130–3
solvent–RNA interactions 273
specific ion RNA binding sites 183–5
specificity (S) domain 284
specificity switch experiments 290
spectroscopic techniques 154–75
  see also individual techniques
  mimics
    ammonium cations 97–100, 132, 135–8
    caesium cations 101, 102, 104, 105
    manganese cations 80–1, 84, 86–90, 141, 164–7, 281
    rubidium cations 101, 130
    thallium cations 24–5, 93–5, 101, 125, 127–8, 132–3, 163
  ribozymal catalytic metal ion–ligand interactions 271–2

Subject Index

sph*n* polyamines 339
spliceosome 285, 287, 293–5
splicing pathways 289–95
stability
  see also thermostability
  G-quadruplex structures 139–41
  metal-coordinated nucleobase pairs 64–5
  RNA
    folding kinetics 221–53
    magnesium cation stabilisation 207–12
    secondary structure elements 223
    stabilisation hierarchy 140
stable gels 118, 119, 145
stacked G-quartets 126, 128, 130, 131, 144
standard primitive model 315–17
statistical analysis 28–32
statistical mechanics 313–15, 331
strontium cations 128–9
structural databases 27–8
subdomains of RNA 247, 248
subtle effects 62–4
sulfur 168
syn–anti–anti–syn pattern 142, 143

Tanford β values 235, 236, 237
TAR *see trans*-activation responsive region
Tat protein 354, 355
tautomer equilibrium constant (K_T) 46–7
tautomeric structures 46–8
TBI *see* tightly-bound ion model
telomeres
  X-ray crystal structure 126, 127
  G-quadruplex polymorphism 141–5
  guanosine-rich sequences 119
  thallium-105 NMR 162–3
temoporfin 377
temperature 119, 326–7
  see also thermo...
terbium ions 173
terminology 309–13
tertiary structure, RNA 3–4, 186, 222, 224–6

Tetrahymena thermophila
  see also P4–P6 domain
  P5abc ribosome 242–6
  ribozyme 210, 228–9, 230, 236, 252
    active site 277–83
    RNA folding 183, 184, 185, 210, 211, 224, 252
  23S-rRNA^{TT} 2
telomeric repeat 141
tetraloop–tetraloop receptor (TL–TLR) 183, 184, 185, 202–4
tetramethylammonium cation 103
tetramolecular structures *see* G-quadruplexes; G-quartets
tetraplatin 369, 370
thallium-105 isotope 163
thallium-205 isotope 132–3
thallium cations
  G-quadruplex X-ray structures 125, 127–8
  major groove localisation mimic 93–5
  minor groove 101
  potassium mimic 24–5
therapeutic agents 353–98
Thermoanaerobacter tengcongensis 184, 185
thermodynamics
  magnesium cation–OP 23–4
  metal–nucleic acid complexes stability 55–6
  polyanion models 335–6
  NLPB theory 319–23
  RNA folding 180–213
  two binding-mode formalism 25–6
thermostability 119, 326–7
thiamine pyrophosphate (TPP)-sensing riboswitch 213
three-helix RNA junctions 234
thrombin-binding aptamer 141
thymidine 354–5
thymine 44
tightly-bound ion (TBI) model 316, 340–2
time-dependent processes *see* kinetics
time-resolved hydroxyl radical footprinting 251–3

time-scales 134–7, 164
titanium agents 379–81
titanocene dichloride 379, 380–1
TL–TLR *see* tetraloop–tetraloop receptor
toxicity
  carcinogenesis causation 351–3
  cisplatin 361, 367–8
  mutagenesis causation 351–3
  photoactivable drugs 376–9, 391
TPP *see* thiamine pyrophosphate
trans-activation responsive (TAR) region 189, 190–1, 209, 211, 354–5
transesterification 295–7
transfer RNA (tRNA) 171, 172, 180
transition metal coordination 39–69
transition state ensemble (TSE) 228, 235–46
transition state models
  general acid–base catalysed RNA cleavage 296
  large ribozyme active sites 283–95, 298
  small ribozyme active sites 261, 295–8
transplatin diam(m)ine platinum(II) complexes 372
trans-platinum complexes 370–4
transport number 264, 265
tricycles 15, 17
tridentate chelation 8, 9, 30
tri-nuclear metal complexes 53
tripartite assays 290, 291
TriplatinNC 375, 376
triple-helical tetracationic cyclinder 395
triplets *see* nucleobase triplets
tRNA *see* transfer RNA
true globular structures 2, 3
TSE *see* transition state ensemble
tug-of-war model 106
two binding-mode formalism 25–7
two-metal ion mechanistic paradigm 261, 283–95
Twort 282–3
two-way RNA junction constructs 233

U167C mutant ribozyme 239
U273A mutant ribozyme 249, 250
U coordination site 132, 138
undulations of charge 310
unfolding, RNA 207–12, 227, 241–6
uridine 354–5

valence, counterions 310, 311, 312–13
vanadyl ions 163–6
very large RNA assemblies 2
vibrational spectroscopy 173–4
V-shaped loops 121, 122

water 101, 102, 104–6
  hydrated magnesium ions 225
  hydration enthalpy 264, 265
  hydration shells 182–3
  measurements 75, 104–6
Watson–Crick base pairs 43, 64–5, 77, 222–3
Web-based databases 273
Wyman linkage relationship 199

XANES *see* X-ray near-edge structure
X-ray absorption spectroscopy (XAS) 166–9
X-ray crystal structure
  alternative cation binding sites in minor groove 107
  cation position determination 24–5
  cisplatin adducts 364, 365, 368
  complex RNA molecules 184–5, 195, 199, 206
  divalent cation localisation 81–5, 88–92
  G-quadruplexes 123–9
  group I ribozymes active site structures 282–3
  human telomere quadruplex 127
  iron helicates 397
  model mixed-nucleobase/mixed-metal complex 46
  monovalent cation localisation 92–6, 101–2
  Oceanobacillus iheyensis group II intron 292

ribozymal catalytic metal ion–
  ligand interactions 270–1
triple-helical tetracationic cyclinder
  395
X-ray near-edge structure (XANES)
  168

Z-DNA 59
zeniplatin 359, 360
zidovudine 353
zinc^{II}–cyclen complexes 353–5
zinc ions 55–6
zip *see* folding